Exposure to Microbiological Agents in Indoor and Occupational Environments

Carla Viegas · Susana Viegas
Anita Quintal Gomes · Martin Täubel
Raquel Sabino
Editors

Exposure to Microbiological Agents in Indoor and Occupational Environments

Springer

Editors
Carla Viegas
GIAS, ESTeSL - School of Health Technology of Lisbon
Polytechnic Institute of Lisbon
Lisbon
Portugal

Center for Research in Public Health, National School of Public Health
University of Lisbon
Lisbon
Portugal

Susana Viegas
GIAS, ESTeSL - School of Health Technology of Lisbon
Polytechnic Institute of Lisbon
Lisbon
Portugal

Center for Research in Public Health, National School of Public Health
University of Lisbon
Lisbon
Portugal

Anita Quintal Gomes
GIAS, ESTeSL - School of Health Technology of Lisbon
Polytechnic Institute of Lisbon
Lisbon
Portugal

Martin Täubel
Department Health Protection
National Institute for Health & Welfare
Kuopio
Finland

Raquel Sabino
Reference Unit for Parasitic and Fungal Infections, Infectious Diseases Department
National Institute of Health Dr. Ricardo Jorge
Lisbon
Portugal

ISBN 978-3-319-61686-5 ISBN 978-3-319-61688-9 (eBook)
DOI 10.1007/978-3-319-61688-9

Library of Congress Control Number: 2017954457

Printed on acid-free paper

This Springer imprint is published by Springer Nature
The registered company is Springer International Publishing AG
The registered company address is: Gewerbestrasse 11, 6330 Cham, Switzerland

Foreword

Microorganisms are indispensable companions to human life.

We have always been faced with the galaxy of microorganisms in our environment, and the science developments during the last decades have unequivocally shown that our "internalized" microorganisms referred to as the microbiome are determining our health and disease.

The new measuring techniques next generation sequencing has opened a whole new area of research, and much is still to be learnt about the interface between humans and their surrounding microorganisms.

Modern living is associated to the movement of people at an unprecedented high level. Such a high mobility leads to new threats in society and new tools to handle these. These issues call for a constant vigilance in the focus on microorganisms in different areas and purposes. The current volume is a great companion for the many people responsible for the management of public health in connection to the diverse effects of microorganisms, from surveillance of moulds in occupational settings to the handling of the microbial environment in hospitals.

This book provides a valuable companion to all who are concerned with diseases, related to microbial exposure at the workplace. The content covers a very diverse turf, from infections and noninfectious effects of microorganisms to the different measuring strategies and special occupational environments with high levels of microorganisms and their endo- and exotoxins.

Whether the reader is concerned with infections, allergies, or with other potential adverse health effects, this book will give valuable background information. By bringing together what is currently known about these conditions, together with the latest information on their detection, monitoring and control, the authors have provided a comprehensive resource for all those concerned with this increasingly important and diverse field of health effects related to microbial exposures. Increased awareness of this field will be needed in order to develop new strategies for intervention and prevention. Given the potential for public health benefit as well as burden of these exposures the book can be highly recommended.

June 17, 2017 Torben Sigsgaard

List of Abbreviations

15-Ac-DON	15-acetyl-deoxynivalenol
16HBE140	human bronchial epithelial cell line
2'R-OTA	2'R-ochratoxin A
3-AcDON	3-acetyl-deoxynivalenol
4R-OH-OTA	4'R-hydroxy-ochratoxin A
A549	human alveolar epithelial type II cell line
AIHA	American Industrial Hygiene Association
AdV	human adenovirus
AFB_1	aflatoxin B_1
AFB1	aflatoxin B1
AFB_1-N^7-guanine	aflatoxin B_1-N^7-guanine
AFB_1-NAC	aflatoxin B_1-N-acetyl-cysteine
AFB_2	aflatoxin B_2
AFG_1	aflatoxin G_1
AFG_2	aflatoxin G_2
AFLP	amplified fragment length polymorphism
AFM_1	aflatoxin M_1
AIDS	acquired immune deficiency syndrome
AM	alveolar macrophages
APS	aerodynamic particle sizer
ASHRAE	American Society of Heating, Refrigerating and Air-Conditioning Engineers
BEAS-2B	human bronchial epithelial cell line
BEN	Balkan endemic nephropathy
Calu3	human airway epithelial cell line
CAM	chorioallantoic membrane
CD14	cluster of differentiation 14 gene
CDC	Centers for Disease Control and Prevention
CFS	chronic fatigue syndrome
CFU	colony forming unit

CMV	cytomegalovirus
CoNTC	concentration of no toxicologic concern
COPD	chronic obstructive pulmonary disease
DBS	dried blood spot
DGGE	denaturizing gradient gel electrophoresis
DGGE	denaturing gradient gel electrophoresis
DMI	sterol dimetilation inhibitors
DNA	deoxyribonucleic acid
DOM-1	deepoxy-deoxynivalenol
DON	deoxynivalenol
DON-15-GlcA	deoxynivalenol-15-O-β-glucuronide
DON-3-GlcA	deoxynivalenol-3-O-β-glucuronide
dsDNA	double strained DNA
ECDC	European Center for Disease Prevention and Control
ECRHS	European Community Respiratory Health Survey
EDC	electrostatic dustfall collector
EDX	energy dispersive X-ray spectroscopy
EF	enhanced Fujita scale
EFSA	European Food Safety Authority
EIA	enzyme immunoassay
ELISA	enzyme linked immunosorbent assay
ELPI	electrical low pressure impactor
EM	electron microscopy
EPS	extracellular polysaccharides
EPSS	electrostatic precipitator with superhydrophobic surface
ERMI	environmental relative moldiness index
EU	endotoxin units
EU	European Union
EU/m^3	endotoin unit per cubic meter
EVs	enteroviruses
FB_1	fumonisin B_1
FDA	Food and Drug Administration
FLD	fluorescence detector
FSSST	fungal spore source strength tester
FUS	fusarenon x
G-	gram negative bacteria
G+	gram positive bacteria
GC	gas chromatography
HAI	hospital acquired infection
HAI	healthcare-associated infections
HBM	human biomonitoring
HBROEL	health-based recommended occupational exposure limit
HCW	healthcare workers
HEPA	high-efficiency particulate arrestance
HEPA	high efficacy particulate airfiltration

HIV	human immunodeficiency virus
HMPV	human metapneumovirus
HPIV	human parainfluenza virus
HPLC	high performance liquid chromatography
HRV	human rhinovirus
HSCT	allogeneic hematopoietic stem cell transplantation
HVAC	heating, ventilation and air conditioning
IA	invasive aspergillosis
IAC	immuno-affinity chromatography
IARC	International Agency for Research of Cancer
IFI	invasive fungal infection
IgA	immunoglobulin A
IgE	immunoglobulin E
IgG	immunoglobulin G
IL	interleukin
INAS	intra-nasal air sampler
IOM	Institute of Medicine
ISO	International Organization for Standardization
ITS	internal transcribed spacer regions
IV	influenza virus
JEFCA	Joint FAO/WHO Expert Committee on Food Additives
LAI	laboratory-acquired infection
LAL	limulus amebocyte lysate
LIF	laser induced fluorescence
LOD	limit of detection
LOQ	limit of quantification
LPS	lipopolysaccharide
LPSN	list of prokaryotic names with standing in nomenclature
m3	cubic metre
MALDI	matrix-assisted laser desorption/Ionization
MALDI-TOF	matrix-assisted laser desorption/ionization time-of-flight
MALDI-TOF-MS	matrix assisted laser desorption/ionization-time of flight mass spectrometry
MDR	multidrug resistance
MEA	malt extract agar
MERS	middle east respiratory syndrome
MERS-CoV	middle east respiratory syndrome coronavirus
MGE	mobile genetic elements
mm	millimetres
Mrna	messenger RNA
MRSA	methicillin-resistant *Staphylococcus aureus*
MS	mass spectrometry
MS/MS	tandem mass spectrometry
MSQPCR	mold-specific quantitative PCR
MSSA	methicillin-resistant *Staphylococcus aureus*

MTT	(3-(4,5-dimethylthiazol-2-yl)-2,5-diphenyltetrazolium bromide
MVOCs	microbial volatile organic compounds
MWF	metalworking fluids
NADPH	nicotinamide adenine dinucleotide phosphate
NAHA	N-acetylhexosaminidase
NCID	National Center for Infectious Diseases
ng/m^3	nanogram per cubic meter
NGS	next generation sequencing
NIOSH	National Institute for Occupational Safety and Health
NIV	nivalenol
NLVs	norwalk-like viruses
NoVs	noroviruses
ODTS	organic dust toxic syndrome
OEL	occupational exposure limit
OPCs	optical particle counters
OTA	ochratoxin A
PBOA	primary biogenic organic aerosols
PCR	polymerase chain reaction
PFGE	pulse field gel electrophoresis
P-FLEC	particle-field and laboratory emission cell
PM	particulate matter
PM10	particulate matter 10 micrometres or less in diameter
PM10	particulate matter
PMN	polymorphonuclear cells
PPE	personal protective equipment
qPCR	quantitative PCR
rDNA	ribosomal DNA
REL	recommended exposure limit
rFC	recombinant factor C
RFLP	restriction fragment length polymorphism
RH	relative humidity
RNA	ribonucleic acid
ROS	reactive oxygen species
rRNA	ribosomal RNA
RSV	respiratory syncytial virus
RT – PCR	reverse transcription polymerase chain reaction
RT-PCR	real time PCR
RV	rhinovirus
RWI	recreational water illness
SARS	severe acute respiratory syndrome
SARS-CoV	severe acute respiratory coronavirus
SAX	anion exchange resin
SCF	scientific committee for food
SEM	scanning electron microscopy
SPE	solid phase extraction

SPP.	species
SRM	single reaction monitoring
ssDNA	single stranded DNA
TCID50	median tissue culture infectious dose
TDI	tolerable daily intake
TGGE	temperature gradient gel electrophoresis
Th cells	T-helper cells of either type 1 (Th1) or type 2 (Th2)
Th-2	type 2 helper cells
TLR	toll-like receptors
Tm	melting temperature
TMC-120A	mycotoxin furo[3,2-h]isoquinoline
TMV	tobacco mosaic virus
TOF	time off light
T_{reg}	regulatory T cells
TRFLP	terminal restriction fragment length polymorphism
TSA	tryptone soy agar
TTC	threshold of toxicological concern
UVAPS	ultra violet aerodynamic particle sizer
VOC	volatile organic compound
VRE	vancomycin-resistant enterococci
VZV	varicella-zoster virus
WGS	whole-genome sequencing
WHO	World Health Organization
XTT	2,3-bis-(2-methoxy-4-nitro-5-sulfophenyl)-2H-tetrazolium-5-carboxanilide
ZAN	zearlanone
ZEN	zearalenone
ZEN-14-GlcA	zearalenone-14-O-β-glucuronide
α-ZAL	α-zearalanol
α-ZEL	α-zearalenol
β-ZAL	β-zearalanol
β-ZEL	β-zearalenol
Δ	difference
µm	micro molar

Contents

Part I
Indoor Exposure to Microorganisms with Emphasis on Occupational Environments

Chapter 1
Occupational Fungal Exposure in the United States

Brett J. Green

Abstract The objective of this book chapter is to provide a review of recent advances in our understanding of fungal exposures encountered in United States occupational environments and the impacts that these exposures have on worker health. Occupational exposure can occur to a broad diversity of fungal bioaerosols that include spores, conidia, hyphae, yeasts, chlamydospores, and submicron fragments. Pulmonary exposure to fungal bioaerosols in the work environment can lead to some respiratory morbidities. In some cases, exposure to dimorphic fungal conidia can also result in a symptomatic pulmonary infection that can disseminate and become life threatening. Transcutaneous penetrating injuries sustained while handling vegetation may additionally implant fungal spores or hyphae and result in a subcutaneous infection. Workers may also be susceptible to dermatophytes that can proliferate in occluded regions of the skin such as interdigital spaces and cause a cutaneous infection. Occupational environments and work related tasks that can lead to fungal exposure are reviewed. Strategies to avoid worker fungal exposures including engineering and administrative controls as well as personal protective equipment are additionally provided.

Keywords Exposure assessment · fungus · dermatophytes · gene sequencing · occupational hazards

1.1 Introduction

Fungi are eukaryotic organisms that lack chlorophyll. Many species are saprophytic and obtain nutrients from live or dead organic material in the environment. Based on these properties, fungi can proliferate within a number occupational settings in the United States, including non-industrial indoor environments that

B.J. Green (✉)
Allergy and Clinical Immunology Branch, Health Effects Laboratory Division, National Institute for Occupational Safety and Health, Centers for Disease Control and Prevention, Morgantown, WV 26505, USA
e-mail: dox6@cdc.gov

C. Viegas et al. (eds.), *Exposure to Microbiological Agents in Indoor and Occupational Environments*, DOI 10.1007/978-3-319-61688-9_1

contain cellulose-based building materials, agricultural settings, or other occupational environments in which contaminated organic material or soil are handled or disturbed. Fungi are a diverse kingdom, and 1.5 million species are estimated to exist (Hawksworth 2001). Compared to prokaryotes described in independent chapters, fungi contain a membrane-bound nucleus, mitochondria, centrioles, and 80S ribosomes (Cannon et al. 2006). The fungal cell wall is rigid and composed of mannose proteins, (1→3)-β-d-glucan, chitin and the sterol, ergosterol (Cannon et al. 2006). Fungi are larger in size than bacteria and produce a variety of structures that range from unicellular yeasts to filamentous hyphae that can include arrangements of sexual spores or asexual conidia. These reproductive particles range in size from as small as 2 μm to greater than 120 μm in size. Biotic or abiotic disturbance can result in the release of spores in concentrations that can exceed 10^8 colony forming units per cubic meter in some industrial environments (Eduard 2009).

Workers can be exposed to a variety of fungal phyla, but three predominate and include the Ascomycota, Basidiomycota, and Zygomycota. Recent metagenomic studies suggest that the Ascomycota and Basidiomycota are the most common fungal phyla detected in indoor, outdoor, and occupational environments (Green et al. 2016; Rittenour et al. 2014). In contrast, some workers may only be exposed to a homogenous fungal source, such as processing workers in a mushroom production facility in which spore concentrations can be as high as 10^6 spores per cubic meter (Sastre et al. 1990). Some of the particles that the U.S. workforce can be exposed to include spores or conidia, unicellular yeasts, chlamydospores, and even fragments of each of these reproductive structures. Hyphae or microscopic structures of the cell wall can also fragment and aerosolize into the breathing zone of a worker (Green et al. 2011).

Fungi can be monomorphic and consist of either a unicellular yeast or a multicellular hyphal form. Dimorphic fungi are a unique group of fungi placed in the phylum Ascomycota and include the primary endemic pathogens that grow in the environment as multicellular filamentous fungi and produce spores that when inhaled by a host, convert into a budding yeast (Gauthier 2015; Klein and Tebbets 2007). Polymorphic fungi include endogenous flora, such as the yeast species *Candida albicans*, that produce budding yeast, pseudohyphae, true hyphae and chlamydospores. Worker exposure to asexual or sexual spores derived from monomorphic fungi such as yeasts or filamentous hyphal forms is the predominant fungal particle exposure in occupational settings. In some cases that involve disturbance to the soil, workers can inhale the spores of dimorphic fungal pathogens that are endemic to specific regions of the U.S., although cases outside of these geographical boundaries have been reported (Marsden-Haug et al. 2012). Exposure to these types of fungi can cause pulmonary or disseminated infection in immunocompetent and immunocompromised hosts. Workers that sustain transcutaneous injuries while handling organic material may also acquire a subcutaneous fungal infection following implantation of fungal spores or hyphae. Worker exposure to infectious dermatophyte arthroconidia in combination with wearing occluded footwear or clothing can result in a fungal infection restricted to the skin, hair or nails. Respiratory morbidity such as hypersensitivity pneumonitis, allergy, and asthma following exposure to fungal particles is also widely reported among working populations (Eduard 2009). Exposure to these particles occurs in a variety of occupational environments especially those that handle or disturb organic material or whose work environment is within a fungal contaminated damp indoor environment.

The objective of this book chapter is to provide a review of recent advances in our understanding of fungal exposures encountered in U.S. occupational environments and the impacts that these exposures have on worker health. Dimorphic fungal pathogens are reviewed, and updates on endemic areas and affected working populations are presented. Fungi that are associated with subcutaneous and cutaneous fungal mycoses in the U.S. workforce are also reviewed. New insights into the diversity of fungi a worker can be exposed to are additionally evaluated. This book chapter will also provide information about preventive and protective measures to avoid each of these fungal hazards in U.S. occupational environments.

1.2 Dimorphic Fungal Pathogens

Most fungi grow as saprophytes in the environment, but some are considered harmful to human health and cause disease in immunocompetent and immunocompromised individuals (Gauthier 2015; Klein and Tebbets 2007). Dimorphic fungi cause endemic mycoses and are considered primary pathogens, placed in a limited group of six phylogenetically related Eurotiomycetes in the Ascomycota order, Onygenales (Thompson and Gomez 2015). Dimorphic fungi can switch between a filamentous hyphal and a yeast phase (Gauthier 2015; Klein and Tebbets 2007). Thermal or non-thermal mechanisms govern this morphological change, a process recently reviewed by Gauthier (Gauthier 2015). In the environment, dimorphic fungi grow as a filamentous hyphal form in soil and organic debris at ambient temperatures and produce infectious spores (Gauthier 2015; Klein and Tebbets 2007). Members of this phylogenetically related group produce unicellular aleuroconidia or arthroconidia (Thompson and Gomez 2015). Abiotic and biotic disturbance to soil results in the aerosolization and inhalation of the infectious spores into the host's lungs. Upon deposition, the spores convert to into a budding yeast as part of the pathogenic phase (Klein and Tebbets 2007).

Dimorphic fungal pathogens are restricted to specific geographic regions and account for several million infections each year (Klein and Tebbets 2007). Three dimorphic fungal pathogens are endemic to the U.S. and include *Histoplasma capsulatum* (Histoplasmosis), *Blastomyces dermatitidis* (Blastomycosis), and *Coccidioides immitis* (Coccidioidomycosis). Histoplasmosis and Coccidioidomycosis are the most frequently encountered endemic mycoses in the U.S., whereas Blastomycosis is less common. Exposure to the infectious fungal conidia derived from these fungi can result in an infection that is often asymptomatic or self-limited, or can present as a symptomatic pulmonary infection that can disseminate and impact multiple organs. Although exposure can occur during recreational activities, there are some recent examples in the published literature where U.S. workers can be exposed to infectious spores while performing work-related tasks.

1.2.1 Histoplasmosis

H. capsulatum is the causal agent of Histoplasmosis, also known as Darlings disease. The species is placed in the Ascomycota order, Onygenales. This species is thermally dimorphic and grows in the soil at ambient temperature (Gauthier 2015; Klein and Tebbets 2007). *H. capsulatum* produces septate hyphae, and two types of asexual spores termed tuberculate macroconidia and microconidia (Fig. 1.1). Tuberculate macroconidia are ornate, and range in size from 8 to 15 μm (Fig. 1.1a), whereas

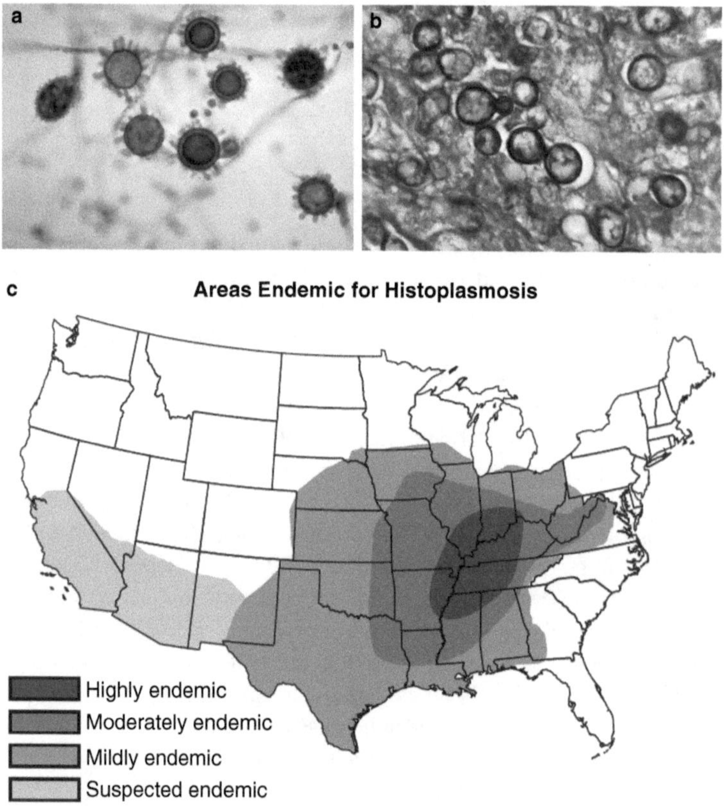

Fig. 1.1 Histoplasmosis, an endemic mycosis, is caused by the inhalation of *Histoplasma capsulatum*. The habitat of the fungus includes nitrogen rich soil that contains large amounts of bird or bat droppings. *H. capsulatum* is a dimorphic fungal pathogen and grows in the environment as a filamentous hyphal form producing tuberculate macroconidia and microconidia (**a**; Photo courtesy of CDC; CDC Public Health Image Library (PHIL) ID#: 299). Inhalation of microconidia results in the conversion to a budding yeast that can proliferate within the reticuloendothelial system (**b**; Photo courtesy of CDC/Dr. Libero Ajello; CDC Public Health Image Library (PHIL) ID#: 4221). Endemic regions of histoplasmosis include central and eastern states, especially areas around the Ohio and Mississippi River valleys (**c**). The endemic regions were determined from histoplasmin skin response surveys and outbreak cases (Map courtesy of the Centers for Disease Control and Prevention). It is important to note that cases can occur outside endemic areas

microconidia are respirable in size (Fig. 1.1b, 2–5 μm) and considered infectious. Biotic or abiotic soil disturbance events within endemic regions aerosolizes microconidia that are then inhaled by the host. Introduced to a higher temperature, suboptimal nutrients, and reduced oxygen, the microconidia then convert to a budding yeast. The yeast can modulate and proliferate within the phagolysosome of macrophages and become an intracellular pathogen within the reticuloendothelial system (Klein and Tebbets 2007). The mechanisms associated with yeast conversion and *H. capsulatum* pathogenesis are reviewed by Klein (Klein and Tebbets 2007) and Gauthier (Gauthier 2015).

Half a million Histoplasmosis cases are reported annually in the U.S. (Gauthier 2015; CDC 2016c). Inhalation of infectious microconidia can result in a lung infection that can be asymptomatic and self-limited, or can present as a symptomatic pulmonary infection that can disseminate and become life threatening. Estimates as high as 45% of hosts are not even aware of an infection and do not seek medical treatment (Lenhart et al. 1997). In symptomatic cases, mild flu-like symptoms with chest radiography resolving patchy pneumonitis that can calcify are experienced within an average of 20 days (Lenhart et al. 1997). Chronic histoplasmosis resembles tuberculosis and can last for extended durations. Disseminated histoplasmosis is the rarest and most severe form of the disease as it can affect multiple organs (Lenhart et al. 1997). Histoplasmosis can occur in both immunocompetent and immunocompromised workers and following infection, the organism can remain latent and may reactivate if the host becomes immunocompromised (Klein and Tebbets 2007). Methods to identify cases of histoplasmosis in clinical samples include viable culture, serological tests, detection of *H. capsulatum* polysaccharide antigen and the histoplasmin skin test. More detailed information related to these methods of detection are presented in detail elsewhere (Lenhart et al. 1997).

Histoplasmin skin test surveys conducted on U.S. Navy recruits in the 1950s (Manos et al. 1956) and outbreak cases are datasets used to map the continental distribution of histoplasmosis (Benedict and Mody 2016). The map presented in Fig. 1.1c outlines endemic areas of Histoplasmosis but it is important to note that cases can occur outside endemic areas. Recently, Benedict and Mody (Benedict and Mody 2016) reviewed U.S. outbreak cases of histoplasmosis between 1938 and 2013. Of the 105 reported outbreaks that included 2, 850 individual cases, over 50% of the cases were localized in Indiana, Ohio, and Iowa and onset occurred between May and November (Benedict and Mody 2016). *H. capsulatum* grows in the soil with high nitrogen content especially in soils enriched with bird droppings or bat guano. Endemic regions occur in eastern states along the Ohio and Mississippi River Valleys (Fig. 1.1c) (Lenhart et al. 1997). Areas of highest contamination include soil of various bird roosting sites, manure, and habitats of pigeons, bats and poultry (Lenhart et al. 1997).

Working environments where bird droppings or bat guano were present were estimated to account for 33% of the reported outbreaks (Benedict and Mody 2016). Other reported workplace sites included buildings, chicken coops, and farms (Benedict and Mody 2016). Occupational tasks performed during the work-related outbreaks included construction, demolition, and maintenance (Benedict and Mody 2016). Disturbance or excavation of soil in areas with bird or bat droppings are additional risk factors for aerosolizing *H. capsulatum* infectious microconidia (Lenhart et al. 1997). In addition to disturbance mediated exposures, laboratory-acquired infections in laboratorians handling *H. capsulatum* cultures have also been reported in North America. A list of occupations at risk of *H. capsulatum* exposure is presented in Table 1.1.

Although personal exposure to microconidia is the source of exposure to *H. capsulatum*, the dose of infectious spores required to elicit symptomatic disease has not been reported. However, the higher the inoculum burden, longer duration of exposure, as well as the immune status and age of the individual are variables that can increase a worker's susceptibility to acquiring symptomatic histoplasmosis (Lenhart et al. 1997). Workers that engage in disturbance activities of nitrogen rich soils containing bird droppings or bat guano, within endemic regions are additional variables that can lead to worker exposures (Huhn et al. 2005). The National Institute for Occupational Safety and Health (NIOSH) and the National Center for Infectious Diseases (NCID) have published a guidance document aimed to increase employer and worker awareness of Histoplasmosis. The guidelines provide strategies to protect workers from occupational exposure to infectious *H. capsulatum* microconidia (Lenhart et al. 1997), particularly in workplaces where bird or bat droppings are present and environmental disruption occurs (Benedict and Mody 2016; Huhn et al. 2005). A similar list of strategies to prevent exposures are published on the Centers for Disease Control and Prevention (CDC) website (CDC 2016c). Listed below are best work practices adapted from the NIOSH/NCID guidance document that aims to prevent worker exposure to *H. capsulatum* in endemic regions.

- Remove bat colonies and birds from buildings.
- Provide health risk warnings and signage in workplaces.
- Suppress dust generation and soil disruptive activities.
- Disinfect areas identified to harbor *H. capsulatum*.
- Workers should wear personal protective equipment (PPE) presented in Table 1.1.

1.2.2 Blastomycosis

Inhalation of infectious spores derived from *B. dermatitidis* (Klein and Tebbets 2007; Klein et al. 1986) can result in Blastomycosis, otherwise referred to as Gilchrist's disease (Saccente and Woods 2010) or Chicago disease (Thompson and Gomez 2015). *B. dermatitidis* is an endemic dimorphic fungal pathogen placed in the phylum Ascomycota. This species grows as septate hyphae in the soil at ambient temperature and is thought to break down leaves and wood. Infectious spores are produced that aerosolize following abiotic or biotic disturbance of the soil (Fig. 1.2a). Compared to the other North American dimorphic fungal pathogens, the natural habitat of *B. dermatitidis* is not as well understood.

Table 1.1 Occupations at risk of exposure to dimorphic primary pathogens and fungi that can cause subcutaneous and cutaneous mycoses. Administrative, engineering, and personal protective controls that are adapted from NIOSH and CDC guidance documents are presented with the aim to minimize worker fungal exposure

Fungal species	Mycosis	Exposure	Detection methods	Occupational exposure	Avoidance strategies	PPE
Histoplasma capsulatum	Histoplasmosis	Microconidia	Viable culture Direct microscopy ITS Gene Sequencing LAMPPCR Serological assays	Bridge inspector/ painter Chimney cleaner Construction worker Demolition worker Poultry worker Farmer Gardener HVAC technician Pest controller Building restoration Roofer Prison Guard Laboratorian	• Avoid bird or bat colonies in endemic regions • Minimize dust generation activities • Disinfect soil • Worksite training • Ventilation systems • Educate and train employees safe work practices	• NIOSH-approved respirator • Disposable protective clothing • Shoe coverings • Gloves
Blastomyces dermatitidis	Blastomycosis	Conidia	Viable culture Direct microscopy ITS Gene Sequencing PCR Serological assays	Forestry worker Construction worker Excavator Rail road worker Diver Laboratorian	• Minimize dust or soil disturbance activities in endemic regions • Educate and train employees safe work practices	• NIOSH-approved respirator • Disposable protective clothing • Shoe coverings • Gloves

(continued)

Table 1.1 (continued)

Fungal species	Mycosis	Exposure	Detection methods	Occupational exposure	Avoidance strategies	PPE
Coccidioides immitis	Coccidioidomycosis	Arthroconidia	Viable culture Direct microscopy Whole genome sequencing ITS Gene Sequencing PCR Serological assays	Geologist Textile Worker Agriculture worker Farmer Construction worker Excavator Military Personnel Archeologist Healthcare worker Oil and Gas worker Telecommunication worker Prison Guard Wildland firefighters Television crew	• Minimize dust or soil disturbance activities • Avoid peak seasons • Wash equipment • Excavate using heavy machinery • Educate and train employees safe work practices • Wet soil during soil disturbance activities	• NIOSH-approved respirator • Disposable protective clothing • Shoe coverings • Gloves
Sporothrix schenckii	Sporotrichosis	Conidia	Viable culture Direct microscopy Fluorescence microscopy ITS Gene Sequencing PCR Serological assays Sporotrichin skin test	Farmer Landscaper Gardener Nursery worker Forestry worker Flower vendor Animal husbandry worker Laboratorian	• Educate and train employees safe work practices • Avoid transcutaneous injuries • Use alternative packaging materials	• Protective clothing that covers lower extremities • Shoe coverings • Gloves • NIOSH-Approved respirator (Laboratorians)

(continued)

Table 1.1 (continued)

Fungal species	Mycosis	Exposure	Detection methods	Occupational exposure	Avoidance strategies	PPE
Fonsecaea pedrosoi *Fonsecaea monophora* *Phialophora verrucosa* *Cladophialophora carrionii*	Chromoblastomycosis	Conidia (muriform or sclerotic cells)	Viable culture Direct microscopy	Farmer Rural worker	• Educate and train employees safe work practices • Avoid transcutaneous injuries	• Protective clothing that covers lower extremities • Shoe coverings • Gloves
Alternaria spp. *Exserohilum* spp. *Exophiala jeanselmei* *Wangiella dermatitidis* *Bipolaris* spp.	Phaeohyphomycosis	Yeast Conidia Hyphae	Viable culture Direct microscopy ITS Gene Sequencing PCR	Farmer Rural worker Boat builder	• Educate and train employees safe work practices • Avoid transcutaneous injuries	• Protective clothing that covers lower extremities • Shoe coverings • Gloves
Mucor spp. *Rhizopus* spp. *Apophysomyces trapeziformis* *Apophysomyces* spp.	Mucormycosis	Sporangiospores Hyphae	Viable culture Direct microscopy ITS Gene Sequencing PCR	First responders – natural disasters Emergency preparedness and response worker	• Educate and train employees safe work practices • Avoid transcutaneous injuries • Minimize dust or soil disturbance activities	• NIOSH-approved respirator • Protective clothing that covers lower extremities • Shoe coverings • Gloves

(continued)

Table 1.1 (continued)

Fungal species	Mycosis	Exposure	Detection methods	Occupational exposure	Avoidance strategies	PPE
Trichophyton spp. *Microsporum* spp. *Epidermophyton* spp.	Dermotophytosis	Arthroconidia	Viable culture Direct microscopy ITS Gene Sequencing PCR	Manual laborer Farmer Miner Military personnel Meat processor Dairy processor Poultry processor Forestry worker	• Educate and train employees safe work practices • Minimize indirect or direct contact with infectious arthroconidia (shared occupational change room facilities) • Minimize wet skin and clothes • Wear non-occluding footwear • Wash hands after handling livestock	

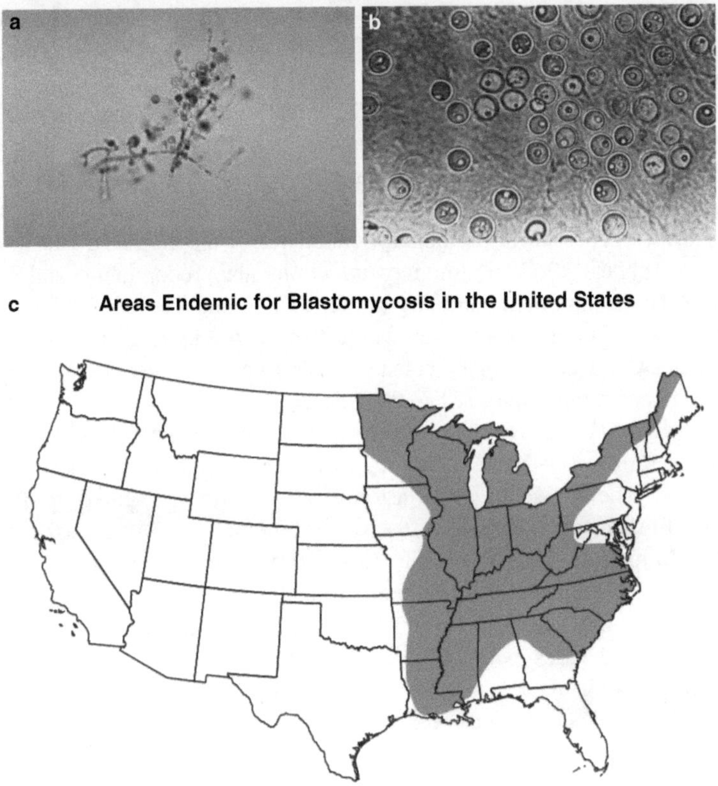

Fig. 1.2 Blastomycosis, an endemic mycosis, is caused by the inhalation of *Blastomyces dermatitidis* conidia. The habitat of the fungus is less well described compared to *H. capsulatum* and includes damp soil near bodies of water such as rivers or lakes. *B. dermatitidis* is a dimorphic fungal pathogen that decomposes organic matter such as leaves and wood as a filamentous hyphal form producing conidia (**a**; Photo courtesy of CDC/Dr. Leanor Haley; CDC Public Health Image Library (PHIL) ID#: 3768). Inhalation of conidia results in the conversion to a large budding yeast with thick refractive walls (**b**; Photo courtesy of CDC/Dr. Lucille K. Georg; CDC Public Health Image Library (PHIL) ID#: 14882). Blastomycosis is sporadic and occurs in southeastern U.S. regions especially the Mississippi and Ohio River valleys as well as the Canadian provinces of Ontario, Quebec, and Manitoba (**c**; Map courtesy of the Centers for Disease Control and Prevention). It is important to note that cases can occur outside endemic areas

Inadequate skin test reagents and serology methods have limited the research community's understanding of Blastomycosis, and much of our knowledge of endemic regions has been based on reported outbreaks (Saccente and Woods 2010). The saprophytic habitat of *B. dermatitidis* includes wet earth and organic debris associated with bodies of waters such as rivers, streams, and lakes (Saccente and Woods 2010). One study located in Northern Wisconsin showed that Blastomycosis cases were related to soil disturbance surrounding a beaver lodge (Klein et al. 1986). Cases of Blastomycosis are sporadic and occur in southeastern U.S. regions, especially the Mississippi and Ohio River valleys, as well as the Canadian provinces of Ontario, Québec, and Manitoba (CDC 2016a). Fig. 1.2c presents the geographic distribution of Blastomycosis cases.

Epidemiological studies suggest that the mode of infection includes the inhalation of conidia following soil disturbance activities (Fig. 1.2a) (Klein et al. 1986). Conversion of the conidia into a broad based-budding yeast cell (8–10 μm) follows respiratory deposition (Fig. 1.2b) (Klein and Tebbets 2007). Exposure to *B. dermatitidis* can result in pulmonary, cutaneous or disseminated Blastomycosis (Klein and Tebbets 2007). Pulmonary infections are predominant, and 50% are reported to be asymptomatic and self-limited (Klein and Tebbets 2007). In symptomatic cases, symptoms appear anywhere from 3 weeks to 3 months following exposure (CDC 2016a). Primary cutaneous blastomycosis has also been reported following a penetrating injury (CDC 2016a).

As *B. dermatitidis* is restricted to wet earth and organic debris, workers that handle or disturb soil within endemic regions are at risk of being exposed to infectious spores. The incidence of Blastomycosis is low compared to the other dimorphic fungal pathogens. The disease is reportable in Arkansas, Louisiana, Michigan, Minnesota, and Wisconsin (CDC 2016a). In a recent study, the annual incidence of Blastomycosis in Quebec, Canada was 0.133 cases per 100,000 individuals (Litvinov et al. 2013), whereas in a Wisconsin country the annual incidence was 40 cases per 100,000 individuals (Saccente and Woods 2010). Occupations at risk of exposure to *B. dermatitidis* in endemic regions are presented in Table 1.1 and include workers that disturb the soil such as professional divers (Kroll and Grossman 2013) and railroad workers (Siemieniuk et al. 2015). In addition to environmental sources of exposure, laboratorians that handle culture specimens can inadvertently aerosolize infectious *B. dermatitidis* spores and acquire blastomycosis (Cote et al. 1997; Saccente and Woods 2010). Based on the low incidence of Blastomycosis in the U.S., methods to prevent exposure are presented by the CDC and are available online (CDC 2016a). Based on the available peer-reviewed literature, the following methods can minimize and help protect workers from exposure in endemic regions.

- Awareness and worksite training before soil disturbance occupational tasks (construction, excavation, and manual digging) in endemic regions.
- Minimize soil disturbance and dust generation on work sites.
- Wear PPE described in Table 1.1.

1.2.3 Coccidioidomycosis

Inhalation of *C. immitis* and *C. posadasii* infectious arthroconidia can result in Coccidioidomycosis or Valley Fever (CDC 2016b). *C. immitis* is thermally dimorphic, and optimal growth occurs in the southwest of the U.S. in the Lower Sonoran Life Zone. Growth is optimal in warm, dry climates with alkaline soil containing salt and borates (Ampel 2009, 2011). The species also extends into

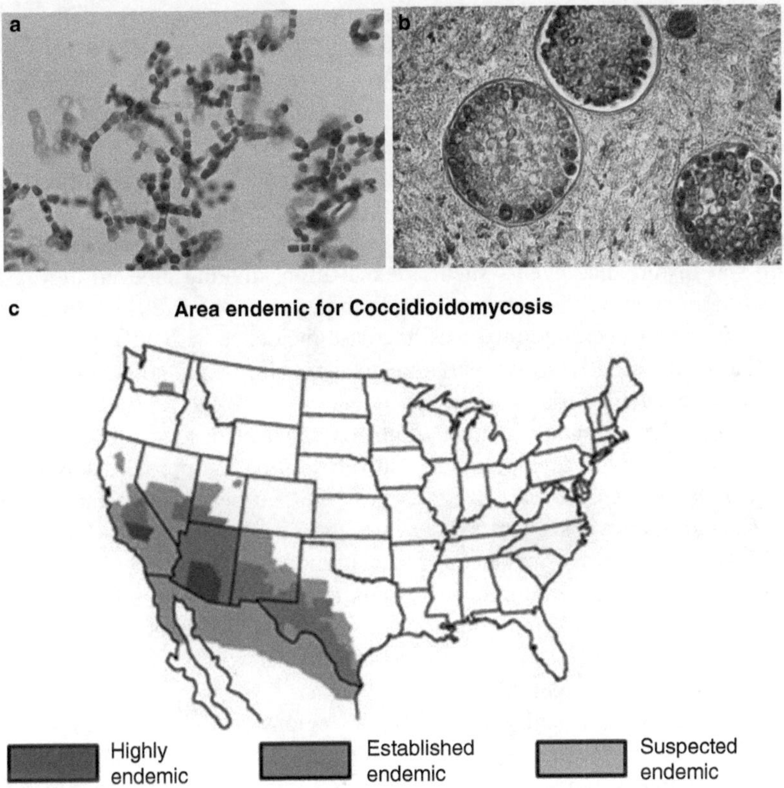

Fig. 1.3 Coccidioidomycosis, also known as Valley Fever, is caused by the inhalation of arthroconidia derived from *Coccidioides* species, such as *C. immitis*. The habitat of the dimorphic fungus includes the soil of semiarid areas of the U.S. where it grows in the environment as a filamentous hyphal form producing arthroconidia (**a**; Photo courtesy of CDC/Dr. Lucille K. Georg; CDC Public Health Image Library (PHIL) ID#: 12196). Inhalation of arthroconidia results in the conversion to a large spherule that contains endospores (**b**; Photo courtesy of CDC/Dr. Lucille K. Georg; CDC Public Health Image Library (PHIL) ID#: 14641). Coccidioidomycosis is endemic in areas such as the southwestern U.S. The mycosis is reportable in endemic regions including California, New Mexico, Arizona and Nevada (**c**; Map courtesy of the Centers for Disease Control and Prevention); however, Coccidioidomycosis is not restricted to these regions as cases have been reported in Washington State, Mexico and South America

Mexico, regions of South America, and has recently been identified as far north as Washington State (Marsden-Haug et al. 2012). Maps of Coccidioidomycosis endemic regions have been produced following a survey of coccidioidin hypersensitivity among U.S. Naval recruits (Edwards and Palmer 1957) and outbreak cases (Ampel 2009, 2011). The regions in the U.S. with the highest coccidioidin hypersensitivity responses occur in the San Joaquin Valley, south-central Arizona, and the lower western Rio Grande Valley in Texas (Ampel 2011). A map showing these endemic regions is presented in Fig. 1.3c.

Inhalation of infectious *Coccidioides* arthroconidia (Fig. 1.3a; 2–5 μm) can result in a pulmonary infection (Stevens et al. 2009) that consists of the conversion to an enlarged spherule (Fig. 1.3b; up to 120 μm) that is resistant to phagocytosis. The spherule contains many endospores (2–4 um) that can be released and spread to surrounding tissue and develop into new spherules repeating the cycle (Fig. 1.3b) (Ampel 2011). Given the ability of arthroconidia to aerosolize and cause respiratory infection, *C. immitis* is a select Biological Safety Level 3 agent (Ampel 2011). Inhalation exposure to arthroconidia occurs during soil disturbance events such as excavation, digging or even during natural disasters such as dust storms and earthquakes (Ampel 2011). In endemic areas that report Coccidioidomycosis, the case incidence in 2011 was 42.6 cases per 100,000 (CDC 2016b), in part due to a mobile aging population (Ampel 2011). Peak endemic periods are in spring and autumn, and the most productive regions of the fungus include Phoenix and Tucson (Ampel 2011). Coccidioidomycosis is asymptomatic and self-limited in approximately 60% of infected persons (Ampel 2011); however, it can further develop into disseminated Coccidioidomycosis, but this is rare and occurs in roughly 1% of cases (CDC 2016b). Primary cutaneous Coccidioidomycosis has been reported following traumatic inoculation, but this is rare compared to inhalation exposures (Ampel 2011). In symptomatic cases, a variety of symptoms are reported including a dry cough, rash, and fatigue that appear within one to three weeks following exposure (Ampel 2011).

Workers that disturb soil within endemic regions of Coccidioidomycosis are susceptible to occupational exposure to infectious arthroconidia. Occupations at risk of *Coccidioides* exposure are presented in Table 1.1 and include agricultural workers, construction crews, and telephone post diggers (Ampel 2011). Workers not located within an endemic region are also at risk, particularly textile workers that handle cotton contaminated with *C. immitis* (Gehlbach et al. 1973). Recent occupational health surveys have identified some additional U.S. occupational populations at risk of *C. immitis* arthroconidia exposure (NIOSH 2016). Construction workers at a solar farm in San Louis Obispo County contracted Coccidioidomycosis following occupational duties that disturbed the soil (Wilken et al. 2015). Dust exposure was identified to be the primary source of exposure for prison employees that acquired Coccidioidomycosis in central California (de Perio et al. 2015). Similarly, dry, dusty conditions, dust from an adjoining mine, and disruptive soil work were all described as potential sources of occupational exposure to *C. immitis* in 5 possible cases among outdoor television workers filming an event in California (Wilken et al. 2014; Wilken et al. 2015). U.S. armed forces personnel are also at risk for acquiring Coccidioidomycosis, especially in workers with an Asian Pacific Islander background. Over a ten year period, 58% of service members developed an extrapulmonary infection (Mease 2012). In California, the highest rate of occupational Coccidioidomycosis was among construction and agricultural workers (Das et al. 2012). Laboratory exposure to *Coccidioides* arthroconidia has also been reported and is one of the major organisms reported to cause laboratory acquired infections (Stevens et al. 2009).

The CDC (CDC 2016b), NIOSH (NIOSH 2016), as well as the California Department of Public Health have outlined strategies to minimize workforce exposures to infectious arthroconidia. A manuscript authored by Das and colleagues (Das et al. 2012) has also identified at risk occupational populations in California and presented procedures to control *Coccidioides* exposure. Stevens et al. have also presented methods to prevent laboratory-acquired infections (Stevens et al. 2009). Below is a list of administrative, engineering, and personal protective controls adapted from the three guidance documents described above that aim to minimize worker *Coccidioides* exposure in endemic regions.

- Educate employees about the risks associated with *Coccidioides* exposure.
- Minimize dust exposure in endemic areas during the peak season.
- Wet soil during soil-disruptive work.
- Excavate soil using heavy machinery and an enclosed cab fitted with HEPA filtration.
- Avoid working during dust storms.
- Wear PPE such as a NIOSH-certified respirator described in Table 1.1 during occupational tasks that involve dust generation.

1.3 Subcutaneous Fungal Infections

Certain occupations may be susceptible to acquiring subcutaneous fungal infections. Occupations at risk of these types of infections include farmers, laborers, landscapers, and gardeners (Chakrabarti et al. 2015; de Lima Barros et al. 2011). Subcutaneous fungal infections require a traumatic inoculation of the viable fungal spore or hyphal fragment into the dermis or subcutaneous tissue. Worker exposure scenarios include sustaining a penetrating injury while handling dead wood, thorned vegetation, or sphagnum moss (Chakrabarti et al. 2015; de Lima Barros et al. 2011; Hajjeh et al. 1997; Powell et al. 1978b). Subcutaneous infections occur in the lower extremities of the body such as the hands, forearm, lower legs and feet. Although not frequently reported in the U.S. workforce, recent literature has reported several cases of Sporotrichosis, Chromoblastomycosis, Phaeohyphomycosis, and Mucormycosis that are briefly summarized below.

Sporotrichosis is a rare occupational infection caused by the dimorphic fungal pathogen, *Sporothrix schenckii*, a species placed in the Ascomycota order, Ophiostomatales (de Lima Barros et al. 2011). In the U.S., the average annual rate of Sporotrichosis hospitalizations is 0.35 per million persons and predominantly affects immunocompromised patients (Gold et al. 2016). The fungus has a broad geographic distribution and is prominent in tropical and subtropical environments (Chakrabarti et al. 2015; de Lima Barros et al. 2011) where it proliferates as filamentous hyphae within the soil and decays organic substrates such as hay, sphagnum moss, roses, dead wood, and corn stalks (Chakrabarti et al. 2015). *S. schenckii* is related to fungal genera associated with bark beetles and have been

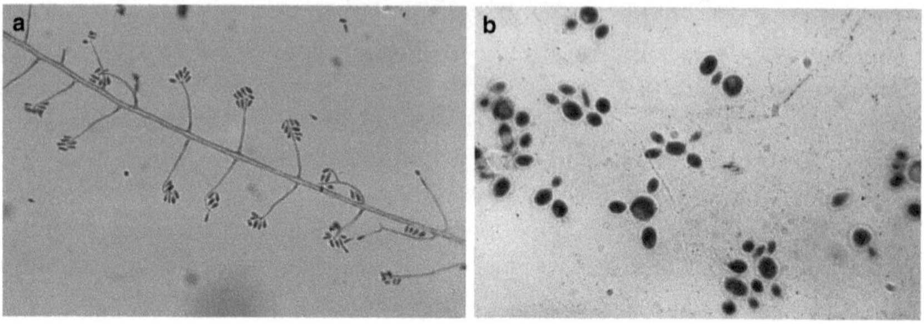

Fig. 1.4 Sporotrichosis is caused by a transcutaneous implantation of spores derived from the dimorphic fungal pathogen, *Sporothrix schenckii*. Implantation of *S. schenckii* clavate spores (**a**; Photo courtesy of CDC; CDC Public Health Image Library (PHIL) ID#: 10603) result in the conversion of the spore to a budding cigar-shaped yeast (**b**; Photo courtesy of CDC/Dr. Lucille K. Georg; CDC Public Health Image Library (PHIL) ID#: 3063)

shown to grow on dead wood but not live plants (Chakrabarti et al. 2015). A transcutaneous penetrating injury result in the implantation of the clavate spores (Fig. 1.4a) into the host followed by conversion into a budding cigar-shaped yeast (Fig. 1.4b; 4–6 μm). Sporotrichosis infections can include a non-painful lesion or chronic cutaneous, subcutaneous, or deeper involvement of the lymphatics (Chakrabarti et al. 2015; de Lima Barros et al. 2011).

Outbreaks of Sporotrichosis reported in the U.S. workforce are rare, and it is not a reportable disease (Chakrabarti et al. 2015). Workers at risk of exposure include those that handle contaminated soil or vegetation such as farmers, landscapers, gardeners, flower vendors, and animal husbandry workers (Chakrabarti et al. 2015; de Lima Barros et al. 2011). In the U.S., workers that handled sphagnum moss have been previously reported to be susceptible to Sporotrichosis (Coles et al. 1992; Dixon et al. 1991; Hajjeh et al. 1997; Powell et al. 1978b). In one outbreak, 14% of nursery workers that handled sphagnum moss developed lymphocutaneous Sporotrichosis (Hajjeh et al. 1997). Powell and colleagues also reported an outbreak in 17 Mississippi forestry workers that handled Wisconsin grown sphagnum moss (Powell et al. 1978b). In this study, the forestry workers handled contaminated sphagnum moss used to package pine tree seedlings (Powell et al. 1978b). Contaminated hay used in an Oklahoma spook house also resulted in an outbreak of Sporotrichosis (Dooley et al. 1997). In addition to environmental exposures, laboratorians are also susceptible to Sporotrichosis, especially those that handle *S. schenckii* (Thompson and Kaplan 1977).

Methods to protect workers from *S. schenckii* exposure are presented in Table 1.1 and include wearing PPE that cover the arms and lower extremities such as gloves and clothing. Protective clothing aids in the prevention of penetrating injuries sustained while handling dead vegetation during work-related tasks in farming, landscaping, and other horticulture occupations (Coles et al. 1992). These preventative measures have been shown to be protective in workers exposed to *S. schenckii* (Powell et al. 1978a). Given the association with sphagnum moss and packaging material, other controls to reduce exposure have included the use of alternative packaging materials such as wood chips or shredded paper (Coles et al. 1992).

In addition to Sporotrichosis, there are several other fungal species presented by Arenas and colleagues (Arenas et al. 2012) that are also capable of causing subcutaneous mycoses in U.S. workers following a transcutaneous injury sustained while performing an occupational task. Among the more widely reported infections caused by melanized fungi is Chromoblastomycosis (Queiroz-Telles et al. 2009). In the U.S. the incidence of Chromoblastomycosis has been estimated at 1 case per 8,625,000 persons and the causal agents are *Fonsecaea pedrosoi, Phialophora verrucosa, Cladophialophora carrionii* and *Fonsecaea monophora* (Queiroz-Telles et al. 2017, 2009). Implantation can occur following contact injuries with vegetation (wood, straw, dead grass), insects (insect stings), and even agricultural tools (Queiroz-Telles et al. 2017, 2009). Commonly associated with tropical and subtropical environments, implantation of infectious conidia (muriform or sclerotic cells) can result in the formation of papules that progress to verrucous lesions to flat plaques (Arenas et al. 2012; Queiroz-Telles et al. 2009; Spiker and Ferringer 2015) that are challenging to manage clinically (Queiroz-Telles et al. 2009). Photomicrographs of *P. verrucosa* and *C. carrionii* are presented in Fig. 1.5. Working populations that are affected include farmers and rural workers (Spiker and Ferringer 2015). Although rarely reported in the U.S., Chromoblastomycosis cases have been reported in immigrant populations (Burks et al. 1995). Cases located in Puerto Rico and the continental U.S. (Florida) have been reported in the peer-reviewed literature (Barwasser 1953; Binford et al. 1944; Burks et al. 1995; Emmons et al. 1941; Queiroz-Telles et al. 2017).

Phaeohyphomycosis is a subcutaneous mycosis caused by pigmented hyphae (Fig. 1.5c). Approximately 100 organisms have been identified to be associated with Phaeohyphomycosis and include *Alternaria* spp., *Exserohilum* spp., *Exophiala jeanselmei, Wangiella dermatitidis* and *Bipolaris* species (Kollipara et al. 2016). Implantation of the organisms occurs via splinters of wood resulting in an inflammatory cyst. The primary sites of Phaeohyphomycosis include the lower extremities such as the hands and feet but can occur in other regions of the body. Like Chromoblastomycosis, Phaeohyphomycosis cases in the U.S. workforce are rarely reported. In one U.S. based case-report, Phaeohyphomycosis due to *Curvularia lunata* was identified following an implantation injury sustained during an explosion at a chemical plant (Grieshop et al. 1993). Phaeohyphomycosis cases have also been reported in a boat builder (Tam and Freeman 1989) and a rural worker (Russo et al. 2010).

Mucormycosis is another subcutaneous infection that could be encountered within U.S. workplaces. Like the examples provided above, soil born fungi placed in the sub-phylum Mucoromycotina can grow and proliferate following an implantation injury and cause soft tissue infection to severe cases that result in a necrotizing fasciitis and some cases, even death (Neblett Fanfair et al. 2012; Snell and Tavakoli 2007). Although occupational examples in the U.S. are limited, Mucormycosis has gained renewed interest following recent natural disasters (Kouadio et al. 2012; Neblett Fanfair et al. 2012; Snell and Tavakoli 2007).

Exposure to the Mucoromycotina occurs following soil disturbance especially during or immediately following a disaster (Benedict and Park 2014; Kouadio et al. 2012), where spores are disseminated into the surrounding environment (Fig. 1.5d). Although not affecting workers, 13 cases of a necrotizing fasciitis were observed in Joplin, Missouri following an EF-5 tornado (Neblett Fanfair et al. 2012). *Apophysomyces trapeziformis* was identified to be the causal agent (Neblett Fanfair et al. 2012). Reports of Mucormycosis have followed flooding events (Bandino et al. 2015), tsunamis (Snell and Tavakoli 2007), volcanic eruption (Patino et al. 1991), and even blast injuries sustained during military operations in Afghanistan (Tribble and Rodriguez 2014). These data demonstrate that first responders and other emergency preparedness staff may be susceptible to acquiring subcutaneous infection following a natural disaster event.

Procedures to prevent worker exposure to Sporotrichosis have been published online by the CDC (CDC 2016e). Although no U.S. guidelines exist for the other subcutaneous fungal infections, methods to minimize exposure should follow those presented for Sporotrichosis. Avoiding transcutaneous injuries while handling

Fig. 1.5 Other subcutaneous fungal infections encountered by U.S. workers can include Chromoblastomycosis, Phaeohyphomycosis, and Mucormycosis. Chromoblastomycosis is caused by a transcutaneous implantation of spores derived from a variety of fungi such as *Phialophora verrucosa* (**a**; Photo courtesy of CDC; Dr. Leanor Haley; CDC Public Health Image Library (PHIL) ID#: 3215) and *Cladophialophora carrionii* (**b**; Photo courtesy of CDC; Dr. Lucille K. Georg; CDC Public Health Image Library (PHIL) ID#: 3061). Implantation of *Exserohilum rostratum* conidia (**c**; Photo courtesy of CDC; CDC Public Health Image Library (PHIL) ID#: 15143) can result in Phaeohyphomycosis. Transcutaneous implantation of Mucormycota sporangiospores (**d**; Photo courtesy of CDC; Dr. Lucille K. Georg CDC Public Health Image Library (PHIL) ID#:3960) can result in Mucormycosis especially following natural disasters

thorned vegetation, wood, or hay will prevent spore implantation. Similarly, wearing PPE such as gloves and protective clothing that cover the arms, lower limbs and feet are among the best preventative strategies for avoiding implantation injuries.

1.4 Cutaneous Fungal Infections

Fungi that colonize, grow, and propagate as either filamentous hyphae or yeasts on the skin and other keratinized structures including hair and nails can cause cutaneous fungal infections (Achterman and White 2013; Cafarchia et al. 2013). Penetration below the granular layer is uncommon (Achterman and White 2012). Fungi associated with cutaneous infections include the dermatophytes, *Scylatidium* spp., *Candida* species, and *Malassezia* spp. The dermatophytes are the most common source of cutaneous fungal infections and comprise three Ascomycota genera including *Trichophyton*, *Microsporum*, and *Epidermophyton* (Achterman and White 2013; Cafarchia et al. 2013). Dermatophyte diseases are varied and the most commonly reported include tinea corporis (ringworm), tinea pedis (athletes foot), tinea cruris (jock itch), and onychomycosis (nail infections) (Achterman and White 2012). There are approximately 25 pathogenic species of dermatophytes (Graser et al. 2008). The species *Trichophyton rubrum* accounts for 80–90% of all dermatophyte infections followed by *Trichophyton mentagrophytes* (Seebacher et al. 2008). Photomicrographs of these species are presented in Fig. 1.6. These fungi are acquired following contact with arthroconidia from geophilic (soil), zoophilic (animal), and anthropophilic (human) sources.(Achterman and White 2013) Anthropophilic species primarily infect humans, and the reoccurrence rate is high following antifungal treatments (Achterman and White 2012).

Dermatophytes secrete the keratinolytic enzyme, keratinase, as well as chymotrypsin to breakdown keratinous tissue (Achterman and White 2013). High molecular

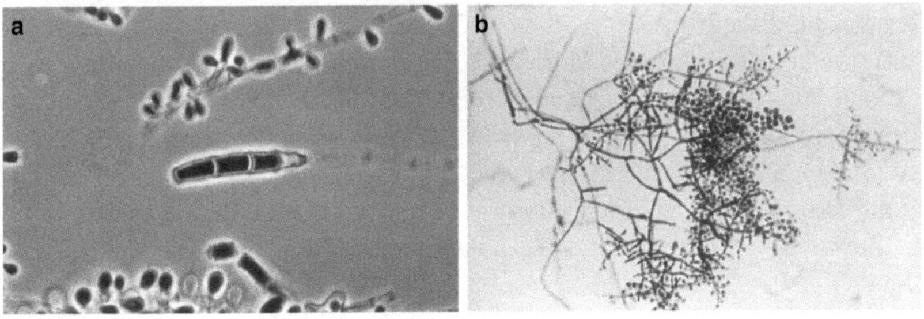

Fig. 1.6 Cutaneous mycoses are caused by fungi that grow and proliferate on the keratinized epithelium including the skin, hair, and nails. *Trichophyton rubrum* is among the most common species and produces filamentous hyphae with macro and numerous microconidia (**a**; Photo courtesy of CDC; Dr. Libero Ajello; CDC Public Health Image Library (PHIL) ID#: 11009). Another common Dermatophyte species, Trichophyton mentagrophytes also produces filamentous hyphae and microconidia (**b**; Photo courtesy of CDC/Dr. Lucille K. Georg; CDC Public Health Image Library (PHIL) ID#: 14955)

weight proteins and enzymes can also be released and result in irritant dermatitis such as the annular rings observed on the trunk of the body in cases of tinea corporis. Dermatophytes are ubiquitous and highly contagious. Dermatophyte infections are prevalent in various sociodemographic groups and a person is likely to experience an infection at least once in their life (Achterman and White 2012). Symptoms range from itching, irritation to hair loss (CDC 2016d). Routes of infection include direct as well as indirect contact with the contaminant; however, the mechanisms that underlie the disease are still the focus of human and animal studies (Achterman and White 2012). Identification of the source is difficult (Graser et al. 2008) as the organism can remain viable for up to a year and there is a high false-negative culture rate from clinical samples (Achterman and White 2012). Recent advances in molecular based techniques have improved the ability to detect and identify dermatophyte species (Achterman and White 2012).

Among the most common forms of human fungal infection are cutaneous mycoses. Recent estimates of therapeutic costs are placed at US$500 million (Achterman and White 2012; Graser et al. 2008). Workers are particularly susceptible to these types of fungal skin diseases, in particular dermatophyte infections (Achterman and White 2012). Recent studies have shown that Latin immigrant poultry processing and manual laborers are vulnerable to dermatophyte infections of the feet (Pichardo-Geisinger et al. 2014). Occluded footwear (rubber boots), vigorous prolonged activity, and wet environmental conditions are variables likely to account for the high incidence of dermatophyte infection (Pichardo-Geisinger et al. 2014). In farming environments, farmers may contract *Trichophyton* species from zoophilic sources such as cattle (Agnetti et al. 2014; Morrell and Stratman 2011). Hot and humid climates are additional variables that can increase a worker's susceptibility to dermatophyte infections (Leite et al. 2014). Other occupational groups that encounter morbidity and are at risk for these infections include miners, military personnel (Achterman and White 2012), as well as workers in meat and dairy processing, forestry and farming (Seebacher et al. 2008).

Procedures aimed to prevent worker exposure to cutaneous fungal infections have been published online by the CDC and are presented in Table 1.1 (CDC 2016d). Methods to minimize indirect or direct contact with infectious arthroconidia should be employed especially in public bathing or shared occupational change room facilities. Employing methods to minimize wet skin and clothes (socks and underwear), wearing non-occluding footwear, and washing hands after handling livestock are additional work practices that employees can take to assist in the prevention of contracting a dermatophyte infection (Prevention 2016d).

1.5 Other Occupational Fungal Hazards

Exposure to fungal particles can exacerbate respiratory morbidities such as hypersensitivity pneumonitis and asthma in U.S. workers (Eduard 2009). Fungal particles can include viable and non-viable spores or conidia, hyphal fragments, and even submicron fragments derived from the fungal cell wall (Afanou et al. 2014; Afanou et al. 2015; Green et al. 2005). Personal exposure to these particles, particularly in damp indoor environments, has been the subject of multiple research studies over the last two decades (Cox-Ganser 2015). The Institute of Medicine (IOM 2004) and the World Health Organization (WHO 2009) published consensus reports that documented associations between respiratory morbidities and damp indoor environments (IOM 2004). Indoor dampness and microbial growth can also result in adverse health effects (WHO 2009). Recent meta-analyses have confirmed these findings and shown associations between exposure to mold (visible mold and mold odor) in indoor damp environments and respiratory morbidities (development, current, and exacerbation) including allergy and asthma (Mendell et al. 2011; Quansah et al. 2012; Jaakkola et al. 2013). However, associations with specific microorganisms were suggestive (Kanchongkittiphon et al. 2015), and prevention of dampness and mold contamination were concluded to minimize health risks (Mendell et al. 2011). Recently, NIOSH (NIOSH Alert 2013) and two professional bodies including The American Society of Heating, Refrigerating, and Air-Conditioning Engineers (ASHRAE) (ASHRAE 2012) and The American Industrial Hygiene Association (AIHA) (AIHA 2013) have published guidance documents for the prevention of personal exposure to fungi in damp non-industrial built environments. In the U.S. alone, the economic burden of respiratory morbidities due to indoor dampness and mold exposure has recently been estimated to be $3.7 billion for allergic rhinitis, $1.9 billion for acute bronchitis, $15.1 billion for asthma morbidity and $1.7 billion for asthma mortality (Mudarri 2016).

The consensus documents, meta-analyses and economic cost studies described above highlight the need for the development of standardized exposure assessment methods to enable improved fungal identification with the aim to understand the potential contributions of specific fungal species to respiratory morbidity (Mudarri 2016). The development of molecular-based assessment methods within the last decade has significantly improved the research community's ability to detect and quantify fungal exposures in a variety of indoor and occupational environments. Studies utilizing these techniques have provided new insight into the contributions of once overlooked fungi derived from the phylum Basidiomycota in indoor, outdoor, and occupational environments (Green et al. 2016; Rittenour et al. 2014; Pashley et al. 2012; Adams et al. 2015; Dannemiller et al. 2014). Dannemiller and colleagues used next-generation DNA sequencing and showed that increased asthma risk in children was associated with both lower *Cryptococcus* diversity and the genus *Volutella* (Dannemiller et al. 2016a; Dannemiller et al. 2014). Higher fungal richness has also been shown to be related to pets, water damage, and urban homes (Dannemiller et al. 2016b) and decreased richness observed with increasing humidity (Dannemiller et al. 2016c). Fungi identified in next-generation DNA sequencing studies include, *Volutella*, *Cryptococcus*, *Kondoa*, *Malassezia*, *Rhodotorula*, *Trichosporon*, *Ustilago*, *Leptosphaerulina*, and *Epicoccum* species (Adams et al. 2015; Dannemiller et al. 2016a, b; Dannemiller et al. 2014; Dannemiller et al. 2016c; Green et al. 2016; Pashley et al. 2012; Rittenour et al. 2014). Species placed in the Basidiomycota class Agaricomycetes that account for over one-fifth of all fungal diversity (Hibbett et al. 2014) have additionally been identified in next generation DNA sequencing studies conducted in U.S. indoor environments (Green et al. 2016; Rittenour et al. 2014).

Other molecular approaches such as mold-specific quantitative polymerase chain reaction (MSQPCR) have been developed within the last decade. This method was developed by Vesper and colleagues at the U.S. Environmental Protection Agency with the aim of developing a national relative moldiness index for U.S. homes (Vesper et al. 2007). The MSQPCR panel includes 36 indicator fungi that consist of 26 fungi commonly cultured in water damaged environments (group 1) and 10 fungi that are commonly identified in outdoor or non-water damaged environments (Vesper et al. 2007). These data can be used to calculate an "Environmental Relative Moldiness Index or ERMI" value that ranges between −10 and 20. This value can be used to estimate mold burden in studies evaluating indoor fungal exposure and respiratory morbidity (Vesper et al. 2007). MSQPCR is utilized in a variety of commercial labs in the U.S. and the United Kingdom (Norback and Cai 2015), and has been used in research studies to estimate mold burden in U.S. indoor environments (Vesper et al. 2009; Vesper et al. 2011). Recent studies have also used MSQPCR in a prospective study of asthma development to investigate the various fungal species present in the homes of asthmatics. Reponen et al. (Reponen et al. 2012) identified *Aspergillus ochraceus*, *A. unguis* and *Penicillium variable*, (both individually and when summed) were in significantly higher concentrations in the homes of infants that developed physician-diagnosed asthma at age 7. The development of MSQPCR and ERMI methods have facilitated improved quantitation of fungi in indoor settings and enabled the identification of associations between members of the Eurotiales and asthma development (Reponen et al. 2012).

Worker exposure to considerably higher concentrations of fungal particles as high as 10^8 colony forming units per cubic meter can occur in some occupational environments (Eduard 2009). Workers that handle, disturb or process organic material such farmers, sawmill workers, waste collectors, mushroom processing workers, or workers that handle grain, hay or straw are especially susceptible to exposure to high concentrations of fungal particles (Eduard 2009). These worksite exposures can result in pulmonary health effects, such as hypersensitivity pneumonitis and asthma and have been previously reviewed by Eduard (Eduard 2009) as well as Nordgren and Bailey (Nordgren and Bailey 2016). Studies that have assessed occupational environments have predominantly utilized traditional methods of assessment including viable culture or microscopic identification of airborne spores. Although these methodological approaches have provided much-needed insight into worker exposures, these methods can underestimate fungi that either cannot grow on standard nutrient medium or cannot be differentiated and identified based on similar morphologies. Using culture-independent approaches, Green et al. showed the contribution of fungi placed in the phylum Basidiomycota was 46-fold higher compared to the species identified in the original culture analysis (Green et al. 2016). Also, fungi that did not grow on nutrient media in the initial culture analysis were resolved including obligate yeasts and xerophilic species, such as *A. halophilicus* (Green et al. 2016). These overlooked fungal exposure sources also contain high molecular weight proteins, secondary metabolites and possibly other immunostimulatory molecules. The results of these preliminary culture-independent analyses suggest that fungal identification may be under quantified using traditional methods of assessment and the results of next-generation sequencing surveys possibly highlight additional fungi that could impact the respiratory health of a worker.

Recent European studies have examined fungal contaminants in a variety of occupational environments including waste processing (Viegas et al. 2015), sawmills (Faerden et al. 2014), swine production (Viegas et al. 2016), and poultry industries (Sabino et al. 2012). Reviews of these working environments are the focus of several chapters presented within this book. In contrast, U.S. occupational health studies published since 2009 have predominantly focused on worker exposures and adverse health effects within damp indoor environments, especially following water damaging events sustained during natural disasters such as hurricanes and floods. In one survey of New York State School teachers where 24% of schools rate indoor air quality as unsatisfactory, current asthma was associated with visible mold, moldy odors, in addition to dust and perfume odors (Kielb et al. 2015). A cross-sectional health and environmental survey of workers in a water-damaged building also showed that subjects with building related rhinosinusitis with higher fungal exposure had a higher odds of developing building-related asthma compared to the lower fungal exposure group (Park et al. 2012). In a water damaged New Orleans school, employees had a higher prevalence of work-related rashes, nasal, and lower respiratory symptoms (Thomas et al. 2012). The data derived from these U.S. school environments further highlight the contribution of fungi to personal exposures within contaminated environments; however, these studies were survey-based and did not provide a detailed synopsis of the diversity of worker fungal exposures.

Several recent U.S. occupational health studies have assessed worker exposures to fungal bioaerosols following natural disasters and flooding events. Water damage and mold were identified as a concern from respondents that were impacted by flood damage following Hurricane Sandy (Burger and Gochfeld 2015) In a study of U.S. Coast Guard responders to Hurricanes Katrina and Rita, personal exposure to fungi was reported from contaminated building materials or water sources as a consequence of their work related tasks (Rusiecki et al. 2014). A positive association between mold exposure and sinus infection was identified among the U.S. Coast Guard responders (Rusiecki et al. 2014). Homes flooded by Hurricane Katrina had a higher burden of fungi compared to non-flooded homes and frequently detected fungi included *Aspergillus*, *Cladosporium* and *Paecilomyces*, *Penicillium* and *Trichoderma* species (Chew et al. 2006; Rao et al. 2007; Bloom et al. 2009). In Cedar Rapids, Iowa, homes undergoing remediation had higher levels of fungi and (1 → 3)-β-d-glucan compared to post-remediated homes (Hoppe et al. 2012). Residents living in homes undergoing remediation also had a higher incidence of doctor diagnosed allergies, self-reported wheeze, and breathing problems (Hoppe et al. 2012). Similarly, flooded homes in Boulder,

Colorado had a higher relative abundance of *Penicillium* species compared to non-flooded homes (Emerson et al. 2015). Differences in fungal diversity between flooded and non-flooded homes were additionally observed (Emerson et al. 2015). Remediation of these environments was ultimately found to improve the quality and respiratory health of the residents in these flooded locations.

The fungal contamination that resulted in homes that sustained flooding following Hurricane Katrina represented a burden for restoration workers (Rando et al. 2014). Rando et al. assessed exposure of New Orleans workers that were conducting disturbance activities such as mold remediation (Rando et al. 2014). Immediately following the Hurricane in 2005, demolition work related tasks resulted in the highest respirable dust exposures of which 17.6% of samples exceeded the permissible exposure limit of 5 mg per m^{-3}. During the same sampling interval, $(1 \rightarrow 3)$-β-d-glucan was as high as 118 μg per m^{-3}. Respirable dust and $(1 \rightarrow 3)$-β-d-glucan measures were also shown to decrease in later sampling collection periods (Rando et al. 2014). The results of these studies indicate that exposure to respirable particles that can include mold during disturbance activities performed as part of recovery operations can be high for first responders, recovery crews, and remediation workers. Based on this growing body of natural disaster literature, occupational exposure to mold has been identified as a growing public and occupational health concern among the disaster medicine community (Johanning et al. 2014).

Methods to minimize fungal exposure in contaminated indoor and occupational environments include removing water intrusion and to dry wet areas. These remediation efforts should occur within the first 48 hours of a water infiltration event. To protect workers from personal exposure within these environments, engineering controls such as HEPA filtration in combination with PPE including respiratory protection and protective clothing will assist in minimizing worker exposures. Guidelines for methods to minimize and reduce worker exposure are presented in Johanning et al. (Johanning et al. 2014)

1.6 Conclusions

The recent literature has shown that exposure to a wide diversity of fungi, including primary, subcutaneous or cutaneous pathogens can occur in various U.S. occupations. The primary workplace exposures result from inhalation of fungal spores and fragments, and these particles can exacerbate respiratory morbidities including but not limited to allergy, asthma, and hypersensitivity pneumonitis. However, due to the limitations and the current lack of standardized methods for assessing mold exposure, consensus threshold levels for fungal contaminants in indoor and occupational environments have not been established. To address this knowledge gap, Eduard reviewed human and animal model studies that evaluated exposure to mold and associated toxicological endpoints. A lowest observed effect level of 1×10^5 culturable fungal spores per m^3 in nonsensitized populations was proposed (Eduard 2009). The utilization of molecular-based methods in exposure assessment studies has significantly improved the ability to detect and quantify fungi in indoor as well as occupational environments in the U.S. The development of standardized qPCR panels for biomarker fungi in specific environments will further improve investigators' ability to assess exposure to specific species in indoor environments as well as detect endemic, subcutaneous, and cutaneous mycoses. However, standardization of extraction methods, amplification, primers, gene regions, and sequencing approaches are all parameters that require further evaluation to allow reproducible comparison between studies. Implementation of these methodologies in exposure assessment and occupational health studies will additionally assist in the generation of data that can be used to derive occupational exposure limits (OEL). These OELs would then have application in diverse occupational workforces and sectors in the U.S.

Based on the review of recent U.S. occupational health studies in this chapter, workers that handle vegetation or organic materials, engage in soil disturb activities, or work in damp indoor environments are at increased risk of exposure to fungi. Workers engaging in these tasks described throughout this chapter should wear PPE, such as a NIOSH approved respirator or protective clothing to protect the worker from inhalation exposures or transcutaneous injuries. In addition to PPE, engineering and administrative controls could also be employed to minimize worker exposure. Engineering controls could include providing improved ventilation within indoor environments or enclosed processing operations or HEPA filtration in the cabs of heavy excavating equipment to protect workers operating in endemic areas of dimorphic fungal pathogens. Administrative controls could also include training programs focused to educate employees about potential exposure to microbial hazards, and encouraging employees to report work-related symptoms to their supervisor and healthcare provider. In addition to these preventative measures, the standardization of methodological approaches will enable improved surveillance of occupational fungal exposure and will provide improved insight into the various occupational tasks and risk factors that result in the aerosolization of spores and other particles derived from fungi.

References

Achterman RR, White TC (2012) A foot in the door for dermatophyte research. PLoS Pathog 8(3):e1002564

Achterman RR, White TC (2013) Dermatophytes. Curr Biol 23(13):R551–R552. doi:10.1016/j.cub.2013.03.026

Adams RI, Bhangar S, Pasut W et al. (2015) Chamber bioaerosol study: outdoor air and human occupants as sources of indoor airborne microbes. PLoS One 10(5):e0128022. doi:10.1371/journal.pone.0128022

Afanou KA, Straumfors A, Skogstad A et al. (2014) Submicronic fungal bioaerosols: high-resolution microscopic characterization and quantification. Appl Environ Microbiol 80(22):7122–7130. doi:10.1128/aem.01740-14

Afanou KA, Straumfors A, Skogstad A et al. (2015) Profile and morphology of fungal aerosols characterized by field emission scanning electron microscopy (FESEM). Aerosol Sci Technol 49(6):423–435. doi:10.1080/02786826.2015.1040486

Agnetti F, Righi C, Scoccia E et al. (2014) Trichophyton verrucosum infection in cattle farms of Umbria (Central Italy) and transmission to humans. Mycoses 57(7):400–405. doi:10.1111/myc.12174

American Industrial Hygiene Association (AIHA) (2013) Position statement on mold and dampness in the built environment. https://www.aiha.org/...affairs/PositionStatements/P-Mold-03-26-13.pdf

American Society of Heating, Refrigerating and Air-Conditioning Engineers ASHRAE (2012) Position document on limiting indoor mold and dampness in buildings. https://www.ashrae.org/.../PositionDocuments/ASHRAE—Limiting-Indo

Ampel NM (2009) Coccidioidomycosis: a review of recent advances. Clin Chest Med 30(2):241–251. doi:10.1016/j.ccm.2009.02.004

Ampel NM (2011) Coccidioidomycosis. In: Kauffman CA, Pappas PG, Sobel JD, Dismukes WE (eds) Essentials of clinical mycology. Springer New York, New York, NY, pp 349-366. doi:10.1007/978-1-4419-6640-7_20

Arenas R, Moreno-Coutiño G, Welsh O (2012) Classification of subcutaneous and systemic mycoses. Clin Dermatol 30(4):369–371. doi:10.1016/j.clindermatol.2011.09.006

Bandino JP, Hang A, Norton SA (2015) The infectious and noninfectious dermatological consequences of flooding: a field manual for the responding provider. Am J Clin Dermatol 16 (5):399–424. doi:10.1007/s40257-015-0138-4

Barwasser NC (1953) Chromoblastomycosis: thirteenth reported case in the United States. J Am Med Assoc153(6):556–556. doi:10.1001/jama.1953.02940230028006f

Benedict K, Mody RK (2016) Epidemiology of histoplasmosis outbreaks, United States, 1938–2013. Emerg Infect Dis 22(3):370–378. doi:10.3201/eid2203.151117

Benedict K, Park BJ (2014) Invasive fungal infections after natural disasters. Emerg Infect Dis 20(3):349–355. doi:10.3201/eid2003.131230

Binford CH, Hess G, Emmons CW (1944) Chromoblastomycosis: report of a case from continental united states and discussion of the classification of the causative fungus. Arch Dermatol Syphilol 49(6):398–402. doi:10.1001/archderm.1944.01510120012002

Bloom E, Grimsley L, Pehrson C et al. (2009) Molds and mycotoxins in dust from water-damaged homes in New Orleans after hurricane Katrina. Indoor Air 19(2):153–158

Burger J, Gochfeld M (2015) Concerns and perceptions immediately following superstorm Sandy: ratings for property damage were higher than for health issues. J Risk Res 18(2):249–265. doi:10.1080/13669877.2014.896401

Burks JB, Wakabongo M, McGinnis MR (1995) Chromoblastomycosis. A fungal infection primarily observed in the lower extremity. J Am Podiatr Med Assoc 85(5):260–264. doi:10.7547/87507315-85-5-260

Cafarchia C, Iatta R, Latrofa MS et al. (2013) Molecular epidemiology, phylogeny and evolution of dermatophytes. Infect Genet Evol 20:336–351. doi:10.1016/j.meegid.2013.09.005

Cannon R, Firth N, Lamont R, et al (2006) Fungi and fungal infections of the oral cavity. Oral Microbiol. Immunol:333–348

Centers for Disease Control and Prevention (CDC) (2016a) Blastomycosis. https://www.cdc.gov/fungal/diseases/blastomycosis/index.html

Centers for Disease Control and Prevention (CDC) (2016b) Coccidioidomycosis. https://www.cdc.gov/fungal/diseases/coccidioidomycosis/index.html

Centers for Disease Control and Prevention (CDC) (2016c) Histoplasmosis. https://www.cdc.gov/fungal/diseases/histoplasmosis/index.html

Centers for Disease Control and Prevention (CDC) (2016d) Ringworm. https://www.cdc.gov/fungal/diseases/ringworm/index.html

Centers for Disease Control and Prevention (CDC) (2016e) Sporotrichosis. https://www.cdc.gov/fungal/diseases/sporotrichosis/index.html

Chakrabarti A, Bonifaz A, Gutierrez-Galhardo MC et al. (2015) Global epidemiology of sporotrichosis. Medi Mycol 53(1):3–14

Chew GL, Wilson J, Rabito FA et al. (2006) Mold and endotoxin levels in the aftermath of Hurricane Katrina: a pilot project of homes in New Orleans undergoing renovation. Environ Health Perspect 114(12):1883–1889

Coles FB, Schuchat A, Hibbs JR et al. (1992) A multistate outbreak of sporotrichosis associated with sphagnum moss. Am J Epidemiol 136(4):475–487

Cote E, Barr SC, Allen C (1997) Possible transmission of Blastomycosis dermatitidis via culture specimen. J Am Vet Med Assoc 210(4):479–480

Cox-Ganser JM (2015) Indoor dampness and mould health effects – ongoing questions on microbial exposures and allergic versus nonallergic mechanisms. Clin Exp Allergy 45(10):1478–1482. doi:10.1111/cea.12601

Dannemiller KC, Gent JF, Leaderer BP, et al. (2016a) Indoor microbial communities: influence on asthma severity in atopic and nonatopic children. J Allergy Clin Immunol. doi:10.1016/j.jaci.2015.11.027

Dannemiller KC, Gent JF, Leaderer BP et al. (2016b) Influence of housing characteristics on bacterial and fungal communities in homes of asthmatic children. Indoor Air 26(2):179–192. doi:10.1111/ina.12205

Dannemiller KC, Mendell MJ, Macher JM et al. (2014) Next-generation DNA sequencing reveals that low fungal diversity in house dust is associated with childhood asthma development. Indoor Air 24(3):236–247

Dannemiller KC, Weschler CJ, Peccia J (2016c) Fungal and bacterial growth in floor dust at elevated relative humidity levels. Indoor Air. doi:10.1111/ina.12313

Das R, McNary J, Fitzsimmons K et al. (2012) Occupational coccidioidomycosis in California: outbreak investigation, respirator recommendations, and surveillance findings. J Occup Environ Med 54(5):564–571. doi:10.1097/JOM.0b013e3182480556

de Lima Barros MB, de Almeida Paes R, Schubach AO (2011) Sporothrix schenckii and Sporotrichosis. Clinic Microbiol Rev 24(4):633–654

de Perio MA, Niemeier RT, Burr GA (2015) Coccidioides exposure and coccidioidomycosis among prison employees, California, United States. Emerg Infect Dis 21(6):1031–1033. doi:10.3201/eid2106.141201

Dixon DM, Salkin IF, Duncan RA et al. (1991) Isolation and characterization of Sporothrix schenckii from clinical and environmental sources associated with the largest U.S. epidemic of sporotrichosis. J Clin Microbiol 29(6):1106–1113

Dooley DP, Bostic PS, Beckius ML (1997) Spook house sporotrichosis. A point-source outbreak of sporotrichosis associated with hay bale props in a Halloween haunted-house. Arch Intern Med 157(16):1885–1887

Eduard W (2009) Fungal spores: a critical review of the toxicological and epidemiological evidence as a basis for occupational exposure limit setting. Crit Rev Toxicol 39(10):799–864. doi:10.3109/10408440903307333

Edwards PQ, Palmer CE (1957) Prevalence of sensitivity to coccidioidin, with special reference to specific and nonspecific reactions to coccidioidin and to histoplasmin. CHEST J 31(1):35–60

Emerson JB, Keady PB, Brewer TE et al. (2015) Impacts of flood damage on airborne bacteria and fungi in homes after the 2013 Colorado Front Range flood. Environ Sci Technol 49 (5):2675–2684. doi:10.1021/es503845j

Emmons CW, Hailey H, Hailey H (1941) Chromoblastomycosis: report of the sixth case from continental United States. J Am Med Assoc 116(1):25–28. doi:10.1001/jama.1941.02820010027006

Faerden K, Lund MB, Mogens Aalokken T et al. (2014) Hypersensitivity pneumonitis in a cluster of sawmill workers: a 10-year follow-up of exposure, symptoms, and lung function. Int J Occup Environ Health 20(2):167–173. doi:10.1179/2049396714y.0000000063

Gauthier GM (2015) Dimorphism in fungal pathogens of mammals, plants, and insects. PLOS Pathog 11(2):e1004608. doi:10.1371/journal.ppat.1004608

Gehlbach SH, Hamilton JD, Conant NF (1973) Coccidioidomycosis. An occupational disease in cotton mill workers. Arch Intern Med 131(2):254–255

Gold JA, Derado G, Mody RK et al. (2016) Sporotrichosis-associated hospitalizations, United States, 2000–2013. Emerg Infect Dis 22(10):1817–1820. doi:10.3201/eid2210.160671

Graser Y, Scott J, Summerbell R (2008) The new species concept in dermatophytesa polyphasic approach. Mycopathologia 166(5–6):239–256. doi:10.1007/s11046-008-9099-y

Green BJ, Lemons AR, Park Y, et al. (2016) Assessment of fungal diversity in a water-damaged office building. J Occup Environ Hyg. doi:10.1080/15459624.2016.1252044

Green BJ, Schmechel D, Summerbell RC (2011) Aerosolized fungal fragments. In: Fundamentals of mold growth in indoor environments and strategies for healthy living, Springer, p 211–243

Green BJ, Sercombe JK, Tovey ER (2005) Fungal fragments and undocumented conidia function as new aeroallergen sources. J Allergy Clin Immunol 115(5):1043–1048. doi:10.1016/j.jaci.2005.02.009

Grieshop TJ, Yarbrough III D, Farrar WE (1993) Case report: phaeohyphomycosis due to Curvularia lunata involving skin and subcutaneous tissue after an explosion at a chemical plant. Am J Med Sci 305(6):387–389

Hajjeh R, McDonnell S, Reef S et al. (1997) Outbreak of sporotrichosis among tree nursery workers. J Infect Dis 176(2):499–504

Hawksworth DL (2001) The magnitude of fungal diversity: the 1.5 million species estimate revised. Mycol Res 105:1422–1432

Hibbett DS, Bauer R, Binder M, et al. (2014) 14 Agaricomycetes. In: McLaughlin JD, Spatafora WJ (eds) Systematics and evolution: part A. Springer, Berlin Heidelberg, pp 373–429. doi:10.1007/978-3-642-55318-9_14

Hoppe KA, Metwali N, Perry SS et al. (2012) Assessment of airborne exposures and health in flooded homes undergoing renovation. Indoor Air 22(6):446–456. doi:10.1111/j.1600-0668.2012.00785.x

Huhn GD, Austin C, Carr M, et al. (2005) Two outbreaks of occupationally acquired histoplasmosis: more than workers at risk. Environ Health Perspect 585–589

Institute of Medicine (IOM) (2004) Damp indoor spaces and health. The National Academies Press, Washington, DC

Jaakkola MS, Quansah R, Hugg TT et al. (2013) Association of indoor dampness and molds with rhinitis risk: a systematic review and meta-analysis. J Allergy Clin Immunol 132(5): 1099–1110.e1018. doi:10.1016/j.jaci.2013.07.028

Johanning E, Auger P, Morey PR et al. (2014) Review of health hazards and prevention measures for response and recovery workers and volunteers after natural disasters, flooding, and water damage: mold and dampness. Environ Health Prev Med 19(2):93–99. doi:10.1007/s12199-013-0368-0

Kanchongkittiphon W, Mendell MJ, Gaffin JM et al. (2015) Indoor environmental exposures and exacerbation of asthma: an update to the 2000 review by the Institute of Medicine. Environ Health Perspect 123(1):6–20. doi:10.1289/ehp.1307922

Kielb C, Lin S, Muscatiello N et al. (2015) Building-related health symptoms and classroom indoor air quality: a survey of school teachers in New York State. Indoor Air 25(4):371–380. doi:10.1111/ina.12154

Klein BS, Tebbets B (2007) Dimorphism and virulence in fungi. Curr Opin Microbiol 10(4): 314–319. doi:10.1016/j.mib.2007.04.002

Klein BS, Vergeront JM, Weeks RJ et al. (1986) Isolation of Blastomyces dermatitidis in soil associated with a large outbreak of blastomycosis in Wisconsin. N Engl J Med 314(9):529–534

Kollipara R, Peranteau AJ, Nawas ZY et al. (2016) Emerging infectious diseases with cutaneous manifestations: fungal, helminthic, protozoan and ectoparasitic infections. J Am Acad Dermatol 75(1):19–30

Kouadio IK, Aljunid S, Kamigaki T et al. (2012) Infectious diseases following natural disasters: prevention and control measures. Expert Rev Anti Infect Ther 10(1):95–104. doi:10.1586/eri.11.155

Kroll RR, Grossman RF (2013) Pulmonary blastomycosis in a professional diver: an occupational risk. Can Respir J 20(5):340–342

Leite Jr. DP, Amadio JV, Simoes Sde A et al. (2014) Dermatophytosis in military in the central-west region of Brazil: literature review. Mycopathologia 177(1–2):65–74. doi:10.1007/s11046-013-9714-4

Lenhart SW, Schafer MP, Hajjeh RA Histoplasmosis: protecting workers at risk. In: US Department of Health and Human Services, Public Health Service, Centers for Disease Control and Prevention, National Institute for Occupational Safety and Health, DHHS (NIOSH) Publication No, 1997. Citeseer

Litvinov IV, St-Germain G, Pelletier R et al. (2013) Endemic human blastomycosis in Quebec, Canada, 1988–2011. Epidemiol Infect 141(6):1143–1147. doi:10.1017/s0950268812001860

Manos NE, Ferebee SH, Kerschbaum WF (1956) Geographic variation in the prevalence of histoplasmin sensitivity. CHEST J 29(6):649–668

Marsden-Haug N, Goldoft M, Ralston C, et al. (2012) Coccidioidomycosis acquired in Washington state. Clin Infect Dis. 1028

Mease L (2012) Pulmonary and extrapulmonary coccidioidomycosis, active component, U.S. Armed Forces, 1999–2011. MSMR 19(12):2–4

Mendell MJ, Mirer AG, Cheung K et al. (2011) Respiratory and allergic health effects of dampness, mold, and dampness-related agents: a review of the epidemiologic evidence. Environ Health Perspect 119(6):748–756. doi:10.1289/ehp.1002410

Morrell J, Stratman E (2011) Primary care and specialty care delays in diagnosing Trichophyton verrucosum infection related to cattle exposure. J Agromed 16(4):244–250. doi:10.1080/1059924x.2011.605715

Mudarri DH (2016) Valuing the economic costs of allergic rhinitis, acute bronchitis, and asthma from exposure to indoor dampness and mold in the US. J Environ Public Health 2016:2386596. doi:10.1155/2016/2386596

National Institute for Occupational Safety and Health (NIOSH) (2016) Coccidioidomycosis (valley fever). https://www.cdc.gov/niosh/topics/valleyfever/prevention.html

National Institute for Occupational Safety and Health (NIOSH) Alert (2013) Preventing occupational respiratory disease from exposures caused by dampness in office buildings, schools, and other non-industrial buildings. Principal contributors: Martin M, Cox-Ganser J, Kreiss K, Kanwal R, Sahakian N. National Institute for Occupational Safety and Health, Department of Health and Human Services, DHHS (NIOSH), 2013–2102

Neblett Fanfair R, Benedict K, Bos J et al. (2012) Necrotizing cutaneous mucormycosis after a tornado in Joplin, Missouri, in 2011. N Engl J Med 367(23):2214–2225. doi:10.1056/NEJMoa1204781

Norback D, Cai GH (2015) Dampness, indoor mould, fungal DNA and respiratory health – molecular methods in indoor epidemiology. Clin Exp Allergy 45(5):840–843. doi:10.1111/cea.12524

Nordgren TM, Bailey KL (2016) Pulmonary health effects of agriculture. Curr Opin Pulm Med 22(2):144–149. doi:10.1097/mcp.0000000000000247

Park JH, Kreiss K, Cox-Ganser JM (2012) Rhinosinusitis and mold as risk factors for asthma symptoms in occupants of a water-damaged building. Indoor Air 22(5):396–404. doi:10.1111/j.1600-0668.2012.00775.x

Pashley CH, Fairs A, Free RC et al. (2012) DNA analysis of outdoor air reveals a high degree of fungal diversity, temporal variability, and genera not seen by spore morphology. Fungal Biol 116(2):214–224

Patino JF, Castro D, Valencia A et al. (1991) Necrotizing soft tissue lesions after a volcanic cataclysm. World J Surg 15(2):240–247

Pichardo-Geisinger R, Mora DC et al. (2014) Comorbidity of tinea pedis and onychomycosis and evaluation of risk factors in Latino immigrant poultry processing and other manual laborers. South Med J 107(6):374–379. doi:10.14423/01.smj.0000450705.67259.26

Powell KE, Taylor A, Phillips BJ et al. (1978a) Cutaneous sporotrichosis in forestry workers. Epidemic due to contaminated Sphagnum moss. JAMA240(3):232–235

Powell KE, Taylor A, Phillips BJ et al. (1978b) Cutaneous sporotrichosis in forestry workers: Epidemic due to contaminated sphagnum moss. JAMA 240(3):232–235. doi:10.1001/jama.1978.03290030050018

Quansah R, Jaakkola MS, Hugg TT et al. (2012) Residential dampness and molds and the risk of developing asthma: a systematic review and meta-analysis. PLoS One 7(11):e47526. doi:10.1371/journal.pone.0047526

Queiroz-Telles F, de Hoog S, Santos DWC et al. (2017) Chromoblastomycosis. Clin Microbiol Revi30(1):233–276

Queiroz-Telles F, Esterre P, Perez-Blanco M et al. (2009) Chromoblastomycosis: an overview of clinical manifestations, diagnosis and treatment. Med Mycol 47(1):3–15. doi:10.1080/13693780802538001

Rando RJ, Kwon CW, Lefante JJ (2014) Exposures to thoracic particulate matter, endotoxin, and glucan during post-Hurricane Katrina restoration work, New Orleans 2005–2012. J Occup Environ Hyg 11(1):9–18. doi:10.1080/15459624.2013.839879

Rao CY, Riggs MA, Chew GL et al. (2007) Characterization of airborne molds, endotoxins, and glucans in homes in New Orleans after Hurricanes Katrina and Rita. Appl Environ Microbiol 73(5):1630–1634

Reponen T, Lockey J, Bernstein DI et al. (2012) Infant origins of childhood asthma associated with specific molds. J Allergy Clin Immunol 130(3):639–644.e635. doi:10.1016/j.jaci.2012.05.030

Rittenour WR, Ciaccio CE, Barnes CS et al. (2014) Internal transcribed spacer rRNA gene sequencing analysis of fungal diversity in Kansas City indoor environments. Environ Sci Process Impact 16(1):33–43. doi:10.1039/c3em00441d

Rusiecki JA, Thomas DL, Chen L et al. (2014) Disaster-related exposures and health effects among US Coast Guard responders to Hurricanes Katrina and Rita: a cross-sectional study. J Occup Environ Med 56(8):820–833. doi:10.1097/jom.0000000000000188

Russo JP, Raffaeli R, Ingratta SM et al. (2010) Cutaneous and subcutaneous phaeohyphomycosis. Skinmed 8(6):366–369

Sabino R, Faisca VM, Carolino E et al. (2012) Occupational exposure to Aspergillus by swine and poultry farm workers in Portugal. J Toxicol Environ Health A 75(22–23):1381–1391. doi:10.1080/15287394.2012.721170

Saccente M, Woods GL (2010) Clinical and laboratory update on blastomycosis. Clin Microbiol Rev 23(2):367–381

Sastre J, Ibanez M, Lopez M et al. (1990) Respiratory and immunological reactions among Shiitake (Lentinus edodes) mushroom workers. Clin Exp Allergy 20(1):13–19

Seebacher C, Bouchara JP, Mignon B (2008) Updates on the epidemiology of dermatophyte infections. Mycopathologia 166(5–6):335–352. doi:10.1007/s11046-008-9100-9

Siemieniuk RA, MacFadden DR, Matukas LM (2015) Disseminated blastomycosis with cutaneous involvement in a Canadian railroad worker. Lancet 385(9971):883. doi:10.1016/s0140-6736(14)61641-x

Snell BJ, Tavakoli K (2007) Necrotizing fasciitis caused by Apophysomyces elegans complicating soft-tissue and pelvic injuries in a tsunami survivor from Thailand. Plast Reconstr Surg 119(1):448–449. doi:10.1097/01.prs.0000233624.34950.f8

Spiker A, Ferringer T (2015) Chromoblastomycosis. Cutis 96(4):224. 267-268

Stevens DA, Clemons KV, Levine HB et al. (2009) Expert Opinion: what to do when there is coccidioides exposure in a laboratory. Clin Infect Dis 49(6):919–923. doi:10.1086/605441

Tam M, Freeman S (1989) Phaeohyphomycosis due to Phialophora richardsiae. Australas J Dermatol 30(1):37–40

Thomas G, Burton NC, Mueller C et al. (2012) Comparison of work-related symptoms and visual contrast sensitivity between employees at a severely water-damaged school and a school without significant water damage. Am J Ind Med 55(9):844–854. doi:10.1002/ajim.22059

Thompson D, Kaplan W (1977) Laboratory-acquired sporotrichosis. Sabouraudia 15(2):167–170

Thompson GR, Gomez BL (2015) Histoplasma, blastomyces, coccidioides, and other dimorphic fungi causing systemic mycoses. In: Manual of clinical microbiology, 11th Edn. Am. Soc. Microbiol., pp 2109–2127

Tribble DR, Rodriguez CJ (2014) Combat-related invasive fungal wound infections. Curr Fungal Infect Rep 8(4):277–286. doi:10.1007/s12281-014-0205-y

Vesper S, McKinstry C, Cox D et al. (2009) Correlation between ERMI values and other moisture and mold assessments of homes in the American Healthy Homes Survey. J Urban Health 86(6):850–860. doi:10.1007/s11524-009-9384-1

Vesper S, McKinstry C, Haugland R et al. (2007) Development of an environmental relative moldiness index for US homes. J Occup Environ Med 49(8):829–833. doi:10.1097/JOM.0b013e3181255e98

Vesper S, Wakefield J, Ashley P et al. (2011) Geographic distribution of environmental relative moldiness index molds in USA homes. J Environ Public Health 2011:242457. doi:10.1155/2011/242457

Viegas C, Faria T, dos Santos M et al. (2015) Fungal burden in waste industry: an occupational risk to be solved. Environ Monit Assess 187(4):199. doi:10.1007/s10661-015-4412-y

Viegas C, Faria T, dos Santos M, et al. (2016) Slaughterhouses fungal burden assessment: a contribution for the pursuit of a better assessment strategy. Int J Environ Res Public Health 13(3). doi:10.3390/ijerph13030297

Wilken JA, Marquez P, Terashita D et al. (2014) Coccidioidomycosis among cast and crew members at an outdoor television filming event—California, 2012. MMWR Morb Mortal Wkly Rep 63(15):321–324

Wilken JA, Sondermeyer G, Shusterman D et al. (2015) Coccidioidomycosis among workers constructing solar power farms, California, USA, 2011–2014. Emerg Infect Dis 21(11):1997–2005. doi:10.3201/eid2111.150129

World Health Organization (WHO) (2009) WHO guidelines for indoor air quality: dampness and mould. WHO Regional Office for Europe, Geneva

Chapter 2
Bacteria in Indoor and Occupational Settings

Mark A. Barnes

Abstract Bacteria are ubiquitously present in our environment. As such, humans are constantly exposed to a multitude of bacterial species include those that are a part of our natural flora, as well as opportunistic and pathogenic bacteria. These constant challenges may potentially result in development of pathogenic diseases, as well as other adverse health effects. Some of these adverse events include development of respiratory diseases such as Farmer's lung, subcutaneous and cutaneous infections, and bacteremia. Environmental exposures have the potential to affect all people equally, such as omnipresent endotoxin which may exert both protective and detrimental health effects, particularly in the context of respiratory health. In contrast, occupational exposures affect specific sections of the population. For example, workers in meat industries tend to exhibit a higher prevalence of colonization with methicillin-resistant *Staphylococcus aureus* than the general population. This chapter will examine various environmental and occupational sources of bacterial exposures, as well as the potential health consequences of those exposures.

Keywords Health effects · occupational exposure · environmental exposure · bacterial metabolites · endotoxin · antibiotic resistance

2.1 Introduction to Prokaryotes

2.1.1 Intro

Bacteria, along with Archaea, make up the organismal group termed prokaryotes. Key cellular properties attribute to the difference in form and function of prokaryotes compared to eukaryotes, which include humans. The total number of

M.A. Barnes (✉)
Allergy and Clinical Immunology Branch, Health Effects Laboratory Division, National Institute for Occupational Safety and Health, Centers for Disease Control and Prevention, Morgantown, WV 26505, USA
e-mail: meu9@cdc.gov

C. Viegas et al. (eds.), *Exposure to Microbiological Agents in Indoor and Occupational Environments*, DOI 10.1007/978-3-319-61688-9_2

prokaryotic cells on the earth has been estimated in the range of 4–6 × 10^{30} as opposed to just 7.3 billion humans, making its total biomass at least equal to, but potentially greater than that of plants (Whitman et al. 1998). Cells typically range in size from 1 to 10 μm, however, there are a number of submicron (*Thermodiscus spp*. and *Mycoplasma spp*.) and macro (*Thiomargarita namibiensis*) prokaryotic organisms (Schulz and Jørgensen 2001). Unlike eukaryotes, prokaryotes lack intracellular membrane-bound structures such as a nucleus and other specified organelles like mitochondria, endoplasmic reticulum, and Golgi apparatus, which are critical to the function and survival of higher organisms (Murat et al. 2010). Additionally, prokaryotes reproduce asexually, forming two or more daughter cells depending on the modes of cellular propagation (Angert 2005). In contrast to complex genetic recombination systems in eukaryotes, prokaryotic organisms gain genetic diversity via horizontal gene transfer methods – that is, by acquiring foreign DNA through viruses such as bacteriopahges, accepting circular, non-chromosomal DNA known as plasmids, or simply taking up free DNA from the environment (Gyles and Boerlin 2014). Despite their differences, prokaryotes share some commonalities with eukaryotes including the use of DNA as genetic material and ribosomes for protein synthesis. Furthermore, evidence suggests that prokaryotes utilize protein-bound microcompartments for storage/sequestration, gas vesicles to assist with buoyancy, and other protein and lipid-bound organelle-like structures for various processes (Murat et al. 2010; Kerfeld et al. 2005).

2.1.2 Archaea

Archaea were historically classified as bacteria and called Archaebacteria, but have since been considered a separate organismal domain due to the structural and genetic differences to bacteria (Woese 1994). However, the Archaea are structurally and genetically related to gram-positive bacteria. For instance, pseudopeptidoglycan molecules in the cell wall of Archaea species offer a form and function similar to peptidoglycan content of gram-positive bacteria while maintaining a different chemical makeup. Archaea and bacteria share similar molecular composition, yet archaea employ gene expression processes more closely resembling those of eukaryotic cells (Koonin et al. 1997). Though initially discovered in habitats with extreme temperature, salt, and pH conditions, Archaea have recently been found in milder environments such as swamps, soils, as well as human skin and gut, thus invalidating the previously used term extremophile as a general characteristic for all Archaea (Bang and Schmitz 2015). Commensalism between archaea and eukaryotes is well appreciated; archaea species have been identified in agricultural settings and pose a potential exposure risk to humans (Nehme et al. 2009; Blais Lecours et al. 2012), however, it is not yet known what, if any, role archaea play in human health and disease as there are currently no known immuno- or pathogenic factors in these species. As such, this section will focus on properties of bacterial species.

2.1.3 Physical Properties of Bacteria

Bacteria come in a few variants of shape at the single cell level but achieve morphological diversity by adopting various arrangements of their cells and cellular structure. The three basic shapes of bacterial cells are rods (bacillus), spherical (coccus), and spiral/helical (Salton and Kim 1996). Bacilli (Fig. 2.1b) arrange as single cells (bacillus), single ovoid cells (coccobacillus) (Fig. 2.1c), in pairs (diplobacilli), and in chain-like forms (streptobacilli); cocci (Fig. 2.1a) are paired (diplococci), link in chains (streptococci), and clump as tetrads or in grape-like clusters (staphylococci). Spiral bacteria (Fig. 2.1d) assume helical, comma-like, corkscrew, and filamentous forms (Young 2006; Salton and Kim 1996). In addition to their basic shapes, bacteria may also present secondary shapes or characteristics such as filamentous appendages known as flagella, shorter hair-like protrusions called pili, bifid protrusions, and cellular membrane protrusions termed

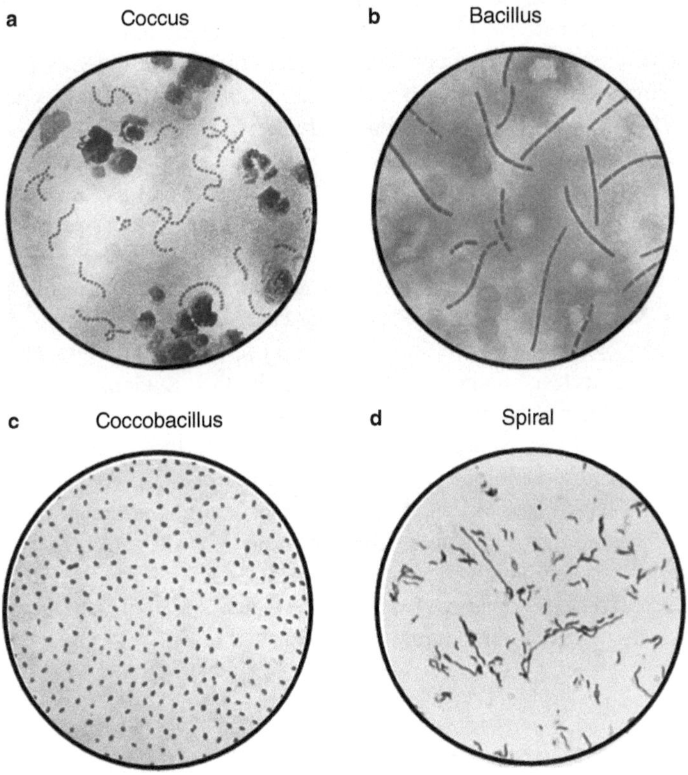

Fig. 2.1 Occupationally relevant cocci, bacillus, coccobacillus, and spiral bacteria. Photomicrographs of (**a**) *Streptococcus pyogenes*, (**b**) *Bacillus anthracis*, (**c**) *Brucella melitensis*, and (**d**) *Campylobacter jejuni*

prosthecae. Endospores, non-vegetative structures that are resistant to many thermal, radiation, and chemical challenges, are produced by some bacteria in the Firmicutes phylum (Poehlein et al. 2013; Thomas 2006; Nicholson et al. 2002). Some bacterial species also form capsules made of a viscous layer of polysaccharides, as well as secrete a slime-like mixture of polysaccharides, proteins, and nucleic acids called biofilm. Biofilms allow bacteria to coalesce and adhere to various surfaces (Hall-Stoodley et al. 2004). These structures variably contribute to survival, nutrient acquisition, motility, and virulence of bacteria.

With the advent of the gram stain, bacteria were classified as either gram-positive or gram-negative (Steinbach and Shetty 2001). This technique takes advantage of a simple generic difference in bacteria: gram-positive species have thick cell walls that contain an abundant amount of tightly-packed peptidoglycan, while gram-negative cell walls contain less peptidoglycans. As microbiologists increase their understanding of bacteria, it becomes evident that more suitable descriptors and advanced methods, such as 16S ribosomal RNA sequencing (Clarridge 2004; James 2010), are required for differentiating bacterial species.

Monoderm and diderm are more apt dichotomous descriptors of bacteria because they do not rely on staining properties of bacterial cell envelopes, rather natural physical properties of bacteria. Monoderm bacteria have one cellular membrane that is surrounded by a cell wall; on the other hand, diderms possess two cellular membranes that are separated by a thin cell wall (Fig. 2.2) (Gupta 2011; Sutcliffe 2010). Monoderm and diderm are the preferred terminology because there are always exceptions to the rule. For example, most species that are placed in the phylum Firmicutes would be classified as gram-positive; however, some bacteria resemble gram-negative bacteria. Moreover, some diderms exhibit gram-positive properties even though they are structurally and genetically similar to other diderms, and would therefore be misclassified if differentiated solely by gram stain. Mono- and diderm as identifying characteristics also allow for logical differentiation of bacterial evolution. Gupta puts forth that diderms developed the outer membrane to resist naturally-occurring antibiotics. Evidence for this mode of evolution is supported by the fact that most known antibiotics are produced by monoderm species, particularly *Streptomyces* spp., and that monoderms are more susceptible to natural antibiotics (Gupta 2011). Environmental selection pressures, including the need to develop antibiotic resistance, have contributed to the overall diversity of bacteria. For example, diversification of bacterial characteristics has led to species with atypical cell wall components, such as mycolic acid found in *Mycobacteria*, diderms without significant endotoxin (eobacateria) (Sutcliffe 2010), and classic endotoxin-expressing diderms.

2.1.4 Bacterial Taxonomy

According to the List of Prokaryotic names with Standing in Nomenclature (LPSN), there are 30 recognized phyla containing thousands of different bacterial

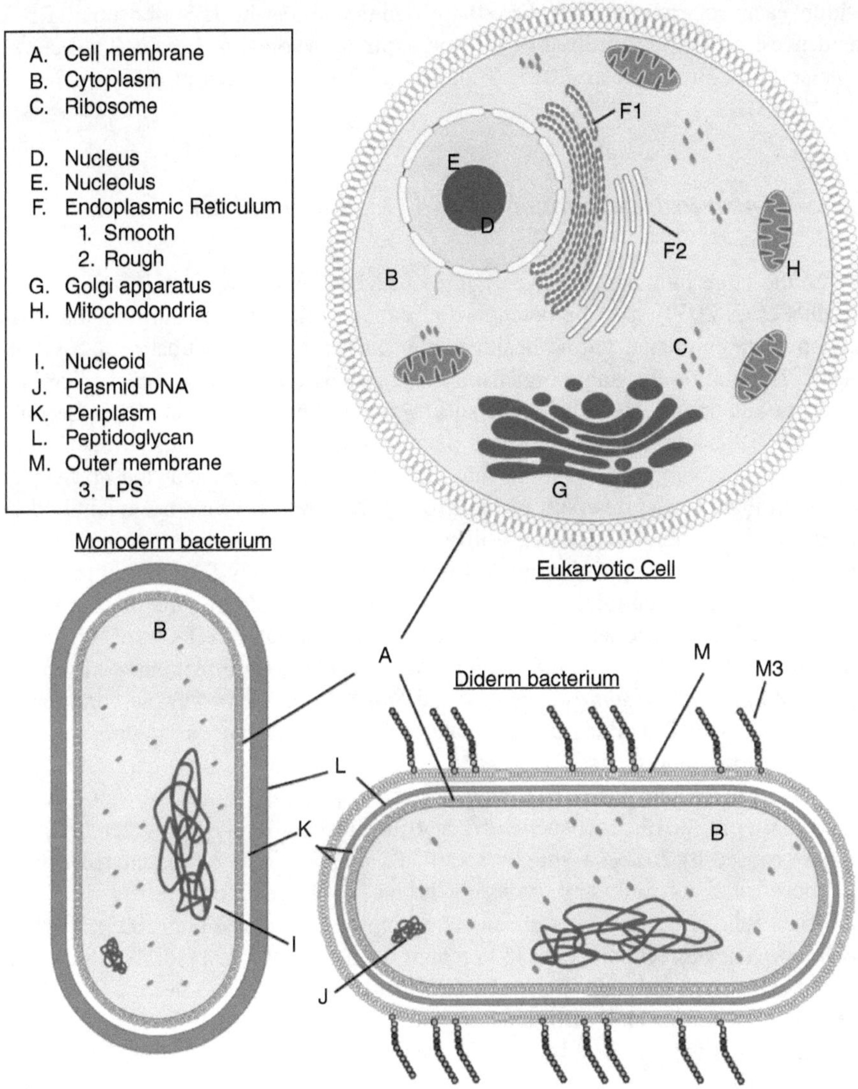

Fig. 2.2 Comparison of eukaryotic and prokaryotic cells. Schematic comparison of cellular differences in eukaryotes and prokaryotes, as well as similarities and differences in monoderm vs diderm bacteria

genera (Euzéby 1997). Even with the vast diversity of species across taxonomic ranks, approximately 40% of all recognized taxa belong to the Firmicutes and Actinobacteria phyla, which primarily consist of gram-positive monoderms, and also include endospore-forming bacteria (Sutcliffe 2010; Poehlein et al. 2013; Thomas 2006; Nicholson et al. 2002). Bacteria can be further grouped into so-called superphyla in addition to the 30 currently recognized phyla. Species within each superphylum are linked by commonalities in 16S ribosomal RNA sequences, as well as similarities in gene expression, proteins, and cell structures (Fuerst 2013; Brown et al. 2015; Yeoh et al. 2016; Sekiguchi et al. 2015; Rinke et al. 2013).

2.1.5 Intro to Environmental and Occupational Exposures

Given the large biomass, the wide range of niches, including natural human flora (Täubel et al. 2009), and the potential for pathogenicity, it is no surprise that bacteria pose a significant public health burden in indoor and occupational environments. Although only viable organisms can cause active infection, other microbial products and fragments also contribute to adverse health effects (Blais-Lecours et al. 2015). Bacterial endotoxin and peptidoglycan can contribute to respiratory morbidity in a variety of settings (Fig. 2.3). Further, species that are non-viable also contribute to the diversity of bioaerosols that are otherwise not quantified in exposure assessment studies that utilize viable culture techniques. Estimations of indoor bacterial contamination range from 10^1 to 10^6 colony forming units (CFU) per cubic meter for culturable species, while non-culturable may be present in the range of 10^3–10^8 bacterial cells/m^3. In addition to respiratory effects, occupational exposures can result in subcutaneous as well as cutaneous infections. Skin infections with group A *Streptococcus* spp. and *Staphylococcus aureus* are common in industries in which workers are at greater risk for transcutaneous injuries such as cuts, abrasions, burns, etc. (Lachapelle 2012). Occupational skin exposures may also give rise to rare infections, and in some cases, rare modes of transmission (Godshall et al. 2013; Lachapelle 2012). Brucellosis, a rarely encountered febrile disease caused by *Brucella* spp., is acquired via exposure to contaminated animal products, such as milk and cheese. Chronic cutaneous norcardiosis has been observed following high pressure injection injury (Lee and Yong 2005). *Listeria monocytogenes* infections occur in pregnant women, neo-nates, elderly adults, and immunocompromised individuals (CDC 2016b). Patient symptoms clinically manifest as bacteremia and infections of the central nervous system. In contrast, Listeria infections in healthy people generally present as gastroenteritis that requires a large inoculum, but can also cause cutaneous infections at a much lower inoculum (Godshall et al. 2013). Here we examine environmental exposures to various bacterial species and products, primarily endotoxin, paying special attention to bioaerosol exposures in indoor and occupational settings.

2.2 Methods to Discover, Analyze, and Enumerate Bacteria

The various methods of collection and analysis of bioaerosols have been thoroughly reviewed by Ghosh et al. (2015), therefore this section will only discuss the most important techniques. The most common methods of bioaerosol

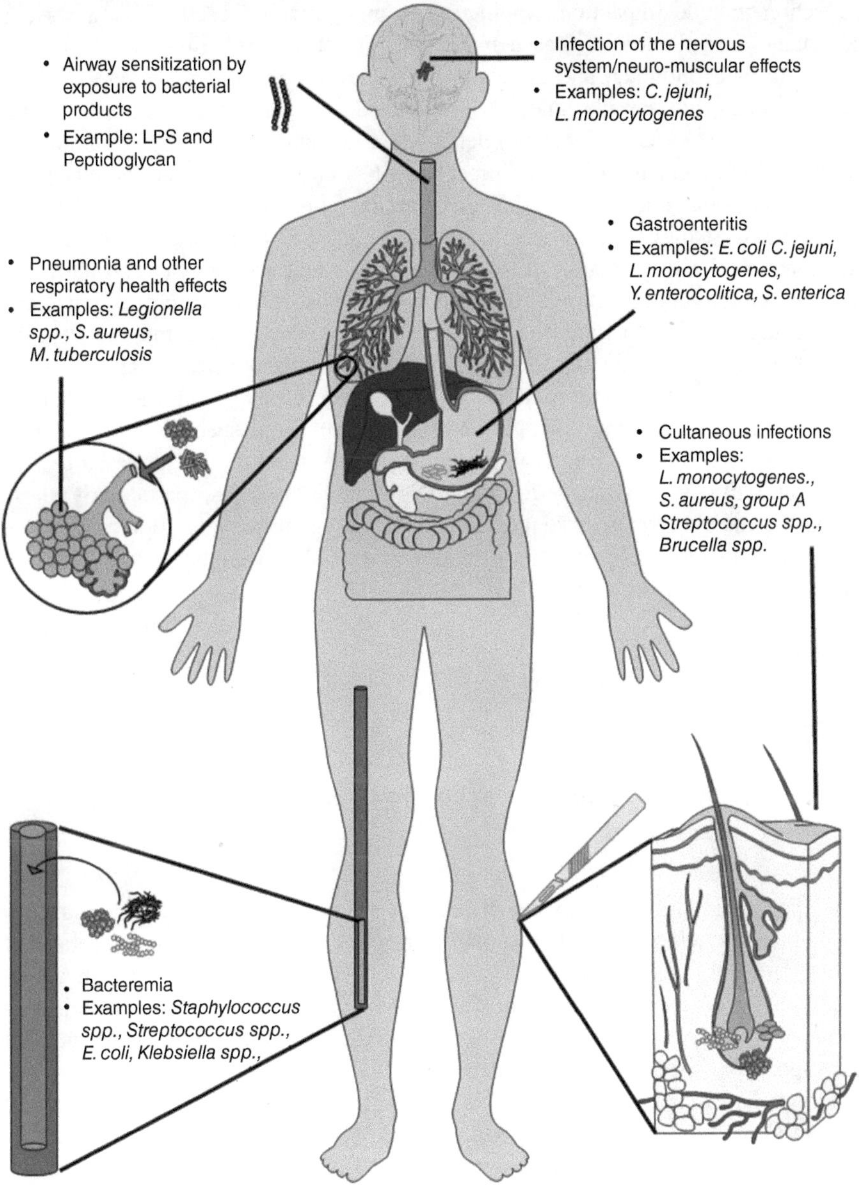

Fig. 2.3 Health effects resultant of environmental and occupational exposure to various bacterial species. Humans experience a variety of adverse health effects which are dependent on the type of bacteria encountered, the route of transmission, size of inoculum, and pre-exposure health status of the individual

collection include impaction, impingement, and filtration, each offering various advantages and challenges (Pasanen 2001; Ghosh et al. 2015; Eduard et al. 2012). Impaction is economical and allows investigators to collect samples directly onto culture media, but sample analysis is restricted to culture-only methods of enumeration and is susceptible to problems when encountering highly contaminated aerosols. Impingement collectors direct microbes into liquid media, thereby reducing the possibility for overloading the sampler. This method is subject to media evaporation, requires post-collection processing, and does not allow for size fractionation. Lastly, filtration methods are simple and economical, and allow for multiple enumeration methods, but also require post-collection processing, have a potential for overloading the sampler, and could decrease the recovery of collected organisms due to desiccation on the filter. Collected samples can be enumerated by culturable methods – light microscopy, matrix-assisted laser desorption/ionization–time-of-flight (MALDI-TOF) mass spectrometry, or non-viable methods – flow cytometry, polymerase chain reaction (PCR) based techniques, and next-generation sequencing. Non-viable methods, particularly those based in gene expression or sequencing, have become increasing popular in recent years to characterize the diversity of endogenous and exogenous bacterial populations (Blais-Lecours et al. 2015).

2.3 Environmental Exposures

2.3.1 Diversity of Bacterial Exposures in the Indoor Environment

The specific environments in which bacteria are encountered vary. For example, one Finnish study showed that human flora contributed to a high percentage of bacteria, mostly gram-positive, in samples collected from indoor dust. In agreement with that work, a study in Boston, MA, USA, also showed that indoor samples were dominated by Firmicutes and Actinobacteria phyla which contain gram-positive species. The most represented genera were *Corynebacterium*, *Lactobacillus*, *Staphylococcus*, and *Streptococcus*, all of which are associated with human flora (Hanson et al. 2016). In contrast, outdoor dust samples were rich in gram-negative Proteobacteria phylum species from the genera Acidovorax and Brevundimonas, and Flavobacterium spp. from the Bacteroidetes phylum. Another study showed that the bacterial composition of air samples in the subway system of New York City, USA, was consistent with outdoor air samples. The authors noted that this was likely due to very efficient air mixing that occurs between the subway system and the outdoor environment (Robertson et al. 2013). Despite the similarities in composition, the investigators did observe an increase of species derived from the Actinobacteria, Bacteroidetes, Firmicutes, and Proteobacteria phyla in the subway system compared to outdoor samples. Exercise facilities also function as an important source of environmental bacterial exposure as they serve as an exogenous source of community-acquired infections. As in the study of New York subways, more species were associated with endogenous human flora and environmental sources derived from dust, soil, and water. However, additional pathogenic species were enumerated (*Bacillus cereus*, *Enterococcus faecalis*, *Klebsiella pneumoniae*, *Pantoea agglomerans*, and *Salmonella enterica*) suggesting that exercise facilities are a potential source for pathogenic bacteria (Mukherjee et al. 2014).

2.3.2 Exposure to Bacterial Products and Metabolites

In addition to potential adverse health effects caused by viable organisms, bacterial products and metabolites such as endotoxin and naturally occurring antibiotics may influence physiologic responses. For example, exposure to endotoxin in homes has been associated with asthma-like symptoms including wheezing and coughing. Symptoms can be made worse dependent on social and environmental factors like presence of pets, carpeting, pests, comorbidities such as smoking, and overall socioeconomic standing (Thorne et al. 2015). Endotoxin exposure has also been linked to development of atopic and nonatopic asthma, and decreased lung functions in asthmatic people (Carnes et al. 2016). Natural antibiotics can be present in moisture-damaged structures, indicating the presence of bacteria, particularly *Streptomyces* spp. These antibiotic metabolites can modulate immune responses, cause cellular apoptosis, and have been linked to both anemia and leukemia. Further, these products often act in a synergistic manner with other microbial species and products in indoor air (Park et al. 2006; Penttinen et al. 2005; Täubel et al. 2011).

2.3.3 *Legionella*

Legionella spp. are important gram-negative pathogens from both an indoor environmental and occupational perspective. These species elicit typical pneumonia type symptoms including cough, fever, headaches, and muscle pains, which may present as the more severe Legionnaires' disease, with an associated 1–4% mortality, or the milder Pontiac fever (Fig. 2.3) (CDC 2016a; Principe et al. 2017). *Legionella* thrive in fresh water sources and rarely cause infection, however, exposure to *Legionella* spp. cultivated in improperly maintained water sources such as cooling systems, hot tubs, and showers of hospitals, hotels and cruise ships may lead to aerosolization and subsequent inhalation. Studies have shown that many cooling systems in particular fail to undergo routine total system cleaning and may not use appropriate amounts of biocide to quell outbreaks (Rangel et al. 2011; Rafiee et al. 2014). The Centers for Disease Control and Prevention offers guidelines for maintenance of building and recreational water systems (CDC 2016c), however, no government regulations currently exist. A review (Principe et al. 2017) of occupational *Legionella* outbreaks from 1949 to 2015 revealed that 795 workers had contracted Legionnaires' disease (211) or Pontiac fever (584) due to exposure to *Legionella pneumophila* and *Legionella feeleii* respectively. Most cases were located in the United States, United Kingdom, and Canada, primarily in industrial settings followed by office buildings, then hospitals. Surprisingly, *Legionella longbeachae*, a soil-adapted species, caused severe pneumonia in two metal recycling workers (Picard-Masson et al. 2016). Being that *L. longbeachae* is typically transmitted in highly organic soils and compost, its occurrence in a low-quality, nutrient-poor soil is fascinating, further highlighting the importance of *Legionella* spp. as environmental and occupational bioaerosols.

2.4 Occupational Exposures

2.4.1 Endotoxin Exposures in Various Occupations

Endotoxin is a major exposure concern for many occupations, but seems to be particularly prevalent in some industries including food, agricultural, and textile professions, as well as occupations in which workers are exposed to metalworking fluids (Basinas et al. 2015; Broadwater et al., 2015; Dahlman-Hoglund et al. 2016; Meza et al. 2013). Here we primarily will focus on agricultural and food occupations. Compared to other bacterial components like peptidoglycan, endotoxin has been shown to be more associated with negative health events in farm workers, and exerts those respiratory effects in the range of moderate exposure, 100–200 EU/m^3, whereas below 100 EU/m^3 is considered low and above 450 EU/m^3 is high (Fig. 2.4b) (Latza et al. 2004; Smit et al. 2010; Basinas et al. 2012). Those health effects include airway hyperresponsiveness, wheezing, chronic obstructive pulmonary disease (COPD), asthma-like symptoms, and decreased lung function (Basinas et al. 2015). In a study examining aerosolized bacterial and endotoxin levels that was conducted in a French cheese factory, workers were exposed to 10–300 EU/m^3 of endotoxin, though no specific health effects were reported (Simon and Duquenne 2014). Endotoxin exposure was also problematic in a facility that produced an agricultural feed component, bioprotein. Workers that were exposed to 130–540 EU/m^3 exhibited improved lung function after one year of no endotoxin exposure, compared to workers exposed to higher levels of endotoxin (>8000 EU/m^3) (Skogstad et al. 2012). The employees in the high exposure group did experience a reduction in inflammation, measured by circulating leukocytes and cytokines. Genetics, in conjunction with environmental factors, also play a role in endotoxin-induced health effects. Agricultural workers with a certain genetic mutation in the gene that produces CD14, a protein that is important in the detection of endotoxin, are more responsive to endotoxin, leading to increased wheezing and asthma-like symptoms

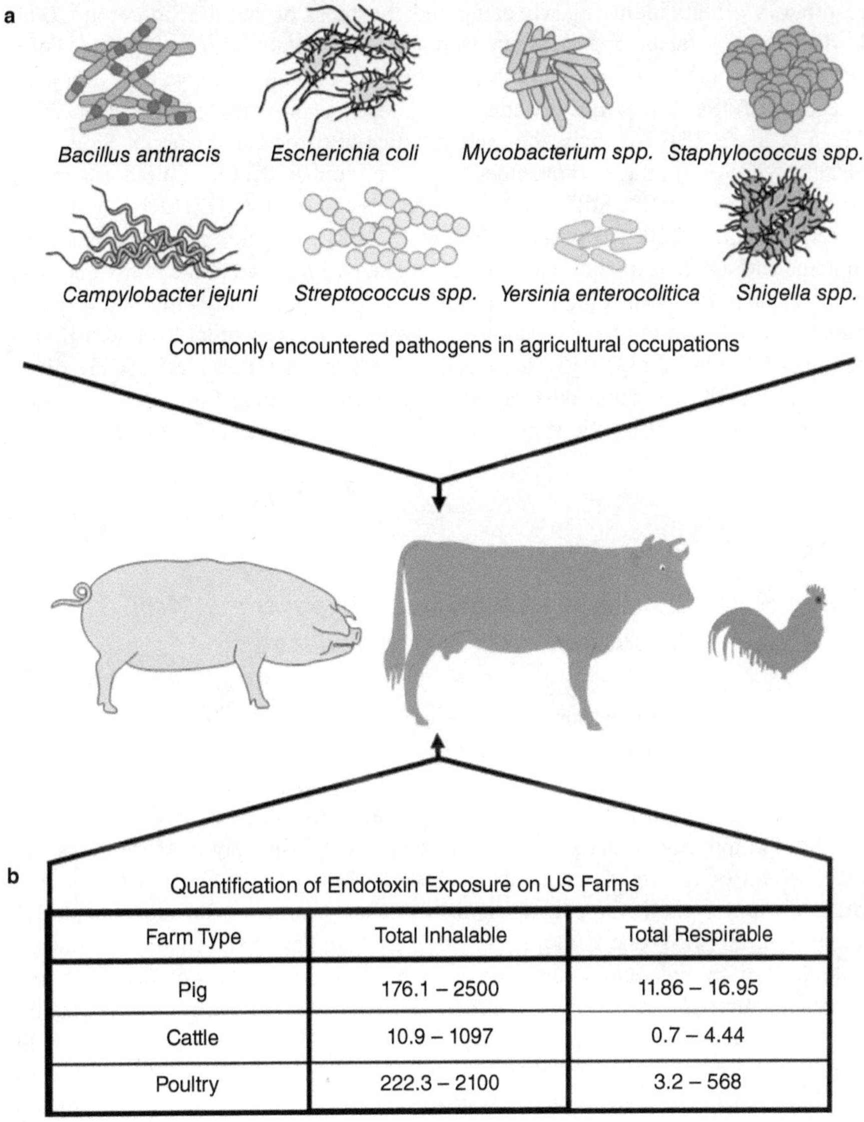

b Quantification of Endotoxin Exposure on US Farms

Farm Type	Total Inhalable	Total Respirable
Pig	176.1 – 2500	11.86 – 16.95
Cattle	10.9 – 1097	0.7 – 4.44
Poultry	222.3 – 2100	3.2 – 568

Fig. 2.4 Agricultural workers on pork, beef, and poultry farms are exposed to a range of bacterial species. (**a**) Workers are encounter various bacterial species and (**b**) varying levels of endotoxin in various meat-handling occupations including animal farms, butchers, and catering. Endotoxin levels are listed as EU/m^3

(Smit et al. 2011). Additionally, investigators discovered an association between negative health effects and high production of cytokines following endotoxin exposure; blood cells collected from farm workers that produce high amounts of Interleukin-10 (IL10) in response to endotoxin show increased adverse respiratory effects including wheezing and shortness of breath (Smit et al. 2009). Endoxtoxin is also prevalent in aerosols of metalworking fluids (MWF). Exposure to endotoxin may lead to development of asthma, as well as dermatitis (Health NIfOSa, 2013; Broadwater et al., 2015; Dahlman-Hoglund et al. 2016; Meza et al. 2013). As such, the National Institute for Occupational Safety and Health (NIOSH) set the recommended exposure limit (REL) for MWF at 0.4 mg/m^3 (Health NIfOSa, 2013). Still, even exposures less than the NIOSH REL may cause adverse health effects in workers (Meza et al. 2013). Endotoxin may be beneficial in some settings as exposure has been associated with protection against lung cancer, sensitization and allergic asthma. For example, early life exposure to endotoxin-rich environment such as a farm, protects people from developing asthma (Schuijs et al. 2015). It should be noted that these effects are likely dose dependent, and other adverse effects – decreased lung function, chest tightness, and organic dust toxic syndrome, remain (Basinas et al. 2015; Lenters et al. 2010).

2.4.2 *Risks Associated with Bacterial Exposure in Meat, Agricultural, and Food Service Industries*

Workers that handle meat are at increasingly greater risk of adverse occupational exposures to bacteria as the occurrence of pathogenic species in animals raised for food increases (Fig. 2.4a) (Castillo Neyra et al. 2012). Extraneous use of antibiotics in farming industries has led to an increase in antibiotic-resistant bacteria, notably through horizontal gene transfer, which will ultimately increase the healthcare burden of not only treatment, but also the risk of occupational exposure to drug-resistant bacteria in patients and healthcare workers (Bezuidt et al. 2011; Love et al. 2011). Workers in a beef slaughterhouse have been found to be nasally colonized with *methicillin-resistant S. aureus* (MRSA), as well as multidrug-resistant *methicillin-susceptible S. aureus* (MSSA) (Leibler et al. 2016). Tasks that require workers to handle live poultry and their products (e.g. eggs) exposes the workers to *Enterococci* and other species in concentrations as high as 1.3×10^4 CFU/m^3 and leads to increased prevalence of negative respiratory events such as wheezing, coughing and sneezing (Brauner et al. 2016; Viegas et al. 2013). Poultry farms also may pose a risk to the neighboring environments as *S. aureus* has been isolated from air samples in areas near the farms (Castillo Neyra et al. 2012). Workers that handle sheep are at risk of developing a flu-like disease, Q fever, due to exposure to *Coxiella burnetii* (Schulz et al. 2005). Handling raw meat is associated with *S. aureus* colonization in food service workers. Employees at catering businesses that handled raw meats in their job duties exhibited increased prevalence of colonization. Even some workers that never handled raw meat were colonized, suggesting aerosolization in their work environment (Ho et al. 2014). Further, the work environment creates a preponderance of transcutaneous penetrative injuries which can lead to *Staphylococcal* spp., *Streptococcal* spp., *Mycobacterium*, or *Bacillus anthracis* infections. Besides the bacteria mentioned above, meat workers may also develop adverse neurological effects following exposure to *Campylobacter jejuni* from poultry or swine, and experience gastroenteritis following exposure to occupationally-derived *S. enterica*, additional *Campylobacter* spp., *Yersinia enterocolitica*, *Escherichia coli*, and *Shigella* spp. (Fig. 2.3) (Davis et al. 2011; Hardy et al. 2011; Helms et al. 2006; Castillo Neyra et al. 2012). One potential consequence of agricultural work is development of Farmer's lung, an inflammatory disease caused by exposure to agricultural dust that contains thermophilic Actinomycetales species including but not limited to *Saccharopolyspora rectivirgula*, *Thermoactinomyces vulgaris*, *Thermoactinomyces viridis*, and *Thermoactinomyces sacchari* (Schäfer et al. 2013; Barrera et al. 2014; Cano-Jimenez et al. 2016). Farmer's lung is a type of hypersensitivity pneumonitis that presents clinically in acute, subacute, and chronic forms, and is characterized by fever-like symptoms and difficulty breathing, and can be associated with a high mortality rate.

2.4.3 Wastewater as a Source of Occupational Bacterial Exposure

Wastewater has been identified as a potential source of community-acquired MRSA, which is an important consideration due to the increase in repurposing waste water for landscaping, ground water recharge, as well as crop irrigation (Rosenberg Goldstein et al. 2012, 2014). Across the US, approximately 6% of wastewater is reused, with the majority being used for landscaping purposes, leading to both environmental and occupational exposure to bacteria in wastewater. A large proportion, over 80%, of influent wastewater test positive for MRSA compared to approximately 8% of effluent samples, indicating that wastewater treatment proves effective in reducing the total burden of MRSA returning to the environment. Still, it has been postulated that the surviving microbes are hardier and could represent multi-drug resistant bacteria (Rosenberg Goldstein et al. 2012). In a study examining exposure to MRSA, *vancomycin-resistant Enterococci* (VRE), and other pathogens from reclaimed wastewater, the investigators found that spray irrigation workers were at increased odds of being colonized with MSSA, multidrug-resistant MSSA, and *vancomycin-susceptible Enterococci* (VSE). Another study showed workers at a wastewater treatment facility were exposed to bacteria in concentrations that could exceed 5×10^4 CFU/m^3, most of which were gram-positive species, which have previously been associated with bioaerosols in wastewater treatment (Prazmo et al. 2003; Teixeira et al. 2013). In addition to the potential for exposure, wastewater workers may experience direct health effects including, but not limited to, adverse respiratory, skin, or allergic reactions (Teixeira et al. 2013).

2.4.4 Bacterial Exposures in Healthcare Settings

Healthcare workers (HCW) are at particular risk for exposure to bioaerosols given the nature and environment of their work. Even with a lack of direct contact with patients workers can be exposed to a wide variety of endogenous and exogenous bacteria. In a study measuring bioaerosols of vacated hospital rooms, workers had the potential to be exposed to bacteria in quantities up to 1.4×10^4 CFU/m^3/hr (Heimbuch et al. 2016). A significant portion of gram-positive and negative species showed resistance to oxacillin, while a small number of gram-positive isolates exhibited resistance to vancomycin. HCW personal protective equipment (PPE) can also act as a source of contamination. In a multi-center study, particular pathogenic species *Clostridium difficile*, VRE, MRSA, and other enterobacteria were discovered on HCW PPE (Mitchell et al. 2015). In one center, over 60% of uniforms of physicians and nurses had detectable levels of bacterial pathogens. As expected, HCW performing routine care for individuals colonized with multidrug-resistant bacteria are at risk of exposure (Morgan et al. 2012). HCW gloves or gowns were contaminated with *Acinetobacter baumannii, Pseudomonas aeruginosa*, VRE, or MRSA in 21% of observed patient interactions. *A. baumannii*, an opportunistic pathogen frequently found in hospitals (Antunes et al. 2014), was the most common contaminant of HCW PPE. HCW themselves could be a source of pathogenic bacteria. Studies of HCW indicate that upwards of 8% exhibit nasal colonization with MRSA (Albrich and Harbarth 2008), while others have suggested the number could be higher, closer to 15% (Hawkins et al. 2011). Collectively, these observations suggest that HCW, and patients alike, must take care to minimize transfer of potentially aerosolizable pathogens. Decreasing the burden of bacteria can be achieved by utilizing clean PPE and adhering to prerequisite hygiene standards for the healthcare setting.

2.4.5 Laboratory-Acquired Infections

Laboratories, both clinical and research, are an additional source of occupational exposure to bacteria due to inhalation of cultured organisms, needle sticks, animal bites, damaged glassware, ingestion, and contact with a contaminated surface (e.g. gloves) with mucous membranes (Coelho and Garcia Diez 2015). Laboratory acquired infections (LAI) totaled more than 5000 between 1930 and 2000, with 190 associated fatalities reported (CDC 2009; Pike 1979, 1976). The most common bacteria associated with LAI infections were *Mycobacterium tuberculosis*, *C. burnetii*, *Salmonella* spp., *Brucella* spp., *Shigella* spp., *Chlamydia psittaci*, and *Francisella tularensis*. However, due to an increase in antibiotic-resistant bacteria that may contribute to and influence LAI, the following species have been recently associated infections in the lab: *C. difficile*, *E. coli*, and *Klebsiella spp* (Fig. 2.3). Clinical labs accounted for approximately 45% of total reported cases between the years 1978–2000 compared to 51% in research labs during that same period. The actual number of LAI may never be known due to a lack of an official reporting system, non-reporting of sub-clinical cases, and fear of laboratory reprimand due to increasing oversight and regulation of biomedical labs. Still, LAIs are much rarer, likely due to vaccinations against potential exposures, improved containment systems, engineering controls, and heavy emphasis and enforcement on a culture of safety and training (CDC 2009; Wurtz et al. 2016; Coelho and Garcia Diez 2015). Even with improved safety training and protocols in labs, LAI can still occur. In one survey based study, the investigators found 15 cases of LAI in 23 biosafety level 3 and 4 labs that were surveyed (Wurtz et al. 2016). Two cases were caused by viruses, and the rest by *M. tuberculosis*, *C. burnetii*, and *Brucella melitensis*. The LAI in these cases were transmitted via inhalation of bioaerosols due to human error. Fortunately, all of the infected recovered and most without sequelae. Continual review and improvement of safety standards in biomedical laboratories is critical to minimizing laboratory worker exposure to pathogenic bacteria and subsequent development of LAIs.

2.5 Future Considerations

Given the sheer volume of bacteria in the environment, preventing exposure is next to impossible. Preventing infection is, however, possible with improved situational awareness, a deeper understanding of the exposures that affect workers, and improvement in methods to combat pathogenic and potentially pathogenic species. Embracing and continually improving upon a culture of safety and training, in conjunction with advances in worker PPE and engineering controls, will help alleviate workplace exposures (Wurtz et al. 2016). Moreover, striving to gain a complete understanding of the make-up and complexities of bacteria in the environment using new approaches such as next-generation sequencing will enhance our understanding of potential health effects of various exposures. For example, *Bacillus thuringiensis* is commonly used in agricultural work as a biopesticide, accounting for as much as 90% of the market share. Still, not much is known about its potential health effects; however, a recent study showed that granulocytes produce massive amounts of reactive oxygen species (ROS) in response to *B. thuringiensis* exposure (Madsen et al. 2014). In addition to improving biosafety and our understanding of bacteria, preventing antibiotic misuse in both humans and animals will help quell the rise in "super bugs" – that is, highly infective, easily transmittable, multidrug-resistant bacteria (Llor and Bjerrum 2014; CDC 2014; Castillo Neyra et al. 2012). Overuse of antibiotics has led to, in addition to increased microbial resistance, increases in: mortality following infection, duration and severity of diseases, erroneous prescriptions for self-limiting infections, and healthcare costs. It is, therefore, critical to discover new antibiotics with no associated resistance, as well as investigate other alternatives. A newly discovered antibiotic, teixobactin, inhibits cell wall synthesis by binding to peptidoglycan and teichoic acid and shows no in resistance various mutants of *S. aureus* and *M. tuberculosis*, both of which have multiple drug-resistant strains (Ling et al. 2015). Additionally, the human microbiota may be a suitable place to search for new, naturally occurring antibiotics (Zipperer et al. 2016). Finally, another alternative is silver. It has demonstrated antimicrobial properties and could be further developed to be incorporated into air filtration systems, PPE, and other sources of contamination in both occupational and environmental settings (Swathy et al. 2014; Lansdown 2006; Paladini et al. 2014).

Acknowledgements The findings and the conclusions in this chapter are those of the author and do not necessarily represent the views of the National Institute for Occupational Safety and Health.

References

Albrich WC, Harbarth S (2008) Health-care workers: source, vector, or victim of MRSA? Lancet Infect Dis 8(5):289–301. doi:10.1016/S1473-3099(08)70097-5

Angert ER (2005) Alternatives to binary fission in bacteria. Nat Rev Micro 3(3):214–224

Antunes LC, Visca P, Towner KJ (2014) Acinetobacter baumannii: evolution of a global pathogen. Pathog Dis 71(3):292–301. doi:10.1111/2049-632x.12125

Bang C, Schmitz RA (2015) Archaea associated with human surfaces: not to be underestimated. FEMS Microbiol Rev 39(5):631–648. doi:10.1093/femsre/fuv010

Barrera C, Millon L, Rognon B et al. (2014) Immunoreactive proteins of Saccharopolyspora rectivirgula for farmer's lung serodiagnosis. Proteomics Clin Appl 8(11–12):971–981. doi:10.1002/prca.201400024

Basinas I, Schlunssen V, Heederik D et al. (2012) Sensitisation to common allergens and respiratory symptoms in endotoxin exposed workers: a pooled analysis. Occup Environ Med 69 (2):99–106. doi:10.1136/oem.2011.065169

Basinas I, Sigsgaard T, Kromhout H et al. (2015) A comprehensive review of levels and determinants of personal exposure to dust and endotoxin in livestock farming. J Expo Sci Environ Epidemiol 25(2):123–137. doi:10.1038/jes.2013.83

Bezuidt O, Pierneef R, Mncube K et al. (2011) Mainstreams of Horizontal Gene Exchange in Enterobacteria: Consideration of the Outbreak of Enterohemorrhagic E. coli O104:H4 in Germany in 2011. PLoS One 6(10):e25702. doi:10.1371/journal.pone.0025702

Blais-Lecours P, Perrott P, Duchaine C (2015) Non-culturable bioaerosols in indoor settings: Impact on health and molecular approaches for detection. Atmos Environ 110:45–53. doi:10.1016/j.atmosenv.2015.03.039

Blais Lecours P, Veillette M, Marsolais D et al. (2012) Characterization of bioaerosols from dairy barns: reconstructing the puzzle of occupational respiratory diseases by using molecular approaches. Appl Environ Microbiol 78(9):3242–3248. doi:10.1128/aem.07661-11

Brauner P, Klug K, Jackel U (2016) Eggshells as a source for occupational exposure to airborne bacteria in hatcheries. J Occup Environ Hyg 13(12):950–959. doi:10.1080/15459624.2016.1200192

Broadwater K de Perio, MA; Brueck, SE et al. (2015) Evaluation of an unpleasant odor at an aircraft ejection seat manufacturer. Centers for Disease Control and Prevention.

Brown CT, Hug LA, Thomas BC, et al. (2015) Unusual biology across a group comprising more than 15% of domain Bacteria. Nature 523(7559):208–211. doi:10.1038/nature14486 http://www.nature.com/nature/journal/v523/n7559/abs/nature14486.html#supplementary-information

Cano-Jimenez E, Acuna A, Botana MI et al. (2016) Farmer's Lung Disease. A review. Arch Bronconeumol 52(6):321–328. doi:10.1016/j.arbres.2015.12.001

Carnes MU, Hoppin JA, Metwali N, et al. (2016) House dust endotoxin levels are associated with adult asthma in a U.S. farming population. *Ann Am Thorac Soc.* doi:10.1513/AnnalsATS.201611-861OC

Castillo Neyra R, Vegosen L et al. (2012) Antimicrobial-resistant bacteria: an unrecognized work-related risk in food animal production. Saf Health Work 3(2):85–91. doi:10.5491/shaw.2012.3.2.85

CDC (2009) Biosafety in Microbiological and Biomedical Laboratories, 5th edn. Centers for Disease Control and Prevention, Atlanta

CDC (2014) Mission critical: preventing antibiotic resistance. Centers for Disease Control and Prevention. http://www.cdc.gov/features/antibioticresistance/. Accessed 9 Dec 2016

CDC (2016a) Legionnaires' Disease. Centers for Disease Control and Prevention. https://www.cdc.gov/legionella/index.html. Accessed 9 Dec 2016.

CDC (2016b) Listeria (Listeriosis). Centers for Disease Control and Prevention. https://www.cdc.gov/listeria/index.html. Accessed 9 Dec 2016

CDC (2016c) Water system maintenance. Centers for Disease Control and Prevention. https://www.cdc.gov/legionella/water-system-maintenance.html. Accessed 12 Jan 2017

Clarridge JE (2004) Impact of 16S rRNA gene sequence analysis for identification of bacteria on clinical microbiology and infectious diseases. Clin Microbiol Rev 17(4):840–862. doi:10.1128/CMR.17.4.840-862.2004

Coelho AC, Garcia Diez J (2015) Biological risks and laboratory-acquired infections: a reality that cannot be ignored in health biotechnology. Front Bioeng Biotechnol 3:56. doi:10.3389/fbioe.2015.00056

Dahlman-Hoglund A, Lindgren A, Mattsby-Baltzer I (2016) Endotoxin in size-separated metal working fluid aerosol particles. Ann Occup Hyg 60(7):836–844. doi:10.1093/annhyg/mew036

Davis MF, Kamel F, Hoppin JA et al. (2011) Neurologic symptoms associated with raising poultry and swine among participants in the agricultural health study. J Occup Environ Med 53 (2):190–195. doi:10.1097/JOM.0b013e31820805d5

Eduard W, Heederik D, Duchaine C et al. (2012) Bioaerosol exposure assessment in the workplace: the past, present and recent advances. J Environ Monit 14(2):334–339. doi:10.1039/c2em10717a

Euzéby JP (1997) List of bacterial names with standing in nomenclature: a folder available on the Internet. Int J Syst Bacteriol 47(2):590–592. doi:10.1099/00207713-47-2-590

Fuerst JA (2013) The PVC superphylum: exceptions to the bacterial definition? Antonie van Leeuwenhoek 104(4):451–466. doi:10.1007/s10482-013-9986-1

Ghosh B, Lal H, Srivastava A (2015) Review of bioaerosols in indoor environment with special reference to sampling, analysis and control mechanisms. Environ Int 85:254–272. doi:10.1016/j.envint.2015.09.018

Godshall CE, Suh G, Lorber B (2013) Cutaneous listeriosis. J Clin Microbiol 51(11):3591–3596. doi:10.1128/jcm.01974-13

Gupta RS (2011) Origin of diderm (Gram-negative) bacteria: antibiotic selection pressure rather than endosymbiosis likely led to the evolution of bacterial cells with two membranes. Antonie Van Leeuwenhoek 100(2):171–182. doi:10.1007/s10482-011-9616-8

Gyles C, Boerlin P (2014) Horizontally transferred genetic elements and their role in pathogenesis of bacterial disease. Vet Pathol 51(2):328–340. doi:10.1177/0300985813511131

Hall-Stoodley L, Costerton JW, Stoodley P (2004) Bacterial biofilms: from the natural environment to infectious diseases. Nat Rev Micro 2(2):95–108

Hanson B, Zhou Y, Bautista EJ et al. (2016) Characterization of the bacterial and fungal microbiome in indoor dust and outdoor air samples: a pilot study. Environ Sci Process Impacts 18(6):713–724. doi:10.1039/c5em00639b

Hardy CG, Lackey LG, Cannon J et al. (2011) Prevalence of potentially neuropathic Campylobacter jejuni strains on commercial broiler chicken products. Int J Food Microbiol 145(2–3):395–399. doi:10.1016/j.ijfoodmicro.2010.12.027

Hawkins G, Stewart S, Blatchford O et al. (2011) Should healthcare workers be screened routinely for meticillin-resistant Staphylococcus aureus? A review of the evidence. J Hosp Infect 77(4):285–289. doi:10.1016/j.jhin.2010.09.038

Health NIfOSa (2013) Metalworking fluids. Centers for Disease Control and Prevention. https://www.cdc.gov/niosh/topics/metalworking/. Accessed 01/12/2017 2017

Heimbuch BK, Wallace WH, Balzli CL et al. (2016) Bioaerosol exposure to personnel in a clinical environment absent patients. J Occup Environ Hyg 13(2):D11–15. doi:10.1080/15459624.2015.1091966

Helms M, Simonsen J, Molbak K (2006) Foodborne bacterial infection and hospitalization: a registry-based study. Clin Infect Dis 42(4):498–506. doi:10.1086/499813

Ho J, O'Donoghue MM, Boost MV (2014) Occupational exposure to raw meat: a newly-recognized risk factor for Staphylococcus aureus nasal colonization amongst food handlers. Int J Hyg Environ Health 217(2–3):347–353. doi:10.1016/j.ijheh.2013.07.009

James G (2010) Universal Bacterial Identification by PCR and DNA Sequencing of 16S rRNA GeneSchuller M, Sloots TP, James GS, Halliday CL, Carter IWJ (Eds.), PCR for Clinical Microbiology: An Australian and International Perspective. Springer, Dordrecht. 209–214.

Kerfeld CA, Sawaya MR, Tanaka S, et al. (2005) Protein structures forming the shell of primitive bacterial organelles. Science 309(5736):936–938. doi:10.1126/science.1113397

Koonin EV, Mushegian AR, Galperin MY et al. (1997) Comparison of archaeal and bacterial genomes: computer analysis of protein sequences predicts novel functions and suggests a chimeric origin for the archaea. Mol Microbiol 25(4):619–637. doi:10.1046/j.1365-2958.1997.4821861.x

Lachapelle J-M (2012) Biologic Causes of Occupational Dermatoses. In: Rustemeyer T, Elsner P, John S-M, Maibach HI (eds) Kanerva's Occupational Dermatology. Springer, Heidelberg, p 231–245. doi:http://dx.doi.org/10.1007/978-3-642-02035-3_23

Lansdown AB (2006) Silver in health care: antimicrobial effects and safety in use. Curr Probl Dermatol 33:17–34. doi:10.1159/000093928

Latza U, Oldenburg M, Baur X (2004) Endotoxin exposure and respiratory symptoms in the cotton textile industry. Arch Environ Health 59(10):519–525. doi:10.1080/00039890409605168

Lee JY, Yong FC (2005) A chronic nocardial infection following conservative treatment of a high pressure injection injury of air. J Hand Surg Asian Pac Vol 10(2-3):255–259. doi:10.1142/s0218810405002796

Leibler JH, Jordan JA, Brownstein K et al. (2016) Staphylococcus aureus nasal carriage among beefpacking workers in a midwestern United States slaughterhouse. PLoS One 11(2): e0148789. doi:10.1371/journal.pone.0148789

Lenters V, Basinas I, Beane-Freeman L et al. (2010) Endotoxin exposure and lung cancer risk: a systematic review and meta-analysis of the published literature on agriculture and cotton textile workers. Cancer Causes Control 21(4):523–555. doi:10.1007/s10552-009-9483-z

Ling LL, Schneider T, Peoples AJ, et al. (2015) A new antibiotic kills pathogens without detectable resistance. Nature 517 (7535):455–459. doi:10.1038/nature14098. http://www.nature.com/nature/journal/v517/n7535/abs/nature14098.html#supplementary-information

Llor C, Bjerrum L (2014) Antimicrobial resistance: risk associated with antibiotic overuse and initiatives to reduce the problem. Ther Adv Drug Saf 5(6):229–241. doi:10.1177/2042098614554919

Love DC, Davis MF, Bassett A et al. (2011) Dose imprecision and resistance: free-choice medicated feeds in industrial food animal production in the United States. Environ Health Perspect 119(3):279–283. doi:10.1289/ehp.1002625

Madsen AM, Zervas A, Tendal K et al. (2014) Exposure and preventive measure to reduce high and daily exposure to Bacillus thuringiensis in potted plant production. Ann Occup Hyg 58(6):664–676. doi:10.1093/annhyg/meu030

Meza F, Chen L, Hudson N (2013) Investigation of respiratory and dermal symptoms associated with metal working fluids at an aircraft engine manufacturing facility. Am J Ind Med 56(12):1394–1401. doi:10.1002/ajim.22253

MitchellA, SpencerM, Edmiston C,Jr (2015) Role of healthcare apparel and other healthcare textiles in the transmission of pathogens: a review of the literature. J Hosp Infect 90(4): 285–292. doi:10.1016/j.jhin.2015.02.017

Morgan DJ, Rogawski E, Thom KA et al. (2012) Transfer of multidrug-resistant bacteria to healthcare workers' gloves and gowns after patient contact increases with environmental contamination. Crit Care Med 40(4):1045–1051. doi:10.1097/CCM.0b013e31823bc7c8

Mukherjee N, Dowd SE, Wise A et al. (2014) Diversity of bacterial communities of fitness center surfaces in a U.S. metropolitan area. Int J Environ Res Public health 11(12):12544–12561. doi:10.3390/ijerph111212544

Murat D, Byrne M, Komeili A (2010) Cell biology of prokaryotic organelles. Cold Spring Harb Perspect Biol 2(10):a000422. doi:10.1101/cshperspect.a000422

Nehme B, Gilbert Y, Letourneau V et al. (2009) Culture-independent characterization of archaeal biodiversity in swine confinement building bioaerosols. Appl Environ Microbiol 75(17):5445–5450. doi:10.1128/aem.00726-09

Nicholson WL, Fajardo-Cavazos P, Rebeil R et al. (2002) Bacterial endospores and their significance in stress resistance. Antonie van Leeuwenhoek 81(1):27–32. doi:10.1023/a:1020561122764

Paladini F, Cooper IR, Pollini M (2014) Development of antibacterial and antifungal silver-coated polyurethane foams as air filtration units for the prevention of respiratory diseases. J Appl Microbiol 116(3):710–717. doi:10.1111/jam.12402

Park JH, Cox-Ganser J, Rao C et al. (2006) Fungal and endotoxin measurements in dust associated with respiratory symptoms in a water-damaged office building. Indoor Air 16(3): 192–203. doi:10.1111/j.1600-0668.2005.00415.x

Pasanen AL (2001) A review: fungal exposure assessment in indoor environments. Indoor Air 11(2):87–98. doi:10.1034/j.1600-0668.2001.110203.x

Penttinen P, Pelkonen J, Huttunen K et al. (2005) Interactions between Streptomyces californicus and Stachybotrys chartarum can induce apoptosis and cell cycle arrest in mouse RAW264.7 macrophages. Toxicol Appl Pharmacol 202(3):278–288. doi:10.1016/j.taap.2004.07.002

Picard-Masson M, Lajoie E, Lord J et al. (2016) Two related occupational cases of legionella longbeachae infection, Quebec, Canada. Emerg Infect Dis 22(7):1289–1291. doi:10.3201/eid2207.160184

Pike RM (1976) Laboratory-associated infections: summary and analysis of 3921 cases. Health Lab Sci 13(2):105–114

Pike RM (1979) Laboratory-associated infections: incidence, fatalities, causes, and prevention. Annu Rev Microbiol 33:41–66. doi:10.1146/annurev.mi.33.100179.000353

Poehlein A, Gottschalk G, Daniel R (2013) First insights into the genome of the gram-negative, endospore-forming organism sporomusa ovata strain H1 DSM 2662. Genome Announc 1(5): e00734–00713. doi:10.1128/genomeA.00734-13

Prazmo Z, Krysinska-Traczyk E, Skorska C et al. (2003) Exposure to bioaerosols in a municipal sewage treatment plant. Ann Agric Environ Med 10(2):241–248

Principe L, Tomao P, Visca P (2017) Legionellosis in the occupational setting. Environ Res 152:485–495. doi:10.1016/j.envres.2016.09.018

Rafiee M, Jahangiri-Rad M, Hajjaran H et al. (2014) Detection and identification of Legionella species in hospital water supplies through polymerase chain reaction (16S rRNA). J Environ Health Sci Eng 12:83. doi:10.1186/2052-336x-12-83

Rangel KM, Delclos G, Emery R et al. (2011) Assessing maintenance of evaporative cooling systems in legionellosis outbreaks. J Occup Environ Hyg 8(4):249–265. doi:10.1080/15459624.2011.565409

Rinke C, Schwientek P, Sczyrba A, et al. (2013) Insights into the phylogeny and coding potential of microbial dark matter. Nature 499 (7459):431–437. doi:10.1038/nature12352. http://www.nature.com/nature/journal/v499/n7459/abs/nature12352.html#supplementary-information

Robertson CE, Baumgartner LK, Harris JK et al. (2013) Culture-independent analysis of aerosol microbiology in a metropolitan subway system. Appl Environ Microbiol 79(11):3485–3493. doi:10.1128/aem.00331-13

Rosenberg Goldstein RE, Micallef SA, Gibbs SG et al. (2012) Methicillin-resistant Staphylococcus aureus (MRSA) detected at four U.S. wastewater treatment plants. Environ Health Perspect 120(11):1551–1558. doi:10.1289/ehp.1205436

Rosenberg Goldstein RE, Micallef SA, Gibbs SG et al. (2014) Occupational exposure to Staphylococcus aureus and Enterococcus spp. among spray irrigation workers using reclaimed water. Int J Environ Res Pub Health 11(4):4340–4355. doi:10.3390/ijerph110404340

Salton MRJ, Kim KS (1996) Structure. In: Baron S (ed) Medical microbiology. University of Texas Medical Branch at Galveston The University of Texas Medical Branch at Galveston, Galveston (TX)

Schäfer J, Klug K, van Kampen V et al. (2013) Quantification of Saccharopolyspora rectivirgula in composting plants: assessment of the relevance of S. rectivirgula. Ann Occup Hyg 57 (7):875–883. doi:10.1093/annhyg/met010

Schuijs MJ, Willart MA, Vergote K et al. (2015) Farm dust and endotoxin protect against allergy through A20 induction in lung epithelial cells. Science 349(6252):1106–1110. doi:10.1126/science.aac6623

Schulz HN, Jørgensen BB (2001) Big bacteria. Ann Rev Microbiol 55(1):105–137. doi:10.1146/annurev.micro.55.1.105

Schulz J, Runge M, Schroder C et al. (2005) [Detection of Coxiella burnetii in the air of a sheep barn during shearing]. DTW Deutsche tierarztliche Wochenschrift 112(12):470–472

Sekiguchi Y, Ohashi A, Parks DH et al. (2015) First genomic insights into members of a candidate bacterial phylum responsible for wastewater bulking. Peer J 3:e740. doi:10.7717/peerj.740

Simon X, Duquenne P (2014) Assessment of workers' exposure to bioaerosols in a French cheese factory. Ann Occup Hyg 58(6):677–692. doi:10.1093/annhyg/meu027

Skogstad M, Sikkeland LI, Ovstebo R et al. (2012) Long-term occupational outcomes of endotoxin exposure and the effect of exposure cessation. Occup Environ Med 69(2):107–112. doi:10.1136/oem.2010.062414

Smit LA, Heederik D, Doekes G et al. (2011) Endotoxin exposure, CD14 and wheeze among farmers: a gene–environment interaction. Occup Environ Med 68(11):826–831. doi:10.1136/oem.2010.060038

Smit LA, Heederik D, Doekes G et al. (2009) Ex vivo cytokine release reflects sensitivity to occupational endotoxin exposure. Eur Respir J 34(4):795–802. doi:10.1183/09031936.00161908

Smit LA, Heederik D, Doekes G et al. (2010) Occupational endotoxin exposure reduces the risk of atopic sensitization but increases the risk of bronchial hyperresponsiveness. Int Arch Allergy Immunol 152(2):151–158. doi:10.1159/000265536

Steinbach WJ, Shetty AK (2001) Use of the diagnostic bacteriology laboratory: a practical review for the clinician. Postgrad Med J 77(905):148–156. doi:10.1136/pmj.77.905.148

Sutcliffe IC (2010) A phylum level perspective on bacterial cell envelope architecture. Trends Microbiol 18(10):464–470. doi:10.1016/j.tim.2010.06.005

Swathy JR, Sankar MU, Chaudhary A, et al. (2014) Antimicrobial silver: an unprecedented anion effect. Scientific Reports 4:7161. doi:10.1038/srep07161. http://www.nature.com/articles/srep07161#supplementary-information

Täubel M, Rintala H, Pitkaranta M et al. (2009) The occupant as a source of house dust bacteria. J Allergy Clin Immunol 124(4):834–840.e847. doi:10.1016/j.jaci.2009.07.045

Täubel M, Sulyok M, Vishwanath V et al. (2011) Co-occurrence of toxic bacterial and fungal secondary metabolites in moisture-damaged indoor environments. Indoor Air 21(5):368–375. doi:10.1111/j.1600-0668.2011.00721.x

Teixeira JV, Miranda S, Monteiro RA et al. (2013) Assessment of indoor airborne contamination in a wastewater treatment plant. Environ Monit Assess 185(1):59–72. doi:10.1007/s10661-012-2533-0

Thomas P (2006) Isolation of an ethanol-tolerant endospore-forming Gram-negative Brevibacillus sp. as a covert contaminant in grape tissue cultures. J Appl Microbiol 101(4):764–774. doi:10.1111/j.1365-2672.2006.02993.x

Thorne PS, Mendy A, Metwali N et al. (2015) Endotoxin exposure: predictors and prevalence of associated asthma outcomes in the United States. Am J Respir Crit Care Med 192(11): 1287–1297. doi:10.1164/rccm.201502-0251OC

Viegas S, Faisca VM, Dias H et al. (2013) Occupational exposure to poultry dust and effects on the respiratory system in workers. J Toxicol Environ Health A 76(4-5):230–239. doi:10.1080/15287394.2013.757199

Whitman WB, Coleman DC, Wiebe WJ (1998) Prokaryotes: the unseen majority. Proc Natl Acad Sci U S A 95(12):6578–6583

Woese CR (1994) There must be a prokaryote somewhere: microbiology's search for itself. Microbiol Rev 58(1):1–9

Wurtz N, Papa A, Hukic M et al. (2016) Survey of laboratory-acquired infections around the world in biosafety level 3 and 4 laboratories. Eur J Clin Microbiol Infect Dis 35(8): 1247–1258. doi:10.1007/s10096-016-2657-1

Yeoh YK, Sekiguchi Y, Parks DH et al. (2016) Comparative genomics of candidate phylum TM6 suggests that parasitism is widespread and ancestral in this lineage. Mol Biol Evol 33(4):915–927. doi:10.1093/molbev/msv281

Young KD (2006) The selective value of bacterial shape. Microbiol Mol Biol Rev 70(3): 660–703. doi:10.1128/MMBR.00001-06

Zipperer A, Konnerth MC, Laux C, et al. (2016) Human commensals producing a novel antibiotic impair pathogen colonization. Nature 535 (7613):511–516. doi:10.1038/nature18634. http://www.nature.com/nature/journal/v535/n7613/abs/nature18634.html#supplementary-information

Chapter 3
Characterization of Viral Exposures in United States Occupational Environments

Tara L. Croston

Abstract Viruses are considered to be the most abundant biological particles and have the capability to infect all forms of life leading to a variety of diseases. American workers in specific occupational environments are threatened by viral exposures, highlighting the importance to recognize the type and risk of exposure, as well as the preventive measures that can be taken to reduce the risk of exposure. For example, healthcare workers can potentially be exposed to air and bloodborne pathogens, such as hepatitis and the human immunodeficiency virus. These types of exposures have led to the development of preventive equipment and regulations intended to reduce viral exposures in occupational settings. This chapter will discuss the characteristics of viruses and the occupationally relevant viruses of which people in varying occupations can potentially encounter. Regulatory guidelines and protective strategies will also be reviewed.

Keywords Virus · viral exposure · occupational exposure

3.1 Introduction

Viruses are abundant microbiological agents (Lawrence et al. 2009) that are capable of infecting all forms of life including but not limited to humans, plants, animals, fungi, bacteria, protozoa, and archaea. The United States workforce, especially health care workers, first responders, industrial workers, and biowaste workers are at risk of occupational exposure to a broad diversity of viruses including hepatitis, human immunodeficiency virus (HIV), Ebola virus, influenza and adenoviruses. Preventative guidelines including administrative and engineering controls, as well as use of personal protective equipment (PPE) have been published by various government agencies, and these avoidance strategies and exposure control plans have

T.L. Croston (✉)
Allergy and Clinical Immunology Branch, Health Effects Laboratory Division, National Institute for Occupational Safety and Health, Centers for Disease Control and Prevention, Morgantown, WV 26505, USA
e-mail: xzu9@cdc.gov

C. Viegas et al. (eds.), *Exposure to Microbiological Agents in Indoor and Occupational Environments*, DOI 10.1007/978-3-319-61688-9_3

been modernized and recently revised (CDC 1988, 2016b, c, v; Kuhar et al. 2013; Garner 1996; Siegel et al. 2007; West and Cohen 1997).

Viruses are acellular and conflict within the scientific community exists as to whether or not viruses are living. Only able to replicate inside living cells, viruses rely on the host cell's reproductive machinery (Sze and Tan 2015). They are generally smaller than bacteria and therefore must be visualized using scanning or transmission electron microscopy. Viral particles range in diameter from 20 nm to approximately 300 nm; approximately one hundredth the size of a bacterium (Sekar and Kathiresan 2013). Stains are used to enhance the visualization of the viral particles, such as fluorescent antibodies that specifically tag antigens specific to the viral particles or uranyl acetate and phosphotungstic acid to enhance the contrast for electron microscopy (Smith and Melnick 1962). Negative staining can also enhance the contrast by staining the background instead of the viral proteins (Kiselev et al. 1990). Complete independent viral particles that have not yet infected, or are in the process of infecting, living cells are called "virions" and consist of two components. The genetic material of the virus, whether it is DNA or RNA, is contained within the protective proteinaceous coat termed the "capsid." Identical protein subunits termed "capsomeres" collectively construct the capsid, which can be enveloped with a layer of lipids formed from the host cell. This envelope, which aids in the infection of a host cell, consists of components from both the viral and the host cell's membranes.

3.1.1 Structure of Viral Capsids

Depending on the virus, virions differ in morphology, size and symmetry. There are four primary morphologies: helical, icosahedral, prolate and enveloped. The majority of viruses contain either a helical or icosahedral capsid (Lidmar et al. 2003; Vernizzi and Olvera de la Cruz 2007) and more commonly will be encountered in an occupational environment. Helical viruses are rigid or flexible rod-shaped (Prasad and Schmid 2012) formed by capsomeres specifically arranged in a helix around a central axis. The genetic material is located either inside the central cavity protected by the capsid or may be attached directly to the capsid. Fig. 3.1 illustrates the longitudinal view of an influenza virion, which is an example of a helical virus (Brown et al. 2010) that can be encountered in healthcare settings. Another example of a helical virus is the tobacco mosaic virus (TMV) discovered in 1898 (Harrison and Wilson 1999). Although humans cannot be infected by TMV, antibodies against TMV can be produced as shown in a study that measured anti-TMV antibodies in cigarette and smokeless tobacco users (Liu et al. 2013).

More commonly, viruses have an icosahedral shape where capsids are composed of 60 repeating identical subunits that give rise to 20 equilateral triangles arranged in a symmetrical manner (Prasad and Schmid 2012). This formation allows for the least amount of genetic material needed to code for the structural proteins that make up the capsid and for a closed capsid to protect the genetic material. An example of a virus with an icosahedral structure is an adenovirus, shown in Fig. 3.2. Other examples of icosahedral viruses include the rhinoviruses,

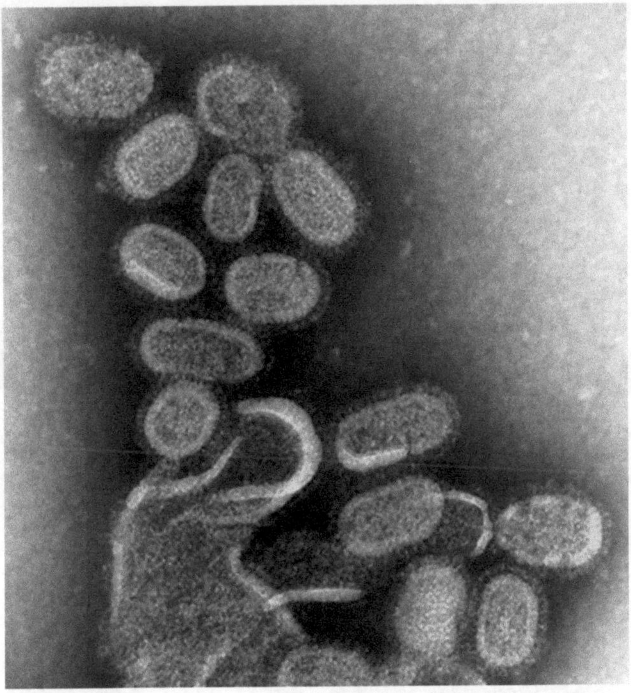

Fig. 3.1 Longitudinal transmission electron microscopic image of influenza virions. A negative-stained transmission electron microscopic (TEM) image illustrating 1918 influenza virions

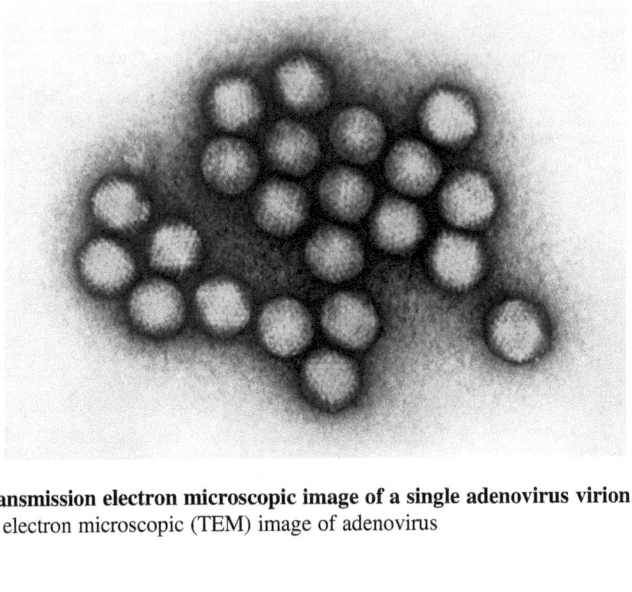

Fig. 3.2 Transmission electron microscopic image of a single adenovirus virion. A colorized transmission electron microscopic (TEM) image of adenovirus

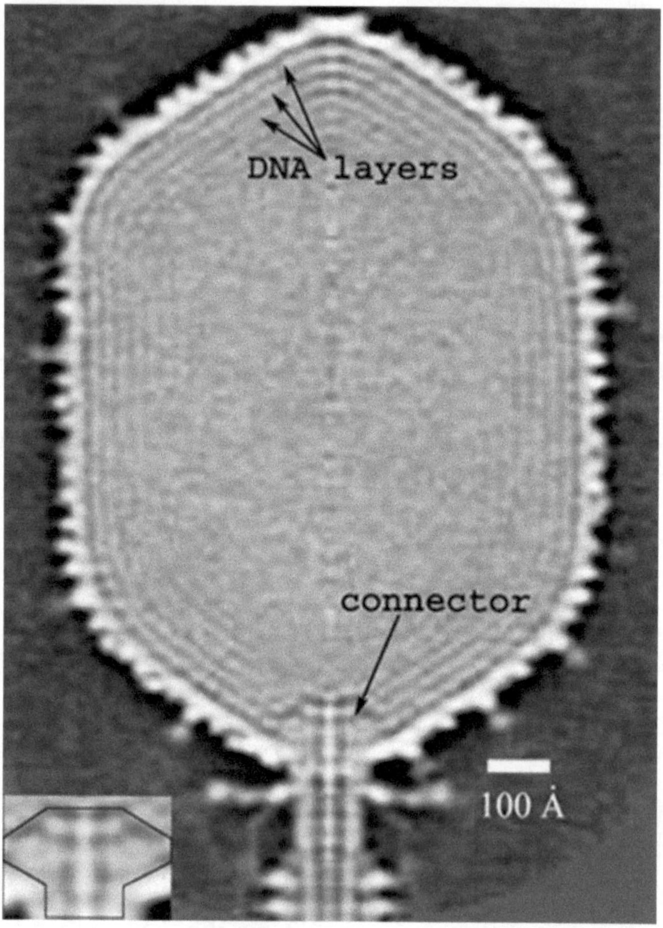

Fig. 3.3 Cryo-electron microscopic image of a T4 bacteriophage. A Cryo-electron microscopic image of the prolate head of bacteriophage T4 (Fokine et al. 2004)

which belong to the family of viruses that cause the common cold, as well as the virus responsible for poliomyelitis, poliovirus (Lodish et al. 2000). Unlike the frequency of acquiring the common cold in the healthcare or service sector, for example (Turner 2007), poliovirus has decreased 99% over the past 20 years, with only 74 cases reported worldwide in 2015 (WHO 2016d) and could be encountered by American emergency responders that visit third-world countries or by infected travelers visiting the United States. Variants to the icosahedral structure are known as "prolate" structures, such as those seen in bacteriophages (Fokine et al. 2004) as depicted in Fig. 3.3. Phage capsids have a variety of sizes and morphologies that are similar to human pathogenic viruses (Ackermann and Prangishvili 2012) and infect different bacterial species. These viruses are used as

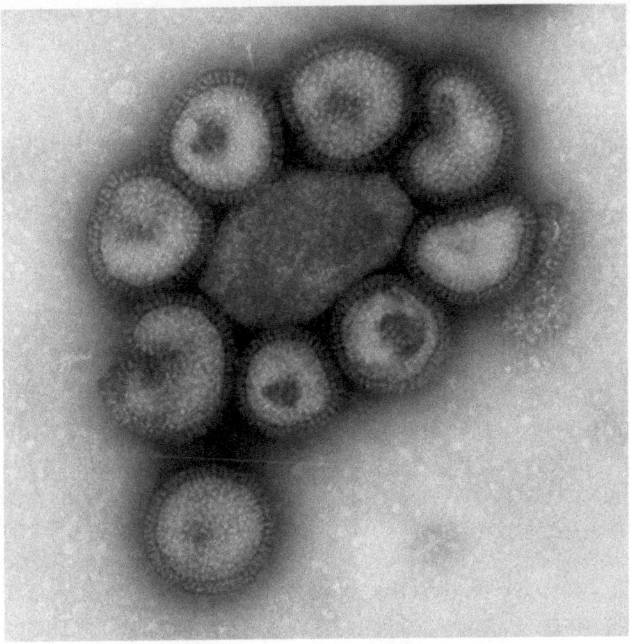

Fig. 3.4 Cross-sectional transmission electron microscopic image of influenza virions. A colorized negative-stained transmission electron microscopic (TEM) image of influenza virus particles

bactericidal agents in "phage therapy" to treat bacterial infections in humans (Abedon et al. 2011).

Viruses such as influenza, herpes virus, and HIV are enveloped viruses that acquire a lipid envelope from the outer or inner membrane of a host cell (Prasad and Schmid 2012). This envelope contains proteins that are coded by both the viral and the host's genome and are required for the infectivity of the virus. Viruses that contain an envelope are more stable under low relative humidity and low temperatures (Duchaine 2016). A cross-sectional image of influenza virions shown in Fig. 3.4 demonstrates the envelope of the virion.

3.2 Classification of Viruses

Viruses are classified according to the Baltimore classification system developed in 1971 (Baltimore 1971). This system classifies viruses based on the type of genome, such as single- or double-stranded and deoxyribonucleic acid (DNA) or ribonucleic acid (RNA) (Prasad and Schmid 2012), as well as the method of replication (Baltimore 1971). The type of genetic material found in a virus depends on the nature and function of the virus. The diversity between the different genomes illustrate the complexity of viruses. Viruses with the largest genomes, including the Mimivirus, Megavirus and Pandoravirus, infect amoebas, with the latter having the largest known genome of all the viruses (Philippe et al. 2013). Having one of the simplest genomes, the circoviruses are the smallest viruses and infect eukaryotic cells (Mankertz et al. 2004).

Viruses contain either RNA or DNA as their genetic material; however, some viruses contain both RNA and DNA at different times during the life cycle. An example of a virus that contains both RNA and DNA is HIV. The replication of the viral genome will also vary, depending on whether the genome is composed of RNA or DNA. For instance, viral RNA replicates primarily in the cell's cytoplasm, whereas DNA viruses replicate in the cell's nucleus.

The viral genome can be linear, circular or segmented in shape. Along with the different shapes of the viral genome, the structure of the nucleic acid will also differ. Viruses can contain single stranded DNA (ssDNA) in which the virus relies on the host's replication machinery, such as the host's DNA polymerase, in order for the ssDNA to be made into double stranded DNA (dsDNA). The dsDNA is then transcribed to messenger RNA (mRNA) after which translation can take place. Single stranded DNA viruses, such as parvovirus, have the ability to infect archaea, bacteria, plants and animals (Koonin et al. 2006). The difference between dsDNA viruses compared to ssDNA viruses is that the virus does not require DNA polymerase to create a complimentary DNA strand to be then transcribed into mRNA because the DNA is already double stranded. Double stranded DNA viruses, such as adenoviruses and herpes viruses, have the ability to infect archaea, bacteria, unicellular eukaryotes and animals (Koonin et al. 2006). Hepatitis A and C, as well as enteroviruses, contain positive-sense ssRNA genomes. A positive-sense RNA genome allows for the genome to be immediately translated in the host cell's cytoplasm. Negative-sense ssRNA viruses, such as Ebola, influenza, measles and mumps, requires an RNA polymerase to produce a positive-sense RNA strand in order for translation to occur. Viruses with dsRNA, such as rotavirus, have the ability to infect bacteria, unicellular eukaryotes, fungi, plants and animals (Koonin et al. 2006).

Lastly, reverse transcribing viruses possess either an ssRNA or a dsDNA genome. HIV contains an ssRNA genome where a reverse transcriptase copies the ssRNA into a complementary DNA strand, and then a DNA polymerase creates a sense DNA strand, making dsDNA. This dsDNA moves into the nucleus of the host cell and integrates into the host's genome. During replication of the host's genome, the dsDNA is transcribed to mRNA from which the viral proteins are then translated. Reverse transcribing viruses that contain dsDNA genome, such as Hepatitis B, use proteins in the host cell to move to the nucleus where the dsDNA serves as a template for mRNA following transcription. The mRNA can then move back into the cytoplasm where it will be reverse transcribed to DNA, allowing for the process to repeat multiple times.

The viral genome can be mutated through different mechanisms to result in new viral strains. First, DNA or RNA bases can mutate to other bases; however, the resultant proteins may be unchanged. Although the proteins may not be initially altered, the accumulation of point mutations may eventually change the viral proteomic profile over time. RNA viruses are more prone to mutations compared to DNA viruses (Domingo et al. 1996). Recombination of the viral genome can also alter the resultant proteins. This type of mutation can render the virus resistant to vaccinations and antiviral drugs, which is detrimental to the preventation against disease development that follow exposures encountered in occupational environments.

3.3 Infection and Effects on Host Cell

Infection generally involves contact between a virus and the host's epithelial surface (Janeway et al. 2001). Viral capsid proteins attach to receptors on the host's outer surface. Glycoproteins or glycolipids serve as the attachment receptors, such as heparin sulfate proteoglycans (de Haan et al. 2005; Vlasak et al. 2005). The attachment process of HIV requires CD4 along with CCR5 or CXCR4 (Grove and Marsh 2011), while rhinoviruses attach to the intercellular adhesion molecule 1 (Norkin 1995). Endocytosis or membrane fusion enables the virus to penetrate the host cell; however, this process may differ depending on the host cell. Penetration into plant or bacterial cell walls is typically more challenging compared to animal cells due to the thickness and complexity of the outer surface composition. Once the virus has entered into the host cell, the capsid is uncoated to release the viral genome. Replication of the viral genome takes place inside the host cell using the host's replication machinery. After the viral components are produced, developed and assembled, they are released from the host cell. This process either occurs by lysis of the host cell when releasing non-enveloped virus or by budding and scission to release enveloped viral particles (Sze and Tan 2015).

Once the virus has migrated out of the host cell, the cytopathic effects caused by the viral infection can result in host cell death through lysis or apoptosis (Roulston et al. 1999). In contrast, there is no visible change to the host cell for some viruses, such as herpes virus (Sissons et al. 2002). These viruses lie dormant inside of the host cell with viral genetic material floating in the nucleus or cytoplasm and can cause chronic infection over long periods of time.

3.3.1 Host Diversity

Viruses can infect a wide diversity of hosts. The infections caused by most viruses are species-specific. On one hand, some viruses can only infect a limited number of hosts and are considered to have a narrow range of hosts. An example of a virus with a narrow range of hosts is the smallpox virus, which only infects humans. On the other hand, other viruses can infect multiple species and are considered to have a broad range of hosts. An example of a virus with a broad range is the rabies virus, which can infect animals and be transmitted to humans. Emerging viral diseases affecting humans, wildlife and agricultural have resulted from viruses infecting new novel hosts, a process known as a "host shift" (Longdon et al. 2014) or a "host switch" (Parrish et al. 2008). These shifts have caused pandemics in the human population, such as HIV from chimpanzees (Sharp and Hahn 2010), severe acute respiratory syndrome from bats (Li et al. 2005), and influenza virus from aquatic birds (Webby and Webster 2001). Although the reservoirs of the influenza virus are in aquatic birds, swine also serve as another zoonotic source as evidenced by the 2009 H1N1 outbreak (Novel Swine-Origin Influenza et al. 2009; Fraser et al. 2009; Gatherer 2009). Rodents are also carriers of viruses, like the hantavirus (CDC 2016e), whereas bats and wild carnivores are zoonotic sources of rabies (CDC 2011; England 2013). Recently, Influenza A (H7N2) virus was detected in an animal shelter worker who was exposed to infected cats (CDC 2016l).

3.4 Viral Transmission Routes

Depending on the origin of the virus, transmission occurs through a variety of mechanisms. Direct transmission occurs through direct physical contact with an infected person's bodily fluids or through indirect transmission by contact with contaminated surfaces or objects (fomites). Viruses, such as HIV and hepatitis, are transmitted by direct contact with bodily fluids, including blood, mucous secretions, urine or feces, from an infected person or through contact with contaminated drug delivery equipment, like needles. Potential exposure to blood-borne diseases can be encountered in occupations involving the use of needles as tools, such as tattooists and piercers, emergency first responders, maintenance and waste workers, or in healthcare environments. Also in a healthcare environment, patients or staff can be exposed to viruses through the use of surgical or medical instruments like catheters or sharp instruments. A sharps injury occurs when a sharp object penetrates the skin, such as a needle, scalpel, or bone fragment. Of all reported sharps injuries, nurses and healthcare assistants accounted for 42%, whereas doctors accounted for 41% (Riddell et al. 2015). Industrial workers are potentially exposed to blood-borne diseases through sharps or a traumatic injury.

To reduce the risk of viral exposures, the Centers for Disease Control and Prevention (CDC) has set guidelines to prevent transmission of infectious agents in healthcare settings (Garner 1996; Siegel et al. 2007). The Universal Precautions guidelines were developed in 1987 as an addition to the 1983 "Guideline for Isolation Precautions in Hospitals" to include the recommendation that blood and body fluid precautions be used for all patients regardless of their blood-borne infection status (CDC 1988). These precautions were set in place to reduce the risk of exposure to blood-borne diseases, specifically for healthcare workers. The Standard Precautions guidelines were proposed in 1996, consisting of precautions that are standard for all patients and includes blood-borne, airborne, and other important pathogens (West and Cohen 1997). The guidelines discuss hand and respiratory hygiene, the use of PPE, such as gloves, gowns, eyewear and face-masks, prevention of sharps injuries, and sanitation practices. By following the recommendations of the guidelines, the risk of viral exposures can be reduced.

Other viruses infect the host via a vector, such as insects like the mosquito. This type of vector-borne transmission can spread through a bite or through contact with the host. This is the route used by the virus involved in the recent disease outbreak caused by the Zika virus. The swine influenza virus is an example of a zoonotic virus that is transmitted from animals to humans. Workers in a meat processing facility and a pig farm have been found to be at greater risk to swine influenza virus compared to people not working in those occupations (Myers et al. 2006; Gray et al. 2007). Hantavirus is another example of a zoonotic virus. As previously mentioned, hantaviruses are carried by rodents and can lead to a rare, but sometimes lethal disease in humans called hantavirus pulmonary syndrome. Contracted through exposure to rodent urine, droppings or saliva, workers in construction, utility, pest control and house cleaning services are at risk of hantavirus pulmonary syndrome exposure (CDC 2016e).

Airborne transmission occurs when a virus is spread by traveling on dust particles or on small respiratory droplets. Viruses can travel on different sizes of airborne particles formed when a person coughs, sneezes, talks or exhales. The larger respiratory droplets that form travel short distances (Mubareka et al. 2009), compared to droplet nuclei that are small and responsible for transmission over long range distances. An example of an airborne virus is the influenza virus, which can be transmitted by direct contact with infected people in a healthcare environment for example or by exposure to fomites that have been contaminated when the droplet nuclei settle out of the air and dry on a surface, a situation that may be encountered in healthcare or academic occupational environments. The viability of the viral particle on the fomites depends on the virus, the environmental conditions surrounding the fomite and whether or not the droplet nuclei is moist or dry. Viral exposure can occur through inhalation of aerosolized droplet nuclei into the respiratory tract of a person, also potentially encountered in healthcare or academic occupational environments. One study determined that professionals in healthcare settings could be exposed to infectious doses of influenza virus within 1.8 meters of an infected patient (Bischoff et al. 2013). Another study detected low concentrations of influenza viral RNA in aerosols located in patients' rooms of which healthcare workers frequently encountered (Leung et al. 2016). The National Institute for Occupational Safety and Health (NIOSH) has developed a bioaerosol sampler that allows for the collection of size-fractionated aerosols (Cao et al. 2011; Lindsley et al. 2010b), which enabled further research into influenza transmission. Subsequent studies utilizing the influenza virus as a model illustrated that infectious viral particles are collected in the smallest particle (airborne) fraction, as well as in the larger (droplet) fraction (Noti et al. 2012) and that more viable influenza particles are detected after coughing than exhaling (Lindsley et al. 2016).

3.5 Viral Exposures in Occupational Settings

Viral exposures are encountered in a wide variety of occupations. An earlier review described methods for sampling airborne viruses in various environments such as hospitals, office buildings, restaurants and schools (Verreault et al. 2008). Most of the studies reviewed occurred before 2008 and do not include recent viral outbreaks. Current examination of viral exposures lead to the identification of viruses that are frequently encountered in occupational environments or have been a recent focus in the United States. Although viral exposures can occur in many different occupational environments, the healthcare sector has one of the greatest risk of exposure to a large variety of infectious viruses.

Environments within the healthcare sector include, but are not limited to, hospitals, medical offices, assisted-living facilities, and dental offices. Viruses in these occupational environments can be transmitted from patient to patient, healthcare worker to patient or patient to healthcare worker as hypothesized in a review on Hepatitis C transmission in a healthcare environment (Pozzetto et al. 2014). One study reported the detection of human cytomegalovirus in the air of hospital rooms, a mode of transmission not previously considered, which demonstrates a potential route of exposure to healthcare workers and patients (McCluskey et al. 1996). Respiratory syncytial virus has also been detected in the air of hospital rooms housing infected patients in which the viral RNA was detected up to 7 m away from the patients for up to 7 days (Aintablian et al. 1998). Severe acute respiratory syndrome (SARS) coronavirus was identified on fomites in a hospital room belonging to a patient with SARS, as well as at a nurses' station. These data suggest generation of aerosol particles containing SARS coronavirus (Booth et al. 2005). Collectively, these studies further highlight the potential exposure for both patients and healthcare workers in a healthcare setting.

Laboratory-associated occupations also serve as potential sources of adverse viral exposures. Researchers that handled clinical samples of the Ebola virus became infected at a government hospital in Sierra Leone (Silver 2015). Individuals that work in biotechnological laboratories, as well as healthcare environments, that use lentiviral vectors for gene delivery systems are also at risk of viral exposure (Howard et al. 2017). Viral exposures may also occur in manufacturing or industrial occupational environments. Human adenovirus was detected in air samples collected from a landfill, a waste recycling plant, an incineration plant, and waste collection vehicles, all serving as environments with a potential risk for viral exposure (Carducci et al. 2013). Modeling systems for airborne viruses used in industry could possibility lead to viral exposures for workers in these facilities. For example, bacteriophages are not only used by the FDA to test the effectiveness of filtration devices (Duchaine 2016), but are also used as surrogates for enteric viruses in wastewater treatment studies (Grabow 2001). Rhinoviruses have been detected in office buildings with low outdoor air supply, suggesting that occupants are at an increased risk of viral exposure from infected colleagues

(Myatt et al. 2004). Other occupational environments where people are susceptible to viral exposures, such as influenza and norovirus, are in the service sector, including schools and daycare settings, as well as agricultural environments (Division of Viral Diseases et al. 2011; CDC 2016h). Areas of construction serve as environments where blood-borne pathogens can be encountered through traumatic or sharp injuries. The following section and Table 3.1 details the viruses that have more recently been characterized in occupational environments.

3.5.1 Viral Hepatitis

In 2014, Hepatitis B affected approximately 850,000 people in the United States (Roberts et al. 2016; Wasley et al. 2010). Hepatitis B and Hepatitis C are transmitted through bodily fluids or by contact with contaminated drug-injection equipment. Sewage workers at waste plants, as well as healthcare workers and patients in healthcare environments can be exposed to both viruses. Hepatitis B (CDC 2016s) is normally an acute infection, but for some, the disease can become chronic. The risk of the infection becoming chronic is related to the age of the infected person. For all viral hepatitis cases, routine blood tests, serological tests, reverse transcription polymerase chain reaction (RT-PCR) to detect viral RNA, and genotyping tests are used to diagnose viral hepatitis. This virus can be prevented by following the recommendations of the Standard Precautions guidelines, including the use of PPE, proper handling of sharps and patient care equipment, avoiding close physical contact and vaccination. Similar to Hepatitis B, Hepatitis C is transmitted by contact with contaminated drug-injection equipment and is also an acute infection, which can become chronic in some. In 2014, 30,500 cases of acute and an estimated 2.7–3.9 million cases of chronic Hepatitis C occurred in the United States (CDC 2016s). Hepatitis C exposure in healthcare environments is less common, but sharps injuries can still occur. Unlike Hepatitis B, a vaccine is not available for Hepatitis C (Riddell et al. 2015), so it is important to avoid sharps injuries or sharing drug-injection equipment with those people infected with Hepatitis C.

Even more uncommon than Hepatitis C exposure in the healthcare environment are exposures to Hepatitis A. Hepatitis A is transmitted by the fecal-oral route or consumption of contaminated food or water. It does not result in a chronic disease and there is no available vaccine (CDC 2016s). Another rare exposure risk in the United States is Hepatitis D. It is transmitted through contact with infectious blood; however, it only occurs in people who are infected with Hepatitis B. Hepatitis D virus is incomplete and therefore, needs Hepatitis B in order to replicate. There is no vaccination for Hepatitis D, so the best prevention strategy is to be vaccinated against Hepatitis B (CDC 2016s). Hepatitis E is also rare in the United States; however, one study reported a significant association between occupational exposure to Hepatitis E and swine (De Schryver et al. 2015). Hepatitis E is transmitted by ingestion of fecal matter through contaminated food or water.

Table 3.1 Prevalent viral exposures encountered in United States occupational environments

Disease	Genome	Transmission	Occupation	Treatment	Virus	Reference
Hepatitis	• Circular dsDNA (Hepatitis B) • (+)ssRNA (Hepatitis C) • Enveloped icosahedral	• Direct contact with bodily fluids from infected person(s) • Contact with contaminated drug-injection equipment	• Healthcare workers	• Vaccine for Hepatitis B, but no vaccine for Hepatitis C	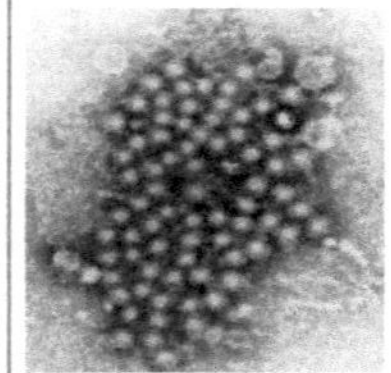	Photo courtesy of CDC/E.H. Cook, Jr.; CDC Public Health Image Library (PHIL) ID#: 8153
HIV	• Linear (+) ssRNA-RT • Viral envelope	• Direct contact with bodily fluids from infected person(s) • Contact with contaminated drug-injection equipment	• Healthcare workers • Research laboratories	• Antiretroviral treatments, but no cure is available	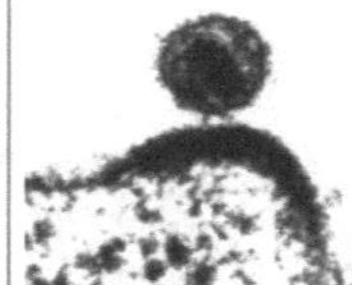	Photo courtesy of National Institute of Allergy and Infectious Diseases (NIAID); PHIL ID#: 18142
Adenovirus	• Linear dsRNA • Non-enveloped icosahedral	• Direct contact with bodily fluids from infected person(s) • Contact with contaminated surfaces	• Healthcare workers • Military	• Available vaccine only approved for military use	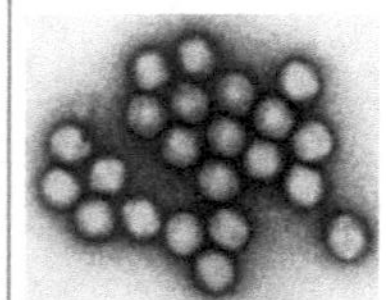	Photo courtesy of CDC/Dr. G. William Gary, Jr.; PHIL ID#: 10010
Influenza	• Segmented • (−)ssRNA • (Influenza A) • Linear (−) ssRNA (Influenza B) • Viral envelope	• Airborne and spreads by coughing and sneezing • Contact with contaminated areas and surfaces	• Healthcare workers • General public • Poultry and meat processing facilities	• Flu vaccine	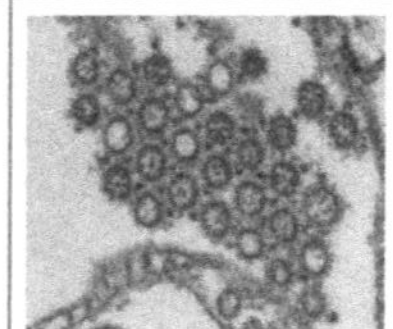	Photo courtesy of CDC/Cynthia Goldsmith; CDC PHIL ID#: 11745

(continued)

Table 3.1 (continued)

Disease	Genome	Transmission	Occupation	Treatment	Virus	Reference
Ebola	• Linear (–) ssRNA • Viral envelope	• Direct contact with bodily fluids from infected person(s)	• Healthcare workers • Response teams • Research laboratories	• No vaccine, but two potential candidates	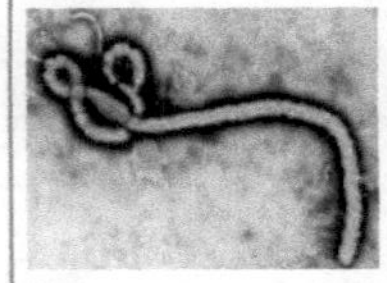	Photo courtesy of CDC/Dr. F. A. Murphy; CDC PHIL ID#: 10815
Norovirus	• Linear (+) ssRNA • Non-enveloped icosahedral	• Contaminated food and water • Direct and indirect contact with infected person(s) or contaminated surfaces • Airborne transmission	• Healthcare workers • Food handling services • Cruise ships • Military	• No vaccine	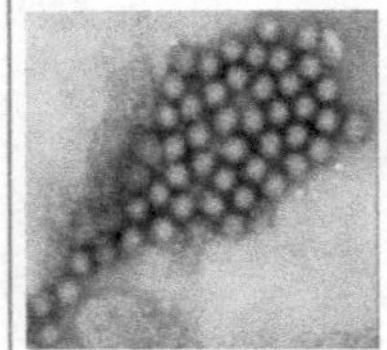	Photo courtesy of CDC/Dr. Charles D. Humphrey; PHIL ID#: 10706
Zika	• Linear (+) ssRNA • Enveloped icosahedral	• Transmitted by mosquitos	• Occupations requiring travel to infected areas • Research laboratories	• No vaccine	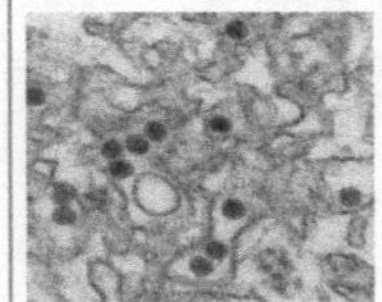	Photo courtesy of CDC/Cynthia Goldsmith; CDC PHIL ID#: 20541

Despite Hepatitis E not resulting in a chronic infection, there is a vaccine available (CDC 2016s). Although these latter types of viral hepatitis are uncommon, each are important to review as lapses in standard precautions and a break in safety protocols can lead to an increased risk of exposure.

To reduce the risk of exposure to viral hepatitis and other blood-borne pathogens, the CDC has set guidelines to prevent transmission of infectious agents in healthcare settings that include standard precautions, which protect the healthcare worker and patient (Garner 1996) as previously mentioned. Guidelines reported in 2007 regarding the prevention of infectious agent transmission in healthcare settings, discuss transmission-based precautions dependent upon the specific disease, as well as type and duration of precautions recommended for selected infections (Siegel et al. 2007). If the virus can be transmitted through bodily fluids, healthcare workers have to be able to assess the risk of exposure by considering the injury type, the bodily fluid involved, the risk for transmission of blood-borne viruses and then test the source patient to determine if blood-borne viruses are present. Although these precautions were recommended to reduce the risk of exposure to blood-borne diseases for healthcare workers, these practices can be applied to prevent exposure in any occupational environment that poses a risk. Exposure Control Plans also need to be executed to ensure the workers have accurate information regarding potential exposures. CDC has also implemented engineering controls and personal protective equipment resources to help reduce exposure to blood-borne infectious diseases, such as viral hepatitis and HIV (CDC 2016b).

3.5.2 Human Immunodeficiency Virus

As of 2010, there were approximately 35,000–38,000 new cases of HIV and 16,000–18,000 deaths per year in the United States (CDC 2016g; Mohebati et al. 2010). In 2015, the CDC reported that approximately 1.2 million people in the United States were living with HIV infection (CDC 2016f). Even though rates are falling (Wyżgowski et al. 2016; CDC 2001), new cases of HIV are reported every year. HIV is transmitted by exposure to infected bodily fluids and is diagnosed through serological testing, as well as detection of the virus through RT-PCR assays. The CDC recommends that all individuals in a healthcare setting aged 13–64 years be tested for HIV, regardless of the exposure risk (Panneer et al. 2014). One review reported 57 documented cases of HIV infections transmitted in an United States healthcare environment after having a negative serology at the time of assumed exposure (Mohebati et al. 2010). HIV is a lentivirus, a subgroup of retroviruses, characterized by long incubation periods. Lentiviruses can integrate large amounts of viral RNA into the DNA of host cells and as a result, lentiviral vectors have become an important research tool as a delivery system for gene therapy strategies. As previously mentioned, workers in biotechnological laboratories or healthcare environments that use these lentiviral vectors are also at risk of viral exposure (Howard et al. 2017). Individuals who work in laboratories that conduct research on HIV are also susceptible to exposure; however, with the protective measures taken to limit exposure to contaminated blood and other bodily fluids, exposure to HIV has decreased over time, specifically in the United States. From 2005 to 2014, HIV infections in the United States have decreased approximately 19% (CDC 2016f). To reduce the risk of exposure, the CDC recommends following the 2007 guidelines for isolation precautions preventing transmission of infectious agents, such as HIV (Siegel et al. 2007). Included in these recommendations are proper use of PPE, such as gloves, safety glasses, and gowns to reduce the risk of exposure. In 2013, an update of the United States Public Health Service guidelines for the management of occupational exposures to HIV was released and included recommendations for the management of healthcare workers who had been exposed to HIV (Kuhar et al. 2013). Administration of antiretroviral drugs to reduce the replication of the virus is also effective following exposure to HIV.

3.5.3 Adenoviruses

Adenoviruses are the most common illness of the respiratory system in healthcare environments and in the military, and can cause a wide range of symptoms (CDC 2015), including common cold-like symptoms, pharyngitis, bronchitis, pneumonia, diarrhea, conjunctivitis or gastroenteritis. Adenoviruses are transmitted by direct personal contact with an infected person or by contact with contaminated surfaces. One study suggested that the crowding of personnel at military stations lead to adenovirus outbreaks, highlighting the importance of vaccinations (Gray et al. 2000). Another study reported the exposure of healthcare workers to Adenovirus Serotype 14 from infected military trainees (Centers for Disease and Prevention 2007). In 2008, adenoviral transmission was documented between healthcare workers and two patients identified as having Adenovirus Serotype 14 (Louie et al. 2008). Good hygiene and sanitation practices, as well as avoiding close contact with infected people or contaminated areas, are prevention strategies to minimize the spread of the virus. Currently, the only available vaccine is for military use, which was approved in 2011 by the Food and Drug Administration (CDC 2015).

3.5.4 Influenza

Influenza virus is a respiratory illness that can affect anyone potentially leading to hospitalization or even death. There are three different types of influenza virus. Influenza type A and B are responsible for the annual influenza outbreaks. Type C also causes the flu, but the symptoms are not as profound as types A and B. Influenza A viruses are found among animals and humans, while influenza B viruses are mainly found in humans. The virus can be transmitted by contact with surfaces or areas where an infected person has touched or sneezed. The virus has also recently been shown to become airborne via coughing (Lindsley et al. 2016). Throughout the world, there are approximately 3–5 million severe influenza cases reported annually with 250,000–500,000 resulting deaths (WHO 2016c). Since 2010, the CDC estimates that 140,000–710,000 influenza cases in the United States have resulted in hospitalizations, resulting in 12,000–56,000 deaths (CDC 2016k).

The influenza virus is diagnosed via several methods including lateral flow and membrane-based immunoassays, as well as RT-PCR assays from nasal swabs. Sensitive real time semi-quantitative PCR assays, such as the viral replication assay developed at NIOSH (Blachere et al. 2011), can detect viable viral RNA influenza quickly and accurately compared to traditional methods (Blachere et al. 2009; Lindsley et al. 2010a; Lindsley et al. 2010b) that cannot distinguish between infectious and non-infectious virus. Prevention of the influenza virus includes proper sanitation practices, such as hand washing and decontamination of infected areas. Guidelines developed by the CDC, such as wearing PPE, will reduce the risk of exposure to the influenza virus (Siegel et al. 2007). An annual flu vaccine is available and is recommended for people older than 6 months (CDC 2016h). The vaccination is composed of inactivated or recombinant strains of influenza virus that research conducted at the collaborating centers of the World Health Organization (WHO) has deemed most probable to be encountered in the upcoming season (CDC 2016h). Trivalent vaccines are most commonly used and the current configuration immunizes workers against two Influenza A (H1N1 and H3N1) viruses and one Influenza B virus. There is also a quadrivalent vaccine that protects against the same three viral strains as the trivalent vaccine, but also includes a second Influenza B virus (CDC 2016h). The National 2009 H1N1 Flu Survey conducted by the CDC from October 2009 through June 2010 found that of 28,710 employed adults, 10.5% of workers employed in real estate and the rental and leasing industry, 10.2% workers in accommodation and food services industry, 11.0% workers in food preparation services, and 8.3% of workers in community and social services reported having an influenza-like illness. It is important to note that the number of workers vaccinated for the seasonal influenza or H1N1 in the aforementioned groups was low (Luckhaupt et al. 2014). A study conducted at NIOSH measured viral particles of influenza in air samples collected from a hospital emergency department and found that over 50% of the virus was detected in the airborne droplets located in waiting rooms, triage rooms and in personal samplers worn by emergency department physicians (Blachere et al. 2009). Even though the CDC, the Advisory Committee on Immunization Practices, and the Healthcare Infection Control Practices Advisory Committee recommend that all United States healthcare workers get vaccinated annually, only 64.3% of healthcare workers were vaccinated during the early 2014–2015 season (CDC 2016j). Changes and updates to the previous recommendations of immunization practices have already been reported for the 2016–2017 influenza season (CDC 2016i).

Workers in the poultry and swine industries are susceptible to zoonotic infections, and exposures have been reported in recent studies (Fragaszy et al. 2016; Gray et al. 2007). Workers in pig farms and their wives have been found to be at a greater risk of exposure to swine influenza virus compared to a control population not exposed to swine (Gray et al. 2007). Veterinarians, farmers and workers who process meat have also been reported to be exposed to swine flu (Myers et al. 2006; O'Brien and Nonnenmann 2016). In December 2016, a New York animal shelter worker was infected with a novel avian lineage influenza (H7N2) virus. The worker was thought to have acquired the virus through close, prolonged unprotected exposure to infected cats (CDC 2016l).

3.5.5 *Ebola*

The Ebola virus disease, also known as Ebola hemorrhagic fever made headlines from December 2013 to January 2016 due to an outbreak in West Africa, including the countries of Guinea, Liberia and Sierra Leone (CDC 2016d). Symptoms included fever, sore throat, muscular pain and headaches followed by vomiting diarrhea and rash. The virus is transmitted into the human population through direct contact with blood or other bodily secretions of infected animals, then transmitted from human to human by contact with bodily fluids of infected individuals (WHO 2016a). To diagnose the Ebola virus, blood samples are tested for viral RNA by RT-PCR and viral proteins by an enzyme-linked immunosorbent assay (ELISA). During the outbreak, approximately 25,000 cases of Ebola exposure were reported, with 41% of those cases ending in death (Suwantarat and Apisarnthanarak 2015). Healthcare-related infections with the Ebola virus accounted for 2.5–12% of total cases and the fatality rate was reported in up to 73% of the emergency responders and healthcare workers in direct contact with infected patients (Suwantarat and Apisarnthanarak 2015). There were five reported exposures that were confirmed in the United States, with four of the cases being healthcare workers (Lyon et al. 2014; Chevalier et al. 2014). The CDC has published guidelines for healthcare workers during the management of patients in United States hospitals during an Ebola virus outbreak (CDC 2016c). The guidelines instruct workers on the selection of the correct PPE, the proper use of PPE, and techniques for donning and doffing PPE. Fig. 3.5 shows a responder wearing PPE during the Ebola outbreak. Preventative measures discussed in the guidelines include strict adherence to PPE recommendations and incubation periods when responders returned to the United States (CDC 2016c). Currently, there are no approved vaccinations against Ebola (CDC 2016d); however, there are potential candidates (WHO 2016b).

3.5.6 *Norovirus*

Gastroenteritis-causing norovirus is a highly contagious virus found in healthcare occupations, food handling services, the military, schools and on cruise ships. The Vessel Sanitation Program at CDC assists with preventing and controlling

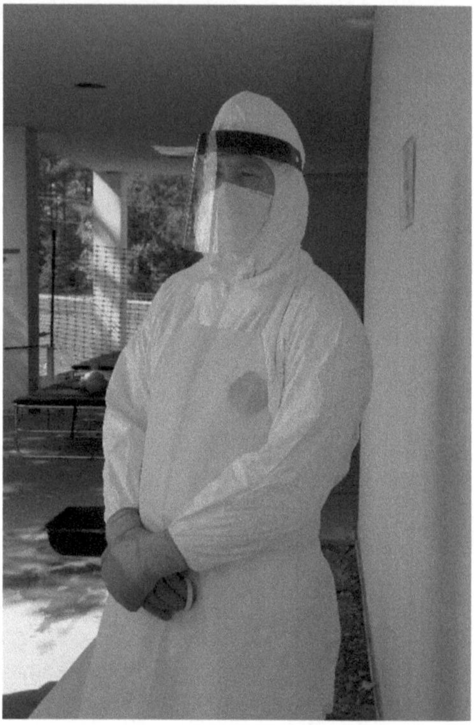

Fig. 3.5 Personal protective equipment. An emergency responder displaying the proper use of personal protective equipment

gastroenteritis illnesses such as the norovirus in the cruise ship industry (CDC 2016r). Norovirus is easily transmitted by contaminated food and water, hands and communal objects, and by direct contact with infected individuals (CDC 2016p). It is suggested that norovirus can be carried through the air to replicate in the host (Duchaine 2016). Recently, molecular analysis of air collected from areas in healthcare facilities revealed evidence of human norovirus (Bonifait et al. 2015). Norovirus is diagnosed through enzyme immunoassays or RT-PCR assays and has no current vaccine. Similar to norovirus, rotavirus causes gastroenteritis and is spread among infants and young children through the fecal-oral route. It is transmitted by contaminated food, water and objects as well as by direct bodily contact. Diagnosis is made through the detection of rotavirus antigen in stool samples. Unlike norovirus, rotavirus has an available vaccine (CDC 2016q). Both viral diseases are prevented through proper hygiene and sanitation practices, as well as avoiding infected and symptomatic people (MacCannell et al. 2011). Washing vegetables and fruits and disinfecting contaminated areas also limits the risk of exposure. Due to the age of the population at risk of rotavirus exposure, it is not common for occupational exposures.

3.5.7 *Zika Virus*

Beginning in 2015, the Zika virus has been a major focus for public health officials. Zika virus is a mosquito-borne illness that was first identified in Uganda (WHO 2016e) and can be transmitted through mosquito bites and possibly blood transfusions, although the latter is unconfirmed (CDC 2016v). In pregnant women, Zika can be transmitted to the fetus and cause birth defects, such as microcephaly. Symptoms that normally last 2–7 days include mild fever, skin rashes, muscle and joint pain, and conjunctivitis. A blood or urine test can confirm the presence of Zika virus. In March 2016, the Food and Drug Administration approved a new Trioplex RT-PCR Assay to test for Zika (CDC 2016o). The CDC suggested that avoiding mosquito bites by wearing protective clothing or applying bug repellents are methods to reduce the risk of exposure. Currently, there is no vaccine against Zika viral infection. In 2016, the CDC responded to an outbreak of Zika in North and South America. Along with NIOSH, the Occupational Safety and Health Administration (OSHA) developed guidelines for protecting workers from occupational exposure to Zika virus, including workers in outdoor environments, business travelers, as well as healthcare and laboratory workers (CDC 2016v). There are reports of locally acquired cases of Zika in Florida that are not associated with travel exposure (CDC 2016w). More recently, five cases of locally acquired Zika were reported in Texas (CDC 2016w). Laboratorians are an additional occupational group at risk of contracting Zika as shown in a case study of a Pittsburgh worker who acquired Zika through laboratory transmission, specifically by a needle stick (CDC 2016w).

3.5.8 *Other Potential Occupational Viral Exposures*

Although not as prevalent in occupational environments, there are other viruses that pose a risk for potential exposure. Viral meningitis, defined as the inflammation of the spinal cord and brain, is the most common type of meningitis. This disease is caused by non-polio enterovirus, mumps, herpes virus, measles, influenza, and arboviruses. Viral meningitis is most common in immunocompromised people and young children (CDC 2016t). The disease can be diagnosed by X-ray of the head, blood collection for culturing, or drawing fluid from around the spinal cord to further test for cellular, sugar and protein levels. There is no specific treatment for viral meningitis. Although viral meningitis may not be commonly found in many occupational environments, the causal agents are found in occupational settings, such as schools (Croker et al. 2015; Sosa et al. 2009). Non-polio enterovirus can be encountered in a healthcare environments (Midgley et al. 2015; Midgley et al. 2014). If symptoms are present, they are similar to the common cold. Non-polio enterovirus is transmitted through feces, or eye, nose and mouth secretions and can be diagnosed through serotype identification and molecular testing. Exposure to the virus can be minimized by washing hands and avoiding direct contact with infected people or contaminated objects. Currently there is no treatment for non-polio enterovirus (CDC 2016a). According to the CDC, mumps can be encountered in healthcare and school environments (CDC 2016n). Transmitted through eye, nose and mouth secretions, mumps is spread through coughing, sneezing or talking and by direct contact with infected people or contaminated objects. The disease is diagnosed using antibody measurements, viral culturing, and RT-PCR assays. A vaccine is available that protects against measles, mumps and rubella (MMR). There is also an MMRV vaccine for children which includes protection against varicella (chickenpox). A third cause of viral meningitis is measles. It is highly contagious and is transmitted through coughing and sneezing. In 2015, there was an outbreak at an amusement park in California exposing both the general public and the park workers (Zipprich et al. 2015), as well as another outbreak in an Ohio Amish community in 2014 (CDC 2016m). Measles is diagnosed using RT-PCR and genetic sequencing and is prevented with the MMR vaccine. Arboviruses, such as the West Nile virus are another group of viruses that can potentially cause viral meningitis. This disease is transmitted to humans by mosquitos; therefore, insect repellent and protective clothing are preventative measures taken to reduce the risk of exposure. West Nile virus is diagnosed by antibody testing, viral culturing or RT-PCR assays to detect viral RNA. Less than 1% of those infected develop life threatening illnesses (CDC 2016u). Currently, there is no treatment available to treat the West Nile virus, but the same precautions for the Zika virus should be taken to reduce exposure to workers in occupations that require travel (CDC 2016v).

3.6 Conclusion

In summary, viral exposures can occur in many occupational environments with the healthcare environment having some of the greatest risk for exposure to a broad diversity of viruses. Table 1 depicts the most prevalent viral exposures encountered in occupational environments, characteristics of those viruses, and current treatment strategies against the viral exposure. The CDC, along with many other governmental agencies, have developed guidelines that are often updated to provide the most current preventative methods to protect workers from potential viral exposures (Garner 1996). Regardless of the virus, the "2007 Guideline for Isolation Precautions: Preventing Transmission of Infectious Agents in Health Care Settings" cover an array of recommendations to protect workers at risk for different viral exposures (Siegel et al. 2007). The CDC offers available resources for workers, especially in the healthcare sector, to assist in limiting potential viral exposures through proper use of selection and the donning of PPE (CDC 2016b, c), as depicted in Fig. 3.5. It is important to follow appropriate prevention practices, such as good hygiene, proper sanitation and disinfection strategies, as well as wearing the correct PPE to limit viral exposure in occupational environments. Knowledge of the available federal and state guideline documents for specific viruses, as well as Exposure Control Plans developed by employers will help reduce occupational exposures to viruses. Taking preventative measures to limit the risk of exposures and following vaccination recommendations can reduce the spread of viral exposure in occupational environments.

Acknowledgements The findings and conclusions in this study are those of the author and do not necessarily represent the views of National Institute for Occupational Safety and Health. The author declares no conflict of interest.

References

Abedon ST, Kuhl SJ, Blasdel BG et al. (2011) Phage treatment of human infections. Bacteriophage 1(2):66–85. doi:10.4161/bact.1.2.15845

Ackermann HW, Prangishvili D (2012) Prokaryote viruses studied by electron microscopy. Arch. Virol 157(10):1843–1849. doi:10.1007/s00705-012-1383-y

Aintablian N, Walpita P, Sawyer MH (1998) Detection of Bordetella pertussis and respiratory syncytial virus in air samples from hospital rooms. Infect Control Hosp Epidemiol 19(12): 918–923. doi:10.2307/30142018

Baltimore D (1971) Expression of animal virus genomes. Bacteriol Rev 35(3):235–241

Bischoff WE, Swett K, Leng I et al. (2013) Exposure to Influenza virus aerosols during routine patient care. J Infect Dis 207(7):1037–1046. doi:10.1093/infdis/jis773

Blachere FM, Cao G, Lindsley WG et al. (2011) Enhanced detection of infectious airborne influenza virus. J Virol Methods 176(1–2):120–124. doi:10.1016/j.jviromet.2011.05.030

Blachere FM, Lindsley WG, Pearce TA et al. (2009) Measurement of airborne influenza virus in a hospital emergency department. Clin Infect Dis 48(4):438–440. doi:10.1086/596478

Bonifait L, Charlebois R, Vimont A et al. (2015) Detection and quantification of airborne norovirus during outbreaks in healthcare facilities. Clin Infect Dis 61(3):299–304. doi:10.1093/cid/civ321

Booth TF, Kournikakis B, Bastien N et al. (2005) Detection of airborne severe acute respiratory syndrome (SARS) coronavirus and environmental contamination in SARS outbreak units. J Infect Dis 191(9):1472–1477. doi:10.1086/429634

Brown JC, Newcomb WW, Wertz GW (2010) Helical virus structure: the case of the rhabdovirus bullet. Viruses 2(4):995–1001. doi:10.3390/v2040995

Cao G, Noti JD, Blachere FM et al. (2011) Development of an improved methodology to detect infectious airborne influenza virus using the NIOSH bioaerosol sampler. J Environ Monit 13(12):3321–3328. doi:10.1039/c1em10607d

Carducci A, Federigi I, Verani M (2013) Virus occupational exposure in solid waste processing facilities. Ann Occup Hyg. doi:10.1093/annhyg/met043

CDC (1988) Update: universal precautions for prevention of transmission of human immunodeficiency virus, hepatitis B virus, and other bloodborne pathogens in health-care settings. MMWR Morb Mortal Wkly Rep 37(24):377–382. 387-378

CDC (2001) HIV and AIDS – United States, 1981–2000. MMWR Morb Mortal Wkly Rep 50(21):430–434

CDC (2011) Rabies. Centers for Disease Control and Prevention. https://www.cdc.gov/rabies/. Accessed November 27, 2016

CDC (2015) Adenoviruses. Centers for Disease Control and Prevention. http://www.cdc.gov/adenovirus/index.html. Accessed November 27, 2016

CDC (2016a) About non-polio enterovirus. Centers for Disease Control and Prevention. https://www.cdc.gov/non-polio-enterovirus/about/index.html. Accessed November 27, 2016

CDC (2016b) Bloodborne infectious diseases: HIV/AIDS, Hepatitis B, and Hepatitis C. Centers for Disease Control and Prevention. https://www.cdc.gov/niosh/topics/bbp/controls.html. Accessed November 27, 2016

CDC (2016c) Ebola (Ebola virus disease). Centers for Disease Control and Prevention. http://www.cdc.gov/vhf/ebola/healthcare-us/ppe/guidance.html. Accessed November 27, 2016

CDC (2016d) Ebola (Ebola virus disease). Centers for Disease Control and Prevention. http://www.cdc.gov/vhf/ebola/index.html. Accessed November 27, 2016

CDC (2016e) Hantavirus. Centers for Disease Control and Prevention. https://www.cdc.gov/hantavirus/. Accessed November 27, 2016

CDC (2016f) HIV in the United States: at a glance. Centers for Disease Control and Prevention. http://www.cdc.gov/hiv/statistics/overview/ataglance.html. Accessed November 27, 2016

CDC (2016g) HIV surveillance reports. Centers for Disease Control and Prevention. http://www.cdc.gov/hiv/library/reports/hiv-surveillance.html. Accessed November 27, 2016

CDC (2016h) Influenza (flu). Centers for Disease Control and Prevention. http://www.cdc.gov/flu/index.htm. Accessed November 27, 2016

CDC (2016i) Influenza (flu). Centers for Disease Control and Prevention. https://www.cdc.gov/flu/professionals/acip/index.htm. Accessed November 27, 2016

CDC (2016j) Influenza (flu). Centers for Disease Control and Prevention. https://www.cdc.gov/flu/healthcareworkers.htm. Accessed November 27, 2016

CDC (2016k) Influenza (flu). Centers for Disease Control and Prevention. https://www.cdc.gov/flu/protect/keyfacts.htm. Accessed November 27. 2016

CDC (2016l) Influenza (flu). Centers for Disease Control and Prevention. https://www.cdc.gov/flu/spotlights/avian-influenza-cats.htm. Accessed January 5, 2017

CDC (2016m) Measles (Rubeola). Centers for Disease Control and Prevention. https://www.cdc.gov/measles/index.html. Accessed November 27, 2016

CDC (2016n) Mumps. Centers for Disease Control and Prevention. https://www.cdc.gov/mumps/. Accessed November 27, 2016

CDC (2016o) New CDC laboratory test for Zika virus authorized for emergency use by FDA. Centers for Disease Control and Prevention. https://wwwdev.cdc.gov/media/releases/2016/s0318-zika-lab-test.html. Accessed November 27, 2016

CDC (2016p) Norovirus. Centers for Disease Control and Prevention. https://www.cdc.gov/norovirus/. Accessed November 27, 2016

CDC (2016q) Rotavirus. Centers for Disease Control and Prevention. https://www.cdc.gov/rotavirus/about/transmission.html. Accessed November 27, 2016

CDC (2016r) Vessel sanitation program. Centers for Disease Control and Prevention. https://www.cdc.gov/nceh/vsp/. Accessed November 27, 2016

CDC (2016s) Viral hepatitis. Centers for Disease Control and Prevention. http://www.cdc.gov/hepatitis/index.htm. Accessed November 27, 2016

CDC (2016t) Viral meningitis. Centers for Disease Control and Prevention. https://www.cdc.gov/meningitis/viral.html. Accessed November 27, 2016

CDC (2016u) West Nile virus. Centers for Disease Control and Prevention. https://www.cdc.gov/westnile/. Accessed November 27, 2016

CDC (2016v) Zika virus. Centers for Disease Control and Prevention. http://www.cdc.gov/zika/index.html. Accessed November 27, 2016

CDC (2016w) Zika virus. Centers for Disease Control and Prevention. http://www.cdc.gov/zika/index.html. Accessed November 27, 2016

Centers for Disease C, Prevention (2007) Acute respiratory disease associated with adenovirus serotype 14 – four states, 2006–2007. MMWR Morb Mortal Wkly Rep 56(45):1181–1184

Chevalier MS, Chung W, Smith J et al. (2014) Ebola virus disease cluster in the United States – Dallas County, Texas, 2014. MMWR Morb Mortal Wkly Rep 63(46):1087–1088

Croker C, Civen R, Keough K et al. (2015) Aseptic meningitis outbreak associated with echovirus 30 among high school football players – Los Angeles County, California, 2014. MMWR Morb Mortal Wkly Rep 63(51):1228

de Haan CAM, Li Z, te Lintelo E et al. (2005) Murine Coronavirus with an extended host range uses heparan sulfate as an entry receptor. J Virol 79(22):14451–14456. doi:10.1128/jvi.79.22.14451-14456.2005

De Schryver A, De Schrijver K, Francois G et al. (2015) Hepatitis E virus infection: an emerging occupational risk? Occup Med (Lond) 65(8):667–672. doi:10.1093/occmed/kqv154

Division of Viral Diseases NCfI, Respiratory Diseases CfDC, Prevention (2011) Updated norovirus outbreak management and disease prevention guidelines. MMWR Recomm Rep 60 (RR-3):1–18

Domingo E, Escarmis C, Sevilla N et al. (1996) Basic concepts in RNA virus evolution. FASEB J 10(8):859–864

Duchaine C (2016) Assessing microbial decontamination of indoor air with particular focus on human pathogenic viruses. Am J Infect Control 44(9, Supplement):S121–S126. doi:10.1016/j.ajic.2016.06.009

England PH (2013) Zoonotic diseases (zoonoses): guidance, data and analysis. https://www.gov.uk/government/publications/list-of-zoonotic-diseases/list-of-zoonotic-diseases. Accessed November 27, 2016

Fokine A, Chipman PR, Leiman PG et al. (2004) Molecular architecture of the prolate head of bacteriophage T4. Proc Natl Acad Sci U S A 101(16):6003–6008. doi:10.1073/pnas.0400444101

Fragaszy E, Ishola DA, Brown IH et al. (2016) Increased risk of A(H1N1)pdm09 influenza infection in UK pig industry workers compared to a general population cohort. Influenza Other Respir Viruses 10(4):291–300. doi:10.1111/irv.12364

Fraser C, Donnelly CA, Cauchemez S et al. (2009) Pandemic potential of a strain of influenza A (H1N1): early findings. Science 324(5934):1557–1561. doi:10.1126/science.1176062

Garner JS (1996) Guideline for isolation precautions in hospitals. The Hospital Infection Control Practices Advisory Committee. Infect Control Hosp Epidemiol 17(1):53–80

Gatherer D (2009) The 2009 H1N1 influenza outbreak in its historical context. J Clin Virol 45 (3):174–178. doi:10.1016/j.jcv.2009.06.004

Grabow WOK (2001) Bacteriophages: update on application as models for viruses in water. Water SA 27(2):251–268

Gray GC, Goswami PR, Malasig MD et al. (2000) Adult adenovirus infections: loss of orphaned vaccines precipitates military respiratory disease epidemics. For the Adenovirus Surveillance Group. Clin Infect Dis 31(3):663–670. doi:10.1086/313999

Gray GC, McCarthy T, Capuano AW et al. (2007) Swine workers and swine influenza virus infections. Emerg Infect Dis 13(12):1871–1878. doi:10.3201/eid1312.061323

Grove J, Marsh M (2011) The cell biology of receptor-mediated virus entry. J Cell Biol 195(7):1071–1082. doi:10.1083/jcb.201108131

Harrison BD, Wilson TM (1999) Milestones in the research on tobacco mosaic virus. Philos Trans R Soc Lond B Biol Sci 354(1383):521–529

Howard J, Murashov V, Schulte P (2017) Synthetic biology and occupational risk. J Occup Environ Hyg 14(3):224–236. doi:10.1080/15459624.2016.1237031

Janeway CJ, Travers P, Walport M et al. (2001) Infectious agents and how they cause disease. Immunobiology: the immune system in health and disease, 5th edn. Garland Science, New York

Kiselev NA, Sherman MB, Tsuprun VL (1990) Negative staining of proteins. Electron Microsc Rev 3(1):43–72. doi:10.1016/0892-0354(90)90013-I

Koonin EV, Senkevich TG, Dolja VV (2006) The ancient Virus World and evolution of cells. Biol Direct 1:29. doi:10.1186/1745-6150-1-29

Kuhar DT, Henderson DK, Struble KA et al. (2013) Updated US Public Health Service guidelines for the management of occupational exposures to human immunodeficiency virus and

recommendations for postexposure prophylaxis. Infect Control Hosp Epidemiol 34(9): 875–892. doi:10.1086/672271
Lawrence CM, Menon S, Eilers BJ et al. (2009) Structural and functional studies of archaeal viruses. J Biol Chem284(19):12599–12603. doi:10.1074/jbc.R800078200
Leung NHL, Zhou J, Chu DKW et al. (2016) Quantification of influenza virus RNA in aerosols in patient rooms. PLoS One 11(2):e0148669. doi:10.1371/journal.pone.0148669
Li W, Shi Z, Yu M et al. (2005) Bats are natural reservoirs of SARS-like coronaviruses. Science 310(5748):676–679. doi:10.1126/science.1118391
Lidmar J, Mirny L, Nelson DR (2003) Virus shapes and buckling transitions in spherical shells. Phys Rev E 68(5):051910
Lindsley WG, Blachere FM, Beezhold DH et al. (2016) Viable influenza A virus in airborne particles expelled during coughs versus exhalations. Influenza Other Respir Viruses 10(5):404–413. doi:10.1111/irv.12390
Lindsley WG, Blachere FM, Davis KA et al. (2010a) Distribution of airborne influenza virus and respiratory syncytial virus in an urgent care medical clinic. Clin Infect Dis 50(5):693–698. doi:10.1086/650457
Lindsley WG, Blachere FM, Thewlis RE et al. (2010b) Measurements of airborne influenza virus in aerosol particles from human coughs. PLoS One 5(11):e15100. doi:10.1371/journal.pone.0015100
Liu R, Vaishnav RA, Roberts AM et al. (2013) Humans have antibodies against a plant virus: evidence from tobacco mosaic virus. PLoS One 8(4):e60621. doi:10.1371/journal.pone.0060621
Lodish H, Berk A, Zipursky S et al. (2000) Molecular Cell Biology. Section 6.3 Viruses: Structure, Function, and Uses, 4th edn. W. H. Freeman, New York
Longdon B, Brockhurst MA, Russell CA et al. (2014) The evolution and genetics of virus host shifts. PLoS Pathogens 10(11):e1004395. doi:10.1371/journal.ppat.1004395
Louie JK, Kajon AE, Holodniy M et al. (2008) Severe pneumonia due to adenovirus serotype 14: a new respiratory threat? Clin Infect Dis 46(3):421–425. doi:10.1086/525261
Luckhaupt SE, Calvert GM, Li J et al. (2014) Prevalence of influenza-like illness and seasonal and pandemic H1N1 influenza vaccination coverage among workers – United States, 2009-10 influenza season. MMWR Morb Mortal Wkly Rep 63(10):217–221
Lyon GM, Mehta AK, Varkey JB et al. (2014) Clinical care of two patients with Ebola virus disease in the United States. New Engl J Med371(25):2402–2409. doi:10.1056/NEJMoa1409838
MacCannell T, Umscheid CA, Agarwal RK et al. (2011) Guideline for the prevention and control of norovirus gastroenteritis outbreaks in healthcare settings. Infect Control Hosp Epidemiol 32(10):939–969. doi:10.1086/662025
Mankertz A, Çaliskan R, Hattermann K et al. (2004) Molecular biology of Porcine circovirus: analyses of gene expression and viral replication. Veterinary Microbiology 98(2):81–88. doi:10.1016/j.vetmic.2003.10.014
McCluskey R, Sandin R, Greene J (1996) Detection of airborne cytomegalovirus in hospital rooms of immunocompromised patients. J Virol Methods 56(1):115–118
Midgley CM, Jackson MA, Selvarangan R et al. (2014) Severe respiratory illness associated with enterovirus D68 – Missouri and Illinois, 2014. MMWR Morb Mortal Wkly Rep 63(36):798–799
Midgley CM, Watson JT, Nix WA et al. (2015) Severe respiratory illness associated with a nationwide outbreak of enterovirus D68 in the USA (2014): a descriptive epidemiological investigation. Lancet Respir Med 3(11):879–887. doi:10.1016/S2213-2600(15)00335-5
Mohebati A, Davis JM, Fry DE (2010) Current risks of occupational blood-borne viral infection. Surg Infect (Larchmt) 11(3):325–331. doi:10.1089/sur.2010.025
Mubareka S, Lowen AC, Steel J et al. (2009) Transmission of influenza virus via aerosols and fomites in the guinea pig model. J Infect Dis 199(6):858–865. doi:10.1086/597073
Myatt TA, Johnston SL, Zuo Z et al. (2004) Detection of airborne rhinovirus and its relation to outdoor air supply in office environments. Am J Respir Crit Care Med 169(11):1187–1190. doi:10.1164/rccm.200306-760OC

Myers KP, Olsen CW, Setterquist SF et al. (2006) Are swine workers in the United States at Increased risk of infection with zoonotic influenza virus? Clin Infect Dis42(1):14–20. doi:10.1086/498977

Norkin LC (1995) Virus receptors: implications for pathogenesis and the design of antiviral agents. Clin Microbiol Rev 8(2):293–315

Noti JD, Lindsley WG, Blachere FM et al. (2012) Detection of infectious influenza virus in cough aerosols generated in a simulated patient examination room. Clin Infect Dis 54(11): 1569–1577. doi:10.1093/cid/cis237

Novel Swine-Origin Influenza AVIT, Dawood FS, Jain S et al. (2009) Emergence of a novel swine-origin influenza A (H1N1) virus in humans. N Engl J Med 360(25):2605–2615. doi:10.1056/NEJMoa0903810

O'Brien KM, Nonnenmann MW (2016) Airborne influenza a is detected in the personal breathing zone of swine veterinarians. PLoS One 11(2):e0149083. doi:10.1371/journal.pone.0149083

Panneer N, Lontok E, Branson BM et al. (2014) HIV and hepatitis c virus infection in the United States: whom and how to test. Clin Infect Dis 59(6):875–882. doi:10.1093/cid/ciu396

Parrish CR, Holmes EC, Morens DM et al. (2008) Cross-species virus transmission and the emergence of new epidemic diseases. Microbiol Mol Biol Rev 72(3):457–470. doi:10.1128/MMBR.00004-08

Philippe N, Legendre M, Doutre G, et al. (2013) Pandoraviruses: amoeba viruses with genomes Up to 2.5 Mb reaching that of parasitic eukaryotes. Science 341(6143):281–286. doi:10.1126/science.1239181

Pozzetto B, Memmi M, Garraud O et al. (2014) Health care-associated hepatitis C virus infection. World J Gastroenterol 20(46):17265–17278. doi:10.3748/wjg.v20.i46.17265

Prasad BVV, Schmid MF (2012) Principles of virus structural organization. In: Rossmann MG, Rao VB (eds) Viral molecular machines. Springer, US, Boston, MA, p 17–47. doi:10.1007/978-1-4614-0980-9_3

Riddell A, Kennedy I, Tong CY (2015) Management of sharps injuries in the healthcare setting. BMJ 351:h3733. doi:10.1136/bmj.h3733

Roberts H, Kruszon-Moran D, Ly KN et al. (2016) Prevalence of chronic hepatitis B virus (HBV) infection in U.S. households: National Health and Nutrition Examination Survey (NHANES), 1988–2012. Hepatology 63(2):388–397. doi:10.1002/hep.28109

Roulston A, Marcellus RC, Branton PE (1999) Viruses and apoptosis. Ann Rev Microbiol 53:577–628. doi:10.1146/annurev.micro.53.1.577

Sekar A, Kathiresan K (2013) Bacterial viruses in marine environment and their ecology role and bioprospecting potential: a review. Int J Curr Microbiol Appl Sci 2(7):151–163

Sharp PM, Hahn BH (2010) The evolution of HIV-1 and the origin of AIDS. Philos Trans R Soc Lond B Biol Sci 365(1552):2487–2494. doi:10.1098/rstb.2010.0031

Siegel JD, Rhinehart E, Jackson M et al. (2007) 2007 Guideline for isolation precautions: preventing transmission of infectious agents in health care settings. Am J Infect Control 35(10 Suppl 2):S65–164. doi:10.1016/j.ajic.2007.10.007

Silver S (2015) Laboratory-acquired lethal infections by potential bioweapons pathogens including Ebola in 2014. FEMS Microbiol Lett 362(1):1–6. doi:10.1093/femsle/fnu008

Sissons JG, Bain M, Wills MR (2002) Latency and reactivation of human cytomegalovirus. J Infect 44(2):73–77. doi:10.1053/jinf.2001.0948

Smith KO, Melnick JL (1962) A method for staining virus particles and identifying their nucleic acid type in the electron microscope. Virology 17(3):480–490. doi:10.1016/0042-6822(62)90143-5

Sosa LE, Gupta S, Juthani-Mehta M et al. (2009) Meningitis in a college student in Connecticut, 2007. J Am Coll Health 58(1):12–14. doi:10.3200/JACH.58.1.12-14

Suwantarat N, Apisarnthanarak A (2015) Risks to healthcare workers with emerging diseases: lessons from MERS-CoV, Ebola, SARS, and avian flu. Curr Opin Infect Dis 28(4):349–361. doi:10.1097/QCO.0000000000000183

Sze CW, Tan Y-J (2015) Viral membrane channels: role and function in the virus life cycle. Viruses 7(6):3261–3284. doi:10.3390/v7062771

Turner RB (2007) Rhinovirus: more than just a common cold virus. J Infect Dis 195(6):765–766. doi:10.1086/511829

Vernizzi G, Olvera de la Cruz M (2007) Faceting ionic shells into icosahedra via electrostatics. Proc Natl Acad Sci 104(47):18382–18386. doi:10.1073/pnas.0703431104

Verreault D, Moineau S, Duchaine C (2008) Methods for sampling of airborne viruses. Microbiol Mol Biol Rev 72(3):413–444. doi:10.1128/MMBR.00002-08

Vlasak M, Goesler I, Blaas D (2005) Human rhinovirus type 89 variants use heparan sulfate proteoglycan for cell attachment. J Virol 79(10):5963–5970. doi:10.1128/jvi.79.10.5963-5970.2005

Wasley A, Kruszon-Moran D, Kuhnert W et al. (2010) The prevalence of hepatitis B virus infection in the United States in the era of vaccination. J Infect Dis 202(2):192–201. doi:10.1086/653622

Webby RJ, Webster RG, (2001) Emergence of influenza A viruses. Philos Trans R Soc Lond B Biol Sci 356(1416):1817–1828. doi:10.1098/rstb.2001.0997

West KH, Cohen ML (1997) Standard precautions – a new approach to reducing infection transmission in the hospital setting. J Intraven Nurs 20(6 Suppl):S7–10

WHO (2016a) Ebola virus disease. World Health Organization. http://www.who.int/mediacentre/factsheets/fs103/en/

WHO (2016b) Final trial results confirm Ebola vaccine provides high protection against disease. World Health Organization. http://www.who.int/mediacentre/news/releases/2016/ebola-vaccine-results/en/

WHO (2016c) Influenza (Seasonal). World Health Organization. http://www.who.int/mediacentre/factsheets/fs211/en/. Accessed November 27, 2016

WHO (2016d) Poliomyelitis. World Health Organization. http://www.who.int/mediacentre/factsheets/fs114/en/. Accessed November 27, 2016

WHO (2016e) Zika virus. World Health Organization. http://www.who.int/mediacentre/factsheets/zika/en/

Wyżgowski P, Rosiek A, Grzela T et al. (2016) Occupational HIV risk for health care workers: risk factor and the risk of infection in the course of professional activities. Ther Clin Risk Manag 12:989–994. doi:10.2147/TCRM.S104942

Zipprich J, Winter K, Hacker J et al. (2015) Measles outbreak – California, December 2014–February 2015. MMWR Morb Mortal Wkly Rep 64(6):153–154

Part II
Sampling and Analysis Approaches in Indoor Microbial Assessments

Chapter 4
Sampling for Microbial Determinations

Tiina Reponen

Abstract Sampling for microbial determination can be conducted by air sampling or by direct source evaluation techniques, such as bulk sampling, surface sampling, and dust sampling. The choice of a sampling method depends on the purpose of the sampling and the method for the subsequent sample analysis. Investigators should be aware of the pros and cons of the methods they are using.

Keywords Bioaerosol · sampling · samplers

4.1 Introduction

Assessment of microbial exposure can be based on qualitative or quantitative methods or a combination of these two. The qualitative evaluation includes visual and olfactory observations of mold, water damage and moldy odor in the context of dampness and mold investigations (Mendell and Kumagai, 2017). The quantitative assessment requires two separate steps - sampling and analysis - or can be conducted with direct reading instruments that are described in Chap. 6 "Analysis Approaches for Fungi in Indoor Environmental Assessments." Sampling is carried out for verifying and quantifying the presence of airborne microorganisms or material contamination, to identify sources, to monitor the effectiveness of control measures and to assess human exposure. Sampling for bioaerosols in indoor and occupational environments can provide useful information necessary to characterize the exposure conditions, determine whether the contamination represents a potential hazard to health, and establish the need for control measures. However, it should be noted that microbial sampling may not be necessary if there is substantial evidence of visible microbial contamination. In these cases, resources should be focused on remediating the microbial contamination. Also, in case of

T. Reponen (✉)
Department of Environmental Health, University of Cincinnati Health, Cincinnati, OH, USA
e-mail: tiina.reponen@uc.edu

C. Viegas et al. (eds.), *Exposure to Microbiological Agents in Indoor and Occupational Environments*, DOI 10.1007/978-3-319-61688-9_4

building investigations for assessing moisture damage and dampness, microbial sampling alone will not be sufficiently informative, but should be integrated in a comprehensive building assessment. If sampling is planned to verify the success of remediation or the efficiency of control measures, it is useful to have a baseline assessment before the control measures.

4.2 General Principles of Microbial Sampling for Exposure Assessment

Sampling for microbial determination is commonly carried out by using air sampling, but sometimes also direct source evaluation techniques, such as bulk sampling, surface sampling, and dust sampling have been used (Table 4.1). Active air sampling is conducted via an active collection of pre-determined air flow and can be conducted both by using stationary and personal samplers or the combination of both methods.

Traditionally, bioaerosol sampling has been done by using stationary samplers. By performing stationary sampling in the main microenvironments of the study subject (e.g., home, workplace, outside air), microenvironmental exposure can be calculated using the time-weighted average of the airborne concentration in the different locations. Personal air sampling, however, represents the overall exposure, as it includes the exposure occurring in all the environments that the subjects spend time. Personal monitoring for bioaerosols has been used in occupational environments with relatively high exposures. Epidemiological studies based on personal sampling of microorganisms in occupational environments have shown exposure-response associations almost twice as often as studies using stationary sampling (Eduard, 2003). Personal samples typically show higher exposures than stationary ones. Toivola et al. (2002) compared personal exposure to fungal spores and bacteria with that measured by stationary samplers in the main microenvironments of teachers, i.e., at home and school. The concentration of fungi was found to be higher in personal samples than in home and school samples. Similar findings have been reported by Green et al. (2006) who used novel personal nasal filters to collect fungal spores in an outdoor setting. Higher concentrations obtained by

Table 4.1 Main sampling principles for microorganisms in built environments

Purpose of sampling	Recommended method
Verify and quantify the presence of bioaerosols	Air and dust samples
Verify and quantify material contamination	Surface and building material samples
Identify sources	Surface and building material samples
Monitor effectiveness of control measures	Air samples (in some cases surface samples)
Assess human exposure	Air and dust samples

personal sampling is a well-known phenomenon with non-biological particles but has also been shown with endotoxin and fungal spores. It is caused by the "personal cloud" effect, i.e., the subject is resuspending settled particles by his/her activities and is often closer to the source than the stationary sampler (Wallace 1996).

Air sampling is considered to be the most representative of respiratory exposure, but there are several problems associated with traditional air sampling methods. Most methods cannot be used for assessing long-term exposure to bioaerosols. The sampling time of commonly used, commercially available instruments (e.g., Burkard, Andersen, Air-O-Cell) is limited to 5–20 min. On the other hand, bioaerosol concentrations are known to have large temporal variation, particularly if an active source is present, such as in mold-problem homes. Hyvärinen et al. (2001) showed that when Andersen 6-stage impactor with 10-min sampling time was used, a sampling campaign over 11 different days is needed to characterize fungal concentrations of a residence with unknown indoor conditions. One of the reasons for the temporal variation is the sporadic release of microorganisms from contaminated surfaces (Górny et al., 2001). Therefore, short-term sampling often fails to detect certain species, which are present in low concentration but may be important for the subsequent health effects.

Direct source evaluation techniques (bulk, surface and dust sampling) allow the investigators to collect extensive information about the source, but cannot measure its aerosolization potential and therefore may not adequately represent the aerosol exposure (Chew et al., 2003; Nelson et al., 1999). Material samples are usually collected during the remediation process when it is easy to collect pieces of contaminated building materials. Another example of bulk sampling is the collection of water from cooling towers in suspected Legionella cases. Surface samples can be collected by using contact agar plates or by swiping the surface with a cotton swab or electrostatic dust cloth. Dust sampling by vacuuming from the floor, bed or other surfaces is commonly used in epidemiological studies that aim at assessing health effects of indoor microbial exposures. This is in part due to the convenience of house dust sampling that can be carried out by building occupants and study participants themselves, restricting the need for field work and allowing sampling campaigns in large population surveys. House dust is believed to represent long-term patterns in exposure, and it allows easy collection of sufficiently large samples for analysis of multiple agents (Casas et al., 2016). Many large-sized allergens (e.g., house dust mite) do not stay airborne for a long time and are, therefore, difficult to detect by air sampling. The disadvantages of dust sampling, especially from the floor, are that the sample may not adequately reflect human inhalation exposure and that the age (i.e. sample accumulation time) of floor dust is typically not known.

Increasingly, researchers have started using a long-term collection of settled dust by placing standardized dust sampling platforms in the breathing zone of the subjects (Wurtz et al., 2005; Noss et al., 2008; Adams et al., 2015). The advantage of collecting settling dust compared to vacuumed dust is that the sampling occurs over a known discrete period, onto a standard surface. Strong correlations were found between five types of passive collectors (plastic Petri dishes and four types of dust cleaning cloths) when used for collecting bacteria and fungi for subsequent analysis with quantitative PCR and high throughput sequencing methods (Adams et al., 2015). The disadvantages of passive air sampling is that the "sampled" air volume is not known and that smaller particles may be underestimated because their gravitation settling can be hindered by air currents.

Another approach is to assess the aerosolization potential of microorganisms from surfaces. Aerosolization chambers have been designed to release biological particles from contaminated surfaces by air currents and vibration (for example, Fungal Spore Source Strength Tester, FSSST and Particle-Field and Laboratory Emission Cell, P-FLEC) (Sivasubramani et al., 2004; Kildesø et al., 2003). Results can be used to estimate the airborne concentration of biological particles that can be released or resuspended from the surface under investigation.

Results from dust and air samples represent different types of potential exposure, and it may be of advantage to use a combination of both, depending on the type of bioaerosol that is of interest (Chew et al., 2003). When designing sampling strategies, the investigators need to consider the bioaerosol type and the particular components that may be causing the relevant exposures. This information will determine the sampling and analytical strategy that is optimal. Also, general considerations for exposure assessment, such as using worst-case, random or systematic approach, need to be considered.

4.3 Methods for Air Sampling of Microorganisms

The general sampling principles for bioaerosol sampling are the same as for the other airborne particles: impaction, impingement, and filtration (Table 4.2). Impaction can be either inertial or centrifugal. Inertial impaction is the most commonly used sampling mechanisms in bioaerosol samplers and occurs when the air is pulled or pushed through a small nozzle, which accelerates air and particles to high speed. The air jet is directed to the impaction surface, where it makes a sharp turn. Larger particles with sufficient inertia continue their movement towards the sampling surface and are impacted. Inertial impactors can have either one stage (single-stage impactors, e.g., Air-O-Cell) or multiple stages (cascade impactors, e.g., Andersen 6-stage impactor). In a centrifugal impactor (e.g., Reuter centrifugal air sampler), air is moving in a circular motion, and centrifugal forces collect larger particles that cannot follow the air streamlines. Impactors collect bioaerosol particles directly onto agar plate, glass slide, tape, or filter.

Impingement (e.g., AGI-30 and BioSampler) utilizes primarily inertial impaction into a liquid and onto the bottom of the collection vessel. Furthermore, diffusion within the bubbles enhances the collection of smaller particles. Wetted wall cyclones (e.g., OMNI 3000) combine centrifugal sampling and impingement as air is moving in a circular motion onto a wet sampling surface. Impingement typically occurs into water-based suspensions such as phosphate buffer or peptone-water. Sometimes also non-water-based collection liquids can be used, e.g., glycerol and mineral oil.

Table 4.2 Examples of commercially available bioaerosol samplers

Sampling principle	Name of the sampler	Manufacturer/supplier	Type of analysis that can follow	Type of microorganism that can be collected
Impaction onto agar				
	Andersen impactor	Thermo Scientific	Cultivation	Bacteria, fungi
	MAS-100	Merck	Cultivation	Bacteria, fungi
	Millipore Air Tester	Millipore Corp.	Cultivation	Bacteria, fungi
	RCS (Reuter Centrifugal Sampler)	Biotest Diagnostics Corp.	Cultivation	Bacteria, fungi
	SAS (Surface Air Sampling System)	International PBI; Scientific Products Corp.; Bioscience International	Cultivation	Bacteria, fungi
Impaction onto glass slide				
	Air-O-Cell	Zefon International;	Microscopic counting	Fungi
	Allergenco	EMS (Environmental Monitoring Systems)	Microscopic counting	Fungi
	Burkard, 7-day recording	Burkard Manufacturing, Co., Ltd	Microscopic counting	Fungi
	Cyclex-D	EMS	Microscopic counting	Fungi
	Micro-5	EMS	Microscopic counting	Fungi
	MoldSNAP	Zefon International	Microscopic counting	Fungi
	Versatrap	SKC, Inc.	Microscopic counting	Fungi
	Via-cell	Zefon International	Cultivation, Microscopic counting	Bacteria, fungi
Impingement				
	AGI-30 and AGI-4	Ace Glass Inc.; Hampshire Glassware; Millipore Corp	Cultivation, Microscopic counting, Other	Bacteria, fungi, viruses
	BioSampler	SKC, Inc.	Cultivation, Microscopic counting, Other	Bacteria, fungi, viruses
	XMC-CV and XMX/102	Dycor Technologies, Ltd	Cultivation, Microscopic counting, Other	Bacteria, fungi, viruses

(continued)

Table 4.2 (continued)

Sampling principle	Name of the sampler	Manufacturer/supplier	Type of analysis that can follow	Type of microorganism that can be collected
Wetted-wall cyclone samplers				
	Microbio MB3	F.W. Parrett Ltd	Cultivation, Microscopic counting, Other	Bacteria, fungi, viruses
	OMNI 3000	Life safety Systems	Cultivation, Microscopic counting, Other	Bacteria, fungi, viruses
	SpinCon	MidWest Research Institute	Cultivation, Microscopic counting, Other	Bacteria, fungi, viruses
	Coriolis	Bertin Instruments	Cultivation, Microscopic counting, Other	Bacteria, fungi, viruses
	SASS 2300	Research International	Cultivation, Microscopic counting, Other	Bacteria, fungi, viruses
Filtration				
	Button-sampler	SKC, Inc.	Cultivation*, Microscopic counting, Other	Bacteria, fungi, viruses
	IOM	SKC, Inc.	Cultivation* Microscopic counting, Other	Bacteria, fungi, viruses
	Sartorius AirPort MD8	Microbiology International	Cultivation*, Microscopic counting, Other	Bacteria, fungi, viruses
	3-piece cassette	SKC, Inc.	Cultivation*, Microscopic counting, Other	Bacteria, fungi, viruses
Electrostatic precipitation				
	Inspirotec	Inspirotec LLC	Other	Bacteria, fungi, viruses

*Cultivation of filter samples recommended only for resistant microorganisms
Other: DNA-based and chemical methods (see Chap. 6 "Analysis Approaches for Fungi in Indoor Environmental Assessments")

Several filter samplers, originally designed for general bioaerosol sampling, have been applied for bioaerosol collection. In filter sampling, air passes through the filter medium, and particles are collected by impaction on the filter, interception with the filter material, diffusion, and electrostatic forces. Examples of commonly used filter samplers are Button sampler (Aizenberg et al., 2000a) and IOM sampler (Kenny et al., 1999). The selection of the filter to be employed in the filter sampling depends on the subsequent analytical method. Capillary pore filters have smooth, flat surfaces and are often chosen for culturing and immunostaining of bioaerosols. Teflon and polycarbonate filters are commonly used if the sample has to be extracted off from the filter as the particles easily wash off from these materials (Schmechel et al., 2003; Burton et al., 2007). Black polycarbonate filters are used if the sample is analyzed through epifluorescence microscopy. Mixed cellulose esterase filter can be made transparent by acetone vapor for subsequent direct microscopic analysis by an optical microscope (Adhikari et al., 2003). Gelatin filter reduces the desiccation of collected microorganisms and quickly dissolves into the water for subsequent analysis (Burton et al., 2007). PVC-filters may be used when there is a need for simultaneous analysis of particle mass.

Electrostatic precipitators are another example of samplers that have been adapted from general particle sampling to bioaerosol sampling (Mainelis et al., 2002). An electrostatic field charges particles in an air stream and forces them to the surface that has an opposite charge than the particles. An Electrostatic Precipitator with Superhydrophobic Surface (EPSS) achieved a concentration rate of 10^6 for latex particles (Han and Mainelis 2008). The EPSS subsequently was found to be compatible with analysis by polymerase chain reaction (PCR) and to reach a collection efficiency of 72% for *P. fluorescens* and *B. subtilis* (Han et al., 2010). Another recently developed electrostatic sampler was shown to have ≥85% efficiency for the collection of bacteria (Roux et al., 2013). A commercially available electrostatic precipitator, Inspirotec, was modified from a portable air cleaner. It was shown to have ~23% collection efficiency for fungal spores when used for collecting samples for quantitative PCR and next generation sequencing analysis (Gordon et al., 2015).

An unconventional personal sampler fits into a test subject's nostrils (Intranasal air sampler, INAS) and has been used to measure allergen and fungal exposures (Green et al., 2006). Another recent development is the use of a laminar-flow water based condensation system (Hering et al., 2005) for the collection of nanoscale viruses. This system has been shown to have high physical collection efficiency while maintaining the infectivity of viruses (Lednicky et al., 2016; Walls et al., 2016).

4.4 Considerations for Air Sampling of Microorganisms

When sampling biological particles, investigators have to consider all the same factors affecting the physical sampling efficiency as when sampling inert particles, i.e., inlet efficiency, transmission efficiency, and collection efficiency (Reponen et al., 2011). The inlet efficiency is not determined for most of the commercially available bioaerosol samplers. Some samplers have specially designed inlets to collect inhalable particles, e.g., a particle that can be inhaled into the human respiratory system (e.g., Button sampler and IOM sampler). A sampling of thoracic (particles that can penetrate through larynx) and respirable particles (that penetrate to alveoli) requires a pre-collector, such as a cyclone, that removes the unwanted larger particles. The IOM sampler can be equipped with polyurethane foams, for sampling either thoracic or respirable particles (Haatainen et al., 2010). Bioaerosols can also be collected according to their aerodynamic size using cascade impactors. Most commonly used is the 6-stage viable Andersen sampler. Other cascade impactors, originally designed for non-biological particles, can also be used if the collection medium is compatible with the analytical method. An example is an Electrostatic Low-Pressure Impactor that has been used in laboratory-based bioaerosol investigations (Cho et al., 2005).

An important consideration is the sampler's collection efficiency, even when using a non-size-selective sampler. Inertial impactors typically have an S-shaped collection efficiency curve, and their efficiency increases with increasing particle size. The size at which the impactor has 50% collection efficiency is called d_{50} cut-size. The sampler should be selected so that the d_{50} cut-size is below the particle size of interest. This limits the use of impactors to larger bioaerosol particles, as impactors are not efficient for particles <0.5 μm, which includes single virions and microbial fragments. High air velocity can cause injury of microorganism during the impaction onto the surface (Stewart et al., 1995; Mainelis and Tabayoyong, 2010). Furthermore, increased jet-to-plate distance has been shown to decrease the culturability of collected bacteria due to increased jet dissipation that desiccates larger fraction of the agar surface (Yao and Mainelis, 2007).

Additional attention has to be paid to the uniformity of the deposition of bioaerosol particles on the collection surface if direct microscopic counting is utilized for the assay (Aizenberg et al., 2000b). Furthermore, the possible bounce of fungal spores when collected by inertial impaction is of concern. Especially extended sampling times may increase the bounce of these larger bioaerosol particles (Trunov et al., 2001). Fungal spores have been shown to bounce off from the sampling surface at lower velocities than non-biological particles (Kuuluvainen et al., 2016). Investigators have used water-based grease or specially designed cyclone samplers to prevent the bounce of fungal spores (Chen et al., 2004; Cho et al., 2005). A two stage sampler designed by investigators at the National Institute for Occupational Safety and Health has been used for the collection of microbial fragments separately from intact bacterial and fungal spores (Singh et al., 2011; Seo et al., 2014).

Filter samplers have a U-shaped collection efficiency curve as they have a minimum efficiency at certain particle size and the efficiency increases for particles that are smaller or larger than this size. Mechanical filters have a minimum efficiency around 0.3 μm (Liu and Lee, 1976; Burton et al., 2007). Membrane filters have electrostatic effects facilitating the collection of small particles, and their efficiency has the minimum efficiency at about 50–70 nm (Bałazy et al., 2006). Filters that have multiple layers, such as Teflon are efficient for a wide range of particle sizes, from nanosized to supermicrometer sized particles. Dehydration of vegetative bacterial cells during filter sampling significantly reduces their viability, but hardy bacterial and fungal spores are not as susceptible to damage (Li et al., 1999; Lin and Li, 1998; Lin and Li 1999). Gelatin filters can reduce the desiccation effect caused by filter sampling, but typically also limit the sampling time due to loss of material properties (Burton et al., 2007). Investigators are commonly using membrane filters to collect bioaerosols in conjunction with culture-independent analyses, e.g., endotoxin and β-glucan, because of the ease of use and high collection efficiency of filters (Meklin et al., 2007; Adhikari et al., 2014) Similarly, filters are used as a sampling surface in inertial impactors, e.g., a non-viable Andersen and a Tisch cascade impactor, when molecular-based methods are used for the analysis (Qian et al., 2012; Yamamoto et al., 2012; Alonso et al., 2017). However, some filters have been shown to have reduced recovery for specific agents, such as endotoxin for which glass fiber filters have been recommended (Douwes et al., 1995).

A recent study evaluated nine different commercially available bioaerosol instruments for the collection of bacteria and viruses (Dybwad et al., 2014). BioSampler was used as a reference sampler and the samples were analyzed by cultivation, microscopic counting and quantitative PCR. Samplers employing dry collection had lower culturable counts for stress sensitive bioaerosols than wet collection methods. Furthermore, non-filter-based samplers had reduced collection efficiencies for 1 μm size bioaerosols. Several samplers underestimated bioaerosol concentration levels when compared to the BioSampler. This was attributed to lower sampling efficiencies, although several samplers obtained more concentrated samples due to having higher concentration factors (Dybwad et al., 2014).

Impingement into liquid can also achieve high efficiency as it utilizes multiple collection mechanisms (impaction, interception, and diffusion). Traditional impingers can be used only with water-based liquids, such as phosphate buffer or peptone water. These evaporate, which limits the sampling time to about 30 minutes. The BioSampler can be used with viscous fluids, such as glycerol and mineral oil, which do not evaporate and allow long-term sampling (Lin et al., 1999).

4.5 Conclusions

As each technique has unique advantages and disadvantages, it is often beneficial to use multiple techniques in an investigation. This way, the limitations of one technique can be are compensated by another one. Sampling for microorganisms in buildings with moisture or mold problems should be designed so that it complements visual assessment in a meaningful way. Practitioners need to understand the underlying limitations of the methods they are using for microbial sampling in order to properly interpret the results. Due to the differences in the performance of various samplers, results between different studies can be compared only if same sampling methods were used.

References

Adams RI, Tian Y, Taylor JW et al. (2015) Passive dust collectors for assessing airborne microbial material. Microbiome 3(1):1

Adhikari A, Martuzevicius D, Reponen T et al. (2003) Performance of the button personal inhalable sampler for the measurement of outdoor aeroallergens. Atmos Environ. 34:4723

Adhikari A, Kettleson EM, Vesper S et al. (2014) Dustborne and airborne Gram-Positive and Gram-Negative bacteria in high versus low ERMI homes. Sci Total Environ. 482–483:92–99

Aizenberg V, Grinshpun SA, Willeke K et al. (2000a) Performance characteristics of the button personal inhalable aerosol sampler. AIHA J 61(3):398–404

Aizenberg V, Reponen T, Grinshpun SA et al. (2000b) Performance of Air-O-Cell, Burkard, and Button samplers for total enumeration of airborne spores. Am Ind Hyg Assoc J. 61:855–864

Alonso C, Olson BA, Goyal S et al. (2017) Comparison of two size-differentiating air samplers for detecting airborne swine viruses under experimental conditions. Aerosol Sci Technol 51:198–205

Bałazy A, Toivola M, Reponen T et al. (2006) Manikin-based performance evaluation of N95 filtering facepiece respirators challenged with nanoparticles. Ann Occup Hyg. 50:259–269

Burton N, Grinshpun SA, Reponen T (2007) Physical collection efficiency of filter materials for bacteria and viruses. Ann Occup Hyg. 51:143

Casas L, Tischer C, Täubel M (2016) Pediatric asthma and the indoor microbial environment. Curr Environ Health Rep 3:238–249

Chen BT, Feather GA, Andrew Maynard A et al. (2004) Development of a personal sampler for collecting fungal spores. Aerosol Sci Technol 38:926–937

Chew GL, Rogers C, Burge HA et al. (2003) Dustborne and airborne fungal propagules represent a different spectrum of fungi with differing relations to home characteristics. Allergy 58:13–20

Cho S-H, Seo S-C, Schmechel D et al. (2005) Aerodynamic characteristics and respiratory deposition of fungal fragments. Atmos Environ. 39:5454–5465

Douwes J, Versloot P, Hollander A et al. (1995) Influence of various dust sampling and extraction methods on the measurement of airborne endotoxin. Appl Environ Microbiol. 61:1763

Dybwad M, Skogan G, Blatny JM (2014) Comparative testing and evaluation of nine different air samplers: end-to-end sampling efficiencies as specific performance measurements for bioaerosol applications. Aerosol Sci Technol 48(3):282–295

Eduard W (2003) The performance of culture-based methods and microscopy for quantification of noninfectious airborne microorganisms in epidemiological studies of highly contaminated work environments. AIHA J. 64:684–689

Gordon J, Gandhi P, Shekhawat G et al. (2015) A simple novel device for air sampling by electrokinetic capture. Microbiome 3:7

Green BJ, O'Meara T, Sercombe J et al. (2006) Measurement of personal exposure to outdoor aeromycota in northern New South Wales, Australia. Ann Agric Environ Med. 13:225–234

Górny RL, Reponen T, Grinshpun SA et al. (2001) Source strength of fungal spore aerosolization from moldy building materials. Atmos Environ 35:4853–4862

Haatainen S, Laitinen J, Linnainmaa M et al. (2010) The suitability of the IOM foam sampler for bioaerosol sampling in Occupational Environments. J Occup Environ Hyg. 7:1–6

Han T, Mainelis G (2008) Design and development of an electrostatic sampler for bioaerosols with high concentration rate. J Aerosol Sci. 39:1066

Han T, An HR, Mainelis G (2010) Performance of an electrostatic precipitator with hydrophobic surface when collecting airborne bacteria. Aerosol Sci Technol. 44:339

Hering SV, Stolzenburg MR, Quant FR et al. (2005) A laminar-flow, water-based condensation particle counter (WCPC). Aerosol Sci Technol. 39(7):659–672

Hyvärinen A, Vahteristo M, Meklin T et al. (2001) Temporal and spatial variation of fungal concentrations in indoor air. Aerosol Sci Technol. 35:688–695

Kenny L, Bowry A, Crook B et al. (1999) Field testing of a personal size-selective bioaerosol sampler. Ann Occup Hyg. 43:393–404

Kildesø J, Würtz KF, Nielsen P et al. (2003) Determination of fungal spore release from wet building materials. Indoor Air 13:148

Kuuluvainen H, Saari S, Mensah-Attipoe J et al. (2016) Triboelectric charging of fungal spores during resuspension and rebound. Aerosol Sci Technol. 50(2):187–197

Lednicky J, Pan M, Loeb J et al. (2016) Highly efficient collection of infectious pandemic influenza H1N1 virus (2009) through laminar flow water based condensation. Aerosol Sci Technol. 50:i–iv. 2016

Li CS, Hao ML, Lin WH et al. (1999) Evaluation of microbial samplers for bacterial microorganisms. Aerosol Sci Technol. 30:100

Lin WH, Li CS (1998) The effect of sampling time and flow rates on the bioefficiency of three fungal spore sampling methods. Aerosol Sci Technol. 28:511

Lin WH, Li CS (1999) Evaluation of impingement and filtration for yeast bioaerosol sampling. Aerosol Sci Technol. 30:119

Lin X, Reponen TA, Willeke K et al. (1999) Long-term sampling of airborne bacteria and fungi into a non-evaporating liquid. Atmos Environ. 33:4291–4298

Liu BYH, Lee KW (1976) Efficiency of membrane and nucleopore filters for submicrometer aerosols. Environ Sci Technol. 10:345

Mainelis G, Willeke K, Adhikari A et al. (2002) Design and collection efficiency of a new electrostatic precipitator for bioaerosol collection. Aerosol Sci Technol. 36:1073

Mainelis G, Tabayoyong M (2010) The effect of sampling time on the overall performance of portable microbial impactor. Aerosol Sci Technol. 44:75

Meklin T, McKinstry C, Reponen T et al. (2007) Comparison of mold concentrations in dust, indoor and outdoor air sampled simultaneously and then quantified by MSQPCR. Sci Total Environ. 382:130–134

Mendell MJ, Kumagai K (2017) Observation-based metrics for residential dampness and mold with dose-response relationships to health: a review. Indoor Air 27:505–517

Nelson HS, Szefler SJ, Jacobs J et al. (1999) The relationships among environmental allergen sensitization, allergen exposure, pulmonary function, and bronchial hyperresponsiveness in the Childhood Asthma Management Program. J Allergy Clin Immunol. 104:775–785

Noss I, Wouters IM, Visser M et al. (2008) Evaluation of a low-cost electrostatic dust fall collector for indoor air endotoxin exposure assessment. Appl Environ Microbiol. 74:5621

Qian J, Hospodsky D, Yamamoto N et al. (2012) Size-resolved emission rates of airborne bacteria and fungi in an occupied classroom. Indoor Air 22(4):339–351

Reponen T, Willeke K, Grinshpun S et al. (2011) Biological particle sampling. In: Kulkarni P, Baron P, Willeke K (eds) Aerosol measurement, principles, techniques, and applications, 3rd edn. John Wiley & Johns, Inc., pp. 549–570.

Roux J-M, Kaspari O, Heinrich R et al. (2013) Investigation of a new electrostatic sampler for concentrating biological and non-biological aerosol particles. Aerosol Sci Technol. 47(5): 463–471

Schmechel D, Górny RL, Simpson JP et al. (2003) Limitations of monoclonal antibodies for monitoring of fungal aerosols using Penicillium brevicompactum as a model fungus. J Immunol Methods 283:235

Seo S, Choung JT, Chen BT et al. (2014) The level of submicron fungal fragments in homes with asthmatic children. Environ Res. 131:71–76

Singh U, Cho KJ, Grinshpun SA et al. (2011) Airborne endotoxin and β-glucan in fine particles in agricultural and home environments. Aerosol Air Qual Res. 11:376–386

Stewart SL, Grinshpun SA, Willeke K et al. (1995) Effect of impact stress on microbial recovery on an agar surface. Appl Environ Microbiol. 61:1232

Sivasubramani SK, Niemeier RT, Reponen T et al. (2004) Assessment of the aerosolization potential for fungal spores in moldy homes. Indoor Air 14:405

Toivola M, Alm S, Reponen T et al. (2002) Personal exposures and microenvironmental concentrations of particles and bioaerosols. J Environ Monit. 4:166–174

Trunov M, Trakumas S, Willeke K et al. (2001) Collection of bioaerosols by impaction: effect of fungal spore agglomeration and bounce. Aerosol Sci Technol. 35:617–624
Wallace L (1996) Indoor particles: a review. J Air Waste Manag Assoc. 46:98–126
Walls HJ, Ensor DS, Harvey LA et al. (2016) Generation and sampling of nanoscale infectious viral aerosols. Aerosol Sci Technol. 50(8):802–811
Wurtz H, Sigsgaard T, Valbjorn O et al. (2005) The dustfall collector – a simple passive tool for long-term collection of airborne dust: a project under the Danish Mould In Buildings Program (DANIB). Indoor Air 15(Suppl 9):33
Yamamoto N, Bibby K, Qian J et al. (2012) Particle-size distributions and seasonal diversity of allergenic and pathogenic fungi in outdoor air. ISME J. 6(10):1801–1811
Yao M, Mainelis G (2007) Analysis of portable impactor performance for enumeration of viable bioaerosols. J Occup Environ Hyg. 4:514

Chapter 5
Analyses Approaches for Bacteria

Edna Ribeiro and Isabel Faria

Abstract Bacterial bioburden assessments, particularly in the context of human environmental/occupational exposure, have major implications in public health risk assessment. Several sampling methods, which must be adapted to the studied environmental context, are currently available. Culture-dependent and culture-independent methodologies have been utilized for the analysis of bacterial communities in various environments. Culture-dependent techniques drove extraordinary advances in microbiology and allow the enhancement of bacterial material to be utilized in supplementary analysis; however, these approaches may underestimate the bacterial bioburden of the studied samples. On the other hand, culture-independent approaches are considered more sensible and efficient with capacity to provide valuable information regarding bacterial diversity and quantity; nevertheless, preferential amplification and poor primers specificity can account for major limitations. In order to perform valuable and efficient assessments of bacterial bioburden booth approaches should be utilized simultaneously.

Keywords Bacterial bioburden · environmental sampling · culture-dependent bacterial analysis · culture-independent bacterial analysis

---

E. Ribeiro (✉)
Research Center LEAF - Linking Landscape, Environment, Agriculture and Food - Instituto Superior de Agronomia, Lisbon University, Lisbon, Portugal

Research Group Environment & Health (GIAS), Lisbon School of Health Technology/Polytechnic Institute of Lisbon, Lisbon, Portugal

Lisbon School of Health Technology/Polytechnic Institute of Lisbon, Av. D. João II, lote 4.69.01, Parque das Nações, Lisbon, Portugal
e-mail: edna.ribeiro@estesl.ipl.pt

I. Faria
Lisbon School of Health Technology/Polytechnic Institute of Lisbon, Av. D. João II, lote 4.69.01, Parque das Nações, Lisbon, Portugal

Centro Hospitalar Lisboa Ocidental, Hospital Egas Moniz, Microbiology and Molecular Biology Laboratory, Lisbon, Portugal
e-mail: isabelfaria@netcabo.pt

C. Viegas et al. (eds.), *Exposure to Microbiological Agents in Indoor and Occupational Environments*, DOI 10.1007/978-3-319-61688-9_5

5.1 Environmental Exposure to Bacteria

Environmental exposure to bacteria and their components is a significant concern because of the impact in human and animal health. Infectious disease, respiratory diseases, cancers, acute toxic allergies are the major adverse health effects associated with bioaerosols (airborne particles of biological origins such us pollen, fungi, bacteria, virus, skin scales or hair of mammals and residues or products of organisms). While also beneficial effects of exposure to environmental bacteria are well documented (see for example Chap. 13) we focus here on documentation of adverse effects of exposure to bacteria in indoor environments.

Bioaerosols occur as solid or liquid particles in the air and include fungi, bacteria, arthropods and protozoa as well as microbial products such as glucans, mycotoxins and endotoxins. Bioaerosols are ubiquitous in nature and their presence in enclosed environment is almost inevitable. The earth's atmosphere is teeming bioaerosols and airborne bacteria are important biological components of bioaerosols. In outdoor environments, most of bacterial aerosols originate from natural and anthropogenic sources such as vegetables, plants, soil and water bodies, sewage sludge, animal feeding, agricultural activities and industrial operations (i.e. waste water treatment, landfills, horticulture facilities, etc.). The knowledge of the type and concentration of microorganisms in air and surfaces of confined environments is of foremost importance to prevent harmful health impacts such as infectious diseases as they can penetrate into the lung or gastrointestinal tract, causing both respiratory and gastro-intestinal symptoms. Prolonged exposure to particular bioaerosols in the indoor environment and in at workplaces may represent a health hazard and potentially result in allergic lung disease or in an infectious disease (Walser et al., 2015).

Microbial contamination of an enclosed area may come from outside or be generated in the area itself. When generated inside, the numbers of occupants of a room, the presence of organic dust, materials stored in the buildings, air inflowing from ventilation and air conditioning systems are some of the environmental variables that may affect the microbial ecology indoors. Aerosolization of microorganisms is also influenced by other factors like temperature and humidity whereby the distribution and concentrations of bacterial aerosols may be modulated by both meteorological conditions and seasons and ventilation and air conditioning systems. The persistence, survival and the change of microbial communities in indoor environments is of immense interest in public health. Several environments been studied over the years with increasing emphasis being given for the detection of bioaerosols in indoor environments such as public buildings (Stanley et al., 2008).

Moreover, in the last decades several studies increased scientific data on non-occupational and occupational exposure to bioaerosols and helped to understand the relationship between exposure and health effects, reviewed in (Ghosh et al., 2015). Sources of bioaerosols include contaminants from waste treatment system, building maintenance, clothes and fomites, skin, among others Currently, bacterial contamination in wastewater treatment (Fuhrimann et al., 2016), public buildings (Stanley et al., 2008), schools (Anderson and Palombo 2009), child care facilities (Lee et al., 2007) and recreation facilities (Kirby-Smith and White, 2006) as well as industrial process (Nehme et al., 2008) is well documented.

Hospital indoor environment is also classified as extremely high risk for bacterial exposure/contamination and has been a focus of intense research. The World Health Organization (WHO) described nosocomial infections caused by airborne agents of microbial origin as one of the major classes of infectious diseases with an enormous associated economic impact worldwide (WHO, 2011). Several studies have been demonstrating the dominant prevalence of bacteria bioburden in hospital settings compared to fungi and viruses. This prevalence is certainly influenced by both patients and visitors of hospital facilities since most of the recovered isolates were similar to skin microbiota (Brady et al, 2007, Mbim et al, 2016). Additionally, in the epidemiology of nosocomial infections, environmental transmission has gained a central role as patients considered more vulnerable to cross contamination often exhibited higher morbidity and mortality (Huang et al., 2013). Relevantly, a recently published study demonstrated that airborne sources as well as inanimate surfaces of hospitals are a major reservoirs of multidrug resistant nosocomial pathogens (Mbim et al, 2016).

5.2 Main Sources of Bacterial Aerosols in Built Environment

Bioaerosols are produced and dispersed through different processes both indoors as outdoors and omnipresent in ambient environment. People can be exposed to contaminants via inhalation, ingestion and dermal contact. The interest in characterizing the microbiota of indoor environment is based on the importance of understanding associations between occupant health and microbial exposure in indoor settings. Because the indoor environment is a predominant habitat for humans in the developed world (modern urbanites allocate approximately 90% of their times indoor) most microorganisms encountered by the most human's lifetime will therefore be those who are present in indoor air and on indoor surfaces. There are several factors that influence the microbiota of indoor environments, and these determinants are discussed in detail in several other chapters in this book (e.g. Chap. 12). Each indoor micro-environment has unique characteristics determinated by the local outdoor air, specific building characteristics, and occupancy and occupant behavior, including indoor activities. For this reason, the different indoor micro-environments to which a person is exposed to and the time spent in each one determine the individual personal exposure. Geographic location (outdoor factor) and the human occupants (human factor) are the two major factors. However, the type of ventilation (natural or mechanic), building design and occupant's movement may also be associated with changes in the microbiota of a particular indoor environment (Leung and Lee, 2016).

Microorganisms can penetrate the indoor environment from outdoor sources. Different geographical locations in addition to climatic factors may be conducive to the presence of specifics of bioaerosols. Furthermore, the natural surrounding areas (urban, rural, agricultural, marine, etc.) impact the background bioaerosol concentrations. Other factors that influence the abundance of some pathogenic microbes are the air temperature and relative humidity as well as the source of ventilation air and occupant density. Mechanically ventilated rooms may shape indoor microbiota differently than do naturally ventilated spaces. Mechanical ventilation can include filters preventing the acquisition of microorganisms and particulate from outdoors while naturally ventilated rooms tend to contain microorganisms more similar to those from adjacent outdoor. On the other hand, human presence and human activities play an important role in indoor microbiota. In fact, microorganisms associated with the human gut, with the human skin and oral and urogenital flora are present in indoor air and surfaces acting as additional sources of the microbial community in the built environment (Leung and Lee, 2016).

Bacterial community in indoor environment can also be affected by building design, such as room type, floor area, floor level, and spatial relationships with neighboring indoor spaces.

Human exposure to bacterial bioburden occurs when an individual comes into contact with a certain concentration and type of bioaerosol during a certain period. The currently existing data reveals a microbial variation in different types of indoor environments such as schools, kindergartens, hospitals, swine barns, wastewater treatment plant, residential homes, among others reviewed in (Ghosh et al., 2015). In order to assess potential exposure hazards bioaerosols should be measured to determine type and quantity of present microorganisms. Air sampling enables the evaluation and quantification of microbial contamination in environments however, the utilized sampling techniques and methods for microbial identification must be adjusted for each context in order to perform valuable assessments.

5.2.1 Environmental Sampling

Microbial sampling of air and surfaces is a fundamental and limiting factor in bioburden analysis utilized in several contexts, namely i) to support investigations of an outbreak of disease or infection when fomites or environmental reservoirs are implicated epidemiologically in disease transmission, ii) to monitor a potentially hazardous environmental condition, iii) to support research, iv) to evaluate the effects of a control-infection protocol. Valid results of both microbial contamination of air and surfaces depend intrinsically on the selection of appropriate sampling and assay techniques. For a more detailed discussion on the topic of microbial sampling the reader is referred to Chap. 4 but we briefly mention key aspects of microbial sampling also here.

5.2.2 Sampling

Currently, several sampling methods are available to perform bioburden assessments of environmental samples such as air and surfaces, and numerous new divergent methodologies are being developed. Considering the challenges presented by environmental samples, no standard protocol is yet available and sampling methods must be adapted to the study design depending upon the sampling objective.

For assessment of air bioaerosol bioburden, active sampling methods, particularly impaction, impingement and filtration, are typically utilized. In these approaches, sampling efficiency is associated with aspiration, transmission and collection efficiency dependent upon particle aerodynamic diameter, wind velocity, direction and inlet characteristic, dependent upon bioaerosol particles conditions. Furthermore passive approaches such as gravity sampling, based on the deposition of airborne microorganisms into agar media, are also frequently utilized (reviewed in Ghosh et al., 2015).

Furthermore, surface sampling allows determining the number and nature of contaminating microorganisms on surfaces and potential environmental reservoirs of pathogens, survival of microorganisms on surfaces and sources of environmental contamination. Surface sampling can be performed trough tape lift imprint, surface swabbing with a culturette swab, or by surface bulk sample. This sampling technique is inexpensive and permits a direct and immediate assessment; however, the presence of microorganisms in surface does not directly confer information regarding airborne bioburden.

5.3 Bacterial Analysis Methods

5.3.1 Culture-Dependent Bacterial Analysis

Since the 19th century the analysis and study of the structure, characteristics, metabolism and resistances of the bacterial communities present in several environments has been performed through culture-dependent approaches, based on conventional culturing followed by phenotypic and/or genotypic (e.g., sequencing, genetic fingerprinting) identification of a selected subset of purified isolates (Al-Awadhi et al., 2013).

In culture-based detection methods a preliminary amplification of microbial population is encompassed, usually accompanied by an enrichment protocol. Using selective and/or differential media, bacteria are isolated from the enrichment media and identification performed based on biochemical properties. Regarding the utilized cultured media, nonselective media allow for an extensive variety of bacteria growth (e.g., plate count agar etc.) while in selective media, which contains specific inhibitors (e.g., bile salts in MacConkey agar), only specific microorganism can grow. Biochemical and immunochemical tests embody the traditional and classical methods of bacterial identification. Currently several commercial kits are available to study the biochemical properties of bacteria in pure culture.

Moreover, in this analysis approach, standard plate count is based in the assumption that each bacteria grows into a single colony. However, since it is unmanageable to determine that each colony formed from individual cells plate counts are reported as number of colony-forming units (CFU). In highly contaminated environments sample serial dilutions must be performed to ensure countable bacteria. Plates can be prepared by either the pour plate method or the spread plate method.

Culture-dependent bacterial analysis has been employed for bioaerosol (Bacteria and Fungi) assessments in several indoor environments worldwide, reviewed in (Ghosh et al., 2015) and was recently utilized in a study that aimed to evaluate the prevalence and antimicrobial susceptibility profile of bacteria bioburden present in Hospital air and surfaces (Mbim et al. 2016).

It is currently acknowledged that although culture-dependent approaches provide microbial "material" that can be used in further studies, these methods present considerable limitations particularly associated with the selectivity of the nutrient media and culture conditions utilized that favors only a fraction of the populating bacterial community and therefore underestimates the microbial prevalence and composition in the samples under study. Despite of the e fact that, in culture-dependent approaches only organisms that can be cultivated *in vitro* are identified, this approach is still utilized to perform determinations of culturable airborne bacteria in environmental samples (Fang et al., 2007).

5.3.1.1 Cell Wall Markers Analysis

In bacteria the cell wall acts as a surface organelle that allows pathogens to interact with their environment, in particular the cells and tissues of the infected host, trough surface proteins and enzymes. These molecules can be used as cell wall markers as it is detailed described and discussed in Chap. 8 "Endotoxins, Glucans and Other Microbial Cell Wall Markers." One of the most studied cell wall markers are endotoxins. One of the most common approaches to analyze endotoxins is the limulous amoebocyte lysate test (LAL test) accepted by the FDA for endotoxin testing of pharmaceuticals and medical devices. This method can be performed by gel clot assay or trough chromogenic methods such as the USP chromogenic method, based on the activation of a protease (coagulase) by the endotoxin, which is the rate-limiting step of the clotting cascade. Several methods of detection such as Mass Spectroscopy (GCMS), Matrix – Assisted Laser Desorption Ionization – Time of Flight Mass spectroscopy(MALDI-TOF-MS) are utilized and recently a published study performed using Raman spectroscopy, a rapid method for chemical structure detection, identified and detected of bacteria and bacterial endotoxin in concentrations as low as 0.0003 Endotoxin unit (EU)/mL and 1 Colony Forming Unit (CFU)/mL, which demonstrated the sensitivity and specificity of cell wall markers (Elsayeh et al., 2016).

5.3.2 Environmental Culture-Independent Bacterial Analysis

Bacterial burden associated with environmental samples has been assessed by characterization of isolated strains, which may only account for a small fraction of total bacterial diversity. For the past years, alternative methods of bacterial analysis based on protein, DNA and/or RNA analysis have been developed in order to increase the efficiency of bacterial identification as well as the assessment of diversity in microbial communities. These methods have been developed and utilized as alternative methodologies which according to (Dahllöf et al., 2000) "provide a new insight into microbial diversity and allow a more rapid, high resolution description of microbial communities than that provided by the traditional approach of isolation of microorganisms." These new molecular approaches include the blend of methodologies such as DGGE fingerprinting with 16S rDNA gene sequencing for bacteria identification and correlation of banding patterns and band numbers on DGGE gels with environmental variables. Several studies have utilized this approach to study the microbial communities in numerous environments, such as soil, water, air, food, plant, humans, and animals, reviewed in (Su et al., 2012). Nevertheless these methodologies also encompass several limitations such as preferential amplification of upcoming species, inadequate specificity of primers used for DNA amplification, production of single bands by multiple strains among others (Al-Awadhi et al., 2013).

5.3.2.1 Mass Spectrometry

Mass spectrometry is utilized as a valuable method to perform measurement of molecular weight and gives information regarding molecular fragmentation in order to determine molecular structures. The implementation of MALDI/TOF (Matrix-assisted Laser Desorption/Ionization) in combination with time offlight (TOF) detectors, lead to the utilization of this methodology in biochemical as well as microbiological analysis. In microbiology, this method has gained prominence in the past years, particularly as a tool for identification of bacteria in samples containing whole cells as a substitute to extracted protein fragments. Although the sensitivity of the method depends on the quality of the data and the nature of the species, published reports have indicated a > 95% success rate, demonstrating that this technology has an effective potential in the field of environmental microbiology. However, despite of the great advantages of the technique in terms of minimal sample preparation, rapid results, and small reagent costs, some limitations are related to higher purchase price of instrumental equipment and importantly evaluation software and database.

5.3.2.2 Nucleic Acid Amplification

New approaches based on the analysis of bacterial genetic information with no previous media cultivation augments the assessment of bacterial diversity in environmental samples. Currently DNA-based molecular methods such as Pulse Field Gel Electrophoresis (PFGE) and Restriction Fragment Length Polymorphism (RFLP) have gain relevance particularly due to their reproducibility, simplicity and high discriminatory capacity.

Polymerase Chain Reaction (PCR)

Polymerase chain reaction (PCR) technique has been used to amplify variable regions of the 16S rRNA gene after DNA extraction. In this methodology, similar sized PCR-products can be separated by successive denaturizing or temperature gradient gel electrophoresis (DGGE or TGGE) and the resultant patterns analyzed and interpreted. In DGGE the electrophoresis occurs under gradually changing conditions of a denaturing gradient of urea, whereas in TGGE the dsDNA is subjected to an increasing time-dependent temperature gradient dependent on the sequence loaded in the acrylamide gel. Currently, with the development of molecular analysis methodologies, one of the most utilized PCR techniques is quantitative real-time PCR (qRT-PCR) which is described in more detail in Chap. 6 "Analysis Approaches for Fungi in Indoor Environmental Assessments." In qRT-PCR target amplification and detection steps, using fluorescent dyes, occur concomitantly, and the copulated software monitors thermal cycler data at each cycle producing a quantitative amplification plot for each reaction which ensures the identification and quantification of specific, amplified DNA fragments.

Restriction Fragment Length Polymorphism Analysis (RFLP)

Restriction Fragment Length Polymorphism designates variances in homologous DNA sequences that can be identified through the detection of fragments with distinct lengths after digestion of the DNA samples with specific restriction endonucleases. RFLP is considered a preliminary typing method that originates restriction profiles of bacterial DNA based on random distribution of restriction sites in the genome. This method is considered as a powerful tool to determine molecular markers due to its specificity to single clone/restriction enzyme combinations. PCR products are further revealed in electrophoresis and the microorganisms present in the samples are classified according to the number and size of obtained fragments. However, RFLP method is still not developed to analyze whole bioburden. Furthermore, Amplified Fragment Length Polymorphism (AFLP) is a variant technique of RFLP method, in which specific adaptors are coupled to enzyme restricted DNA which is successively amplified using specific primers from the adaptor and restriction site-specific sequences.

Terminal Restriction Fragment Length Polymorphism (TRFLP)

TRFLP is a molecular methodology primarily developed to characterize bacterial community fingerprinting in mixed-species samples. This method analyses DNA patterns of PCR amplification products, of fluorescent labeled DNA, digested using restriction enzymes and visualized in a DNA sequencer. The resultant patterns can be analyzed trough peaks counts and comparisons in the TRFLP profile or trough matching peaks from one or more TRFLP runs to a database of identified species.

Pulse Field Gel Electrophoresis (PFGE)

PFGE methodology is based on *in situ* lysis of bacterial whole-cells in agarose plugs that are subjected to electrophoresis. This method is able to analyze large DNA fragments (10 to 800 kb in size) and the DNA fragments are visualized on the gel following staining.

Denaturing Gradient Gel Electrophoresis (DGGE)

DGGE is a specific type of gel electrophoresis in which a constant heat and an increasing concentration of denaturing chemicals are used to force DNA molecules to unwind. This methodology is coupled with PCR technique (PCR-DGGE) in which DNA amplification by genus-specific primers that target 16S rDNA sequences is performed, followed by differentiation on DGGE gel. This approach can be utilized for screening complex ecosystems including the identification of uncultivable microorganisms and has the ability to analyze diverse environmental samples in a time effective manner (Nehmé et al., 2009).

DNA Sequencing

Currently, DNA sequencing is considered one of the most accurate and reliable methodologies for describing microbial communities and for identification of microorganisms. With applicability in contexts were hasty and precise identifications are required, the identification of microorganisms trough 16S rRNA gene sequencing, classified as one of the best target genes, presents a most valuable method. It is acknowledged that 16S rRNA sequence comprises both variable and conserved regions. For full 16S rRNA gene sequencing, universal primers are chosen from conserved region while the variable region is used for comparative taxonomy. However, the inability of classical DNA sequencing methods (e.g., Sanger) to sequence fragments superior to 1000 bp, the associated high prices and the high purity requirements of DNA analysis presents some important limitations. Nevertheless, new sequencing technologies have been developed (*next generation sequencing*) with the aim to allow a fast/efficient technique for analysis of bacterial bioburden (Štursa et al., 2009). For more discussion on this topic see also Chap. 6 "Analysis Approaches for Fungi in Indoor Environmental Assessments."

Genotyping

Genotyping allows for the identification of genetic variants and can be performed through chips or arrays, depending on the variants of interest and available resources. The identifications of bacterial species have gained valuable advances with the development of genotypic methods such as 16S rRNA gene sequencing and DNA-DNA hybridization. Currently bacterial species are defined as a group of strains sharing 70% or more DNA-DNA affinity with 5°C or fewer ΔTm value (Tm is the melting temperature of the hybrid) among members of the group. In 16S rRNA gene sequencing methodology bacterial strains are considered to be the members of different species when presenting more than 3% sequence divergence.

5.3.2.3 Granulocyte Assay

The assessment of microbial contamination can also be achieved through a cell line (HL-60 cells) based granulocyte luminol-dependent chemiluminometric assay, which was reported to be extremely sensitive to a wide range of microorganisms including fungi. This methodology was proven to be appropriate to evaluate the inflammatory potential of bioaerosols or similar environmental samples (Timm et al., 2009).

5.3.3 Polyphasic Approach

Presently, the concept of microorganism hierarchical classification based on single gene or a cluster of genes at times is considered obsolete in the evolution scenario. Polyphasic approach, which comprises genotypic, chemotypic and phenotypic data, is used to distinguish bacterial species based on morphological and biochemical data supplemented with information obtained from molecular techniques. Considering that genes are subjected to deletions, duplications, mutations, recombination and lateral gene transfer, including conserved genes, the genomic complexity of the microorganisms is remarkably dynamic and encloses multiple parameters. Thus, the accessibility of complete genome sequencing of more than 1000 bacteria allows for the assessment of phylogenetic evolution. Additionally, new lines of comprehensive evolution that considers the combination of vertical and horizontal gene transfer has been considered for bacterial phylogeny analysis in order to further assess bacterial evolution in the concept of polyphasic approach (Prakash et al., 2007).

5.4 Concluding Notes

The assessment of bacterial bioburden in environmental samples encounters serious challenges and limitations mostly associated with sample contamination and the often vast bacterial diversity. Moreover, bacteria identification can differ depending on the utilized approaches as exemplified by the study performed by (Martin and Jackel, 2011). The codominant use of culture-dependent and culture-independent methodologies can overcome the limitations associated with both approaches and warrant more valuable, efficient and precise results regarding bacterial diversity and predominance.

References

Al-Awadhi H, Dashti N, Khanafer M et al. (2013) Bias problems in culture-independent analysis of environmental bacterial communities: a representative study on hydrocarbonoclastic bacteria. SpringerPlus 2:369. doi:10.1186/2193-1801-2-369

Anderson G, Palombo EA (2009) Microbial contamination of computer keyboards in a university setting. Am J Infect Control 37(6):507–509. doi:10.1016/j.ajic.2008.10.032

Brady RRW, Kalima P, Damani NN et al. (2007) Bacterial contamination of hospital bed control handsets in a surgical setting: A potential marker of contamination of the health care environment. Ann R Coll Surg Engl 89(7):656–660. doi:10.1308/003588407X209347

Dahllöf I, Baillie H, Kjelleberg S (2000) rpoB-based microbial community analysis avoids limitations inherent in 16S rRNA gene intraspecies heterogeneity. Appl Environ Microbiol 66(8):3376–3380. doi:10.1128/AEM.66.8.3376-3380.2000.Updated

Elsayeh MM, Kandil AH, Elsayeh M (2016) Detection and identification system of bacteria and bacterial endotoxin based on raman spectroscopy. doi:10.14569/IJACSA.2016.070328

Fang Z, Ouyang Z, Zheng H, Wang X, Hu L et al. (2007) Culturable airborne bacteria in outdoor environments in Beijing, China Microb Ecol 54(3):487–496. doi:10.1007/s00248-007-9216-3

Fuhrimann S, Pham-Duc P, Cissé G et al. (2016) Microbial contamination along the main open wastewater and storm water channel of Hanoi, Vietnam, and potential health risks for urban farmers. Sci Total Environ 566:1014–1022. doi:10.1016/j.scitotenv.2016.05.080

Ghosh B, Lal H, Srivastava A (2015) Review of bioaerosols in indoor environment with special reference to sampling, analysis and control mechanisms. Environ Int 85:254–272. doi:10.1016/j.envint.2015.09.018

Huang PY, Shi ZY, Chen CH et al. (2013) Airborne and surface-bound microbial contamination in two intensive care units of a medical center in central Taiwan. Aerosol Air Qual Res 13(3): 1060–1069. doi:10.4209/aaqr.2012.08.0217

Kirby-Smith WW, White NM (2006) Bacterial contamination associated with estuarine shoreline development. J Appl Microbiol 100(4):648–657. doi:10.1111/j.1365-2672.2005.02797.x
Lee L, Tin S, Kelley ST (2007) Culture-independent analysis of bacterial diversity in a child-care facility. BMC Microbiol 7(1):27. doi:10.1186/1471-2180-7-27
Leung MHY, Lee PKH (2016) The roles of the outdoors and occupants in contributing to a potential pan-microbiome of the built environment: a review. Microbiome 4(1):21. doi:10.1186/s40168-016-0165-2
Martin E, Jackel U (2011) Characterization of bacterial contaminants in the air of a duck hatchery by cultivation based and molecular methods. J Environ Monit 13(2):464–470. doi:10.1039/c0em00272k
Mbim E, Mboto C, Agbo B (2016) A review of nosocomial infections in sub-Saharan Africa. Br Microbiol Res J 15(1):1–11. doi:10.9734/BMRJ/2016/25895
Mbim E, Mboto C, Edet U (2016) Prevalence and antimicrobial susceptibility profile of bacteria isolated from the environment of two tertiary hospitals in Calabar Metropolis, Nigeria. J Adv Med Pharm Sci 10(4):1–15. doi:10.9734/JAMPS/2016/29316
Nehmé B, Gilbert Y, Létourneau V et al. (2009) Culture-independent characterization of archaeal biodiversity in swine confinement building bioaerosols. Appl Environ Microbiol 75(17): 5445–5450. doi:10.1128/AEM.00726-09
Nehme B, Létourneau V, Forster RJ et al. (2008) Culture-independent approach of the bacterial bioaerosol diversity in the standard swine confinement buildings, and assessment of the seasonal effect. Environ Microbiol 10(3):665–675. doi:10.1111/j.1462-2920.2007.01489.x
Prakash O, Verma M, Sharma P et al. (2007) Polyphasic approach of bacterial classification - An overview of recent advances. Ind J Microbiol 47(2):98–108. doi:10.1007/s12088-007-0022-x
Stanley NJ, Kuehn TH, Kim SW et al. (2008) Background culturable bacteria aerosol in two large public buildings using HVAC filters as long term, passive, high-volume air samplers. J Environ Monit 10(4):474–81. doi:10.1039/b719316e
Štursa P, Uhlík O, Kurzawová V et al. (2009) Approaches for diversity analysis of cultivable and non-cultivable bacteria in real soil. Plant, Soil Environ 55(9):389–396
Su C, Lei L, Duan Y et al. (2012) Culture-independent methods for studying environmental microorganisms: methods, application, and perspective. Appl Microbiol Biotechnol 93(3): 993–1003. doi:10.1007/s00253-011-3800-7
Timm M, Madsen AM, Hansen JV et al. (2009) Assessment of the total inflammatory potential of bioaerosols by using a granulocyte assay. Appl Environ Microbiol 75(24):7655–7662. doi:10.1128/AEM.00928-09
Walser SM, Gerstner DG, Brenner B et al. (2015) Evaluation of exposure-response relationships for health effects of microbial bioaerosols - A systematic review. Int J Hyg Environ Health 218(7):577–589. doi:10.1016/j.ijheh.2015.07.004
WHO World Health Organization. (2011). Report on the Burden of Endemic Health Care-Associated Infection Worldwide. Clean Care is Safer Care. Retrieved from http://apps.who.int/iris/bitstream/10665/80135/1/9789241501507_eng.pdf

Chapter 6
Analysis Approaches for Fungi in Indoor Environmental Assessments

Jacob Mensah-Attipoe and Martin Täubel

Abstract The challenge of fungal measurements in indoor environments is complex. Almost all studies that have used several methods for the assessment of fungal exposure have only observed moderate or weak correlation between them. These variations can be explained by the fungal life cycle with differences in spore release and the variation in the characteristics of spores of different species, and with differences in the target molecules used by the various fungal exposure assessment methods. Therefore, the use of different analysis methods will provide a different perspective on the stages of fungal growth and quantity.

Keywords Fungi · indoor · biomass markers · glucan · ergosterol · quantitative PCR · NAHA · microscopy · cultivation · real-time monitoring

6.1 Introduction

Since we spend more than 90% of our time in indoor environments and breathe about 10 m^3 of air every day (Dacarro et al., 2003) our proximity and interaction with indoor microbes, including fungi, is considerable. It has been established that indoor air quality is one of the most important factors that influence our general quality of life. Indoor air pollution can result in health problems and even in an increase in human mortality (Kanchongkittiphon et al., 2015, Mendell et al., 2011, Heseltine and Rosen, 2009). With respect to fungal contamination of indoor environments, the issue of moisture damage and dampness problems in buildings and the associated microbial proliferation and health problems observed in building occupants is the central issue.

J. Mensah-Attipoe (✉)
Department of Environmental and Biological Sciences, University of Eastern Finland, Kuopio, Finland
e-mail: jacob.mensah-attipoe@uef.fi

M. Täubel
Environmental Health Unit, National Institute for Health and Welfare, Kuopio, Finland
e-mail: martin.taubel@thl.fi

C. Viegas et al. (eds.), *Exposure to Microbiological Agents in Indoor and Occupational Environments*, DOI 10.1007/978-3-319-61688-9_6

The interaction between humans and indoor microbes is complex, stemming from the fact that the microbes' amounts, activity, physiology and diversity depend on both human activities and operational conditions of the indoor environment. To deal with the complexity of the diverse effects of mold on health, various different measurement strategies have been developed to assess indoor fungal contamination. Depending on the purpose of the investigation, indoor samples, including air, dust and surface materials, are analyzed to detect either fungal particles or specific fungal compounds. Fungal spores and/or fungal fragments may be detected with the aim of assessing fungal exposure as such, while cell components and metabolites of molds known to be associated with adverse health effects may be specifically quantified to assess health risks.

6.2 Analytical Methods for Measuring Fungal Concentrations

Measurements of fungal concentrations in indoor air and on materials are important in determining the sources and nature of fungal contamination which in turn helps in estimating the risks associated with exposures to fungal particles in moisture damaged buildings. Fungi grow differently on different building materials and surfaces and are affected by the diverse and changing conditions in the indoor environment. Methods that accurately estimate their amounts and also provide qualitative information on the nature of the fungi present are of interest.

The quantitative measurements assess how many fungal cells or how much fungal biomass are present on material surfaces or in the air, being determined by techniques such as the culture-based methods, microscopic counts and molecular methods, the latter targeting chemical cell wall markers of fungal biomass or fungal DNA. These methods determine the extent of growth, sporulation and total number of cells, or biomass (Krause et al., 2003). Total fungal biomass has been used as a surrogate of the overall fungal exposure and can be determined using chemical markers that are found in the fungi. Such chemical markers include ergosterol (Szponar et al., 2003), N-acetylhexosaminidase (NAHA), (Reeslev et al., 2003, Rylander et al., 2010), 1 → 3-β- glucan (Foto et al., 2005) and extracellular polysaccharides (EPS) (Douwes et al., 1999; Noterman and Soentoro, 1986). Each one of the methods is thought to provide a different perspective of fungal quantities since they evaluate specific responses of the various stages of fungal growth.

Summary of methods used for fungal determination

Measurement	Principle	Short description
Cultivation method	Determination of the number and type of colony forming units (CFUs) on agar media. Taxonomic identification of fungi with microscopy.	Determines the viable (culturable and "alive") spectrum of fungi and bacteria. In indoor applications mostly used for determination of viable fungi and bacteria from building materials or short-term active air samples.

(continued)

(continued)

Measurement	Principle	Short description
Epifluorescence Microscopy	Determination of the total number of fungal spores using dyes that stain the DNA of the fungal and bacterial cells.	Determines both viable and non-viable fungi in samples. This method is mostly used for determining total spore counts in samples.
Cell wall components and biomass measurements	Measure surrogates of total biomass of fungi. These methods quantify the total amount of cellular constituents in a sample instead of counting cells.	Measures the total biomass of microbes by quantifying their cellular constituents, mostly cell wall components. These include, for example, ergosterol, beta-glucans, fungal extracellular polysaccharides, N-acetylhexosaminidase. Almost all sample types can be used (dust, air, surface/material samples).
DNA-based methods	Detection, profiling and quantifying microbes targeting specific and suitable sequence regions in their DNA.	Detection is based on DNA, therefore, not dependent on the viability of the microbe (detects alive, dormant and dead cells as long as their DNA is present and detectable). Profiling is done with DNA fingerprinting methods, such as DGGE, TGGE, and more recently using various sequencing approaches. Quantification is done with qPCR, using primers and probes designed for the detection of a given species, genus or groups of taxa.
Real-time detection of fungi	Detects and enumerates fungal particles in real-time. These methods employ either the use of light scattering to detect any particles or the use of biochemical components of the cells to conclude on biological or non-biological origin of particles.	Direct measurements of fungal particles during active sampling of indoor air, avoiding separate steps for analysis before the concentration of fungal particles can be determined. Optical measurements involve the use of light scattering that detect and count all particles. The use of laser induced fluorescence (LIF) devices, targeting biochemical constituents, enables the differentiation of the origin of the particles into either biological or non-biological.

6.2.1 *Cultivation Method for Determining Fungal Growth*

Traditional quantification of fungi is based on the determination of the number and type of colony forming units (CFU), providing quantitative and qualitative data on viable and culturable fungi from different types of samples. This method is useful in identifying fungi down to the genus level using microscopy of individual colonies (Burge et al., 1999). It may help in providing useful information to confirm an environmental source for an outbreak investigation. Counting culturable microorganisms not only allows for a quantitative, but also qualitative assessment of exposure by identifying the genus of fungi since not all fungi pose the same hazard. After cultivation, fungi may be identified to the genus or species level using morphological criteria and microscopy, Matrix Assisted Laser Desorption/Ionization-Time of Flight Mass Spectrometry (MALDI-TOF-MS) (De Carolis et al., 2012) or DNA based methodologies.

Cultivation method has its inherent limitations, such as the inability of particular media to satisfy the specific growth requirements of certain fungal species (Douwes et al., 2003). The choice of growth media used can contribute to substantial variability of the types and amounts of species that are cultured. For example, malt extract agar (MEA) has high sugar content and water activity allowing fast growing fungal species to flourish on its surface. Dichloran glycerol 18 on the other hand allows detection of a more diverse fungal flora, however excluding fungi which require high water activity (Chao et al., 2002, Wu et al., 2000). Cultivation method lacks the ability to detect non-culturable and dead microorganisms, cell debris and microbial components, although all of those may be of health relevance (Green et al., 2005). It has been demonstrated that by using the cultivation method, concentrations of viable fungal cells detected in relation to the total amount of fungal cells present vary widely, depending on the type of sample (<1–100%) (Lee et al., 2006, Meklin et al., 2004, Toivola et al., 2002). The total quantity and diversity of fungal cells are usually drastically underestimated by culture-based methods (Bridge and Spooner, 2001, Douwes et al., 2003). In addition to these limitations, the cultivation method is laborious and requires a long incubation time (minimum of 1 week) to detect fungal growth (Douwes et al., 2003). The method shows poor precision and is variable among replicate samples (Eduard and Halstensen, 2009, Mensah-Attipoe et al., 2016a). Since culture-based methods measure only a fraction of sampled fungi, CFU counted cannot completely characterize the fungal spectrum that might influence human health and well-being. For all of these reasons, other, culture independent approaches have been developed to identify and enumerate fungi in indoor samples (Viegas et al., 2012, Eduard and Halstensen, 2009).

6.2.2 *Microscopic Spore Counting*

To circumvent some of the limitations associated with culture-based techniques, methods that detect both viable and non-viable cells using microscopic counting have been developed (Palmgren, 1986, Bauer et al., 2008, Ho et al., 2005, Sattler et al., 2001). Simple light microscopy may be used to count microorganisms, but counting is based only on morphological recognition, which may result in severe measurement errors (Douwes et al., 2003). For example, identification of the fungal spores is often difficult: only a small number of fungal spore types can be identified with confidence at genus level, and important genera, such as *Aspergillus* and *Penicillium,* cannot be differentiated. In epifluorescence microscopy, dyes such as acridine orange which stain the spores' DNA, are used to help in counting spores. The dye makes the spore fluoresce when viewed under the microscope (Thorne et al., 1994, Palmgren, 1986). The epifluorescence microscopic spore count method has its own limitations which include the masking of spores by large particles and the inability of some spores to absorb the dye (Burge, 1995). In addition, the spore counting approach is time consuming, laborious and not trivial to perform. The method is, however, relatively cheap to perform and may provide a general indication of atypical indoor fungal growth (Douwes et al., 2003).

Electron microscopy (EM) or scanning EM can also be used and it provides a better determination of spore counts and concentrations (Eduard et al., 1988, Karlsson and Malmberg, 1989). Like the simple light microscopy, identification and enumeration is done by morphology. There is an increased interest in identifying and differentiating particles from different sources but having similar appearances which often lead to difficulties in their quantification (Wittmaack et al., 2005). A good technique developed to circumvent this problem is the use of SEM microscopy coupled with energy dispersive X-ray spectroscopy (EDX). This analysis is based on determining the elemental composition of the particles after they have been identified with SEM. The elemental composition of biological particles differs from other particles and they also behave differently from non-biological particles. Based on this property, criteria for determining primary biogenic organic aerosols (PBOA) in atmospheric samples have been developed (Matthias-Maser and Jaenicke, 1991, 1994). These were based on the detection of minor amounts of K, P, S, Na and Ca (usually $<10\%$ of relative element of X-ray intensity of the particle). This criterion was recently adopted (Coz et al., 2010) to characterize PBOA in the atmosphere. Also a recent study (Mensah-Attipoe et al., 2016b) applied this method to differentiate fragments of biological origin from those of non-biological origin.

6.2.3 DNA-Based Methods

The last decades have seen a surge in the development of several culture-independent, molecular, DNA based techniques supplying many advantages over the traditional cultivation technique (Amann et al., 1995). The DNA-based method, like the total spore count method, detects both culturable and unculturable spores. DNA-based methods in addition, measure mycelial cells (Meklin et al., 2004, Gonzalez and Saiz-Jimenez, 2004, Herrera et al., 2009, Yamamoto et al., 2010). Those methods require an initial step of extracting DNA from an environmental sample prior to subsequent analysis. The molecular methods most often used in fungal studies include conventional or quantitative PCR (qPCR) specific for fungal species or groups (Haugland et al., 2004, Zeng et al., 2006), as well as ribosomal DNA amplicon sequencing or metagenome analysis (Tringe et al., 2008, Frohlich-Nowoisky et al., 2009, Liu et al., 2012, Adams et al., 2013a, Adams et al., 2013b, Dannemiller et al., 2014b, Yamamoto et al., 2014, Dannemiller et al., 2014a, Pitkaranta et al., 2008). In the pre-next generation sequencing (NGS) era, DNA fingerprinting methods were based on universal fungal PCR combined with denaturing or temperature gradient gel electrophoresis (DGGE, TGGE) (Gonzalez and Saiz-Jimenez, 2004), and terminal or conventional restriction fragment length polymorphism analysis (Buttner et al., 2007). Other methods include the molecular tracer methods (Elbert et al., 2007)

The advantages of using the DNA-based methods for detecting and quantifying fungi instead of cultivation-based methods and microscopic counts are the speed, accuracy, and analytical sensitivity of this approach and the possibility to detect and identify also dead or dormant microorganisms (Ettenauer et al., 2014). Detection is based on DNA, and therefore, not dependent on the viability of the microbe. The ability of DNA-based methods to detect dead or dormant cells is important in indoor studies since the main exposure hazards relating to indoor microbial contamination may not require viability. Thus, these techniques have enabled a reliable assessment of fungal communities associated with different materials such as wood, concrete, mineral wool, paper, or dust (Ettenauer et al., 2012, Piñar and Sterflinger, 2009). Quantitative PCR has provided valuable information on the occurrence and levels of the most common indoor fungi, and exhibits great potential at being able to provide quickly quantitative data on the occurrence of the studied organisms (Meklin et al., 2004, Pietarinen et al., 2008, Pitkäranta et al., 2011). Primers and probes are generally designed for the detection of a given genus (genus-specific primers), groups of genera (group-specific), or for the detection of a single species (species-specific primers). The 18S ribosomal RNA gene and internal transcribed spacer regions (ITS) can be used in the design of specific primers and probes because they contain sequences that are highly conserved between members of the same species or genus, for example, but are variable among different species or genera (Haugland et al., 2004).

There are a studies describing the exploitation of qPCR analytical methods in estimating the concentrations of individual species or groups of fungi in indoor dust and air samples (Vesper, 2007, Haugland et al., 2004, Meklin et al., 2004, Kaarakainen et al., 2009) and building materials (Pitkäranta et al., 2011, Pietarinen et al., 2008). For example, real-time PCR methods have been utilized to detect and quantify *Cladosporium* (Zeng et al., 2006) and *Aspergillus* (Goebes et al., 2007) at the genus level. Similar methods have been developed for targeting species, groups of species or genera of common indoor fungi such as *Aspergillus, Cladosporium, Penicillium* and *Alternaria* (Vesper et al., 2005, Meklin et al., 2007, Haugland et al., 2004). Applying multiple PCR assays make it possible to assess the presence and amounts of large groups of microorganisms. For instance, a quantitative PCR approach has been developed integrating measurement of 36 fungal species commonly associated with damp houses and background species, used to define an "environmental relative mouldiness index" (ERMI) for houses in the United States (Vesper, 2007). It is also important to mention in this context that PCR and qPCR are *targeted* approaches that detect only what the primer set is designed for, unlike cultivation or next generation sequencing methods, which are largely untargeted methods.

There are well known limitations inherent to DNA based approaches referring to biases in DNA extraction and PCR amplification that can be preferential of one over another species or taxon (see more detail below). Quantitation in qPCR can also have biases due to variation in target gene copy numbers in different fungal species and genera (Herrera et al., 2009). However, results may not be affected by these bias when primers and probes are used that target the ITS gene region and keep the copy numbers fairly stable across varying conditions (Herrera et al., 2009). In addition, when fungal spore suspensions of known concentrations are used in creating standard curves for individual species rather than genera or groups, the issue of target gene copy numbers is less pronounced.

Next generation sequencing (NGS) approaches today are widely used in fungal ecological studies in different environments. NGS refers to a suite of different methods carried out on different sequencing platforms, and these methods include: sequencing of 16S rRNA and internal transcribed spacer region (ITS) amplicons for studies of the bacterial and fungal communities, respectively (amplicon sequencing); whole-genome sequencing for understanding an organism's function; and metagenomics sequencing to understand the functioning of microbial communities (Cox et al., 2013). More recently, NGS – and here thus far almost exclusively amplicon sequencing of the fungal ITS regions – has been introduced to studies of fungal communities in indoor environments. Given constant improvements in sample throughput, resolution, costs per analysis and bioinformatics that tackle the large data amounts typically produced in NGS surveys (Metzker, 2010), the number of studies using NGS in indoor assessments is steadily increasing. Early efforts utilizing DNA extracted from house dust samples revealed a hitherto unknown richness and diversity of the indoor fungal flora (Amend et al., 2010, Pitkaranta et al., 2008). Further indoor NGS studies have dealt with the ecology and sources of fungi determined from indoor samples and also attempted to study health implications of fungal exposure (Adams et al., 2013b, 2013c, 2015, Dannemiller et al., 2014a, 2016a, 2016b, Lymperopoulou et al., 2016). Amplicon sequencing has obvious advantages over other assessment methods: it is a non-targeted method that permits detection of theoretically all fungal taxa present in a sample, independent of viability and culturability; the method is characterized by high resolution that allows determination of hundreds of different fungal taxa from indoor samples; and, in case of ITS sequencing, taxonomic allocation of the fungal sequences detected upon database comparisons is often possible to the species or at least the genus level. Investigations that have interest in knowing which fungal cells in a sample are metabolically active and alive versus dead, may need to use other approaches than amplicon sequencing, as this method can at current not efficiently distinguish what is dead and alive in a sample. Another issue is absolute quantification, desirable in e.g. determining human exposure levels. NGS data are typically presented as relative abundance of a given fungal taxon rather than total amount of the respective fungal cells in a sample. There are further technical limitations in amplicon sequencing that refer generally to DNA and PCR based approaches and that need to be carefully considered. Selectivity in DNA extraction as well as in the PCR amplification prior sequencing, leading to preferential detection of some taxa over others, and variation in target sequence copy number can introduce biases into NGS studies (Amend et al., 2010, Huber et al., 2009, Rastogi et al., 2009, von Wintzingerode et al., 1997). Nonetheless, next generation sequencing approaches have already revolutionized our view and interpretation of the role of bacterial and fungal communities in human health and disease, and similar advances can be expected from studies applying these approaches in indoor microbial assessments.

6.2.4 *Methods Measuring Cell Wall Components and Other Indicators of Biomass*

Microbial cell wall components, including fungal compounds, are discussed in Chap. 8 "Endotoxins, Glucans and Other Microbial Cell Wall Agents" and the reader is referred to this part of the book for more detailed information. There is a variety of approaches that assay cellular constituents, usually microbial cell wall agents, instead of counting culturable and/or non-culturable microbial particles. These methods typically are used to measure surrogates of total fungal biomass; however, in some cases the targeted compound itself may be of interest due to potential health implications upon exposure. More or less commonly used markers for the assessment of fungal biomass include ergosterol, (Miller et al., 1988, Szponar et al., 2003), fungal extracellular polysaccharides (EPS) and β-glucan (Sonesson et al., 1988, Douwes et al., 2000). Measuring the activity of N-acetylhexosaminidase enzyme (Rylander et al., 2010, Reeslev et al., 2003) is another way to estimate total fungal biomass. Microbial volatile organic compounds (MVOCs) produced by fungi have been proposed as markers of active fungal growth (Dillon et al., 2007, Moularat et al., 2008). Also measuring mycotoxins, i.e. toxic fungal secondary metabolites, from indoor samples is being done, mostly in a search for exposing agents involved in provoking adverse health effects in occupants of moisture damaged building. These latter two categories – MVOCs and mycotoxins – will not be discussed further here, as separate chapters in this book have been dedicated to these agents.

There are several advantages common to these assay methods. These include: the stability of most of the measured components, allowing – among others – longer sampling times for airborne measurements; and not restricting the assessment to only "alive" fungal material. Storing samples frozen prior to analyses is typically not an issue, unlike for example in cultivation based approaches. Standards can be used in many of the methods allowing sound quantification, and typically, sensitivity and specificity of the measurement approaches is high, even though there often is a tradeoff between those two (i.e. high sensitivity comes to the expense of lower specificity and vice versa). Major limitations are, however, that these methods do not allow for an identification of the fungal taxa present in a sample, but rather provide information on total fungal material present. Furthermore, these methods do not leave fungal isolates for further investigation where needed, a disadvantage common to all methods but cultivation.

6.2.5 Ergosterol

Ergosterol is the major sterol present within cell membranes of fungal spores and hyphae and is considered an indicator for total fungal biomass (Szponar et al., 2003). This agent has been assayed in indoor dust (Saraf et al., 1997), in building materials (Szponar et al., 2003, Gutarowska and Piotrowska, 2007), and indoor air (Park and Cox-Ganser, 2011). As just mentioned as a common limitation when measuring cell wall agents, also quantifying ergosterol does not provide information about the individual fungal species present in a sample. Ergosterol contents are measured by gas chromatography-mass spectrometry (Miller et al., 1988, Szponar et al., 2003). The amount of ergosterol measured from a particular fungal isolate depends on its surface area and growth conditions. It has been shown that ergosterol is somewhat labile and thus its concentration declines after the death of fungal spores and hyphae (Mille-Lindblom et al., 2004, Gutarowska and Piotrowska, 2007), though it is not well understood how long ergosterol stays stable within dead fungal cells or in different sample materials. Some studies have shown that levels of ergosterol correlate well with total spore counts (Mensah-Attipoe et al., 2016a). Although this method is highly specific and sensitive giving accurate estimate of fungal amounts, ergosterol determination in samples is not done routinely, as it is a rather expensive methodology requiring specific and costly infrastructure and expertise and time to process and analyze samples.

6.2.6 Fungal Extracellular Polysaccharides

These are stable carbohydrates that are produced and excreted during fungal growth and are suggested as a marker for fungal exposure. Measurements are usually done with immunoassays and as the polysaccharides confer antigenic specificity and differ somewhat between major fungal taxa it is in theory possible to target fungi at the genus level. For example, Douwes et al. (2003) quantified EPS from *Aspergillus* and *Penicillium* fungi (EPS-Asp/Pen) in house dust. The authors of that paper found a correlation between the EPS measurements and culturable fungal spore counts. EPS-Asp/Pen in house dust serves as a marker of a somewhat specific fungal exposure, but EPS as such is not suspected to be causally related with poor respiratory or other ill health in children or adults. The method is not widely used today, as the required immunoassays have been developed for research purposes only and are not commercially available.

6.2.7 Beta-Glucan

Beta-glucans are polysaccharides found in the outer cell membrane of fungi, higher plants and some bacteria. In the fungal cell wall, glucans comprise a three-dimensional network of (1 → 3) and (1 → 6)-β D-linked anhydroglucose repeat units that are connected to other carbohydrates, proteins and lipids. (1 → 3)-β-D-glucan polymers can exist as a stable complex of three polymer strands forming a triple helix. The triple helical structure is generally considered to be the preferred form in nature (Young and Castranova, 2005). From a quantitative point of view, (1 → 3)-β-D-glucans are the main constituent, accounting for between 47% and 60% by weight of the cell wall (Young and Castranova, 2005). Measurements of (1 → 3)-β-D-glucan in indoor air have been done with a glucan-sensitive preparation of the Limulus amebocyte lysate (LAL) assay (Iossifova et al., 2009, 2007). While the LAL test is highly sensitive, it's specificity for fungi is somewhat limited, as it reacts to some extent also to plant or bacterial material. Other methods for the analysis of fungal glucans have been based on antibodies. For example, Douwes et al. (1996, 1998) developed an inhibition enzyme immunoassay (EIA), which is specific to the (1 → 3)-glycosidic linkage and to water-insoluble glucans, but is less sensitive than the LAL test. Glucan immunoassays that are more specific to fungal material by targeting (1 → 3)-β glucan have been developed (Low et al., 2009). Generally, glucan content in fungal cells has been shown to vary according to the fungal species and dependent on the spore surface area (Iossifova et al., 2008), but appears to be relatively independent of growth conditions (Foto et al., 2004). Assessment of exposure to glucan is done both in homes and in occupational context in work places. For the latter, several epidemiological studies have reported glucans to have strong immunomodulating and inflammatory effects in occupational, high exposure settings. Douwes (2005) asserted from his review that the biological effects observed are not dependent on viability and that (1 → 3)-β-D-glucans from dead organisms may thus be equally relevant in causing potential health effects.

There have been mixed observation with exposures to (1 → 3)-β-glucan and the health effects, pointing also towards a beneficial role of glucan exposure. Some studies have suggested beneficial impacts on the development of immune system of infants (Iossifova et al., 2007, Schaub et al., 2006), lower prevalence of allergic sensitization in 2–4 year olds when exposed to (1 → 3)-β-glucan from mattress dust (Gehring et al., 2007) and inverse association with wheezing symptoms in children (Iossifova et al., 2007, 2009). Furthermore, (Tischer et al., 2011) found a difference in (1 → 3)-β-glucan effects between countries.

6.2.8 NAHA Enzyme Activity

Measurement of the activity of β-N-acetylhexosaminidase (NAHA) in fungi is another way of measuring fungal total biomass (Rylander et al., 2010, Reeslev et al., 2003). NAHA is present in both the growth and stationary phases of fungal growth (Rast et al., 2003, Reeslev et al., 2003) and its activity is reported to be relatively stable under appropriate storage conditions (Rylander, 2015). The activity measurement is based on a fluorescence labeled substrate which is cleaved by the enzyme that is only present in fungi. The amount of fluorescence detected is proportional to the amount of enzyme/biomass present, and this measurement approach has been translated into a commercially available product (Reeslev et al., 2003). By using enzyme activity as an indicator for fungal biomass, fungal growth present on a building material surface or the amount of fungal biomass in air can be determined. Since the method is fast and can be done onsite, it allows determining the extent of mould-affected materials and the efficacy of cleaning after remediation efforts in the field. Beta-*N*-acetylhexosaminidase activity has been shown to correlate well with the fungal molecules such as ergosterol and the phospholipid fatty acid 18:2ω6 in soil samples (Miller et al., 1988) and building materials (Mensah-Attipoe et al., 2016a). Significant correlations have been reported between NAHA and total spore counts in dust (Madsen, 2003, 2009) and building materials (Mensah-Attipoe et al., 2016a); fungal biomass (by gravimetric weight) of fungal species grown on nutrient agar; and ergosterol content of gypsum boards (Reeslev et al., 2003) and mineral wool contaminated by fungi (Mensah-Attipoe et al., 2016a). In a recent study (Mensah-Attipoe et al., 2015), the authors found a good correlation between NAHA enzyme activity and cultivation method.

The NAHA method is not able to differentiate between fungal species. Furthermore, levels of NAHA enzymes detected in air samples are usually very low and do not represent well the fungal amounts measured. To be able to achieve a high enough concentration in air, movement and agitation during sampling (usually in the form of "aggressive" blowing, i.e., resuspension of material) is required.

6.3 Real-Time Detection of Airborne Fungal Particles

It has been stated that a thorough understanding of the significance of microbial exposure in indoor environments is impaired by the methodological difficulties in identifying and enumerating various microbial components (Green et al., 2006). Traditional bioaerosol detection methods such as the Andersen impactor and filter sampling require a separate step for analysis after sampling and before concentration can be determined, which results in relatively low time resolution (Reponen et al., 2011, Górny et al., 2002). These methods are well-established for culture-based and microscopic analyses, and more recently also for DNA-based analyses in the case of filter sampling. Sample collection usually only lasts for a short period of time and thus samples collected by these methods reflect the concentration of the target organism only at the specific time of sampling (Reponen et al., 2007), while the great temporal as well as spatial variation in airborne fungal concentrations are well known. Detection of target bioaerosols in real time would allow understanding of emissions and the temporal variation of airborne concentrations. Therefore, real time detection techniques are needed in various fields, e.g., bioprocess monitoring (Ganzlin et al., 2007), health related applications (Elston, 2001), and in environmental, defense and public health (Davitt et al., 2005, Kanaani et al., 2007, Sivaprakasam et al., 2004, Méjean et al., 2004).

Direct reading instruments that measure particles in real-time have been used in several laboratory studies, e.g., optical particle counters (OPCs), which measure particle concentration in the size range of 0.3–20 μm based on their light scattering properties. The electrical low pressure impactor (ELPI), a multistage impactor, classifies aerosol samples into size fractions over a size range of 0.07–10 μm and the Aerodynamic Particle Sizer (APS) measures the dynamic size distribution of particles in the size range of 0.5–20 μm by determining the time-of-flight of individual particles in an accelerating flow field (Volckens and Peters, 2005). Particle concentrations in the size range of 0.02–1 μm can be measured using the P-Trak. This instrument measures the number concentration of particles by saturating them with either water or alcohol vapour and cooling them so that their enlarged particle sizes can be detected by optical methods. Although very useful in laboratory-based studies where other particles can be eliminated, especially particles within a certain size range, the above described instruments have limited utility in the assessment of bioaerosol exposures because they are not very specific since the process of distinguishing microbial and non-microbial particulate matter is complex (Green et al., 2011).

The Laser Induced Fluorescence (LIF) technique enables real-time detection of biological aerosol particles. LIF techniques have been developed to detect biological warfare agents (Hairston et al., 1997). The use of LIF techniques can give insight into the origin of the fluorescent spectral features and contribute to the interpretation of data obtained using other fluorescence-based techniques.

The best known and most widely used LIF device is the Ultra Violet aerodynamic particle sizer (UVAPS). Other devices include the BioScout and Wide Issue Bioaerosol Sensor (WIBS-3) and the waveband integrated bioaerosol sensor (WIBS-4). The basis by which all of these devices detect relies on the fluorescence of compounds such as reduced nicotinamide adenine dinucleotide phosphate (NAD(P)H), flavins, melanin, carotenoids, phenols, terpenoids, and DNA (Raimondi et al., 2009, O'Connor et al., 2011, Pöhlker et al., 2012, Saari et al., 2013, Frohlich-Nowoisky et al., 2009, Després et al., 2007), present in all living cells but at different proportions and measured at selective wavelengths. Thus, LIF enables the differentiation of bioaerosols from other particles because of their fluorescence capabilities (Hill et al., 2013, Pöhlker et al., 2012).

The UVAPS measures both aerodynamic particle size and autofluorescence of a single particle. The WIBS, on the other hand, measures optical size and the autofluorescence of bioaerosol particles (Gabey et al., 2010, Healy et al., 2014) by utilizing two excitation wavelengths and by detecting two bands of fluorescence. The BioScout measures both autofluorescence and optical particle size of single particles using continuous wave laser diode and light scattering. Saari et al. (2014) have used both BioScout and UVAPS to measure fungal spores in the laboratory and found the former device to be more sensitive. This indicates that the LIF devices have varying capabilities in detecting biological particles based on the wavelengths of light used and the type and amount of fluorescent compounds present in the samples. Data obtained from the use of these devices have usually been on fungal spores and not on small-sized fragments. This is because very little or no fluorescence is emitted from fungal fragments compared to larger spores (Kanaani et al., 2008, Saari et al., 2014).

Different fungal species have characteristic structures and more or less differing biochemical compositions which in turn could influence their autofluorescence. A variety of factors may affect the fluorescent properties of fungi under various conditions. These factors include the type of the fungal species under consideration, growth substrate, air velocity and age of the culture. A better understanding of the effects of these factors on the fluorescent properties of fungal spores measured with different LIF devices side-by-side would help with instrument calibration and ease the interpretation of LIF-based field results.

6.4 Conclusion

Depending on the purpose of a measurement, an analysis approach serving that purpose need to be chosen. For example, cultivation method accounts only for viable spores and cells that can grow on the culture media. DNA based methods and NGS give in-depth information of the fungi being assessed while ergosterol content, NAHA enzyme activity and other cell wall components display the total biomass of the fungi and indirectly estimate total exposure. Since the life cycle of the fungi is dynamic, the different methods employed will give insight into the different stages of the fungal growth.

References

Adams RI, Amend AS, Taylor JW et al (2013a) A unique signal distorts the perception of species richness and composition in high-throughput sequencing surveys of microbial communities: a case study of fungi in indoor dust. Microb Ecol 66(4):735–741

Adams RI, Bhangar S, Pasut W et al (2015) Chamber bioaerosol study: outdoor air and human occupants as sources of indoor airborne microbes. PLoS One 10(5):e0128022

Adams RI, Miletto M, Taylor JW et al (2013b) Dispersal in microbes: fungi in indoor air are dominated by outdoor air and show dispersal limitation at short distances. ISME J 7(7): 1262–1273

Adams RI, Miletto M, Taylor JW et al (2013c) The diversity and distribution of fungi on residential surfaces. PLoS One 8(11):e78866

Amann RI, Ludwig W, Schleifer K (1995) Phylogenetic identification and in situ detection of individual microbial cells without cultivation. Microbiol Rev 59(1):143–169

Amend AS, Seifert KA, Bruns TD (2010) Quantifying microbial communities with 454 pyrosequencing: does read abundance count? Mol Ecol 19(24):5555–5565

Bauer H, Schueller E, Weinke G et al (2008) Significant contributions of fungal spores to the organic carbon and to the aerosol mass balance of the urban atmospheric aerosol. Atmos Environ 42(22):5542–5549

Bridge P, Spooner B (2001) Soil fungi: diversity and detection. Plant Soil 232(1-2):147–154
Burge H, Otten J, Fungi JM et al (1999) Bioaerosols: assessment and control. In: Anonymous American Conference of Governmental Industrial Hygienists (ACGIH), vol 19., p 1–13
Burge HA (1995) Bioaerosols. CRC Press
Buttner MP, Cruz P, Stetzenbach LD et al (2007) Evaluation of two surface sampling methods for detection of Erwinia herbicola on a variety of materials by culture and quantitative PCR. Appl Environ Microbiol 73(11):3505–3510
Chao HJ, Milton DK, Schwartz J et al (2002) Dustborne fungi in large office buildings. Mycopathologia 154(2):93–106
Cox MJ, Cookson WO, Moffatt MF (2013) Sequencing the human microbiome in health and disease. Hum Mol Genet 22(R1):R88–94
Coz E, Artíñano B, Clark LM et al (2010) Characterization of fine primary biogenic organic aerosol in an urban area in the northeastern United States. Atmos Environ 44(32):3952–3962
Dacarro C, Picco A, Grisoli P et al (2003) Determination of aerial microbiological contamination in scholastic sports environments. J Appl Microbiol 95(5):904–912
Dannemiller KC, Gent JF, Leaderer BP et al (2016a) Indoor microbial communities: influence on asthma severity in atopic and nonatopic children. J Allergy Clin Immunol 138(1):76–83-e1
Dannemiller KC, Mendell MJ, Macher JM et al (2014a) Next-generation DNA sequencing reveals that low fungal diversity in house dust is associated with childhood asthma development. Indoor Air 24(3):236–247
Dannemiller KC, Reeves D, Bibby K et al (2014b) Fungal High-throughput Taxonomic Identification tool for use with Next-Generation Sequencing (FHiTINGS). J Basic Microbiol 54(4):315–321
Dannemiller KC, Gent JF, Leaderer BP et al (2016b) Influence of housing characteristics on bacterial and fungal communities in homes of asthmatic children. Indoor Air 26(2):179–192
Davitt K, Song Y, Patterson III W et al (2005) 290 and 340 nm UV LED arrays for fluorescence detection from single airborne particles. Opt Express 13(23):9548–9555
De Carolis E, Posteraro B, Lass-Flörl C et al (2012) Species identification of Aspergillus, Fusarium and Mucorales with direct surface analysis by matrix-assisted laser desorption ionization time-of-flight mass spectrometry. Clin Microbiol Infect 18(5):475–484
Després V, Nowoisky J, Klose M et al (2007) Characterization of primary biogenic aerosol particles in urban, rural, and high-alpine air by DNA sequence and restriction fragment analysis of ribosomal RNA genes. Biogeosciences 4(6):1127–1141
Dillon HK, Boling DK, Miller JD (2007) Comparison of detection methods for Aspergillus fumigatus in environmental air samples in an occupational environment. J Occup Environ Hyg 4(7):509–513
Douwes J, van der Sluis B, Doekes G et al (1999) Fungal extracellular polysaccharides in house dust as a marker for exposure to fungi: relations with culturable fungi, reported home dampness, and respiratory symptoms. J Allergy Clin Immunol 103:494–500
Douwes J (2005) (1→3)-β-D-glucans and respiratory health: a review of the scientific evidence. Indoor Air 15(3):160–169
Douwes J, Doekes G, Heinrich J et al (1998) Endotoxin and β (1 → 3)-Glucan in House Dust and the Relation with Home Characteristics: A Pilot Study in 25 German Houses. Indoor Air 8(4):255–263
Douwes J, Zuidhof A, Doekes G et al (2000) (1 → 3)-β-D-glucan and endotoxin in house dust and peak flow variability in children. Am J Resp Crit Care Med 162(4):1348–1354
Douwes J, Doekes G, Montijn R et al (1996) Measurement of beta (1→3)-glucans in occupational and home environments with an inhibition enzyme immunoassay. Appl Environ Microbiol 62(9):3176–3182
Douwes J, Thorne P, Pearce N et al (2003) Bioaerosol health effects and exposure assessment: progress and prospects. Ann Occup Hyg 47(3):187–200

Eduard W, Sandven P, Johansen BV et al (1988) Identification and quantification of mould spores by scanning electron microscopy (SEM): analysis of filter samples collected in Norwegian saw mills. Ann Occup Hyg 32(inhaled particles VI):447–455

Eduard W, Halstensen AS (2009) Quantitative exposure assessment of organic dust. Scand J Work Environ Health. Supplement (7):30.

Elbert W, Taylor P, Andreae M et al (2007) Contribution of fungi to primary biogenic aerosols in the atmosphere: wet and dry discharged spores, carbohydrates, and inorganic ions. Atmos Chem Phys 7(17):4569–4588

Elston DM (2001) Fluorescence of fungi in superficial and deep fungal infections. BMC Microbiol 1:21

Ettenauer J, Piñar G, Tafer H et al (2014) Quantification of fungal abundance on cultural heritage using real time PCR targeting the β-actin gene. Front Microbiol 5.

Ettenauer JD, Pinar G, Lopandic K et al (2012) Microbes on building materials – evaluation of DNA extraction protocols as common basis for molecular analysis. Sci Total Environ 439:44–53

Foto M, Vrijmoed L, Miller J et al (2005) A comparison of airborne ergosterol, glucan and Air-O-Cell data in relation to physical assessments of mold damage and some other parameters. Indoor Air 15(4):257–266

Foto M, Plett J, Berghout J et al (2004) Modification of the Limulus amebocyte lysate assay for the analysis of glucan in indoor environments. Anal Bioanal Chem 379(1):156–162

Frohlich-Nowoisky J, Pickersgill DA, Despres VR et al (2009) High diversity of fungi in air particulate matter. Proc Natl Acad Sci U S A 106(31):12814–12819

Gabey A, Gallagher M, Whitehead J et al (2010) Measurements and comparison of primary biological aerosol above and below a tropical forest canopy using a dual channel fluorescence spectrometer. Atmos Chem Phys 10(10):4453–4466

Ganzlin M, Marose S, Lu X et al (2007) In situ multi-wavelength fluorescence spectroscopy as effective tool to simultaneously monitor spore germination, metabolic activity and quantitative protein production in recombinant Aspergillus niger fed-batch cultures. J Biotechnol 132(4):461–468

Gehring U, Heinrich J, Hoek G et al (2007) Bacteria and mould components in house dust and children's allergic sensitisation. Eur Respir J 29(6):1144–1153

Goebes MD, Hildemann LM, Kujundzic E et al (2007) Real-time PCR for detection of the Aspergillus genus. J Environ Monitor 9(6):599–609

Gonzalez JM, Saiz-Jimenez C (2004) Microbial diversity in biodeteriorated monuments as studied by denaturing gradient gel electrophoresis. J Separ Sci 27(3):174–180

Górny RL, Reponen T, Willeke K et al (2002) Fungal fragments as indoor air biocontaminants. Appl Environ Microbiol 68(7):3522–3531

Green BJ, Millecchia LL, Blachere FM et al (2006) Dual fluorescent halogen immunoassay for bioaerosols using confocal microscopy. Anal Biochem 354(1):151–153

Green BJ, Schmechel D, Summerbell RC (2011) Aerosolized fungal fragments. In: Anonymous fundamentals of mold growth in indoor environments and strategies for healthy living. Springer, p. 211–243

Green BJ, Sercombe JK, Tovey ER (2005) Fungal fragments and undocumented conidia function as new aeroallergen sources. J Allergy Clin Immunol 115(5):1043–1048

Gutarowska B, Piotrowska M (2007) Methods of mycological analysis in buildings. Build Environ 42(4):1843–1850

Hairston PP, Ho J, Quant FR (1997) Design of an instrument for real-time detection of bioaerosols using simultaneous measurement of particle aerodynamic size and intrinsic fluorescence. J Aerosol Sci 28(3):471–482

Haugland RA, Varma M, Wymer LJ et al (2004) Quantitative PCR analysis of selected *Aspergillus*, Penicillium and *Paecilomyces Species*. Syst Appl Microbiol 27(2):198–210

Healy D, Huffman J, O'Connor D et al (2014) Ambient measurements of biological aerosol particles near Killarney, Ireland: a comparison between real-time fluorescence and microscopy techniques. Atmos Chem Phys 14(15):8055–8069

Herrera ML, Vallor AC, Gelfond JA et al (2009) Strain-dependent variation in 18S ribosomal DNA Copy numbers in Aspergillus fumigatus. J Clin Microbiol 47(5):1325–1332

Heseltine E, Rosen J (2009) WHO guidelines for indoor air quality: dampness and mould. WHO Regional Office Europe.

Hill SC, Pan Y, Williamson C et al (2013) Fluorescence of bioaerosols: mathematical model including primary fluorescing and absorbing molecules in bacteria. Opt Express 21 (19):22285–22313

Ho H, Rao CY, Hsu H et al (2005) Characteristics and determinants of ambient fungal spores in Hualien, Taiwan. Atmos Environ 39(32):5839–5850

Huber JA, Morrison HG, Huse SM et al (2009) Effect of PCR amplicon size on assessments of clone library microbial diversity and community structure. Environ Microbiol 11(5): 1292–1302

Iossifova Y, Reponen T, Daines M et al (2008) Comparison of two analytical methods for detecting (1-3)-β-D-glucan in pure fungal cultures and in home dust samples. Open Allergy J 1:26–34

Iossifova YY, Reponen T, Ryan PH et al (2009) Mold exposure during infancy as a predictor of potential asthma development. Ann Allergy Asthma Immunol 102(2):131–137

Iossifova Y, Reponen T, Bernstein D et al (2007) House dust (1–3)-β-d-glucan and wheezing in infants. Allergy 62(5):504–513

Kaarakainen P, Rintala H, Vepsäläinen A et al (2009) Microbial content of house dust samples determined with qPCR. Sci Total Environ 407(16):4673–4680

Kanaani H, Hargreaves M, Ristovski Z et al (2007) Performance assessment of UVAPS: Influence of fungal spore age and air exposure. J Aerosol Sci 38(1):83–96

Kanaani H, Hargreaves M, Smith J et al (2008) Performance of UVAPS with respect to detection of airborne fungi. J Aerosol Sci 39(2):175–189

Kanchongkittiphon W, Mendell MJ, Gaffin JM et al (2015) Indoor environmental exposures and exacerbation of asthma: an update to the 2000 review by the Institute of Medicine. Environ Health Perspect 123(1):6–20

Karlsson K, Malmberg P (1989) Characterization of exposure to molds and actinomycetes in agricultural dusts by scanning electron microscopy, fluorescence microscopy and the culture method. Scand J Work Environ Health 353–359

Krause JD, Hammad YY, Ball LB (2003) Application of a fluorometric method for the detection of mold in indoor environments. Appl Occup Environ Hyg 18(7):499–503

Lee T, Grinshpun SA, Martuzevicius D et al (2006) Culturability and concentration of indoor and outdoor airborne fungi in six single-family homes. Atmos Environ 40(16):2902–2910

Liu CM, Kachur S, Dwan MG et al (2012) FungiQuant: a broad-coverage fungal quantitative real-time PCR assay. BMC Microbiol 12:255

Low SY, Hill JE, Peccia J (2009) DNA aptamers bind specifically and selectively to (1 → 3)-β-d-glucans. Biochem Biophys Res Commun 378(4):701–705

Lymperopoulou DS, Adams RI, Lindow SE (2016) Contribution of vegetation to the microbial composition of nearby outdoor air. Appl Environ Microbiol 82(13):3822–3833

Madsen A (2003) NAGase activity in airborne biomass dust and relationship between NAGase concentrations and fungal spores. Aerobiologia 19(2):97–105

Madsen AM, Schlunssen V, Olsen T et al (2009) Airborne fungal and bacterial components in PM1 dust from biofuel plants. Ann Occup Hyg 53(7):749–757

Matthias-Maser S, Jaenicke R (1991) A method to identify biological aerosol particles with radius> 0.3 μm for the determination of their size distribution. J Aerosol Sci 22:S849–S852

Matthias-Maser S, Jaenicke R (1994) Examination of atmospheric bioaerosol particles with radii> 0.2 μm. J Aerosol Sci 25(8):1605–1613

Méjean G, Kasparian J, Yu J et al (2004) Remote detection and identification of biological aerosols using a femtosecond terawatt lidar system. Appl Phys B 78(5):535–537

Meklin T, Haugland RA, Reponen T et al (2004) Quantitative PCR analysis of house dust can reveal abnormal mold conditions. J Environ Monitor 6(7):615–620

Meklin T, Reponen T, McKinstry C et al (2007) Comparison of mold concentrations quantified by MSQPCR in indoor and outdoor air sampled simultaneously. Sci Total Environ 382(1):130–134

Mendell MJ, Mirer AG, Cheung K et al (2011) Respiratory and allergic health effects of dampness, mold, and dampness-related agents: a review of the epidemiologic evidence. Environ Health Perspect 119(6):748

Mensah-Attipoe J, Reponen T, Salmela A et al (2015) Susceptibility of green and conventional building materials to microbial growth. Indoor Air 25(3):273–284

Mensah-Attipoe J, Reponen T, Veijalainen A et al (2016a) Comparison of methods for assessing temporal variation of growth of fungi on building materials. Microbiol 162(11):1895–1903

Mensah-Attipoe J, Saari S, Veijalainen A et al (2016b) Release and characteristics of fungal fragments in various conditions. Sci Total Environ 547:234–243

Metzker ML (2010) Sequencing technologies – the next generation. Nature Rev Genet 11(1):31–46

Mille-Lindblom C, von Wachenfeldt E, Tranvik LJ (2004) Ergosterol as a measure of living fungal biomass: persistence in environmental samples after fungal death. J Microbiol Methods 59(2):253–262

Miller J, Laflamme A, Sobol Y et al (1988) Fungi and fungal products in some Canadian houses. Internat Biodeterior 24(2):103–120

Moularat S, Robine E, Ramalho O et al (2008) Detection of fungal development in closed spaces through the determination of specific chemical targets. Chemosphere 72(2):224–232

Noterman S, Soentoro PS (1986) Immunological relationship of extra-cellular polysaccharide antigens produced by different mould species. Antonie Van Leeuwnhoek 52:393–401

O'Connor DJ, Iacopino D, Healy DA et al (2011) The intrinsic fluorescence spectra of selected pollen and fungal spores. Atmos Environ 45(35):6451–6458

Palmgren U (1986) Collection of airborne micro-organisms on Nuclepore filters, estimation and analysis – CAMNEA method. J Appl Bacteriol Oxford 61(5):401–406

Park J, Cox-Ganser JM (2011) Mold exposure and respiratory health in damp indoor environments. Front Biosci E 3:575–571

Pietarinen V, Rintala H, Hyvärinen A et al (2008) Quantitative PCR analysis of fungi and bacteria in building materials and comparison to culture-based analysis. J Environ Monitor 10(5):655–663

Piñar G, Sterflinger K (2009) Microbes and building materials. Building materials: properties, performance and applications. Nova Publishers, New York, p 163–188

Pitkäranta M, Meklin T, Hyvärinen A et al (2011) Molecular profiling of fungal communities in moisture damaged buildings before and after remediation – a comparison of culture-dependent and culture-independent methods. BMC Microbiol 11(1):1

Pitkaranta M, Meklin T, Hyvarinen A et al (2008) Analysis of fungal flora in indoor dust by ribosomal DNA sequence analysis, quantitative PCR, and culture. Appl Environ Microbiol 74(1):233–244

Pöhlker C, Huffman J, Pöschl U (2012) Autofluorescence of atmospheric bioaerosols – fluorescent biomolecules and potential interferences. Atmos Meas Tech 5(1):37–71

Raimondi V, Agati G, Cecchi G et al (2009) In vivo real-time recording of UV-induced changes in the autofluorescence of a melanin-containing fungus using a micro-spectrofluorimeter and a low-cost webcam. Opt Express 17(25):22735–22746

Rast DM, Baumgartner D, Mayer C et al (2003) Cell wall-associated enzymes in fungi. Phytochemistry 64(2):339–366

Rastogi R, Wu M, Dasgupta I et al (2009) Visualization of ribosomal RNA operon copy number distribution. BMC Microbiol 9:208. doi:10.1186/1471-2180-9-208

Reeslev M, Miller M, Nielsen KF (2003) Quantifying mold biomass on gypsum board: Comparison of ergosterol and beta-N-acetylhexosaminidase as mold biomass parameters. Appl Environ Microbiol 69(7):3996–3998

Reponen T, Seo S, Grimsley F et al (2007) Fungal fragments in moldy houses: a field study in homes in New Orleans and Southern Ohio. Atmos Environ 41(37):8140–8149

Reponen T, Willeke K, Grinshpun S et al (2011) Biological particle sampling. In: Aerosol measurement: principles, techniques, and applications, 3rd Edn. 549–570

Rylander R (2015) β-N-Acetylhexosaminidase (NAHA) as a marker of fungal cell biomass – storage stability and relation to β-glucan. Int J Monitor Anal 3(4):205–209

Rylander R, Reeslev M, Hulander T (2010) Airborne enzyme measurements to detect indoor mould exposure. J Environ Monitor 12(11):2161

Saari S, Reponen T, Keskinen J (2014) Performance of two fluorescence-based real-time bioaerosol detectors: BioScout vs. UVAPS. Aero Sci Technol 48(4):371–378

Saari S, Putkiranta M, Keskinen J (2013) Fluorescence spectroscopy of atmospherically relevant bacterial and fungal spores and potential interferences. Atmos Environ 71:202–209

Saraf A, Larsson L, Burge H et al (1997) Quantification of ergosterol and 3-hydroxy fatty acids in settled house dust by gas chromatography-mass spectrometry: comparison with fungal culture and determination of endotoxin by a Limulus amebocyte lysate assay. Appl Environ Microbiol 63(7):2554–2559

Sattler B, Puxbaum H, Psenner R (2001) Bacterial growth in supercooled cloud droplets. Geophys Res Lett 28(2):239–242

Schaub B, Lauener R, von Mutius E (2006) The many faces of the hygiene hypothesis. J Allergy Clin Immunol 117(5):969–977

Sivaprakasam V, Huston A, Scotto C et al (2004) Multiple UV wavelength excitation and fluorescence of bioaerosols. Opt Express 12(19):4457–4466

Sonesson A, Larsson L, Fox A et al (1988) Determination of environmental levels of peptidoglycan and lipopolysaccharide using gas chromatography with negative-ion chemical-ionization mass spectrometry utilizing bacterial amino acids and hydroxy fatty acids as biomarkers. J Chromatog B: Biomed Sci Appl 431:1–15

Szponar B, Szponar A, Larsson L (2003) Direct assessment of microbial colonisation in damp houses by chemical marker analysis. Indoor Built Environ 12(4):251–254

Thorne PS, Lange JL, Bloebaum P et al (1994) Bioaerosol sampling in field studies: can samples be express mailed? Am Ind Hyg Assoc 55(11):1072–1079

Tischer C, Chen CM, Heinrich J (2011) Association between domestic mould and mould components, and asthma and allergy in children: a systematic review. Eur Respir J 38(4):812–824

Toivola M, Alm S, Reponen T et al (2002) Personal exposures and microenvironmental concentrations of particles and bioaerosols. J Environ Monitor 4(1):166–174

Tringe SG, Zhang T, Liu X et al (2008) The airborne metagenome in an indoor urban environment. PLoS One 3(4):e1862

Vesper S, Wymer L, Meklin T et al (2005) Comparison of populations of mould species in homes in the UK and USA using mould-specific quantitative PCR. Lett Appl Microbiol 41(4):367–373

Vesper S (2007) Development of an Environmental Relative Moldiness Index for US Homes. J Occup Environ Med 49(8):829

Viegas C, Viegas S, Monteiro A et al (2012) Comparison of indoor and outdoor fungi and particles in poultry units. WIT Transactions on Ecology and the Environment 162

Volckens J, Peters TM (2005) Counting and particle transmission efficiency of the aerodynamic particle sizer. J Aerosol Sci 36(12):1400–1408

von Wintzingerode F, Gobel UB, Stackebrandt E (1997) Determination of microbial diversity in environmental samples: pitfalls of PCR-based rRNA analysis. FEMS Microbiol Rev 21(3):213–229

Wittmaack K, Wehnes H, Heinzmann U et al (2005) An overview on bioaerosols viewed by scanning electron microscopy. Sci Total Environ 346(1):244–255

Wu P, Su HJ, Ho H (2000) A comparison of sampling media for environmental viable fungi collected in a hospital environment. Environ Res 82(3):253–257

Yamamoto N, Dannemiller KC, Bibby K et al (2014) Identification accuracy and diversity reproducibility associated with internal transcribed spacer-based fungal taxonomic library preparation. Environ Microbiol 16(9):2764–2776

Yamamoto N, Kimura M, Matsuki H et al (2010) Optimization of a real-time PCR assay to quantitate airborne fungi collected on a gelatin filter. J Biosci Bioeng 109(1):83–88

Young S, Castranova V (2005) Toxicology of 1-3-beta-glucans: glucans as a marker for fungal exposure. CRC Press

Zeng Q, Westermark S, Rasmuson-Lestander Å et al (2006) Detection and quantification of Cladosporium in aerosols by real-time PCR. J Environ Monitor 8(1):153–160

Chapter 7
Viruses Present Indoors and Analyses Approaches

Edna Ribeiro, Céu Leitão, Elisabete Cristovam and Ana Dias

Abstract Through human history viruses have shown enormous epidemiological and pandemic potential as the occurrence and spread of viruses in pandemic dimensions poses a threat to the health and lives of seven billion people worldwide. Scientific evidence has associated harmful health effects to indoor air hazards recognizing the existence of a vital concern in public health sector. Thus the assessment of human exposure to biological aerosols and droplets indoor became an imperative requirement of investigation. Environmental bioburden assessment of viruses relies in both culture-dependent approaches that comprise classical methodologies, still prominent and vital in the field of modern biotechnology, and culture-independent approaches based on nucleic acid amplification techniques, which are considered the gold standard in clinical virology. The main factor influencing indoor microbiology is the human being and their activities. Indoor environments to be considered are those regularly occupied by humans: residences, offices, schools, industrial buildings, health care facilities, farming activities and other settings occupied all the time, or in which occupant density is high. It's well known that approximately 60% of total human respiratory and gastrointestinal infections are acquired indoor, since viruses have a rapid spread in the community and can be transmitted easily, especially in crowded and poorly ventilated environments, causing high morbidity and decline in quality of life and productivity. Studies have shown that respiratory syncytial virus, rhinovirus, metapneumovirus, influenza and parainfluenza virus, and

E. Ribeiro (✉)
Research Center LEAF – Linking Landscape, Environment, Agriculture and Food – Instituto Superior de Agronomia, Lisbon University, Lisbon, Portugal

Research Group Environment & Health, Lisbon School of Health Technology/Polytechnic Institute of Lisbon, Lisbon, Portugal
e-mail: edna.ribeiro@estesl.ipl.pt

C. Leitão
Lisbon School of Health Technology/Polytechnic Institute of Lisbon Lisbon Portugal, Lisbon, Portugal

E. Cristovam · A. Dias
Centro Hospitalar Lisboa Ocidental, Hospital Egas Moniz, Microbiology and Molecular Biology Laboratory, Rua da Junqueira, Lisbon, Portugal

C. Viegas et al. (eds.), *Exposure to Microbiological Agents in Indoor and Occupational Environments*, DOI 10.1007/978-3-319-61688-9_7

human enterovirus infections may be associated with virus-induced asthma, leading to diseases such as pneumonia. Gastroenteritis infectious (about 30±40% of cases) is attributable to viruses. Rotavirus, Astrovirus, Norwalk-like viruses and other caliciviruses are responsible for 48% of all reported outbreaks of infectious intestinal disease. Safe working conditions are essential for healthy living, that's why the programmes conceived as a result of strategic and preventive policy maintenance, in refrigeration and ventilation systems, are the determining factor for the control of biological pollutants. Moreover, the development of highly sensitive and specific detection and identification methodologies with capacity to be used in diverse applications, such as diagnosis, public health risk assessment, research and for the implementation of preventive measures and protocols are imperative.

Keywords Viruses indoor · air transmission · culture-dependent virus analysis · culture-independent virus analysis · environmental assessment

7.1 Environmental Exposure

Indoor air pollution is a major global public health threat requiring increased hard work, linking research to policy-making. The evidence of the effects of physical and chemical pollutants on human health, present in the external and internal environment has no superior aggression than the existing bioaerosols. Health effects from indoor air pollutants may be experienced soon after exposure or, possibly, years later, according to the nature of air contaminants, being classified, in an abbreviated approach, as physical, chemical or biological (Nazaroff, 2016).

Several studies have related hostile health effects to Indoor air hazards and adequate assessment of human exposure to biological aerosols has been recognized as an imperative requirement and a very important concern in the area of public health (Douwes et al., 2003). Scientific evidence has showed that the air within buildings can be more seriously polluted than the outdoor air, even in developed and industrialized countries, and it is well known that indoor environments occupied by humans, contain abundant material of microbial origin, consequently, the risks to human health may be greater due to exposure to air pollution indoors (Fisk et al., 2007).

On the other hand, investigation indicates that people spend approximately 90% of their time indoors, with mechanical heating, cooling and ventilation systems, influencing irrefutably the quality of life (Klepeis et al., 2001; Nazaroff, 2016). Complex syndromes arise associated with indoor air quality, such as sick building syndrome, building-related disease leading to loss of productivity and absence at work. These syndromes are associated with increased incidence and prevalence of asthma and other chronic diseases worldwide. Exposure to bioaerosol material can cause or can contribute to several relevant diseases.

Human occupancy and activities are major factors influencing indoor microbiology. Humans are important primary sources of certain bacteria and viruses. The recent development of quantitative polymerase chain reaction (PCR) and other RNA/DNA-based measurement technologies has permitted studies that measure pathogenic material in indoor air (Lax et al., 2014; Nazaroff, 2016).

7.1.1 *Droplets vs Airborne*

Droplet transmission is not to be confused with airborne transmission. Aerosols are suspensions in air (or in a gas) of solid or liquid particles, small enough to remain airborne for a prolonged period of time because of their low settling velocity. Droplets do not remain suspended in the air. Airborne transmission depends on viruses from evaporated droplets or dust particles that can remain suspended in the air for long periods. Droplet transmission occurs when viruses travel on relatively large respiratory droplets (>10 μm) that people sneeze, cough, or exhale during conversation or breathing (primary aerosolization) (La Rosa et al., 2013). Respiratory droplets initially all move forward with the exhaled air jet; very large droplets leave the jet quickly and fall on the ground and small droplets completely desiccate within the jet (Tellier, 2009).

The transport and the settling of a bioaerosol are affected by its physical properties and the environmental parameters that it encounters. Size, density and shape of droplets or particles, air currents, relative humidity and temperature, determine the capacity of generation of airborne bioaerosols from liquid suspensions, undergo desiccation, whereas those generated as dusts or powders partially rehydrate. The presence of moulds indicates a problem with water penetration or high humidity. Bioaerosols can be transmitted at long distances. Small particle aerosols, as shown during endotracheal intubation, are transmitted to persons in the immediate area near the patient. Viruses' inductors of Severe Acute Respiratory Syndrome (SARS), influenza and norovirus are transmitted from patients primarily by contact and/or droplet routes, while airborne transmission occurs over a limited distance (Srikanth et al., 2008). Indoor–outdoor air exchange (mechanical ventilation), penetration (air filter), deposition, sources and aerosol resuspension, are extremely relevant for spread contamination.

Once aerosolized, aerosols viruses' particles may travel significant distances through buildings before being captured and retained by HVAC filters, or they pass through as well, because most HVAC filters are not 100% efficient in capturing particles. The rapidity with which airborne viruses are inactivated during transport or after filter capture is uncertain and merits additional study. Analysis of ventilation filters certainly could play a role in the epidemiology of infectious diseases caused by pathogens released into the environment, suiting all the different situations of confinement (Goyal et al., 2011).

7.2 Viruses Indoor

The main source of indoor viruses is the human being. Viruses are spread by air currents after resuspension of material scattered by aerosols droplets or saliva. We can say that viral infections are probably the most common acquired diseases indoor that affect man, knowing approximately a thousand types of different viruses involved. It is estimated about 60% of total human respiratory and gastrointestinal infections, with a rapid spread in the community, being a cause of high morbidity and decline in quality of life and productivity, since viruses can be easily transmitted, especially in crowded and poorly ventilated environments. It is well-known that viruses are shed in large numbers, with transmission routes extraordinary diverse, including direct contact with infected persons, faecal-oral transmission (through contaminated food and water), droplet and airborne transmission, and can survive for long periods on surfaces or fomites, emphasizing the possible role of surfaces in the transmission of viruses (Barker et al., 2001; La Rosa et al., 2013).

For instance, faeces can contain up to 10^{12} viruses particles per gram and vomit up to 10^{7} per millilitre, so the potential transfer contamination from hands to surfaces is frighteningly considerable. The most important source of potentially pathogenic viral aerosol is other humans and other means, as the flushing of a toilet that can aerosolize significant concentrations of airborne viruses. Viruses' survival on fomites is influenced by temperature, humidity, pH and exposure to ultraviolet light.

Particle size, depth of penetration and the minimum dosage of the agent capable of causing disease are implicated in the infectivity. In addition, it is also important to be conscious of the risk groups, the most susceptible to contracting infection when exposed to microorganisms, conditioned by factors such as a weakened immune system, the children, the elderly, the pregnant women, the chronically ill, especially those suffering from respiratory or cardiovascular disease. Chronic obstructive pulmonary disease (COPD) and acute exacerbations are frequent complications, thought to be caused by interactions between host factors, bacteria, viruses and changes in air quality producing increased inflammation in the lower airway with a long-lasting adverse influence on health status (Celli and Macnee, 2004; Celli and Barnes, 2007). Approximately of 50% of acute exacerbations of COPD are associated with symptoms of viral infections of the respiratory tract by rhinovirus, respiratory syncytial virus and influenza. Studies have shown that respiratory syncytial virus (RSV), human rhinovirus (HRV), human metapneumovirus (HMPV), influenza and parainfluenza virus (HPIV), and human enterovirus infections may be associated with virus-induced asthma, leading to diseases such as pneumonia or death (Tsukagoshi et al., 2013).

Respiratory and enteric viruses are opportunistic pathogens transmitted mainly via other routes are able to spread via droplet nuclei or dust in certain circumstances (La Rosa et al., 2013). Numerous studies identify the factors that are involved in the transmission of infection by aerosol indoor with the correlation between the pulmonary mechanism, the different human activities and the critical concentration of particles expelled. A virus (H3N2) had a higher transmissibility and uncontrollable potential than the A (H1N1) and B viruses (Chen et al., 2009; Chen and Liao, 2010).

7.2.1 Respiratory Viruses

In developing countries, acute respiratory infections, due to various viruses, is not only the most common cause of upper respiratory infection in humans but is also a major cause of morbidity and mortality, leading to diseases such as bronchiolitis, asthma and pneumonia (Tsukagoshi et al., 2013; Paba et al., 2014).

Despite all the progress of the last decades in the prevention and care of health, respiratory infections represent one of the major causes of disease in humans. This group remains the leading cause of outpatient consultation as well as antibiotic prescription and work absence. The chronic infections, causing disability, have the most important impact on quality of life. The main effect of the inadequate quality of indoor air is in respiratory system. Most studies focus on the influenza virus, but in general, we can say that a total concentration of viruses, in a variety of environments such as classrooms, health institutions, restaurants, offices and others corresponds to an average $4.7 \pm 2.5 \times 10^5$ particles per cubic meter, without significant differences between the different indoor environments. Also, is estimated that the viral particles inhaled daily indoor correspond approximately to 5×10^6. Respiratory viruses can be transported over considerable distances by air currents and be inhaled, penetrating deep into the respiratory system (Prussin II et al., 2015).

Droplet transmission occurs when viruses travel on relatively large respiratory droplets (>10 μm) that people sneeze, cough, or exhale during conversation or breathing, primary aerosolization. A single cough can release hundreds of droplets, up to 40,000, at speeds of up to 50–200 miles per hour, each droplet containing millions of viral particles. Aerosol droplets travel only short distances (1–2 m) before settlings on surfaces, where viruses can remain infectious for hours or days. Secondary aerosolization can occur when air displacements disperse the viruses back into the air from contaminated surfaces (La Rosa et al., 2013). Transmission occurs through air droplets, aerosol and fomites that may come into contact with nasal and conjunctival epithelium. The etiological viral agents involved include: influenza virus types A and B, parainfluenza viruses types 1, 2, 3, and 4, respiratory syncytial virus, adenovirus, and rhinoviruses/enteroviruses (Paba et al., 2014). The acute respiratory infections are the result of active multiplication of microbiological agents in the respiratory system when favourable conditions of the host exist. Viral or bacterial ethology, predisposing factors related to anatomical factors, immune changes, colonization of the naso and oropharynx, and the spread of these infections is favoured by the continuity of the epithelia of the respiratory system and a continuum between upper and lower airways. The upper and lower respiratory infections are in most cases (60%) of viral origin.

Adenovirus (type 4), the first virus to be isolated from indoor aerosol, was identified in 1966 in aerosol samples from the quarters of military recruits infected with Acute Respiratory Disease (Artenstein and Miller, 1966). Since then, human infections due to viral aerosol (or contact with contaminated surfaces) have been studied in various environments, including office building, hospitals, restaurants, transport systems and schools. In the last few years, other respiratory viruses have been discovered and linked to the upper and lower respiratory tract infections: human metapneumovirus, SARS coronavirus, HKU1 coronavirus, NL63 coronavirus, MERS coronavirus and bocavirus.

In 2007, two novel human polyomaviruses named KIPyV and WUPyV were discovered in the respiratory secretions of patients with acute respiratory symptoms.

The outbreak of the influenza A virus (H1N1) infection in 2009 has reminded us again of the importance of monitoring and controlling airborne microorganisms in public facilities (Lee et al., 2012). Unlike conventional viral cell cultures, with the introduction of the real-time PCR assay, the diagnosis of respiratory infections improved greatly. It is possible to search for up to 21 different respiratory pathogens including viruses and bacteria. In addition, it is possible to detect co-infections that may have implications on disease severity or therapeutic strategies (Paba et al., 2014).

For bioaerosol particles, maybe the most important exposure pathway is inhalation followed by deposition in the respiratory tract. The probability of deposition varies with particle size, with lung morphology, and with breathing characteristics. Airborne exposure indoors makes a meaningful contribution to the occurrence or spread of disease.

7.2.2 *Respiratory Viruses Associated with Indoor Air Transmission*

7.2.2.1 Rhinovirus

Rhinovirus (RV) is a small RNA virus belonging to the Picornaviridae family. More than 100 immunologically distinct serotypes have been identified and new serotypes are continuously emerging. This virus is responsible for more than 50% of cases of common cold, being the virus with the highest morbidity among respiratory diseases. In children causes bronchitis. Although the method of transmission of RVs is in doubt, they are thought to be mainly transmitted via large droplets, but indirect contact with contaminated fomites has also shown to transmit infection or by aerosols. It's known that RV need only 10 s of contact (hand-to-hand) for infecting another person. The recurrence of viral infections is predominantly associated with an immune response to a serotype that does not provide immunity to reinfection with another serotype for the same virus (Ballow, 2008).

Rhinoviruses can survive on environmental surfaces for several hours. Infectious viruses have been recovered from naturally contaminated objects in the surroundings of persons with RV colds. Aerosols are generated by coughing, talking, sneezing and even simply breathing (Huynh et al., 2008). Rhinovirus outbreaks in health care facilities, capable of determining severe infections and also death have been documented (MacIntyre et al., 2012). RVs have also been detected in transport vehicles (La Rosa et al., 2013). Even in the common cold, rhinoviruses and coronaviruses predominate (Barker et al., 2001; Heikkinen et al., 2003), but the role of rhinoviruses is the most prominent, in particular are responsible for outbreaks of the common cold in the general community such as schools, day care centres and hospitals (Barker et al., 2001; Heikkinen et al., 2003; La Rosa et al., 2013). Rhinoviruses and coronaviruses also cause a greater disease burden in elderly people living at home, compared with influenza or syncytial viruses.

The infections by rhinovirus may be triggering factors of exacerbations of asthma, probably because they induce inflammation in airways that are already damaged and sensitized. Rhinoviruses were detected by RT-PCR among infants (von Mutius, 2004), in the acute exacerbation of asthma when infection was prolonged (Tsukagoshi et al., 2013). Some authors describe that bronchial epithelial patient cells with asthma are more effectively infected with RV than normal epithelia. It was shown that viral are upregulated in the epithelial cells of patients with allergies (Canonica et al., 1995; Tantilipikorn and Auewarakul, 2011).

7.2.2.2 Respiratory Syncytial Virus

Respiratory syncytial virus (RSV) is a single-stranded RNA virus belonging to Paramixoviridae family. RSV infections occur all over the world and outbreaks are common in the cold season in temperate climates and in the rainy season in tropical climates. Is one of the most common viruses which can cause a type of potentially lethal pneumonia in older people and is a major cause of respiratory illness in young children producing bronchitis and pneumonia worldwide, affecting about 90% of children by the age of 2 years (24), inducting a preparative stage in the development of asthma (Tsukagoshi et al., 2013).

School-aged children often carry RSV to their homes and extent infection to younger brothers as well as when admitted to hospital, tend to shed the virus abundantly and for prolonged periods permitting sufficient opportunity for spread in adults. Sometimes this virus produces a Flu-like syndrome indistinguishable from influenza (Barker et al., 2001).

RSV is highly contagious and transmission can occur when infectious material comes into contact with mucous membranes of the eyes, mouth or nose, and possibly through the inhalation of droplets generated by a sneeze or cough. Also result from contact with contaminated environmental surfaces, the commonest mode of transmission in school classrooms and day care centres. Transmission with fomites predominates over droplet contact can be transmitted by the airborne route. RSV RNA is detected in air samples from the hospital rooms of infected patients at large distances from the patient's bedside. Particles containing RSV RNA were detected in airborne throughout a health care facility, particles small enough to remain in the air for an extended period and to be inhaled deeply into the respiratory tract (6). Evidence shows that direct and indirect contact is a key factor in transmission hands, touching surfaces contaminated with fresh secretions from RSV-infected infants (Barker et al., 2001).

7.2.2.3 Influenza Virus

Influenza virus (IV) is one of the most common and highly contagious infectious diseases and can occur in people of any age. Influenza or flu produces an acute infection of the respiratory system with high transmissibility and global distribution. Is a RNA virus belonging to the Orthomixoviridae family which is subdivided into three distinct antigenic serotypes: A, B and C, causing moderate to severe acute febrile illness, resulting in variable degrees of systemic symptoms, ranging from mild fatigue to respiratory failure and death. Strains A and B have greatest potential epidemic, causing more versions of the flu. The flu vaccine only protects against viruses A and B. Influenza C viruses are antigenically stable, cause subclinical disease and do not cause epidemics. The influenza A and B have multiple subtypes that still suffer mutations, emerging new strains with consequent increased risk of epidemics or pandemics.

The continuing threat of pandemics by the human influenza virus suggests an urgent and constant surveillance. Despite the vaccination continues to be considered the intervention of containment of infection, the viruses A (H1N1), A (H3N2) and influenza B persist with a global distribution and with such a power of infectivity that Influenza activity management has been undertaken by the Centres for Diseases Control. Influenza A (H3N2) has a uncontrolled and potential infectivity demonstrating a much greater transmissibility than A (H1N1) and B viruses (Chen and Liao, 2010). Influenza affects all age groups, but it is the elderly and persons with underlying health problems who are at particular risk from complications of influenza and are more likely to require hospitalization. Both influenza virus A and B revealed to survive on hard surfaces such as stainless steel and plastic for 24 ± 48 h and on absorbent surfaces such as cloth, paper and tissues for up to 12 h (Barker et al., 2001). Asymptomatic patients shed virus and can transmit the disease, thus creating a reservoir for the virus. Influenza virus is transmitted by droplets, through the coughing and sneezing of infected persons, but it can also be transmitted by airborne droplet nuclei as well as by contact, either through direct skin-to-skin contact or through indirect contact with contaminated environments (La Rosa et al., 2013).

Influenza viruses have been detected in different indoor environments, homes, schools, offices and others public buildings. Places such as hospitals, where the presence of a susceptible population is often combined with a high population density, may harbour high concentrations of pathogens and therefore pose a considerable risk for the transmission of the virus, with potentially fatal consequences for hospitalized patients. Schools are known to have an important role in influenza transmission in the community since children have a higher influenza attack rate than adults (children get the flu twice as often as adults) (Zhao et al., 2007).

On buildings ventilation systems Influenza A and B were detected (along with other groups of viruses), meaning that contamination exists in the surrounding environment. Several reviews consider three modes of transmission of influenza, not mutually exclusive, by large droplets, self-inoculation of the nasal mucosa by contaminated hands and the aerosol transmission, the mode of the greatest impact for infection, since it requires specialized personal protective equipment. Influenza virus RNA was directly detected in aerosol particles generated by normal breathing in patients with influenza and collected through an orinasal facemask; particles of 5 mm or less have a significant penetration into the respiratory tract all the way to the alveolar region. Increasing evidences point towards a role for aerosol transmission in the spread of influenza, at least over short distance where exposure to both aerosol and large droplets occurs. In most settings where there is adequate ventilation, long-range transmission does not appear to occur so frequently (Tellier, 2009).

The relative importance of the aerosol transmission route for influenza remains controversial. To determine the potential for influenza to spread via the aerosol route, the authors 6x measured the size distribution of airborne influenza A virus. Collected size-segregated aerosol samples during the 2009–2010 flu season in a health centre, a day-care facility and on board of aeroplanes. Filter extracts were analysed using quantitative reverse transcriptase polymerase chain reaction. Half of the 16 samples were positive, and their total virus concentrations ranged from 5800 to 37,000 genome copies m23. On average, 64% of the viral genome copies were associated with fine particles smaller than 2.5 mm, which can remain suspended for hours. Modelling of virus concentrations indoors suggested source strength of $1.6 + 1.2 \times 10^5$ genome copies m23 air h21 and a deposition flux onto surfaces of 13 + 7 genome copies m22 h21 by Brownian motion. Over 1 h, the inhalation dose was estimated to be 30 +18 median tissue culture infectious dose (TCID50), adequate to induce infection. These results provide quantitative support for the idea that the aerosol route can be an important mode of influenza transmission (Yang et al., 2011). Parainfluenza viruses (PIVs) are a further major group of respiratory pathogens. They cause severe colds, croup, bronchitis and pneumonia in children and adults and in infants the virus can cause life-threatening disease. Infection is probably spread by aerosols in addition to direct contact with contaminated surfaces. The persistence of PIV on hospital surfaces contaminated with patients' secretions was noted as a potential source of transmission (Barker et al., 2001).

Goals for epidemiological surveillance of Influenza include supervise the currently circulating virus subtypes and offer quick response to spread of new subtypes; follow the tendency of the morbidity and mortality to plan strategies to reduce the burden of the disease in public health; define strategies to reduce the occurrence of deaths; monitor the severity standard of the disease by detecting any virulence changes.

7.2.2.4 Severe Acute Respiratory Syndrome

Severe acute respiratory syndrome (SARS) is a respiratory illness caused by a type of coronavirus from the Coronaviridae family, which can cause mild to moderate upper respiratory illness, such as the common cold or develop into potentially severe pneumonia. This virus is known as SARS-CoV. The SARS epidemic broke out in 2002–2003 in southern China, and spread to other regions of Asia and also to Europe and North America. It caused more than 8000 infections worldwide with an approximately 10% fatality rate, along with enormous economic losses. One or a few trivial mutations at the receptor-binding surface of a virus may lead to dramatic epidemic outcomes by facilitating cross-species infections and human-to-human transmission of the virus (Li, 2013). SARS is a condition associated with substantial morbidity and mortality with patterns suggesting droplet or contact transmission (Poutanen et al., 2003).

The earliest symptom is a sudden onset of high fever. Some patients may also have chills and headaches followed by pneumonia; others showed respiratory distress (severe breathing difficulty) and sometimes death (La Rosa et al., 2013).

The most common mode of transmission is contamination by warm air indoor, by water droplets generated by coughs or sneezes, but may be transmitted through the airborne route as well. Transmission in an aircraft from an infected person to passengers located seven rows of seats ahead had been described. Aerosol generated by the building's sewage systems is also responsible. Many health care workers were infected after endotracheal intubation and bronchoscopy procedures which often involve aerosolization. These observations indicate the possible role of more remote modes of transmission, including airborne spread by small droplet nuclei, and emphasize the need for adequate respiratory protection in addition to strict contact and droplet precautions when managing SARS patients. Contaminated fomites or hospital surfaces might contribute to spread. It is known that these viruses may live on hands, tissues and other surfaces for up to 6 h and up to 3 h after droplets have dried. Airborne spread of the virus appears to explain the happened large community outbreaks of SARS (Ignatius et al., 2004; Hui and Chan, 2010; La Rosa et al., 2013).

A novel coronavirus, MERS-CoV (NCoV, HCoV-EMC/2012), originating from the Middle-East, has been discovered. Incoming data reveal that the virus is highly virulent to humans. The members of this group (C) are likely to persist in the environment for a longer period of time and possess the highest oral-fecal components but relatively low respiratory transmission components. Oral-urine and saliva transmission are also highly possible (Goh et al., 2013).

7.2.2.5 Adenovirus

Human adenovirus (AdV) is a non-enveloped, icosahedral virus of the genus Mastadenovirus, family Adenoviridae. There are more than 60 types classified into seven strains from A to G, defined via biological and molecular characteristics. Clinical manifestations are highly heterogeneous, ranging from upper and lower respiratory tract infections to gastroenteritis, pneumonia, urinary tract infection, conjunctivitis, hepatitis, myocarditis and encephalitis. The burden of disease manifests as pneumonia, bronchiolitis, otitis media, conjunctivitis, and tonsillitis. The adenoviral detection rates indicate the potential contamination of the environment, with adverse effects on public health. Adenoviruses can cause severe or life-threatening illness, particularly in immunocompromised patients, children and the elderly. Some types are capable of establishing persistent asymptomatic infections in tonsils, adenoids, and intestines of infected hosts, and shedding can occur for months or years (Lessa et al., 2009; Osuolale and Okoh, 2015). Adenoviruses can occur anytime throughout the year but adenoviral respiratory infections are most common in the late winter, spring, and early summer.

Modes of transmission are also diverse, primarily spread by the respiratory route, person-to-person contact, fomites, and occasionally by airborne aerosols.

Since AdVs are able to infect a wide range of tissues, they can be excreted in large numbers in different body fluids during the acute illness, including oral secretions and faeces. The spread by the fecal-oral route happens through the ingestion of contaminated food or water. Small doses of AdV in aerosols resulted in infection accompanied by febrile acute respiratory disease, sometimes with pneumonia. Humidity affects the viability and dispersal of AdVs in aerosol. These viruses tend to survive best at high relative humidities. Aerosols seem to be very resistant to UV air disinfection. Adenovirus outbreaks have been documented in different indoor environments, including health care facilities, schools, military hospitals and barracks, throughout the year (La Rosa et al., 2013).

7.2.3 Gastrointestinal Viruses Associated with Indoor Air Transmission

In developed countries, it is estimated that 30% ± 40% of infectious gastroenteritis cases are attributable to viruses. Rotavirus, Astrovirus, Norwalk-like viruses, also known as small round structured viruses and other caliciviruses are responsible for 48% of all reported outbreaks of infectious intestinal disease.

7.2.3.1 Rotavirus

Rotaviruses belong to the Reoviridae family; they are segmented bicatenary RNA viruses, which explain their genetic variability. Is the most common cause of severe diarrhea among children, resulting in the death of over 500,000 children annually worldwide. Rotaviral gastroenteritis is a serious public health problem in both developed and developing countries.

The disease occurs most often in the winter, with annual epidemics occurring from December to June. The highest rates of illness occur among infants and young children. A large proportion of hospital admissions due to gastroenteritis in children less than 5 years old are caused by rotavirus. Some adults acquired rotavirus infections a few days after their children's illnesses, suggesting that the children rather than the parents brought infection into the home, though disease tends to be mild. Immunity after infection is incomplete, but repeated infections tend to be less severe than the original infection.

Rotavirus is shed in large numbers from an infected person by animate and non-porous inanimate surfaces. Excretion of the virus can persist for up to 57 days after diarrhea has stopped in symptomatic patients, contributing to an increase of the number of environmental surfaces contaminated with rotavirus. It has been suggested that low humidity and people spending more time indoors contribute to the spread of rotavirus infections. Handwashing is a very important means of preventing the spread of rotavirus (Barker et al., 2001; Anderson et al., 2004; Bernstein, 2009).

7.2.3.2 Norwalk-Like Virus

Norwalk-like viruses (NLVs), RNA virus belonging to Caliciviridae family, also known as small round structured viruses or caliciviruses, are an important cause of gastroenteritis outbreaks and are spread frequently through contaminated food or water. Projectile vomiting associated with NLVs is probably a major source of cross-infection because it is estimated that 3×107 particles are distributed as an aerosol into the environment during a vomiting attack. Carpets can also serve as reservoirs of infection.

Aerosols produced by vomiting can be inhaled or can contaminate hands or work surfaces, with the potential for subsequent transfer to foods or direct hand-to-mouth transfer. The importance of airborne transmission was demonstrated in a recent outbreak in a restaurant where no food source was detected but an analysis of the attack rate showed an inverse correlation with the distance from a person who had vomited. It was found that the risk of gastroenteritis amongst workers and customers who shared toilet facilities was twice of those who had a private bathroom (Green et al., 2000; Barker et al., 2001).

7.2.3.3 Enterovirus

Enteroviruses (EVs) are members of the Picornaviridae family, a large and diverse group of small RNA viruses present worldwide. In humans, EVs target a variety of different organs causing gastrointestinal, respiratory and myocardial and central nervous system diseases. In temperate climates, enteroviral infection occurs primarily in the summer and early fall. Although the majority of infections are asymptomatic or result in a self-limited illness, fatalities do occur, especially in neonates or individuals with B-cell immunodeficiency. Enterovirus outbreaks in neonatal units and school nurseries, reflects the susceptibility of infants to EVs infection, leading to extensive discussion on control measures and interventions.

Faecal-oral transmission is the major mode of transmission. Other important routes of transmission are person-to-person contact and the inhalation of airborne viruses in respiratory droplets. Infectious coxsackievirus, a member of the EV genus, in large droplets and droplet nuclei generated by coughs and sneezes as well as in the air of rooms contaminated by such discharges, transmit this viral infection by the airborne route. Aerosol transmission is suspected of having contributed significantly to the EV epidemic which infected up to 300,000 children and caused 78 deaths in Taiwan in 1998 (Chang et al., 2004; La Rosa et al., 2013).

7.2.3.4 Norovirus

Noroviruses (NoVs), emerging as the leading cause of epidemic gastroenteritis, are RNA viruses belonging to the family Caliciviridae, currently subdivided into five genogroups. NoVs are responsible for nearly half of all gastroenteritis cases and for more than 90% of non-bacterial infection epidemics worldwide. The illness can be severe and sometimes fatal, especially among vulnerable populations – young children, the elderly and the immunocompromised – and is a common cause of hospitalization. Encephalopathy, disseminated intravascular coagulation, convulsions, necrotizing enterocolitis, post-infectious irritable bowel syndrome, and infantile seizures highly contagious with a low infectious dose, occurs repeatedly.

Faecal-oral spread is the primary transmission mode and the foodborne and waterborne transmission. Airborne transmission of NoV is also a cause of acute viral gastroenteritis. Sources of contaminated aerosol are diverse. Droplets being inhaled can be deposited in the upper respiratory tract, and subsequently be swallowed along with respiratory mucus. Aerosol droplets produced during vomiting could settle onto indoor surfaces that might then be transferred to hands of exposed individuals through physical contact, or deposited on the floor from which they can be resuspended by human movement and turbulence. Aerosol droplets can also be generated from toilet flushing. Transmission via fomites is documented. The viruses were identified in indoor environments such as hospitals, schools, kindergartens, restaurants, care facilities, hotels and concert halls as well as airplanes, buses and cruise ships (Morillo and Timenetsky Mdo, 2011; La Rosa et al., 2013).

7.2.3.5 Enteric Adenovirues

Human adenoviruses (AdV) are classified into 47 serotypes and six subgenera (A–F) with different tropisms are associated with outbreaks of gastroenteritis in schools, paediatric hospital and nursing homes. They may be second to rotavirus as a cause of gastroenteritis in young children, especially newborns, mainly caused by serotypes AdV 40 and AdV 41 of subgenus F. The clinical characteristics include watery diarrhea accompanied by vomiting, low grade fever and mild dehydration. Institutionalized persons, immunocompromised persons and Transplant recipients seem to be among the most severely affected, with mortality rates as high as 60%. Respiratory symptoms are infrequent but some studies have suggested that adenovirus infections may be involved with chronic airway obstruction, pulmonary dysplasia, myocarditis and dilated cardiomyopathy, mononucleosis-like syndromes, sudden infant perinatal death and, perhaps most intriguingly, the development of obesity. Although these associations may or may not be causal, understanding adenovirus transmission seems to be the key to their further study. Adenoviruses are the most UV-resistant viruses, and their detection is now a key indicator of water quality (Uhnoo et al., 1990; Zlateva et al., 2005; Gray, 2006).

7.3 Healthcare Facilities Infections

Indoor air quality proves to be of great importance in hospitals due to the spread of air microorganisms maximizing nosocomial infections. Reports about infections correlated with the presence of viral aerosols in indoor air remain scarce. During and after illness, viruses are shed in large numbers in body secretions, including blood, faeces, urine, saliva, and nasal fluid (La Rosa et al., 2013). The microbial load in hospital indoor air is highline nuanced by the number of occupants, their activity and the ventilation. Occupants are a potential source of microorganisms as they shed the microorganisms from the skin squamous and the respiratory tract. Ventilation causes dilution thus reducing the microbial load. Sinks, wash-basins and drains, nebulisers, humidiphiers, and cooling towers are the potential sources which colonize on the moist surfaces. Sweeping of floors, changing of bed linens and entry into the hospital buildings through ventilation ducts also can be the sources of airborne microorganisms. Since exposure levels are high, this may be an issue for immunocompromised patients (Srikanth et al., 2008).

During the 2009–2010 flu seasons, size-resolved particle samples were collected on filters in a day care center and a health center. Influenza A virus was identified in 50% of the samples with concentrations ranging from 5800 to 37,000 genomes per cubic meter and a substantial proportion of the detected viruses was associated with fine particles (<2.5 μm) that can remain airborne for extended periods and that can also penetrate and deposit deeply in the respiratory tract when inhaled (Barker et al., 2001; Nazaroff, 2016).

Establishing how viruses are transmitted under different circumstances, and whether transmission requires close contact, is of great importance as such information will affect the choice of infection control measures in health-care settings (La Rosa et al., 2013).

In a hospital paediatric unit when there was an increase in the number of children suffering from rotavirus gastroenteritis, on the surfaces in direct contact with children (thermometers, play mats and toys) rotavirus was detected in 63% of samples compared with 36% for surfaces without direct contact (telephones, door handles and washbasins). Rotavirus was also found in hand washings of 19% of attendants of patients with non-rotavirus diarrhoea, indicating that they may have come into contact with other attendants and patients in adjacent beds. This highlights the potential for contaminated hands to spread the infection (Soule et al., 1999).

Norwalk-like virus gastroenteritis in an elderly care unit spread rapidly within and between wards, affecting both patients and staff. Although infectious aerosols were probably the main route of dissemination of infection within a particular cohort of guests, contact with contaminated fomites was the most likely factor responsible for maintaining the outbreak by forming the link between successive cohorts (Barker et al., 2001).

7.4 Farming Activities

Farming, one of the oldest professions of mankind, is by far the one that employs the largest number of individuals worldwide. Although outdoor country work is supposedly healthy, farmers are at risk of respiratory diseases because of their work environment. It is very well established that chronic and acute respiratory diseases have been associated with work in confinement operations, like underfloor manure pits. Respirable dust and asphyxiating gases such as hydrogen sulphide, carbon dioxide, methane, and ammonia in high concentrations were measured on these area stations (Reeve et al., 2013; Manbeck et al., 2016). But bioaerosols are recognized as a serious threat on these environments. Mice are prevalent on farms. Faeces of deer mice can be contaminated with the Hantavirus, which can cause a devastating infection in humans, an influenza-like syndrome that often leads to respiratory failure. Farmers should wear protective respiratory equipment when cleaning building areas where mouse droppings are present. Farms are also the usual sources of influenza outbreaks. Influenza viruses infect pigs or poultry that can be transferred from animals to humans even without a mutation. This has so far been the case of the Influenza A virus subtype H5N1 that, to date, only infected farmers in close contact with birds (Cormier, 2007).

In animal slaughterhouses handling out cattle and sheep is hypothesized that these environments would contain significant amounts of bioaerosol due to the mechanical processes used to kill and process animals, a high degree of splashing and fluid handling, and also a high relative humidity of the environment. Workers on these conditions and meatworkers having exposure to a number of significant zoonotic diseases including leptospirosis, parapoxvirus, human papillomavirus subtypes HPV2, HPV4 and HPV7 are known to be at occupational health risk. These workers are also affected by a higher-rate of malignancies of the lung compared to the general population (Hall et al., 2013). Farmers and consumers of fresh farm products from farms irrigated with river water may be at risk of infection from adenoviruses. The findings highlight the lurking dangers of using contaminated surface water and the need for routine monitoring of such waters for protection of public health (Sibanda and Okoh, 2012). Workers in concentrated animal feeding operations are at risk of adverse respiratory outcomes from exposures to indoor contaminants. Is indispensable an optimal management of indoor air quality, preventing the transmission of infectious respiratory disease to workers and animals (Kim et al., 2005).

7.5 Analyses Approaches for Virus

7.5.1 Virus Analysis Methods

7.5.1.1 Culture-Dependent Analysis

Considering that viruses are obligate intracellular parasites, the use of culture-dependent methods is achieved with the use of suitable hosts, such as whole animals or cultured cells. Currently the most commonly used methods for virus cultures are inoculation of viruses into embryonated eggs and tissue cultures.

Inoculation of Embryonated Eggs

Virus started to be propagated in whole animals or embryonated chicken eggs before the use of cell culture methods. Virus cultivation in embryonated eggs is intrinsically dependent on the utilized egg, which must be sterile and the shell should be intact and healthy. The inoculation of the samples is made by injection into the fluid of the egg through a hole drilled in the shell. Viral growth and multiplication is revealed by embryo death, cell damage or through the formation of typical pocks or lesions on the membranes. The selection of the sites of viral inoculation in embryonated eggs is dependent of the studied virus, as each virus has different preferential location for growth and replication.

Chorioallantoic Membrane (CAM): Virus growth and replication is indicated by visible lesions (pocks) derived, under optimal conditions, from a single virion and a grey white area in transparent CAM.

Allantoic Cavity: Usually utilized for growth and replication of virus for vaccine production, provide a rich yield of influenza, some paramyxoviruses and avian viruses isolation.

Amniotic Cavity: Virus growth and replication of virus can be analyzed by haemagglutination assay.

Yolk Sac: Frequently utilized for growth, multiplication and isolation of mammalian viruses.

Currently, embryonated egg inoculation is conventionally considered the “gold standard” method for isolation and propagation of virus such as the influenza virus (Jianqiang Zhang, 2014).

Cell Culture

Mammalian cell culture technology has become a prominent and fundamental field in modern biotechnology, especially in the area of human health and has replaced embryonated eggs as the preferred methodology for virus growth and replication. Cellular cultures rely on techniques such as media changes, passaging, and transfection under aseptic conditions to avoid contaminations (e.g. bacteria, yeast, among others). Presently, numerous valuable cell monolayers are commercially available, and are regularly utilized in clinical laboratories for the diagnosis of virus infections. Some of the most utilized cell are HeLa derivative (Hep2), rhesus monkey kidney cells (RhMK), human lung fibroblasts (MRC-5), human lung carcinoma cells (A549), among others. These cell lines are selected for their ability to support the replication of a wide variety of clinically relevant viruses due to their ability to express cell type-specific factors that contribute to pathology during viral infection. For instance A549 cell line is considered representative of the Alveolar Type II pneumocytes of the human lung as it exhibits features of an ATII epithelial cell phenotype. Applying such methodology enables the isolation of diverse viruses, such as adenovirus, CMV, RSV, influenza A & B, parainfluenza viruses types 1 to 3, VZV, as well as the Ebola virus, severe acute respiratory coronavirus (SARS-CoV), and human metapneumovirus (hMPV).

The recent adaptation of cellular cultures to shell vials with subsequent direct or indirect immunofluorescence technique has dramatically decreased the time of diagnostics for clinical samples from weeks to less than 48 h by staining for early antigens of viral infections. Currently, this methodology is the most sensitive, non-molecular viral detection method, utilized for the identification of viruses such as respiratory viruses, enterovirus and adenovirus among others. Although embryonated chicken egg inoculation is still considered the "gold standard" method for influenza virus isolation and propagation, several primary cells as well as continuous cell lines have also been developed for influenza virus isolation and replication (Jianqiang Zhang, 2014).

7.5.1.2 Culture-Independent Virus Analysis

Nucleic Acid Amplification

The development of nucleic acid amplification techniques has endorsed the advance of molecular tools for virus identification using low specimen quantity with higher sensitivity and specificity, at the same time dramatically decreasing the time for identification. Currently these methodologies are accepted as the gold standard for clinical virology and have been utilized for the identification of different viruses in environmental samples from diverse contexts.

Single Target Nucleic Acid Amplification

The fundamental principles of nucleic acid amplification techniques are based on the thermostable polymerase-based target nucleic acid amplification which results in the production of millions of copies of the targeted sequence. These amplification products are then analyzed through diverse techniques. The advances in molecular analysis techniques lead to the development of the real-time PCR where the target amplification and detection steps, using fluorescent dyes, occur concomitantly. This methodology uses copulated software that monitors the thermal cycler data at every cycle and produces a quantitative amplification plot for each reaction. Thus, RT-PCR allows the performance of viral load assays to quantitatively assess the amount of virus in a sample. Furthermore, a modification of polymerase chain reaction, nested polymerase chain reaction (nested PCR), has been developed in order to decrease non-specific binding in products. This methodology encompasses two sets of primers, used in two successive PCR runs, whereas the second set intended to amplify a secondary target within the first run product. Currently, nucleic acid amplification techniques are considered the gold standard for rapid and accurate detection of viruses, compared to methodologies such as shell vial cell cultures.

Multiplex Nucleic Acid Amplification

Multiplex PCR assays were developed based on single target nucleic acid amplification methodologies with the aim to quantify multiple nucleic acid targets using specific probes to diverse viral targets in a single PCR reaction, allowing the assessment of viral co-infections. Currently, for respiratory viruses three FDA-approved platforms for multiplex PCR assays are available.

Multiplex PCR combined with liquid-phase bead-based array technology.
Multiplex RT-PCR followed by electrochemical detection of hybridized capture probes on gold-plated electrodes.
Multiplex-PCR preceded by nested RT-PCR with detection trough melt curve analysis.

Sequencing and Genotyping

Currently, DNA sequencing is considered one of the most valuable, accurate and consistent methodologies for microorganisms identification and is applied in contexts were quick and precise identifications are required. This method is used to determine the exact sequence of a certain stretch of DNA. Several new sequencing technologies have been developed (next generation sequencing) with the aim to provide fast/efficient techniques for analysis of microorganisms bioburden. Genotyping assays utilize a combination of PCR and nucleic acid sequencing to identify viral genotypes. Genotyping methodology allows for the identification of genetic variants and can be performed via genotyping chips or arrays, depending on the variants of interest and resources available. Viral genotyping is mostly utilized to provide relevant clinical data to predict therapeutic responses to antiviral drugs and/or epidemiologic comparison.

7.5.2 Viral Contamination Assessments in Environmental Samples

Currently, it is a consensus among researchers that methodologies based on nucleic acid amplification techniques offer advantages compared with traditional methods such as inoculation on embryonated eggs and cell culture, due to higher sensitivity, specificity and fast results, which will be further discussed. A comparison between the different viral identification methods is presented in Table 7.1. The efficacy of sampling methodologies for viruses in environmental samples is still a matter of intense debate. A meta-analysis study performed with the aim to assess the efficacy of virus concentration methods associated to the molecular detection of adenovirus demonstrated that for detection in environmental samples qPCR or Nested-PCR should be prioritized over PCR; in water samples (e.g. rivers or lakes) ultracentrifugation should be associated with nested-PCR and that microfiltration membrane, ultrafiltration, and qPCR must be associated for assessment of treated and untreated sewage samples (Silva and Melo, 2010).

Table 7.1 Comparison of viral identification methods

Methodologies	Time	Advantages	Disadvantages
Embryonated eggs	Days to weeks	High sensitivity, allows viral growth, multiplication and isolation	High cost, low efficiently, limited number of virus
Cell culture	Days to weeks	Identification of unknown virus	Poor sensibility, high labor cost
Shell vial	2 days	Improved time to identification and sensibility in comparing to Cell culture	Limited number of viruses
Single target nucleic acid amplification	hours	High sensitivity, viral quantification, rapid results	High costs, only identifies target virus
Multiplex nucleic acid amplification	hours	High sensitivity, viral quantification, rapid results, cost-effective	Decreased sensitivity in relation to single target PCR
Sequencing and Genotyping	hours	High sensitivity, rapid results	Require prior knowledge of the variants to analyze

7.5.2.1 Viral Contamination in Water

It is currently acknowledged that viral contamination of recreational water presents a high risk of infection and is considered a significant public health hazard. Published studies, summarized in Table 7.2, have addressed the assessment of virus bioburden in water environmental samples. The most commonly found viruses in aquatics environments are enteric viruses, such as Enterovirus, Rotavirus and Norovirus, Adenovirus and Hepatitis Virus A and C. Begier and coworkers have correlated swimming in polluted seawater with enterovirus infection (Begier et al., 2008). For enteroviruses, the most common mode of transmission is the fecal-oral route, although aerosol transmission has also been reported. In a similar manner, drinking water contaminated with virus, such as Norovirus, also presents a risk for human health, highlighting the importance of good hygiene practices with respect to water storage, Shin and Sobsey suggested that water chlorination could inactivate enteric viruses (Shin & Sobsey, 2008). Sample concentration is a critical step in viral diagnosis, since the number of viral particles in water is generally very low, which often results in false results if samples are tested directly using PCR (Katayama et al., 2002). Therefore, some authors choose an adsorption–elution method, followed by ultrafiltration, to concentrate the viruses, before nested-PCR or quantitative Real Time PCR. In severally contaminated environments such as hospital wastewater treatment plants, Prado and coworkers utilized PCR/RT-PCR, quantitative real-time PCR (qPCR) and genome sequencing, after sample concentrations techniques, to assess the presence of viruses associated with human pathologies such as acute gastroenteritis and

Table 7.2 Virus contamination assessment in environmental water samples

Setting	Sampling	Methods	Results	References
Wastewater plants	Elution-abortion; Ultrafiltration; Concentration	PCR/RT-PCR; qPCR	Rotavirus A; adenoviruses; norovirus genogroup I and II; hepatitis A viruses	(Prado et al., 2011)
		genome sequencing		
Receiving waters	Concentration	Cell culture	Adenovírus Norovirus	(Rodríguez et al., 2012)
		PCR/RT-PCR		
Urban waters	Adsorption-elution	RT nested – PCR	Enterovirus	(Zhang et al., 2008)
Surface water lagoon	Elution; Filtration; Concentration	RT-PCR	Adenovirus, polyomavirus; hepatitis A virus; rotavirus A.	(Fongaro et al., 2012)

hepatitis (Prado et al., 2011). In intermediate contaminated environments, such as urban and bathing waters the assessment of microbial hazards with potential public health risk associated with viral contamination also preferentially utilizes nucleic acid amplification methodologies, after sample concentration.

Some authors chose to utilize the combination of culture dependent and independent methodologies for virus identification from environmental samples. In a study performed by Roberto A. Rodríguez and coworkers that aimed to assess the effects of sewer overflows to the viral contamination of receiving waters, concentrated samples were assayed for total culturable viruses using the PLC/PRF/5 cell line with associated confirmation by PCR/RT-PCR (Rodríguez et al., 2012).

7.5.2.2 Viral Contamination on Surfaces

Viruses can be transmitted directly from individual to individual via sneezing, coughing and touching, or indirectly via the environment. The prevalence of pathogenic viruses in healthcare settings potentially transmitted by airborne, droplet and contact represents a significant threat for both workers' and patients' health. Thus, most of the presently performed studies regarding virus contamination of surfaces have focused on clinical settings, as demonstrated in Table 7.3. The data resultant from these studies is of foremost importance since it can allow the identification of critical locations and direct the selection of infection control measures for health-care settings to decrease viral associated nosocomial infections. Protocols regarding transmission based precautions in both hospital personnel and patients are continuously updated at the international level (La Rosa et al., 2013).

Table 7.3 Virus contamination assessment from surface samples

Setting	Sampling	Methods	Results	References
Toys in daycare nurseries	Swabs	qRT-PCR	Coronavirus; Bocavirus; Adenovirus; Rhinovirus;	(Ibfelt et al., 2015)
Personal protective equipment; laboratory instruments	Swabs	qRT-PCR	Rhinovirus	(Ling et al., 2016)
Intensive care unit	Swabs	Nested RT-PCR	Rotavirus	(Ganime et al., 2012)
Invasive medical instruments	Swabs	Nested RT-PCR	HBV	(Zhou et al., 2006)
Hospital	Swabs	Nested RT-PCR	Adenovirus; Norovirus; Torque teno virus	(Carducc et al., 2011)
Pediatric primary immunodeficiency unit	Swabs	RT-PCR	Norovirus, Astrovirus, and Rotavirus	(Gallimore et al., 2008)
General pediatric ward				
Pediatric primary immunodeficiency unit	Swabs	RT-PCR	Norovirus, Astrovirus, and Rotavirus	(Gallimore et al., 2006)
Pediatric hospital	Swabs	qPCR	Norovirus, Adenovirus, Respiratory syncytial Virus, Rotavirus, human metapneumovirus, Cytomegalovirus and Torque teno virus	(D'Arcy et al., 2014)

7.5.2.3 Viral Aerosol Contamination

Airborne diseases are caused by pathogens transmitted through the air, that may be spread through coughing, sneezing, resuspension of dust, spraying of liquids, or similar activities likely to generate aerosol particles or droplets. The World Health Organization (WHO) has suggested that aerosol exposure may be the major route for spreading viral diseases (World Health Organization, 2010). Environmental monitoring of airborne viruses could provide an early indicator of dangerous viruses in the air. Filtration methods that may utilize gelatin membrane filters, impaction collectors and impingers are usually used to collect viral aerosols. Sampling using air filters for several hours with posterior extraction and nucleic acid amplification can increase the sensitivity of virus detection. Currently, well-established filter/real-time qPCR assays are performed to successfully measure viral aerosols. The prevalence of respiratory tract infections is intrinsically associated with respiratory viruses which account for high percentage of pediatric as well as adult infections. A study performed in a pediatrics department of a medical center, detected influenza A virus, human adenovirus, and enterovirus in aerosol samples trough filtration method combined with real-time quantitative polymerase chain reaction (qPCR) technique (Tseng et al., 2010). In hospital settings, torque teno virus has been detected in samples collected with an impactor sampler with posterior culture in Tryptone Soy Agar (TSA) and identification trough nested RT-PCR (Carducci et al., 2011) or trough Burkard C90M cyclone sampler and qRT-PCR (D'Arcy et al., 2014).

Nevertheless, culture-dependent methods and nucleic acid amplification techniques can also be used simultaneously. As an example, a study performed by Goyal and coworkers utilized embryonated eggs and tissue cultures of Vero, MDCK, and RK-13 cell lines combined with nucleic acid extraction, PCR, RT-PCR and Nucleic acid sequencing to assess contamination of respiratory viruses (VSR, Influenza A and B, Parainfluenza 1, 2 and 3, Rinovirus, Enterovirus, Coronavirus, Filoviruses, Adenovirus and Orthopoxvirus) and viruses with bioterrorism potential in ventilation filters from two large public buildings (Goyal et al., 2011).

7.5.3 State of the Art and Future Directions

The assessment of viral contamination and/or infections play a key role in reducing global viral disease associated burden. It is currently acknowledged that the advances and development of molecular assays such as PCR, qRT-PCR, genotyping and multiplex assays endorsed important improvements in viral assessments regarding sensitivity, specificity, speed, simplicity and cost-effectiveness. Nevertheless, the developments of new approaches that can provide alternatives to evade the limitations of nucleic acid amplification techniques are of foremost importance. In this context, currently nanobiosensors represent a new promising tool for virus detection. This methodology, still at research stage, encompasses the technology of viral disease biosensors with nanoparticles and nanomaterials, focused on the development of miniaturized biosensors with high sensitivity, specificity, and stability. This state of the art technology aims to deliver an alternative tool for effective and rapid viral disease diagnosis with no requirement of highly trained personnel or heightened laboratory facilities (Kizek et al., 2015).

7.6 Concluding Notes

During the last two decades, there has been increasing concern within the scientific community over the effects of indoor air quality on human health. Everything looked definitive when vaccines were produced, diseases had the end announced, but the truth is that infections insist on persisting in our common environment although the better knowledge.

Changes in building design formulated to improve energy efficiency have meant that modern homes and offices are frequently more airtight than older structures. Indoor pollutants can emanate from a range of sources and we know much less about the health risks from indoor air pollution than we do about those attributable to the contamination of outdoor air (Jones, 1999).

The indoor environments to be considered are those ordinarily and commonly occupied by humans: residences, offices, schools, industrial buildings, hospitals and other settings occupied a high proportion of the time, or in which occupant density is high (Hanski et al., 2012). Biological hazards to man arise from exposure to high concentrations forms of bio-aerosols and three major groups of diseases associated with bio-aerosol exposure are infectious diseases, respiratory diseases and cancer. Current knowledge is unclear regarding risk to cancer whether these excess risks occur from exposures to biological agents or are due to various chemicals used in industries (Srikanth et al., 2008).

Acclimated environments have an artificially multitude of chemical compounds (toxic, carcinogenic, radioactive) and biological (pathogenic) issued by a variety of sources, depending on the physical conditions (air humidity, air temperature, inadequate ventilation) of the environment. The air recirculation phenomenon is responsible for the increase of pathogenic microorganisms in the order of 1000 to 100,000 times in relation to the external air (Lee et al., 2006). Incorrect cleaning filters and ducts of air conditioning provide the development of microbial particles including viruses that may lead the occupants of air-conditioned environments contracting respiratory infections or allergic diseases. Viruses can persist in sufficient number to act as sources of infection for several hours, weeks or even months.

The level of information and awareness of agricultural health and safety risks, disease, and injury prevention among the dairy farmers is low. Training on health and safety in agriculture field is urgently needed.

Safe working conditions are essential for healthy living. The lack of a preventive policy maintenance programmes in refrigeration and ventilation systems is the determining factor for the occurrence of biological pollutants. Take corrective actions preventing the spread of pathogens by the airborne route requires the use of special air handling and ventilation systems, such as airborne infection isolation rooms to contain and then safely remove the infectious agents. In addition, respiratory protection with validated and certified equipment is recommended (Fisk et al., 2007).

Environments have been studied more extensively than other issues due to their greater clinical significance. However, more work is still needed to provide a clearer picture regarding the rates of viral diseases transmission, airborne transmission in particular, in closed environments, and potential ways for reducing the levels of indoor viral pollution and transmission must be investigated. Studies including homes, non-industrial workplaces and public buildings, are scarce (Prussin II et al., 2015).

Ventilation filters of two large public buildings were sampled to determine the presence of human respiratory viruses by polymerase chain reaction and reverse-transcription polymerase chain reaction. Nine of the 64 filters tested were positive for influenza A virus, two filters were positive for influenza B virus, and one filter was positive for parainfluenza virus 1. Filters are installed in HVAC systems of buildings to protect ventilation equipment and maintain healthy indoor air quality. These filters process enormous volumes of air. Building HVAC filters may be used as a method of detection for airborne viruses. They may yield valuable information on the epidemiology and aerobiology of viruses in air that can report to the development of methods to prevent airborne transmission of viruses (Goyal et al., 2011). Regrettably, several investigations have revealed that many HVAC installations have a lot of operational and maintenance problems. Numerous practical recommendations for design and operation of HVAC systems are needed. Following the recommendations will result in less pollution and increased indoor environmental quality (Hanssen, 2004). Basic strategies of source control should not to make indoor air sterile but keep indoor environments dry, maintain good hygienic conditions in ventilation systems, apply effective filtration on mechanical supply ventilation, and use masks in the event of respiratory illness. The importance of hands in the transmission of viruses is well recognized and many of the studies relate specifically to handwashing.

The number of asthma patients in most industrial countries has greatly increased, resulting in a morbidity rate of around 10–15% of the population. May be aerosol transmission is responsible for the most severe cases of disease involving viral infection of the lower respiratory tract (Tellier, 2009; Tsukagoshi et al., 2013).

References

Anderson EJ, Weber SG (2004 Feb) Rotavirus infection in adults. Lancet Infect Dis 4(2):91–9

Artenstein MS, Miller WS (1966) Air sampling for respiratory disease agents in army recruits. Bacteriol Rev 30(3):571–572

Ballow M (2008) Approach to the patient with recurrent infections. Clin Rev Allergy Immunol 34(2):129–140

Barker J, Stevens D, Bloomfield SF (2001) Spread and prevention of some common viral infections in community facilities and domestic homes. J Appl Microbiol 91:7–21

Begier EM, Oberste MS, Landry ML et al (2008) An outbreak of concurrent echovirus 30 and coxsackievirus A1 infections associated with sea swimming among a group of travelers to Mexico. Clin Infect Dis 47(5):616–623. doi:10.1086/590562

Bernstein DI (2009) Rotavirus overview. Pediatr Infect Dis J 28(3):50–53

Canonica GW, Ciprandi G, Pesce GP et al (1995) ICAM-1 on epithelial cells in allergic subjects: a hallmark of allergic inflammation. Int Arch Allergy Immunol 107(1–3):99–102

Carducci A, Verani M, Lombardi R et al (2011) Environmental survey to assess viral contamination of air and surfaces in hospital settings. Journal of Hospital Infection 77(3):242–247. doi:10.1016/j.jhin.2010.10.010

Celli BR, Macnee W (2004) Standards for the diagnosis and treatment of patients with COPD: a summary of the ATS/ERS position paper. Eur Respir J 23:932–946

Celli BR, Barnes PJ (2007) Exacerbations of chronic obstructive pulmonary disease. Eur Respir J. 29(6):1224–1238

Chang LY, Tsao KC, Hsia SH et al (2004) Transmission and clinical features of enterovirus 71 infections in household contacts in Taiwan. JAMA 291(2):222–227

Chen SC, Chio CP, Jou LJ et al (2009) Viral kinetics and exhaled droplet size affect indoor transmission dynamics of influenza infection. Indoor Air 19:401–413

Chen SC, Liao CM (2010) Probabilistic indoor transmission modeling for influenza (sub) type viruses. J Infect 60(1):26–35

Cormier Y (2007) Respiratory health and farming: an essay. Can Respir J 14(7):419–422

D'Arcy N, Cloutman-Green E, Klein N et al (2014) Environmental viral contamination in a pediatric hospital outpatient waiting area: implications for infection control. Am J Infect Control 42(8):856–860. doi:10.1016/j.ajic.2014.04.014

Douwes J, Thorene P, Pearce N et al (2003) Bioaerosol health effects and exposure assessment: progress and prospects. Ann Occup Hyg 47:187–200

Fisk WJ, Lei-Gomez Q, Mendell MJ (2007) Meta-analyses of the associations of respiratory health effects with dampness and mold in homes. Indoor Air 17(4):284–296

Fongaro G, Nascimento MA, Viancelli A et al (2012) Surveillance of human viral contamination and physicochemical profiles in a surface water lagoon. Water Sci Technol. 66(12):2683–2687. doi:10.2166/wst.2012.504

Gallimore CI, Taylor C, Gennery AR et al (2006) Environmental monitoring for gastroenteric viruses in a pediatric primary immunodeficiency unit. J Clin Microbiol 44(2):395–399. doi:10.1128/JCM.44.2.395

Gallimore CI, Taylor C, Gennery AR et al (2008) Contamination of the hospital environment with gastroenteric viruses: comparison of two pediatric wards over a winter season. J Clin Microbiol 46(9):3112–3115. doi:10.1128/JCM.00400-08

Ganime AC, Carvalho-Costa FA, Mendonça MCL et al (2012) Group A rotavirus detection on environmental surfaces in a hospital intensive care unit. Am J Infect Control 40(6):544–547. doi:10.1016/j.ajic.2011.07.017

Goh GK, Dunker AK, Uversky V (2013) Prediction of intrinsic disorder in MERS-CoV/HCoV-EMC supports a high oral-fecal transmission. PLoS Curr 13(1):1–63

Goyal SM, Anantharaman S, Ramakrishnan MA et al (2011) Detection of viruses in used ventilation filters from two large public buildings. Am J Infect Control 39(7):e30–e38. doi:10.1016/j.ajic.2010.10.036

Gray GC (2006) Adenovirus transmission – worthy of our attention. J Infect Dis 194(7): 871–873

Green KY, Ando T, Balayan MS et al (2000) Taxonomy of the caliciviruses. J Infect Dis 181(2): 322–330

Hall RJ, Mily LM, Wang J et al (2013) Metagenomic detection of viruses in aerosol samples from workers in animal slaughterhouses. PLoS One 8(8):1–8

Hanski I, von Hertzen L, Fyhrquist N et al (2012) Environmental biodiversity, human microbiota, and allergy are interrelated. Proc Natl Acad Sci U S A 109(21):8334–8339

Hanssen SO (2004) HVAC – the importance of clean intake section and dry air filter in cold climate. Indoor Air 14(7):195–201

Heikkinen T, Jarvinen A (2003) The common cold. Lancet 361:51–59

Hui DS, Chan PK (2010) Severe acute respiratory syndrome and coronavirus. Infect Dis Clin North Am 24(3):619–638

Huynh KN, Oliver BG, Stelzer S et al (2008) A new method for sampling and detection of exhaled respiratory virus aerosols. Clin Infect Dis 46(1):93–95

Ibfelt T, Engelund EH, Schultz AC et al (2015) Effect of cleaning and disinfection of toys on infectious diseases and micro-organisms in daycare nurseries. J Hosp Infect 89(2):109–115. doi:10.1016/j.jhin.2014.10.007

Ignatius TS, Yu, Li Y et al (2004) Evidence of airborne transmission of the severe acute respiratory syndrome virus. N Engl J Med 350:1731–1739

Jianqiang Zhang PCG (2014) Isolation of swine influenza virus in cell cultures and embryonated chicken eggs. Methods Mol Biol. 1161:265–276. doi:10.1007/978-1-4939-0758-8_22

Jones AP (1999) Indoor air quality and health. Atmos Environ 33(28):4535–4564

Katayama H, Shimasaki A, Katayama H et al (2002) Development of a virus concentration method and its application to detection of enterovirus and Norwalk Virus from Coastal Seawater. Appl Environ Microbiol 68(3):1033–1039. doi:10.1128/AEM.68.3.1033

Kim KY, Ko HJ, Lee KJ et al (2005) Temporal and spatial distributions of aerial contaminants in an enclosed pig building in winter. Environ Res 99(2):150–157

Kizek R, Krejcova L, Michalek P et al (2015) Nanoscale virus biosensors: state of the art. Nanobiosensors Dis Diagn 4:47. doi:10.2147/NDD.S56771

Klepeis NE, Nelson WC, Ott WR et al (2001) The National Human Activity Pattern Survey (NHAPS): a resource for assessing exposure to environmental pollutants. J Expo Anal Environ Epidemiol 11:231–252

La Rosa G, Fratini M, Della Libera S et al (2013) Viral infections acquired indoors through airborne, droplet or contact transmission. Ann Ist Super Sanità 49(2):124–132

Lax S, Smith DP, Hampton-Marcell J et al (2014) Longitudinal analysis of microbial interaction between humans and the indoor environment. Science 345:1048–1052

Lee ByungUk, Hong InGi, Lee DaeHee et al (2012) Bacterial bioaerosol concentrations in public restroom environments. Aerosol Air Qual Res 12:251–255

Lee T, Grinshpun SA, Martuzevicius D et al (2006) Relationship between indoor and outdoor bioaerosols collected with a button inhalable aerosol sampler in urban homes. Indoor Air 16(1):37–47

Lessa FC, Gould PL, Pascoe N et al (2009) Health care transmission of a newly emergent adenovirus serotype in health care person nel at a military hospital in Texas. J Infect Dis 200(11):1759–1765

Li Fang (2013) Receptor recognition and cross-species infections of SARS coronavirus. Antiviral Res 100(1):246–254

Ling WX, Juan S, Qin SQ et al (2016) Viral contamination source in clinical microbiology laboratory. Biomed Environ Sci 29(8):609–611. doi:10.3967/bes2016.082

MacIntyre CR, Ridda I, Seale H et al (2012) Respiratory viruses transmission from children to adults within a household. Vaccine 30(19):3009–3014

Manbeck HB, Hofstetter DW, Murphy DJ et al (2016) Online design aid for evaluating manure pit ventilation systems to reduce entry risk. Fronti Public Health 4(108):1–16

Morillo SG, Timenetsky Mdo C (2011) Norovirus: an overview. Rev Assoc Med Bras 57(4):453–458

Nazaroff WW (2016) Indoor bioaerosol dynamics. Indoor Air 26(1):61–78

Osuolale O, Okoh A (2015) Incidence of human adenoviruses and hepatitis A virus in the final effluent of selected wastewater treatment plants in Eastern Cape Province, South Africa. Virol J 12(98):1–8

Paba P, Farchi F, Mortati E et al (2014) Screening of respiratory pathogens by Respiratory Multi Well System (MWS) r-gene™ assay in hospitalized patients. New Microbiologica 37:231–236

Poutanen SM, Low DE, Henry B et al (2003) Identification of severe acute respiratory syndrome in Canada. N Engl J Med 348:1995–2005

Prado T, Silva DM, Guilayn WC et al (2011) Quantification and molecular characterization of enteric viruses detected in effluents from two hospital wastewater treatment plants. Water Res 45(3):1287–1297. doi:10.1016/j.watres.2010.10.012

Prussin II AJ, Garcia EB, Marr LC (2015) Total virus and bacteria concentrations in indoor and outdoor. Environ Sci Technol 2(4):84–88

Reeve KA, Peters TM, Anthony TR (2013) Wintertime factors affecting contaminant distribution in a swine farrowing room. J Occup Environ Hyg 10(6):287–296

Rodríguez RA, Gundy PM, Rijal GK et al (2012) The impact of combined sewage overflows on the viral contamination of receiving waters. Food Environ Virol 4(1):34–40. doi:10.1007/s12560-011-9076-3

Shin GA, Sobsey MD (2008) Inactivation of norovirus by chlorine disinfection of water. Water Res 42(17):4562–4568. doi:10.1016/j.watres.2008.08.001

Sibanda T, Okoh AI (2012) Assessment of the incidence of enteric adenovirus species and serotypes in surface waters in the Eastern Cape Province of South Africa: Tyume River as a case study. ScientificWorldJournal 949216:1–9

Silva HD, Melo MR (2010) Artículo original Avaliação de métodos de concentração e detecção molecular de adenovírus em águas não tratadas – uma metanálise, 65–71.

Soule H, Genoulaz O, Gratacap-Cavallier B et al (1999) Monitoring rotavirus environmental contamination in a paediatric unit using polymerase chain reaction. Infect Control Hosp Epidemiol 20:432–434

Srikanth P, Sudharsanam S, Steinberg R (2008) Bio-aerosols In indoor environment: composition, health effects and analysis. Indian J Med Microbiol 26(4):302–312

Tantilipikorn P, Auewarakul P (2011) Airway allergy and viral infection. Asian Pac J Allergy Immunol 29:113–119

Tellier R (2009) Aerosol transmission of influenza A virus: a review of new studies. J R Soc Interface 6:783–790

Tseng CC, Chang LY, Li C (2010) Detection of airborne viruses in a pediatrics department measured using real-time qPCR coupled to an air-sampling filter method. J Environ Health 73(4):22–28

Tsukagoshi H, Ishioka T, Noda M et al (2013) Molecular epidemiology of respiratory viruses in virus-induced asthma. Front Microbiol 4(278):1–10

Uhnoo I, Svensson L, Wadell G (1990) Enteric adenoviruses. Baillieres Clin Gastroenterol 4(3):627–642

von Mutius E (2004) Influences in allergy: epidemiology and the environment. J Allergy Clin Immunol 113(3):373–379

World Health Organization. (2010). Limiting spread. Retrieved from http://www.who.int/influenza/resources/research/research_agenda_influenza_stream_2_limiting_spread.pdf

Yang W, Elankumaran S, Marr LC (2011) Concentrations and size distributions of airborn e influenza A viruses measured indoors at a health centre, a day-care centre and on aeroplanes. J R Soc Interface 8:1176–1184

Zhang CM, Wang XC, Liu YJ et al (2008) Characteristics of bacterial and viral contamination of urban waters: a case study in Xi'an, China. Water Sci Technol. 58(3):653–660. 10.2166/wst.2008.438

Zhao H, Joseph C, Phin N (2007) Outbreaks of influenza and influenza-like illness in schools in England and Wales, 2005/06. Euro Surveill 12(5):3–4

Zhou LF, Zhu HH, Lin J et al (2006) Surveillance of viral contamination of invasive medical instruments in dentistry. J Zhejiang Univ Sci B 7(9):745–748. doi:10.1631/jzus.2006.B0745

Zlateva KT, Maes P, Rahman M et al (2005) Chromatography paper strip sampling of enteric adenoviruses type 40 and 41 positive stool specimens. Virol J 2(6):1–5

Part III
Microbial Cell Wall Components, Secondary Metabolites and Health Implications upon Exposure

Chapter 8
Endotoxins, Glucans and Other Microbial Cell Wall Agents

Ioannis Basinas, Grethe Elholm and Inge M. Wouters

Abstract During the last decades an increasing interest in microbial cell wall agents has been established, since exposure to these agents has been linked to a wide range of adverse and beneficial health effects. The term microbial cell wall agents refers to a group of molecules of different composition that are integral structural components of microorganisms like gram-negative and gram positive bacteria and fungi. The available information on exposure characteristics for these cell wall agents within indoor environments and their associated health effects is summarized in this chapter.

Large variation in exposure levels of microbial cell wall agents in indoor occupational environments is documented, whereas actual airborne levels of exposures and determinants of residential indoor air are lacking. Standardisation of methods for determination is highly recommended for future studies.

Endotoxins, cell wall agents of gram-negative bacteria, are well studied and involved in the development of adverse and protective health effects, but for cell wall agents of fungi, like glucans the evidence is more limited and inconclusive. For other microbial cell wall agents, like muramic acid, EPS and ergosterol, studies have been sparse and very diverse in their design and applied methods.

Future recommendations include studies in large populations with a longitudinal design involving both exposure assessment and health effects assessment of

I. Basinas (✉)
Centre for Human Exposure Science, Institute of Occupational Medicine, Edinburgh, United Kingdom
e-mail: ioannis.basinas@iom-world.org

G. Elholm
Department of Public Health, Section for Environment, Occupation and Health, Danish Ramazzini Centre, Aarhus University, Aarhus, Denmark
e-mail: gelh@ph.au.dk

I.M. Wouters
Division of Environmental Epidemiology, Institute for Risk Assessment Sciences (IRAS), Utrecht University, Utrecht, The Netherlands
e-mail: i.wouters@uu.nl

C. Viegas et al. (eds.), *Exposure to Microbiological Agents in Indoor and Occupational Environments*, DOI 10.1007/978-3-319-61688-9_8

distinct microbial cell wall agents and co-existent microbes, which is needed to understand the role of individual and combined exposures in health.

Keywords Cell wall agents · endotoxins · glucans

8.1 Introduction

A variety of potentially hazardous agents can be found in indoor air. Generally dusty and moist indoor environments are unpleasant to most people, but determining which components in the air are significantly associated with specific health outcomes is very challenging. Some indoor air exposures have already been found to have a negative impact on human health, others are still only under suspicion and yet some appear to even have beneficial effects. Micro-organisms, such as bacteria and fungi, and agents from a microbiological origin have been widely studied in relation to indoor air-related health outcomes. Agents composing the cell walls of microbes, such as endotoxins, glucans and extracellular polysaccharides are often implicated as risk factors or used as markers for exposure to microbiological agents. This chapter aims to summarise the available information on the exposure characteristics for these cell wall agents within indoor environments and their associated health effects.

8.2 What Are Microbial Cell Wall Agents?

Microbial cell wall agents are a group of molecules of different composition that are integral structural components of microorganisms (Fig. 8.1). They are released into the environment following replication, apoptosis, lysis or death of the microbial cell. Depending on their origin, fungal, gram positive or gram negative bacteria, microbial cell walls consist of different types of polysaccharides, proteins and acids. Although similar structures may also be present in outer layers of cereals and plant tissues, they are mostly considered to represent microbial exposures. Microbial cell wall agents are an important constituent of the so called "organic dust" arising from microbial, plant and animal origin. During the last decades an increasing interest in microbial cell wall agents has been established, since exposure to these agents has been linked to a wide range of adverse and beneficial health effects.

8.3 Why Are Microbial Cell Wall Agents Important?

Several symptoms and diseases have been associated with exposure to cell wall agents. These include systemic reactions (e.g. inflammation, fever and chills), allergies, acute respiratory symptoms, chronic respiratory disorders such as chronic bronchitis and asthma, as well as cancer (Smit et al., 2006; Madsen et al., 2012; Basinas et al., 2012a; Gladding et al., 2003; Li et al., 2006; Fang et al., 2013;

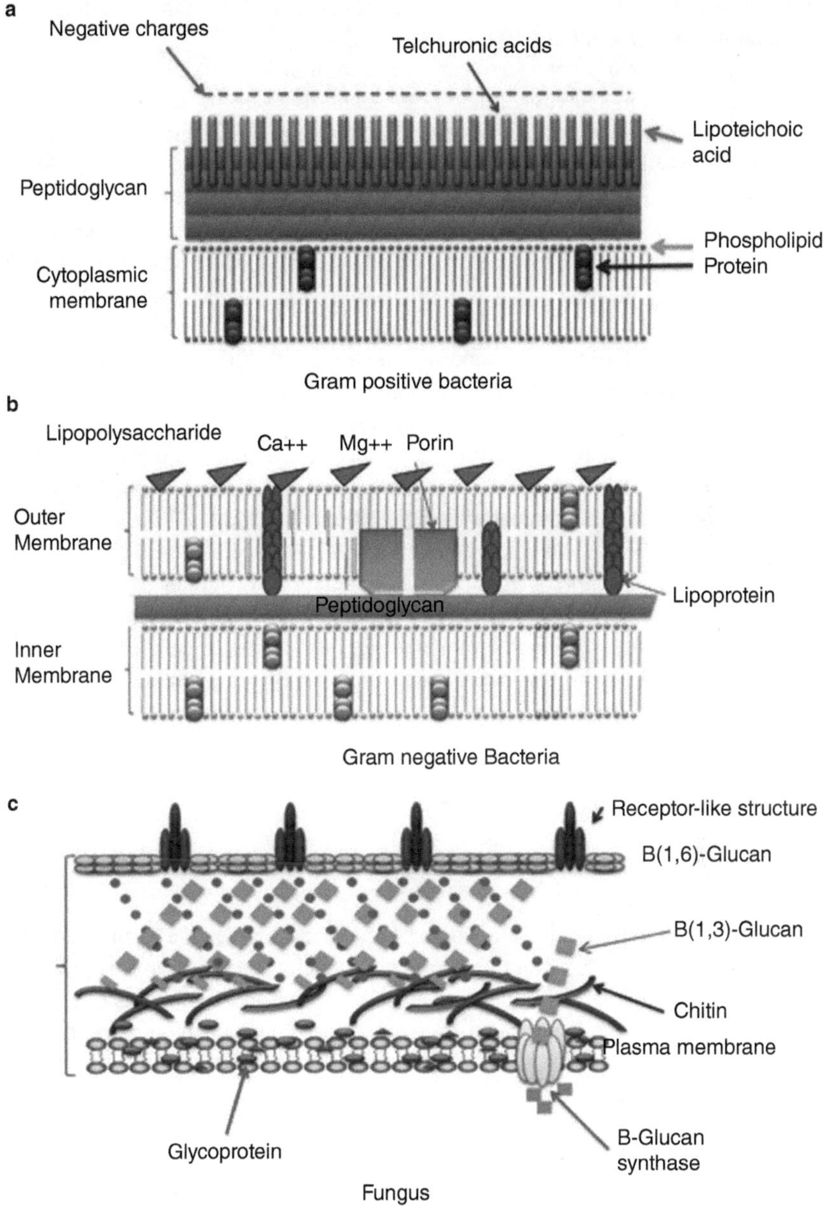

Fig. 8.1 Cell wall structures of three different types of microbial organisms. (**a**) Gram-positive bacteria which have an outer cell wall containing a thick layer of peptidoglycan. (**b**) Gram-negative bacterial cell walls which contain a thin layer of peptidoglycan and a lipid bilayer containing lipopolysaccharide. (**c**) Fungal cell walls which are composed of beta-glucan structures and chitin. (From Vatansever et al., 2013)

Eduard et al., 2004; Rylander et al., 1999; Vogelzang et al., 2000; Eduard et al., 2009; Braun-Fahrlander et al., 2002).

Bacterial endotoxins, peptidoglycans (incl. muramic acid), the fungal sourced β-D-glucans and fungal extracellular polysaccharides are all microbial cell wall agents that are either considered to have a key role in associations with health effects or, as not all necessarily have human antigenic and/or inflammatory properties themselves, being used as markers for exposure to microbes.

Once released and aerosolised, microbial cell wall agents can enter the human body mainly through inhalation. Exposure through other routes has not been thoroughly studied yet. Generally the potential for dermal absorption can be considered as rather small because of a high molecular agent weight (Bos and Meinardi, 2000), whereas direct or inadvertent (i.e. through hand to mouth contact and eating in contaminated areas) exposure by ingestion can occur (Cherrie et al., 2006; Gorman et al., 2012) but it is likely of lesser importance for respiratory diseases. After entering the human body some microbial cell wall agents may trigger a line of different receptors which evoke an increase in the release of cytokines, chemokines, adhesion molecules, and other mediators resulting in an inflammatory reaction (Reed and Milton, 2016).

The main microbiological cell wall agents that have been studied either as independent agents or as markers of exposures in relation to human health outcomes are summarized below. It is important to note that other cell wall agents of microbial origin (e.g. various types of proteins) exist but at present their immunological importance is either considered rather small or remains unknown.

8.4 Microbial Cell Wall Agents

8.4.1 Endotoxins

Endotoxins are commonly also known as Lipopolysaccharides (LPS) in reference to their purified derivative and chemical structure, which typically comprises of a long polysaccharide complex chain bound to a lipid A component (Douwes et al., 2003; Williams, 2007b). They are located at the external cell wall membrane of gram-negative bacteria and are released to the environment primarily following cell replication, death or lysis (Williams, 2007a). Endotoxin and their purified derivatives are present in the oral and nasal cavity and throughout the gastrointestinal tract of mammals, and are found ubiquitously on plant surfaces, animals, and soil (Bos et al., 2007). They are considered as one of the main and biologically most active pro-inflammatory constituents of organic dusts (Sigsgaard et al., 2010).

8.4.2 Glucans

The (1→3)-β-D-glucans are glucose polymers which are part of the cell wall structure of fungi (and of some bacteria), yeasts and mushrooms (Douwes et al., 2003;

Sigsgaard et al., 2005; Williams, 1997). They can also be present in the bran of some cereal (e.g. oat and barley) and be produced as a result of plant synthesis in response to tissue wounds (Finkelman et al., 2005; Lazaridou and Biliaderis, 2007). Their physicochemical properties vary depending on their source. Generally, they are stable molecules, non-soluble in water, and composed of a β-D-linked linear backbone containing anhydroglucose repeat units linked with a glycosidic bond between the 1 and 3 positions and sometimes bearing side chains at position 6 (Williams et al., 2005). In fungi they form the cell wall through a linkage to mannoproteins (i.e. fungal proteins linked with chains of up to several hundred mannoses), proteins, lipids and chitin and the (1→6)-β-side-branches (Miura, 2005). Their exact primary structure, solubility, degree of branching, and molecular weight play an important role in glucans biological activity (Zeković et al., 2005). Glucans are mainly studied for their immunomodulatory properties.

8.4.3 Peptidoglycans and Muramic Acid

Peptidoglycans are composed of amino acids and sugar polymers and form the backbone of the cell walls of bacteria (Figure 8.1). They are present in both gram-positive and gram-negative bacteria (Fig. 8.1). Within gram-positive bacteria, peptidoglycans form the core of the cell wall membrane comprising up to 70% of the composition, whereas in gram-negative bacteria they form only a minor part of the cell wall. Therefore peptidoglycans are considered to be a marker of exposure to gram-positive bacteria. Peptidoglycans are formed by alternating N-acetyl-muramic and N-acetylglucosamine acid residues linked by β -1→4 bonds with a pentapeptide attached to the d-lactoyl group of each combination residue (Vollmer et al., 2008; Meroueh et al., 2006). The N-acetyl-muramic acid constituent is an amino saccharide which is also commonly known as muramic acid and is measured as a marker for the presence and quantification of peptidoglycans (Poole et al., 2010; Van Strien et al., 2004; Lappalainen et al., 2012; Karvonen et al., 2014). Peptidoglycans are known to induce an inflammatory response.

8.4.4 Extracellular Polysaccharides and Ergosterol

Extracellular Polysaccharides (EPS) are stable carbohydrates that dominate the cell wall and periphery of fungal structures including septa, spores and hyphens, whereas ergosterol is a steroid alcohol (sterol) compound of the fungal cell membrane. While their immunomodulatory value is considered rather small, both ergosterol and EPS are considered as good markers for fungal exposures. Particularly EPS from *Aspergillus* and *Penicillium* spp. have been shown to correlate well with the biomass of viable fungi in house dust. On the other hand ergosterol is considered a good marker for both viable and non-viable fungal biomass (Douwes et al., 1999; Casas et al., 2016).

8.5 Methods of Quantification

Overviews of exposure measurement techniques of biological agents including microbial cell wall agents have been described previously by Douwes et al. (2003) and Casas et al. (2016). In short, quantification of microbial cell wall agents relies on the collection of dust followed by subsequent laboratory analysis of the agents within the dust. For an airborne exposure route the preferable sampling method is active airborne sampling: air is sucked though a sampling head by means of a (portable) pump in which dust is captured through filtration. Based on the sampling characteristics of the sampler specific size fractions of dust may be captured. Generally, in occupational studies the inhalable dust fraction is sampled. Alternatively passive sampling methods capturing settling airborne dust may be employed, e.g through air exposure of petri dishes, "pizza boxes" or electrostatic collectors (Frankel et al. 2012b). Instead of airborne sampling many epidemiological studies in the past have relied on dust samples of floor dust samples representing settled dust or mattress dust samples. Those dust samples are collected using a combination of regular vacuum cleaners fitted with specialised sampling devices like nozzles with collection filters or specially designed bags).

Endotoxins contained in the dust are generally measured by the Limulus Amebocyte Lysate (LAL) assay. The LAL is a biological assay which makes use of an enzyme reaction process from the horseshoe crab, Limulus Polyphymus to quantify non-cell bound endotoxins. Results are expressed in Endotoxin Units (EU), a standardized metric introduced to account for differences in biological activity (potency) per mass unit between endotoxins. The assay is very sensitive and available in several formats from which the kinetic colorimetric ones are considered as the most precise, and thus are most commonly used. Inter-laboratory variations have been described, mainly sourcing from differences in sampling and analytical methodologies between laboratories (Chun et al., 2006). To overcome the problem of batch to batch differences and interference, and to protect the horseshoe crab from extinction, an endotoxin assay has recently become available that uses recombinant Factor C (rFC) reagent produced from the cDNA of the Mangrove horseshoe crab (*Cacinoscorpius rotundicauda*) (Ding et al., 1995). Studies in livestock facilities and houses showed good correlation between results from the recombinant Factor C (rFC) assay compared to the LAL assay (Thorne et al., 2010; Alwis and Milton, 2006). However, little is still known on interference of other agents on the rFC assay results. It can be expected that the recombinant assay will be applied more and more in future studies.

Endotoxins can also be measured chemically through gas chromatography / mass spectrometry (GC/MS) to identify and quantify 3-hydroxy fatty acids (3-OHFAs) in the lipid A of endotoxin (Saraf et al., 1997). The method quantifies both cell bound and non-cell bound endotoxin with results expressed in mass concentrations, and thus cannot be compared directly to results obtained with the LAL assay. It has not been widely applied and associations with human health endpoints remain to be fully studied.

Several different assays have been applied in studies investigating (1→3)-β-D-glucans, including assays based on a modification of the Limulus amebocyte lysate (LAL) assay in which only active factor G is present. Earlier this method was referred to as the LAL assay, whereas later a commercially available Glucatell assay became available based on the same principle (Rylander, 1997; Cherid et al., 2011). A number of immunoassays to detect glucans have been developed and applied as well. Initially an inhibition immunoassay was developed (Douwes et al., 1996), which had relatively low sensitivity. More recently, several laboratories have developed more sensitive sandwich Enzyme Immunoassays (EIAs) (Noss et al., 2010b; Sander et al., 2008; Milton et al., 2001). Few data are available comparing outcomes of different (1→3)-β-D-glucans assays where results are typically expressed in units of mass. An interlaboratory comparison study showed that results of different methods were comparable in relative terms as most methods correlated moderately well with each other. Yet direct comparison of results between laboratories and assays is compromised, due to discrepancies in applied standards and extraction procedures resulting in major differences in absolute levels (Brooks et al., 2013). Available comparison data is yet to scarce to provide reliable conversion factors.

Peptidoglycans are determined through GC/MS analysis by quantification of their composite muramic acid (Poole et al., 2010; Van Strien et al., 2004; Lappalainen et al., 2012; Karvonen et al., 2014). The muramic acid content is regarded to be a measure of exposure to gram-positive bacteria. Similarly, ergosterol, which can be determined through GC/MS analyses, is a measure of fungal biomass (Saraf et al., 1997; Miller and Young, 1997). Fungal extracellular polysaccharides (EPS) are considered fungal biomarkers as well, although they allow for a certain level of differentiation of mould genera present. They are measured through a specific sandwich enzyme immunoassay (Douwes et al., 1999).

8.6 Exposure Limits

A number of countries have established occupational exposure limits for exposure to organic dust, which are commonly used as guidelines for advising and protecting workers from overexposure to microbial agents. Generally, these limits have been established based on the available information on exposure levels within certain industries and vary considerably from country to country. For example, the occupational exposure limit (OEL) for organic dust is 3mg/m^3 of ''total'' dust in Denmark and 5mg/m^3 in Norway and Sweden (Arbeidstilsynet, 2011; Arbejdstilsynet, 2007). In the US, the Occupational Safety and Health Administration (OSHA) has since 1989 advised a permissible exposure limit of 10mg/m^3 for total grain dust (OSHA, 1995). Whereas the National Health Council of the Netherlands has recommended a Health-Based OEL (HBROEL) of 1.5mg/m^3 of inhalable grain dust (DECOS, 2011).

However, despite the well-recognised strong inflammatory capability, thus far no agent- and environment-specific (i.e. residential or occupational) health-based limit values for exposure to microbial cell wall agents have been established. The only exception, to our knowledge, is the limit for endotoxin that was established by the National Health Council of the Netherlands in conjunction with the Nordic research council (DECOS, 2010). They jointly proposed a HBROEL of 90 EU/m^3, largely based on acute respiratory effects.

8.7 Exposures in Indoor and Occupational Environments

Despite the broad recognition of different cell wall agents playing a part in the development of respiratory symptoms and other health disorders, relatively little is actually known with respect to their airborne exposure levels and prevalence. Most exposure information is available for endotoxin and (1→3)-β-D-glucans airborne concentrations and an overview of measured airborne levels for these two agents across different occupational and residential environments is provided in Table 8.1. It should be noted that most data from residential environments relate to floor dust and/or mattress dust rather than airborne exposure levels. However, the focus of the current overview is on airborne levels as those are considered to be more representative of inhalation exposures.

In general, the levels of exposure to endotoxins and glucans are very varied across both occupational and residential environments. In occupational settings, levels are clearly dependent on the presence or absence of an exposure source such as manure, composted waste, animals, and/or plant materials. For endotoxin the highest levels of exposure commonly occur among workers in primary agricultural workplaces such as poultry, dairy and pig farms and among those involved in cotton processing and grain handling. Average personal concentrations measured within these industries are reported to typically range between a few hundred to many thousands of EU/m^3 (Table 8.1). Other workplaces with considerably high exposures to endotoxin include waste collection and handling, seed and paper processing and veterinary practices. The levels of exposure within these environments can be several orders of a magnitude higher than those reported within residential and office environments.

Similarly, (1→3)-β-D-glucans exposures appear to be an issue mainly in workplaces of agricultural production, waste collection and management, paper processing as well as podiatry clinics. Direct comparisons between these results however cannot be made because measured concentrations for glucans largely depend on the type and inherent sensitivity of the quantification assay applied within a study (see methods of quantification section above). The higher sensitivity of the LAL assay (Sander et al., 2008; Douwes, 2005) may, at least partly, explain the reported lower levels of exposure in studies that use this methods compared with those using the inhibition enzyme immunoassays (EIA). Other parameters such as the extraction medium, or the type of filter used and its storage or transport

Table 8.1 Overview (non-exhaustive list) of results from studies of airborne endotoxin and (1→3)-β-D-glucans levels within occupational and residential environments

Type of environment	Endotoxin (EU/m^3)					(1 → 3)-β-D-glucan (ng/m^3)				
	Measurement type	Analytical method	Range of means	Range of individual concentrations	References	Measurement type	Analytical method	Range of means	Range of individual concentrations	References
Primary animal production										
Dairy farming	P	KC/T-LAL, rFC	220–1570	<LOD–8290	(Basinas et al., 2012b; Samadi et al., 2012; Garcia et al., 2013; Spaan et al., 2006; Smit et al., 2008; Saito et al., 2009; Burch et al., 2010)	P	SI-EIA	10,300	150–232,000	(Samadi et al., 2012)
Pig farming	P	KC/T-LAL, rFC	400–6600	<LOD–374,000	(Basinas et al., 2012b; Smit et al., 2008; O'Shaughnessy et al., 2010; Simpson et al., 1999; Szadkowska-Stańczyk et al., 2010; Radon et al., 2002)	P	Glucatell	223	6–5208	(Szadkowska-Stańczyk et al., 2010)
						P	SI-EIA	4340	200–38,490	(Douwes et al., 1996)
						S	SI-EIA	NR	33–410	(Sander et al., 2008)
						S	Glucatell	NR	18–96	(Sander et al., 2008)
Poultry farming, general		KC/T-LAL	2576	190–16,348	(Radon et al., 2002)	S	Glucatell	NR	13–5000	(Sander et al., 2008)
						S	SI-EIA	NR	2–972	(Sander et al., 2008)

(continued)

Table 8.1 (continued)

Type of environment	Endotoxin (EU/m³)					(1 → 3)-β-D-glucan (ng/m³)				
	Measurement type	Analytical method	Range of means	Range of individual concentrations	References	Measurement type	Analytical method	Range of means	Range of individual concentrations	References
Poultry farming, layers	P	KC/T-LAL, rFC	694–7517	1162–19,745	(Basinas et al., 2012b; Spaan et al., 2006; Senthilselvan et al., 2011; Arteaga et al., 2015)					
Poultry farming, broilers	P	KC/T-LAL	596–9609	61–8120	(Spaan et al., 2006; Senthilselvan et al., 2011)					
Mink farming	P	KC/T-LAL	214	93–1050	(Basinas et al., 2012b)					
Mixed livestock production farming	P	KC/T-LAL	448	<LOD–2910	(Basinas et al., 2012b)					
Horse keeping/ farming	P	KC/T-LAL	742	92–9846	(Samadi et al., 2009)	P	SI-EIA	9500	<LOD–631,000	(Samadi et al., 2009)
Plant cultivation										
Field crops (arable)	P	KC/T-LAL	63–2700	96–41,200	(Spaan et al., 2006; Smit et al., 2008)					
Mushrooms	P	KC/T-LAL	110	10–4450	(Simpson et al., 1999)					
Flowers, greenhouses	P	KC/T-LAL	27–140	0.84–1097	(Thilsing et al., 2015; Spaan et al., 2006)					
Vegetables, greenhouses	P	KC/T-LAL	13–1180	5.4–4020	(Spaan et al., 2006; Madsen et al., 2009)					

(continued)

Table 8.1 (continued)

Type of environment	Endotoxin (EU/m^3)					(1 → 3)-β-D-glucan (ng/m^3)				
	Measurement type	Analytical method	Range of means	Range of individual concentrations	References	Measurement type	Analytical method	Range of means	Range of individual concentrations	References
Industrial processing of agriculture products										
Abattoirs	P	KC/T-LAL	28–310	27–6230	(Spaan et al., 2006)					
Seed processing, grass and cereals	P	KC/T-LAL	1160–12,869	9.1–79,900	(Madsen et al., 2012; Spaan et al., 2008a)	P	LAL	3.83	2.82–4.84	(Madsen et al., 2012)
Seed processing, vegetables	P	KC/T-LAL	22–770	25.6–42,200	(Spaan et al., 2008a)					
Fruit and vegetable preservation	P	KC/T-LAL	61	4.9–1200	(Spaan et al., 2006)					
Grain handling and animal feed industry	P	KC/T-LAL	270–628	11–80,500	(Spaan et al., 2008a; Halstensen et al., 2013)	P	SI-EIA	7400	200–1,290,000	(Halstensen et al., 2013)
Waste collection and management										
Domestic waste collection	P	KC/T-LAL	40	<4–7182	(Wouters et al., 2006)	P	SI-EIA	1220	<260–52,500	(Wouters et al., 2006)
	S	KC/T-LAL	5–7		(Thorn et al., 1998)	S	LAL	9.2–19.1		(Thorn et al., 1998)

(continued)

Table 8.1 (continued)

Type of environment	Endotoxin (EU/m^3)					(1 → 3)-β-D-glucan (ng/m^3)				
	Measurement type	Analytical method	Range of means	Range of individual concentrations	References	Measurement type	Analytical method	Range of means	Range of individual concentrations	References
Power plants (biofuel/mass)	P	KC/T-LAL	9-200	<3–2104	(Wouters et al., 2006)	P	SI-EIA	<100–290,900	<100–290,900	(Wouters et al., 2006)
Composting, domestic waste	P	KC/T-LAL	17–1038	<3–37,043	(Wouters et al., 2006)	P	SI-EIA	<600–4930	<150–206,600	(Wouters et al., 2006)
Composting, green waste	P	KC/T-LAL	6–32	<3–345	(Wouters et al., 2006)	P	SI-EIA	<600–530,000	<600–2850	(Wouters et al., 2006)
waste transferral	P	KC/T-LAL	36–520	16–3536	(Wouters et al., 2006)					
Sewage treatment	P	KC/T-LAL	15.4	0.7–214	(Cyprowski et al., 2015b)					
Wood and paper processing										
Sawmills	P	KC/T-LAL	130	10–1870	(Simpson et al., 1999)					
Sawmills	P	EC-LAL	43	1.9–784	(Mandryk et al., 1999)	P	LAL	1.37	0.16–11.74	(Mandryk et al., 1999)
Joineries	P	EC-LAL	11–24.1	1–279	(Mandryk et al., 1999; Harper and Andrew, 2006)	P	LAL	0. 43	0.11–3.6	(Mandryk et al., 1999)
Wood chipping	P	EC-LAL	32.7	20–487	(Mandryk et al., 1999)	P	LAL	2.32	0.13–10.4	(Mandryk et al., 1999)
Paper processing factories	S	KC/T-LAL	20-977	0–2200	(Rylander et al., 1999)	S	LAL	10–240	49–366	(Rylander et al., 1999)

(continued)

Table 8.1 (continued)

Type of environment	Endotoxin (EU/m^3)					(1 → 3)-β-D-glucan (ng/m^3)				
	Measurement type	Analytical method	Range of means	Range of individual concentrations	References	Measurement type	Analytical method	Range of means	Range of individual concentrations	References
Textile manufacturing and processing										
Cotton mills	P	KC/T-LAL	70–6316	10–26,300	(Simpson et al., 1999; Mehta et al., 2007; Paudyal et al., 2011)					
	S	EC-LAL	10–7500	10–17,000	(Christiani et al., 1993; Christiani et al., 1994)					
	S	KC/T-LAL	37–4556	2–18,344	(Mehta et al., 2007; Marchand et al., 2007)					
Wool mill	P	KC/T-LAL	960	10–3045	(Simpson et al., 1999)					
Hemp	P	KC/T-LAL	19,569	4734–59,801	(Fishwick et al., 2001)					
Other workplaces										
Metal working/ machining plants	P	KC/T-LAL	2	1–31	(Cyprowski et al., 2015a)					
	S	EC-LAL	25.3	<LOD–183	(Gilbert et al., 2010)					
Veterinary clinics, companion animals	P	KC/T-LAL	4.4	<LOD–75	(Samadi et al., 2011)	P	SI-EIA	3.39	<LOD–111.5	(Samadi et al., 2011)

(continued)

Table 8.1 (continued)

Type of environment	Endotoxin (EU/m^3)					(1 → 3)-β-D-glucan (ng/m^3)				
	Measurement type	Analytical method	Range of means	Range of individual concentrations	References	Measurement type	Analytical method	Range of means	Range of individual concentrations	References
Veterinary clinics, farm animals	P	KC/T-LAL	520–1498	60–49,846	(Samadi et al., 2011)	P	SI-EIA	3.10	<LOD–46.1	(Samadi et al., 2011)
Podiatry clinics	P	KC/T-LAL	9.6	0.5–32.6	(Coggins et al., 2012)					
Laboratories with animals						S	Glucatell	NR	13–5,000	(Sander et al., 2008)
						S	SI-EIA	NR	16–38	(Sander et al., 2008)
Public and social service workplaces										
Office buildings	S	EC-LAL	0.5–3		(Reynolds et al., 2001; Rylander et al., 1992)	S	LAL	<0.1–3.2		(Rylander et al., 1992; Wan and Li, 1999)
Schools without sources	S	KC/T-LAL	9.34	<2.83–>225	(Holst et al., 2015b)	S	LAL	2.9	0–6.9	(Rylander et al., 1998)
Schools with sources	S	KC/T-LAL	2.1–2.6		(Rylander et al., 1992)	S	LAL	0.49–15.3	9.2–27.4	(Rylander et al., 1998; Rylander et al., 1992)

(continued)

Table 8.1 (continued)

Type of environment	Endotoxin (EU/m³)					(1 → 3)-β-D-glucan (ng/m³)				
	Measurement type	Analytical method	Range of means	Range of individual concentrations	References	Measurement type	Analytical method	Range of means	Range of individual concentrations	References
Daycare centres	S	EC-LAL	24.3		(Rylander et al., 1992)	S	LAL	0.2–5.7		(Rylander et al., 1992; Wan and Li, 1999)
	S	KC/T-LAL	1.6		(Wan and Li, 1999)					
Dwellings										
Residence, general	S	KC/T-LAL	0.36–6.5	<0.005–389.2	(Noss et al., 2008; Wan and Li, 1999; Frankel et al., 2012a; Park et al., 2000; Singh et al., 2011; Dassonville et al., 2008)	S	Glucatell	1.96	0.002–41.91	(Singh et al., 2011; Thorn and Rylander, 1998)
						S	LAL	3.7		(Wan and Li, 1999)
Residence with sources	S	KC/T-LAL	22.8–64	4–256	(Semple et al., 2010; Adhikari et al., 2010)	S	LAL	3.1–15.9		(Adhikari et al., 2010)
Farm residence	S	KC/T-LAL	1.04		(Noss et al., 2008)					

EU/m³ = Endotoxin Unit per cubic meter; ng/m³=nanogram per cubic meter; P=personal sampling; S=Stationary/areal sampling; LAL= Limulus amebocyte lysate assay; EC-LAL=Endpoint chromogenic LAL assay; KC/T-LAL= Kinetic and/or Turbidimetric chromogenic LAL assay; rFC= recombinant Factor C Assay; SI-EIA=Specific Inhibition Enzyme-linked ImmunoAssay; Glucatell= Glucatell modification of the LAL assay; LOD=Limit of Detection; NR=Not Reported.

conditions may also play a role as has been described for endotoxin (Noss et al., 2010a; Spaan et al., 2007). However, such analytical errors are unlikely to be a major contributor to the total variability of exposure for these agents as differences in intra-laboratory variations are generally small.

Besides exposure sources, other important determinants of endotoxin exposure include the dustiness of materials handled, the production in bulk (i.e. in large quantities), and the cyclical nature of the process (Spaan et al., 2008a). Personal levels of exposure largely depend on the activities performed by the workers as well as the environmental conditions and workplace characteristics. For example, among livestock workers practices related to ventilation, animal feeding, distribution of bedding and improved building hygiene have been demonstrated as important determinants for exposure to endotoxins and (1 → 3)-β-D-glucans (Basinas et al., 2015; Samadi et al., 2009; Thilsing et al., 2015). Similarly, in sewage treatment plants higher exposures have been reported among workers performing activities related to cleaning and maintenance (Spaan et al., 2008b).

Within residential environments the level of airborne endotoxin exposure has been reported to average between 0.36-6.5 EU/m^3 in absence of an obvious exposure source (Table 8.1) which is similar to that reported for the general environment (Madsen, 2006). However, in other settings where a direct source of exposure is present, such as the burning of biomass, airborne endotoxin levels may increase to 64.0 EU/m^3 (Table 8.1). Similar differences in exposure patterns have been reported for glucan exposures with burning of biomass (Semple et al., 2010) and with the presence of moisture/mould problems within the building (Adhikari et al., 2010). The importance of mould as an exposure source for residential and public environments is well documented also from exposure studies in schools and office buildings (Rylander et al., 1998) and this is broadly supported by results from studies that utilised samples of settled house dust, like floor dust and mattress dust (Douwes et al., 1999; Douwes et al., 1998; Schram et al., 2005; Gehring et al., 2001). Very little information is available concerning other determinants of airborne levels of these agents within home environments. However, results from studies on settled house dust suggest that keeping pets, the number of occupants in the home, the flooring type, whether or not the house is a farm residence, the season and the heating system are important factors in determining the dust composition in these environments (Douwes et al., 1998; Schram et al., 2005; Giovannangelo et al., 2007; Casas et al., 2013; Abraham et al., 2005; Holst et al., 2015a).

The other microbial cell wall agents which have been reported to be elevated in settled dust from indoor environments of houses and farms include muramic acid and ergosterol (Poole et al., 2010; Van Strien et al., 2004) as well as EPS (Giovannangelo et al., 2007; Casas et al., 2013). However, little is known about actual airborne levels of these agents (Dales et al., 2006; Adhikari et al., 2014). Furthermore, it has to be noted that collection of samples and analysis of settled house dust, primarily from floor and mattresses, has been the most common approach for determination of microbial cell wall agent concentrations in epidemiological studies in the home environment. This is due to the increased cost-efficiency of these sampling strategies compared to active airborne dust sampling, as they allow collection of dust to be performed by the participants themselves. As deposited dust is time-integrated, it is less vulnerable to short term variation in exposure and allows relative ranking of exposure levels (Douwes, 2005; Tischer et al., 2011). Nevertheless, results obtained through these methods are unlikely to be fully representative of actual airborne levels and personal exposure within indoor home environments (Adhikari et al., 2010; Adhikari et al., 2014; Noss et al., 2008; Samadi et al., 2010). Recently a simple and rather inexpensive method for passive collection of airborne dusts, the Electrostatic Dustfall Collectors (EDCs), has become available which is proving rather promising with regard to sampling efficiency for endotoxin and glucans (Noss et al., 2010a; Noss et al., 2008; Samadi et al., 2010; Frankel et al., 2012b; Jacobs et al., 2014).

8.8 Health Effects

8.8.1 Endotoxin Exposure and the Janus Faced Effect on Health

Endotoxin is a well-established pro-inflammatory agent with a broad range of health effects documented in epidemiological, toxicological, and experimental studies in humans. It is considered one of the main causes of respiratory disease in populations highly exposed to organic dusts such as farmers, cotton and grain workers (Rylander, 2006). Endotoxin can cause both acute and chronic effects. Endotoxin exposure has been linked to acute symptoms such as wheezing, dyspnea, irritation of the nose and throat, chest tightness, dry cough, fever, headache, and acute airway obstruction and inflammation (Douwes et al., 2003; Rylander, 2006; Bakirci et al., 2007; Castellan et al., 1987). High endotoxin exposure has been shown to cause organic dust toxic syndrome (ODTS) and to increase the risk of chronic respiratory diseases, including extrinsic allergic alveolitis (i.e. Famer's lung), chronic bronchitis, accelerated lung function decline, asthma and asthma-like syndrome. Endotoxin can also simply increase disease severity by causing lung function adverse effects and promoting inflammatory responses (Smit et al., 2006; Donham et al., 2000; Sigsgaard et al., 2004; Wang et al., 2002; Liu, 2002). Positive associations between endotoxin and malignant disease such as nasopharyngeal cancers have also been reported among cotton workers(Li et al., 2006; Fang et al., 2013). In contrast, more recently a protective effect of endotoxin exposure against lung cancer has also been proposed (Lenters et al., 2010). However, evidence supporting this association remains limited primarily to studies among cotton workers (Astrakianakis et al., 2007; McElvenny et al., 2011). Respiratory symptoms and bronchial hyperresponsiveness have been demonstrated among workers and healthy volunteers to initiate with exposure levels in the range of 100 to 200 EU/m^3 (Basinas et al., 2012a; Smit et al., 2008; Castellan et al., 1987; Larsson et al., 1994; Smit et al., 2010; Latza et al., 2004).

During recent decades evidence has become available for an inverse association between endotoxin exposure and atopy, allergic rhinitis and/or atopic asthma. These protective effects from endotoxin have been observed particularly among children (Braun-Fahrlander et al., 2002; Gereda et al., 2000; Douwes et al., 2006; Schram-Bijkerk et al., 2005; Von Mutius et al., 2000) but also among adults, in workers such as farmers (Eduard et al., 2004; Portengen et al., 2005) agriculture workers (Basinas et al., 2012a; Smit et al., 2008; Smit et al., 2010) and even for residential endotoxin exposures (Gehring et al., 2004; Bakolis et al., 2012). Among the adult population the protective effects of endotoxin against atopy and atopic sensitization were always observed in conjunction with a significant increase in risk for non-allergic respiratory morbidity (Basinas et al., 2012a; Eduard et al., 2004; Smit et al., 2008; Smit et al., 2010; Portengen et al., 2005) suggesting a Janus-faced (i.e. dual) role for endotoxin on the development of health symptoms among humans. For example, in a pooled analysis of four epidemiological studies from the Netherlands and Denmark including workers in farming, agricultural processing and power plants using biofuel as well as students in veterinary medicine, an inverse dose-dependent association between measured endotoxin exposure and allergic sensitization and hay fever (i.e. allergic rhinitis) was observed (Basinas et al., 2012a). However, in the same population increased endotoxin exposure was associated with an increased risk for organic dust toxic syndrome and chronic bronchitis when exposure exceeded 100 EU/m^3 (Fig. 8.2).

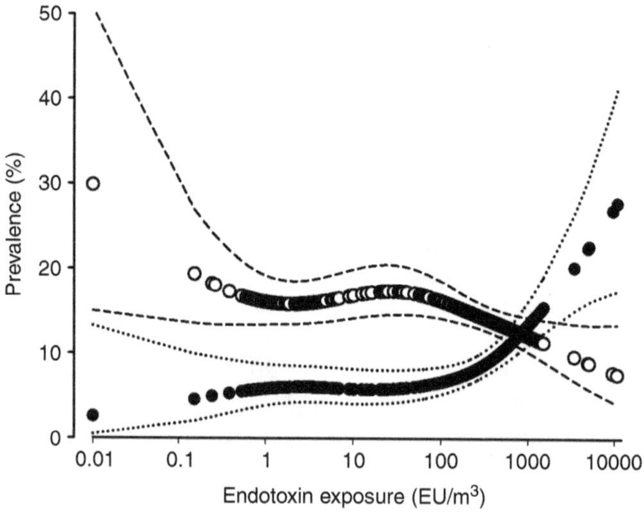

Fig. 8.2 The association between endotoxin exposure and prevalence of hay fever (circles) and chronic bronchitis (filled circles) in a population of 3883 Dutch and Danish employees in veterinary medicine, power plants using biofuel, agricultural processing, and farming. (From Basinas et al., 2012a)

These findings are in line with the hygiene hypothesis (see below) and suggest that some individuals may be more susceptible to endotoxin exposure than others. Though initial interpretation of these findings was fairly cautious, because of the cross-sectional nature of the research studies, emerging results from longitudinal studies among Danish farmers and Dutch agricultural workers seem to confirm the protective effects of adult endotoxin exposure on atopy and atopic sensitization (Elholm et al., 2011; Spierenburg et al., 2016). The individual immunological response to endotoxin exposure is determined by the interaction between dose and timing of exposure, other environmental factors and genetic predisposition (Vandenbulcke et al., 2006).

8.8.2 A Proposed Immunological Mechanism Supporting the Hygiene Hypothesis

The hygiene hypothesis suggests that exposure to microbial components like endotoxin promotes the development of a healthy immune system. The adaptive immune response is thus modified by prior events like infection (Liebers et al., 2008). The initial proposed mechanism associated with the hygiene hypothesis was that an increased microbial exposure induces a shift from atopic T-helper type 2 (Th2) responses to Th1-dominated responses through stimulation of the innate immune system. In addition, it has emerged that regulatory T cells (T_{reg}) play a crucial role in suppressing allergic and non-allergic immune responses (Schaub et al., 2006; Renz et al., 2006; Sigsgaard and Heederik, 2005). Toll-like receptors (TLRs) present on the cell surface of innate immune cells recognize microbial motifs called microbial-associated molecular patterns (MAMPs) (Sabroe et al., 2003). Following entry to the body through the airways, endotoxins/LPS will encounter alveolar macrophages carrying CD14 and LPS binding receptors (Ingalls et al., 1999). The binding of LPS to CD14 is mediated by LPS binding protein (LBP). Via toll-like receptors (TLR-3 and TLR-4) (Beutler, 2004) the alveolar macrophages will be activated, leading to the production and release of proinflammatory cytokines (Reed and Milton, 2016). Cytokines associated with endotoxin exposure are TNF-α, interleukin (IL) 1-β, IL-6, and IL-8, as well as metabolites of arachidonic acid. These cytokines will then recruit and activate neutrophils, resulting in local and systemic inflammation with leukocytosis and neutrophilia. This effect can also be seen experimentally or observationally: swine dust, cotton dust, or grain dust exposure is found to increase IL-1β, IL-6, IL-8, TNF-α, and circulating neutrophils in the airways and causes airway obstruction and methacholine responsiveness (Li et al., 1995; Schwartz et al., 1995; Wang et al., 1999; Wang et al., 1997; Senthilselvan et al., 1997; Malmberg and Larsson, 1993; Forteza et al., 1994; Jorna et al., 1994; Rylander and Bergstrom, 1993). Impairment of TLR4 has also been found to be associated with a history of atopic disease (Prefontaine et al., 2010).

8.8.3 The Role of T Regulatory Cells (T_{reg})

Lack of functional T_{reg} cells, due to a defect in T_{reg} activation is associated with insufficient repression of both Th1 and Th2 immune responses and has been found to be associated with atopic disease (Savilahti et al., 2010; Braga et al., 2011; Braga et al., 2012; O'Garra and Vieira, 2004). $Treg_s$ are a subpopulation of T cells which modulate the immune system, maintain tolerance to self-antigen, and prevent autoimmune disease. T regulatory cells are a T cell subset that produces IL-10 and TGF-β. T_{reg} cells may act both by cytokine production and by cell–cell contact signals, as programmed death-1, glucocorticoid-induced TNF receptor, membrane TGF-β, and cytotoxic T lymphocyte-associated antigen 4 (CTLA-4). T_{reg} cells contribute to the control of allergen-specific immune responses in five major ways: (1) T_{reg} cells suppress antigen-presenting cells that support the generation of effector Th2 and Th1 cells. (2) They suppress Th2 and Th1 cells. (3) They regulate B cells by suppression of allergen-specific Immunoglobulin E (IgE) antibodies and induction of Immunoglobulin G4 (IgG4), A (IgA), or both. (4) They suppress mast cells, basophils, and eosinophils. (5) They interact with resident tissue cells and remodeling (Braga et al., 2012).

8.8.4 Microbial Diversity vs. The Effect of Single Agents

Besides endotoxin, Ege et.al (Ege et al., 2011) recently argued that most likely it is the diversity and wider range of types of microbes offered by the farming environment that contributes to beneficial effects of farming exposure, rather than a single agent such as endotoxin. Other studies have tried to determine the effect of specific microorganisms on the development of allergies, and recently the effect of exposure to Acinetobacter lwoffii F78 and Lactococcus lactis G121 was investigated (Debarry et al., 2007). These two bacteria are in particular found on cattle farms. Both bacteria showed an ability to reduce allergic reactions in mice, to activate mammalian cells in vitro, and to induce a Th1-polarizing program in dendritic cells (Brand et al., 2011). Findings like these suggest that exposure to other components than cell wall agents may affect health as well, however the specific role and contribution to the health effects of the various microbial agents as well as their potential synergic effects with cell wall agents is still to be established.

8.8.5 Diverse Microbial Exposure and TLR Expression

Research has shown that prenatal and/or early life exposure to the rich microbial environment of traditional farms induces an up-regulation of innate immunity receptors that is both robust and long-lasting (Stern et al., 2007). Exposure of the mother during pregnancy to inhalant allergens is less likely to result in sensitization in the child than exposure of the child in early infancy (Kihlstrom et al., 2003; Szepfalusi et al., 2000). It has been seen that peripheral blood cells from farm children expressed significantly higher levels of *CD14*,Toll-like receptor 2 (*TLR2*) and Toll-like receptor 4 (*TLR4*) than cells from non-farm children. Furthermore, it was indicated that it was farming exposure of the pregnant mothers that were associated with the enhanced expression (Ege et al., 2006; Lauener et al., 2002). Additionally reduced maternal T_{reg} numbers and increased *Th2* cytokine production during pregnancy has been found to influence the allergy risk of the child (Hinz et al., 2010). There is evidence that among children of farmers genetic variation in *TLR2* is a major determinant of the susceptibility to asthma and allergies (Eder et al., 2004).

8.8.6 (1 → 3)-β-D-Glucan Exposure and Known Health Effects

Indoor exposure to fungi has been associated with the development of respiratory symptoms, though the mechanisms are far from clear (Douwes, 2005). It has been shown that (1 → 3)-β-D-glucan can initiate a wide range of biological responses in vertebrates including stimulation of the mononuclear phagocyte system (Di Luzio, 1979), activation of neutrophils (Zhang and Petty, 1994), macrophages (Adachi et al., 1994; Lebron et al., 2003), complement (Saito et al., 1992) and possibly eosinophils (Mahauthaman et al., 1988). These potent biological properties of (1 → 3)-β-D-glucan are relevant irrespective of originating from either live or dead organisms. However, clarifying the health effect of (1 → 3)-β-D-glucan exposure has so far been very challenging and largely inconclusive as many studies have reported conflicting results. Some of the health effects which have been evaluated include lung function [forced expiratory volume in 1 s (FEV1) and peak flow (PEF) variability], nasal congestion, airway hyperreactivity, atopy, symptoms (upper and lower respiratory symptoms, eye irritations, head ache, fatigue/tiredness, joint pains, skin symptoms, flu-like symptoms, nausea, gastro-intestinal symptoms), inflammation characterized by inflammatory cells (T-lymphocytes, neutrophils, eosinophils, macrophages), and cytokines and other inflammatory markers –i.e interleukin (IL)-1ß, IL-4, IL-6, IL-8, IL-10, Interferon (INF)-c, Tumour necrosis factor (TNF)-a, Eosinophil cationic protein (ECP), Myeloperokidase (MPO), C-reactive protein (CRP), albumin- in blood, sputum and nasal lavage (Douwes, 2005).

In an epidemiological context positive associations with glucan exposures have been reported among both adults and children in relation to symptoms of upper airway irritation and inflammation, airway responsiveness, increased peak expiratory flow variability, systemic reactions and atopy (Gladding et al., 2003; Rylander et al., 1999; Thorn et al., 1998; Thorn and Rylander, 1998; Douwes et al., 2000; Bønløkke et al., 2006). Interpretation of the study findings though need to be made cautiously as population sizes were rather small, study designs were cross-sectional and in some cases potential interactions with other co-existing exposures were not taken into account. In a number of studies strong correlations between endotoxin and (1 → 3)-β-D-glucan levels have been reported and previously experimental studies in animals have suggested inflammatory responses to enhance in response to combinations of glucans and endotoxin exposures (Douwes, 2005). More research studies with improved and standardised exposure assessments in longitudinal designs are warranted to provide insight on the actual health effects of exposure to glucans.

8.8.7 Health Effects of Other Cell Wall Agents

As mentioned earlier, to date only a limited number of studies addressed the health effects of cell wall agents other than endotoxins and (1 → 3)-β-D-glucans. There is some evidence for a potential and maybe even independent role for muramic acid and ergosterol in the development of health symptoms. Specifically, in a case comparison study of symptomatic and non-symptomatic workers of an office building with a history of water damage Park et al. (2008) examined the association between house dust measured fungi, ergosterol and endotoxin levels and asthma. The authors reported increased levels of ergosterol and total fungi to be associated with an increased prevalence of current asthma (Park et al., 2008). A similar association has also been reported in a cross-sectional analysis of the 1996 follow up of the European Community Respiratory Health Survey (ECRHS) cohort (Dharmage et al., 2001). However, cross-sectional studies from Canada reported no association between ergosterol and respiratory symptoms and cough among elementary school children (Dales et al., 1999), whereas neither ergosterol nor indoor moulds seem to influence the illness-associations with endotoxin exposure in infants (Dales et al., 2006). In contrast to these findings, among school-aged farm children from Austria, Germany, and Switzerland, increased levels of muramic acid were found to be associated with lower prevalence of wheezing but not with atopic sensitization (Van Strien et al., 2004). An inverse association between increased levels of muramic acid in classroom dust and the prevalence of wheeze and daytime breathlessness has been reported also among Chinese school children (Zhao et al., 2008). Based on these findings muramic acid like endotoxin has been suggested to serve as an independent marker of microbial exposure (Van Strien et al., 2004). Similar inverse associations have been found between EPS exposure in mattress dust in German school children and doctor-diagnosed asthma and rhinitis(Tischer et al., 2011). More recently, chitin, one of the earliest identified and most abundant extracellular polysaccharides in nature, has been hypothesised as playing a role in the development of asthma and allergies but the actual supporting evidence to date remains rather small (Brinchmann et al., 2011).

8.9 Conclusions and Future Directions

We spend a large proportion of our time indoors, and it is needless to say that our indoor environment will affect us for better or for worse. Indoor and occupational exposures to microbial cell wall agents and their associated health effects are far from elucidated. It is therefore of great importance to continue to improve our understanding of cell wall component agents that contaminate our indoor air and how they affects us. It is clear that the well-studied endotoxins are involved in the development of the adverse and protective health effects, but for glucans the evidence is more limited and inconclusive. There is some evidence that other microbial cell wall agents are involved in the development of the adverse and/or protective health effects as well. However, relevant studies have been sparse and very diverse in their design and applied methods.

In addition, the literature shows large variation in exposure to microbial cell wall agents in indoor occupational environments, and we still simply lack studies of actual airborne levels of exposures and determinants of residential indoor air. The fact that many different assays and sampling methods have been deployed for evaluation of exposures and levels complicates comparison of results and affects the establishment of proper exposure limits to protect workers from excess exposure to these agents. Standardisation in methods of determination is highly recommended for future studies as well as a broader adaptation of the recently available passive airborne dust sampling methods (e.g. EDCs or dustfall collectors) for residential exposures. It has recently been suggested that both PM10 and PM>10 size fractions elicit a pro-inflammatory response in airway epithelial cells (Hawley et al., 2015), which means that dust size fractions should be taken into consideration when assessing potential risks from exposure to agricultural dusts and other microbial agents which could be found in the indoor environment.

Next to direct effects of cell-wall agents, other components and/or microbial diversity might be important with respect to both detrimental and beneficial health effects. The development and application of molecular techniques in exposure assessment – as reviewed by Casas et al. (2016) – will aid to study the role of microbial diversity and specific microbes in future studies, and may help to understand the role of the individual and combined exposures in health. Such knowledge is highly needed both for the development of targeted prevention strategies and the establishment of adequate exposure limits especially within workplaces. Further research, in particular studies in large populations with a longitudinal design involving the assessment of the health effects of both distinct microbial cell wall agents and co-existent microbes is needed to provide more in-depth insight.

References

Abraham JH, Gold DR, Dockery DW et al (2005) Within-home versus between-home variability of house dust endotoxin in a birth cohort. Environ Health Perspect 113:1516–21

Adachi Y, Okazaki M, Ohno N, Yadomae T (1994) Enhancement of cytokine production by macrophages stimulated with (1–>3)-beta-D-glucan, grifolan (GRN), isolated from Grifola frondosa. Biol Pharm Bull 17:1554–60

Adhikari A, Kettleson EM, Vesper S et al (2014) Dustborne and airborne Gram-positive and Gram-negative bacteria in high versus low ERMI homes. Sci Total Environ 482–483:92–9. doi:10.1016/j.scitotenv.2014.02.110

Adhikari A, Lewis JS, Reponen T et al (2010) Exposure matrices of endotoxin, (1→3)-β-d-glucan, fungi, and dust mite allergens in flood-affected homes of New Orleans. Sci Total Environ 408:5489–98. doi:10.1016/j.scitotenv.2010.07.087

Alwis KU, Milton DK (2006) Recombinant factor C assay for measuring endotoxin in house dust: comparison with LAL, and (1 –> 3)-beta-D-glucans. Am J Ind Med 49:296–300. doi:10.1002/ajim.20264

Arbejdstilsynet (2007) At-vejledning. Grænseværdier for stoffer og materialer. [Danish Working Environment Authority. Limit values for substances and materials]. Publication no. C.0.1. The Danish Working Environment Authority, Copenhagen, Denmark. https://arbejdstilsynet.dk/da/regler/at-vejledninger/g/c-0-1-graensevaerdi-for-stoffer-og-mat. Accessed 05 Jan 2016

Arbeidstilsynet (2011) Veiledning om administrative normer for forurensning i arbeidsatmosfære. [The Norwegian Labour Inspection Authority. Guidance for administrative standards for contamination of the work environment]. Manual no. 361. Trondheim, Norway. Available via http://www.arbeidstilsynet.no/binfil/download2.php?tid=77907. Accessed 05 Jan 2016

Arteaga V, Mitchell D, Armitage T et al (2015) Cage versus noncage laying-hen housings: respiratory exposures. J Agromed. 20:245–255. doi:10.1080/1059924X.2015.1044681

Astrakianakis G, Seixas NS, Ray R et al (2007) Lung cancer risk among female textile workers exposed to endotoxin. J Natl Cancer Inst 99:357–364. doi:10.1093/jnci/djk063

Bakirci N, Kalaca S, Francis H et al (2007) Natural history and risk factors of early respiratory responses to exposure to cotton dust in newly exposed workers. J Occup Environ Med 49:853–61. doi:10.1097/JOM.0b013e3180dca598

Bakolis I, Doekes G, Heinrich J et al (2012) Respiratory health and endotoxin: associations and modification by CD14/-260 genotype. Eur Respir J 39:573–81. doi:10.1183/09031936.00164410

Basinas I, Schlunssen V, Heederik D et al (2012a) Sensitisation to common allergens and respiratory symptoms in endotoxin exposed workers: a pooled analysis. Occup Environ Med 69:99–106. doi:10.1136/oem.2011.065169

Basinas I, Sigsgaard T, Heederik D et al (2012b) Exposure to inhalable dust and endotoxin among Danish livestock farmers: results from the SUS cohort study. J Env Monit 14:604–614. doi:10.1039/c1em10576k

Basinas I, Sigsgaard T, Kromhout H et al (2015) A comprehensive review of levels and determinants of personal exposure to dust and endotoxin in livestock farming. J Expo Sci Env Epidemiol 25:123–137. doi:10.1038/jes.2013.83

Beutler B (2004) Inferences, questions and possibilities in Toll-like receptor signalling. Nature 430:257–263

Bønløkke JH, Stridh G, Sigsgaard T et al (2006) Upper-airway inflammation in relation to dust spiked with aldehydes or glucan. Scand J Work Environ Health 32:374–82

Bos JD, Meinardi MM (2000) The 500 Dalton rule for the skin penetration of chemical compounds and drugs. Exp Dermatol 9:165–9

Bos MP, Robert V, Tommassen J (2007) Biogenesis of the Gram-Negative bacterial outer membrane. Ann Rev Microbiol 61:191–214. doi:10.1146/annurev.micro.61.080706.093245

Braga M, Quecchia C, Cavallucci E et al (2011) T regulatory cells in allergy. Int J Immunopathol Pharmacol 24:55S–64S

Braga M, Schiavone C, Di Gioacchino G et al (2012) Environment and T regulatory cells in allergy. Sci Total Environ 423:193–201. doi:10.1016/j.scitotenv.2010.08.015

Brand S, Teich R, Dicke T et al (2011) Epigenetic regulation in murine offspring as a novel mechanism for transmaternal asthma protection induced by microbes. J Allergy Clin Immunol 128:617–618. doi:10.1016/j.jaci.2011.04.035

Braun-Fahrlander C, Riedler J, Herz U et al (2002) Environmental exposure to endotoxin and its relation to asthma in school-age children. N Engl J Med 347:869–877. doi:10.1056/NEJMoa020057

Brinchmann BC, Bayat M, Brøgger T et al (2011) A possible role of chitin in the pathogenesis of asthma and allergy. Ann Agric Environ Med 18:7–12

Brooks CR, Siebers R, Crane J et al (2013) Measurement of β-(1,3)-glucan in household dust samples using Limulus amebocyte assay and enzyme immunoassays: an inter-laboratory comparison. Environ Sci Process Impacts 15:405–11. doi:10.1039/c2em30749a

Burch JB, Svendsen E, Siegel PD et al (2010) Endotoxin exposure and inflammation markers among agricultural workers in Colorado and Nebraska. J Toxicol Env Heal A 73:5–22. doi:10.1080/15287390903248604

Casas L, Tischer C, Täubel M (2016) Pediatric asthma and the indoor microbial environment. Curr Environ Heal Reports 3:238–249. doi:10.1007/s40572-016-0095-y

Casas L, Tischer C, Wouters IM et al (2013) Endotoxin, extracellular polysaccharides, and beta (1-3)-glucan concentrations in dust and their determinants in four European birth cohorts: results from the HITEA project. Indoor Air 23:208–218. doi:10.1111/ina.12017

Castellan RM, Olenchock SA, Kinsley KB, Hankinson JL (1987) Inhaled endotoxin and decreased spirometric values. An exposure-response relation for cotton dust. N Engl J Med 317:605–610. doi:10.1056/NEJM198709033171005

Cherid H, Foto M, Miller JD (2011) Performance of two different Limulus amebocyte lysate assays for the quantitation of fungal glucan. J Occup Environ Hyg 8:540–3. doi:10.1080/15459624.2011.601994

Cherrie JW, Semple S, Christopher Y et al (2006) How important is inadvertent ingestion of hazardous substances at work? Ann Occup Hyg 50:693–704. doi:10.1093/annhyg/mel035

Christiani DC, Velazquez A, Wilcox M, Olenchock SA (1993) Airborne endotoxin concentrations in various work areas within a cotton mill in Central America. Environ Res 60:187–192. doi:10.1006/enrs.1993.1026

Christiani DC, Ye TT, Wegman DH et al (1994) Pulmonary function among cotton textile workers. A study of variability in symptom reporting, across-shift drop in FEV1, and longitudinal change. Chest 105:1713–21. doi:10.1378/chest.105.6.1713

Chun DT, Bartlett K, Gordon T et al (2006) History and results of the two inter-laboratory round robin endotoxin assay studies on cotton dust. Am J Ind Med 49:301–306. doi:10.1002/ajim.20266

Coggins MA, Hogan VJ, Kelly M et al (2012) Workplace exposure to bioaerosols in podiatry clinics. Ann Occup Hyg 56:746–753. doi:10.1093/annhyg/mer124

Cyprowski M, Ławniczek-Wałczyk A, Górny RL (2015a) Airborne peptidoglycans as a supporting indicator of bacterial contamination in a metal processing plant. Int J Occup Med Environ Health 29:427–437. doi:10.13075/ijomeh.1896.00594

Cyprowski M, Sobala W, Buczyńska A, Szadkowska-Stańczyk I (2015b) Endotoxin exposure and changes in short-term pulmonary function among sewage workers. Int J Occup Med Environ Health 28:803–811. doi:10.13075/ijomeh.1896.00460

Dales R, Miller D, Ruest K et al (2006) Airborne endotoxin is associated with respiratory illness in the first 2 years of life. Environ Health Perspect 114:610–614. doi:10.1289/ehp.8142

Dales RE, Miller D, White J (1999) Testing the association between residential fungus and health using ergosterol measures and cough recordings. Mycopathologia 147:21–7

Dassonville C, Demattei C, Vacquier B et al (2008) Indoor airborne endotoxin assessment in homes of Paris newborn babies. Indoor Air 18:480–7. doi:10.1111/j.1600-0668.2008.00549.x

Debarry J, Garn H, Hanuszkiewicz A et al (2007) Acinetobacter lwoffii and Lactococcus lactis strains isolated from farm cowsheds possess strong allergy-protective properties. J Allergy Clin Immunol 119:1514–1521. doi:10.1016/j.jaci.2007.03.023

DECOS (2010) Endotoxins: health based recommended exposure limit. A report of the Health Council of the Netherlands, publication no. 2010/04OSH. Health Council of the Netherlands, The Hague.

DECOS (2011) Grain dust: Health-based recommended occupational exposure limit. A report of the Health Council of the Netherlands, publication no. 2011/13. Health Council of the Netherlands, The Hague.

Dharmage S, Bailey M, Raven J et al (2001) Current indoor allergen levels of fungi and cats, but not house dust mites, influence allergy and asthma in adults with high dust mite exposure. Am J Respir Crit Care Med 164:65–71. doi:10.1164/ajrccm.164.1.9911066

Di Luzio NR (1979) Lysozyme, glucan-activated macrophages and neoplasia. J Reticuloendothel Soc 26:67–81

Ding JL, Navas MA, Ho B (1995) Molecular cloning and sequence analysis of factor C cDNA from the Singapore horseshoe crab, Carcinoscorpius rotundicauda. Mol Mar Biol Biotechnol 4:90–103

Donham KJ, Cumro D, Reynolds SJ, Merchant JA (2000) Dose-response relationships between occupational aerosol exposures and cross-shift declines of lung function in poultry workers: recommendations for exposure limits. J Occup Environ Med 42:260–269

Douwes J (2005) (1–>3)-Beta-D-glucans and respiratory health: a review of the scientific evidence. Indoor Air 15:160–169. doi:10.1111/j.1600-0668.2005.00333.x

Douwes J, Doekes G, Heinrich J et al (1998) Endotoxin and β(1→3)-glucan in house dust and the relation with home characteristics: a pilot study in 25 german houses. Indoor Air 8:255–263. doi:10.1111/j.1600-0668.1998.00006.x

Douwes J, Doekes G, Montijn R et al (1996) Measurement of beta(1–>3)-glucans in occupational and home environments with an inhibition enzyme immunoassay. Appl Env Microbiol 62:3176–3182

Douwes J, Thorne P, Pearce N, Heederik D (2003) Bioaerosol health effects and exposure assessment: progress and prospects. Ann Occup Hyg 47:187–200. doi:10.1093/annhyg/meg032

Douwes J, van der Sluis B, Doekes G et al (1999) Fungal extracellular polysaccharides in house dust as a marker for exposure to fungi: relations with culturable fungi, reported home dampness, and respiratory symptoms. J Allergy Clin Immunol 103:494–500. doi:10.1016/S0091-6749(99)70476-8

Douwes J, van Strien R, Doekes G et al (2006) Does early indoor microbial exposure reduce the risk of asthma? The prevention and incidence of asthma and mite allergy birth cohort study. J Allergy Clin Immunol 117:1067–1073. doi:10.1016/j.jaci.2006.02.002

Douwes J, Zuidhof A, Doekes G et al (2000) (1–>3)-beta-D-glucan and endotoxin in house dust and peak flow variability in children. Am J Respir Crit Care Med 162:1348–54. doi:10.1164/ajrccm.162.4.9909118

Eder W, Klimecki W, Yu L et al (2004) Toll-like receptor 2 as a major gene for asthma in children of European farmers. J Allergy Clin Immunol 113:482–8. doi:10.1016/j.jaci.2003.12.374

Eduard W, Douwes J, Omenaas E, Heederik D (2004) Do farming exposures cause or prevent asthma? Results from a study of adult Norwegian farmers. Thorax 59:381–386

Eduard W, Pearce N, Douwes J (2009) Chronic bronchitis, COPD, and lung function in farmers: the role of biological agents. Chest 136:716–725. doi:10.1378/chest.08-2192

Ege MJ, Bieli C, Frei R et al (2006) Prenatal farm exposure is related to the expression of receptors of the innate immunity and to atopic sensitization in school-age children. J Allergy Clin Immunol 117:817–823. doi:10.1016/j.jaci.2005.12.1307

Ege MJ, Mayer M, Normand A-C et al (2011) Exposure to environmental microorganisms and childhood asthma. N Engl J Med 364:701–709. doi:10.1056/NEJMoa1007302

Elholm G, Omland Ø, Sigsgaard T et al (2011) Endotoxin exposure protects against new onset of pollen sensitisation. Eur Respir J 38:p3315

Fang SC, Mehta AJ, Hang JQ et al (2013) Cotton dust, endotoxin and cancer mortality among the Shanghai textile workers cohort: a 30-year analysis. Occup Environ Med 70:722–9. doi:10.1136/oemed-2012-100950

Finkelman MA, Tamura H (2005) Detection and measurement of (1→3)-beta-D-glucan with Limulus amebocyte lysate-based reagents. In: Young S-H, Castranova V (eds) Toxicology of 1→3-Beta-Glucans: Glucans as a Marker for Fungal Exposure. CRC Press, Taylor & Francis Group, Boca Raton, USA, pp 179–198

Fishwick D, Allan LJ, Wright A, Curran AD (2001) Assessment of exposure to organic dust in a hemp processing plant. Ann Occup Hyg 45:577–83

Forteza R, Lauredo IT, Burch R, Abraham WM (1994) Extracellular metabolites of Pseudomonas aeruginosa produce bronchoconstriction by different mechanisms. Am J Respir Crit Care Med 149:687–693. doi:10.1164/ajrccm.149.3.8118638

Frankel M, Bekö G, Timm M et al (2012a) Seasonal variations of indoor microbial exposures and their relation to temperature, relative humidity, and air exchange rate. Appl Environ Microbiol 78:8289–8297. doi:10.1128/AEM.02069-12

Frankel M, Timm M, Hansen EW, Madsen AM (2012b) Comparison of sampling methods for the assessment of indoor microbial exposure. Indoor Air 22:405–414. doi:10.1111/j.1600-0668.2012.00770.x

Garcia J, Bennett DH, Tancredi D et al (2013) Occupational exposure to particulate matter and endotoxin for California dairy workers. Int J Hyg Environ Health 216:56–62. doi:10.1016/j.ijheh.2012.04.001

Gehring U, Bischof W, Schlenvoigt G et al (2004) Exposure to house dust endotoxin and allergic sensitization in adults. Allergy 59:946–952. doi:10.1111/j.1398-9995.2004.00551.x

Gehring U, Douwes J, Doekes G et al (2001) Beta(1–>3)-glucan in house dust of German homes: housing characteristics, occupant behavior, and relations with endotoxins, allergens, and molds. Environ Health Perspect 109:139–44

Gereda JE, Leung DY, Thatayatikom A et al (2000) Relation between house-dust endotoxin exposure, type 1 T-cell development, and allergen sensitisation in infants at high risk of asthma. Lancet 355:1680–1683. doi:S014067360002239X. [pii]

Gilbert Y, Veillette M, Mériaux A et al (2010) Metalworking fluid-related aerosols in machining plants. J Occup Environ Hyg 7:280–289. doi:10.1080/15459621003680227

Giovannangelo M, Gehring U, Nordling E et al (2007) Determinants of house dust endotoxin in three European countries – the AIRALLERG study. Indoor Air 17:70–79. doi:10.1111/j.1600-0668.2006.00461.x

Gladding T, Thorn J, Stott D (2003) Organic dust exposure and work-related effects among recycling workers. Am J Ind Med 43:584–91. doi:10.1002/ajim.10220

Gorman NgM, Semple S, Cherrie JW et al (2012) The relationship between inadvertent ingestion and dermal exposure pathways: a new integrated conceptual model and a database of dermal and oral transfer efficiencies. Ann Occup Hyg 56:1000–12. doi:10.1093/annhyg/mes041

Halstensen AS, Heldal KK, Wouters IM et al (2013) Exposure to grain dust and microbial components in the norwegian grain and compound feed industry. Ann Occup Hyg 57:1105–1114. doi:10.1093/annhyg/met036

Harper M, Andrew ME (2006) Airborne endotoxin in woodworking (joinery) shops. J Env Monit 8:73–78. doi:10.1039/b508065g

Hawley B, Schaeffer J, Poole JA et al (2015) Differential response of human nasal and bronchial epithelial cells upon exposure to size-fractionated dairy dust. J Toxicol Environ Heal Part A 78:583–594. doi:10.1080/15287394.2015.1015699

Hinz D, Simon JC, Maier-Simon C et al (2010) Reduced maternal regulatory T cell numbers and increased T helper type 2 cytokine production are associated with elevated levels of immunoglobulin E in cord blood. Clin Exp Allergy 40:419–426. doi:10.1111/j.1365-2222.2009.03434.x

Holst G, Host A, Doekes G et al (2015a) Determinants of house dust, endotoxin, and beta-(1–>3)-D-glucan in homes of Danish children. Indoor Air 25:245–259. doi:10.1111/ina.12143

Holst GJ, Høst A, Doekes G, et al (2015b) Allergy and respiratory health effects of dampness and dampness-related agents in schools and homes: a cross-sectional study in Danish pupils. Indoor Air 1–12. doi:10.1111/ina.12275

Ingalls RR, Heine H, Lien E et al (1999) Lipopolysaccharide recognition, CD14, and lipopolysaccharide receptors. Infect Dis Clin North Am 13:341–53. vii

Jacobs JH, Krop EJM, Borras-Santos A et al (2014) Endotoxin levels in settled airborne dust in European schools: the HITEA school study. Indoor Air 24:148–57. doi:10.1111/ina.12064

Jorna TH, Borm PJ, Valks J et al (1994) Respiratory symptoms and lung function in animal feed workers. Chest 106:1050–1055

Karvonen AM, Hyvärinen A, Rintala H et al (2014) Quantity and diversity of environmental microbial exposure and development of asthma: a birth cohort study. Allergy 69:1092–101. doi:10.1111/all.12439

Kihlstrom A, Lilja G, Pershagen G, Hedlin G (2003) Exposure to high doses of birch pollen during pregnancy, and risk of sensitization and atopic disease in the child. Allergy 58:871–877

Lappalainen MHJ, Hyvärinen A, Hirvonen M-R et al (2012) High indoor microbial levels are associated with reduced Th1 cytokine secretion capacity in infancy. Int Arch Allergy Immunol 159:194–203. doi:10.1159/000335596

Larsson KA, Eklund AG, Hansson LO et al (1994) Swine dust causes intense airways inflammation in healthy subjects. Am J Respir Crit Care Med 150:973–977. doi:10.1164/ajrccm.150.4.7921472

Latza U, Oldenburg M, Baur X (2004) Endotoxin exposure and respiratory symptoms in the cotton textile industry. Arch Env Heal 59:519–525. doi:10.1080/00039890409605168

Lauener RP, Birchler T, Adamski J et al (2002) Expression of CD14 and Toll-like receptor 2 in farmers' and non-farmers' children. Lancet 360:465–466. doi:10.1016/S0140-6736(02)09641-1

Lazaridou A, Biliaderis CG (2007) Molecular aspects of cereal β-glucan functionality: physical properties, technological applications and physiological effects. J Cereal Sci 46:101–118. doi:10.1016/j.jcs.2007.05.003

Lebron F, Vassallo R, Puri V, Limper AH (2003) Pneumocystis carinii cell wall beta-glucans initiate macrophage inflammatory responses through NF-kappaB activation. J Biol Chem 278:25001–8

Lenters V, Basinas I, Beane-Freeman L et al (2010) Endotoxin exposure and lung cancer risk: a systematic review and meta-analysis of the published literature on agriculture and cotton textile workers. Cancer Causes Control 21:523–555. doi:10.1007/s10552-009-9483-z

Li D, Zhong YN, Rylander R et al (1995) Longitudinal study of the health of cotton workers. Occup Environ Med 52:328–31

Li W, Ray RM, Gao DL et al (2006) Occupational risk factors for nasopharyngeal cancer among female textile workers in Shanghai, China. Occup Environ Med 63:39–44. doi:10.1136/oem.2005.021709

Liebers V, Raulf-Heimsoth M, Brüning T (2008) Health effects due to endotoxin inhalation (review). Arch Toxicol 82:203–210. doi:10.1007/s00204-008-0290-1

Liu AH (2002) Endotoxin exposure in allergy and asthma: reconciling a paradox. J Allergy Clin Immunol 109:379–392

Madsen AM (2006) Airborne endotoxin in different background environments and seasons. Ann Agric Environ Med 13:81–86

Madsen AM, Hansen VM, Nielsen SH, Olsen TT (2009) Exposure to dust and endotoxin of employees in cucumber and tomato nurseries. Ann Occup Hyg 53:129–138. doi:10.1093/annhyg/men073

Madsen AM, Tendal K, Schlunssen V, Heltberg I (2012) Organic dust toxic syndrome at a grass seed plant caused by exposure to high concentrations of bioaerosols. Ann Occup Hyg 56:776–788. doi:10.1093/annhyg/mes012

Mahauthaman R, Howell CJ, Spur BW et al (1988) The generation and cellular distribution of leukotriene C4 in human eosinophils stimulated by unopsonized zymosan and glucan particles. J Allergy Clin Immunol 81:696–705

Malmberg P, Larsson K (1993) Acute exposure to swine dust causes bronchial hyperresponsiveness in healthy subjects. Eur Respir J Off J Eur Soc Clin Respir Physiol 6:400–404

Mandryk J, Alwis KU, Hocking AD (1999) Work-related symptoms and dose-response relationships for personal exposures and pulmonary function among woodworkers. Am J Ind Med 35:481–90

Marchand G, Lalonde M, Beaudet Y et al (2007) Documentation of the endotoxins present in the ambient air of cotton fiber textile mills in Québec. J Env Monit 9:869. doi:10.1039/b704087c

McElvenny DM, Hurley MA, Lenters V et al (2011) Lung cancer mortality in a cohort of UK cotton workers: an extended follow-up. Br J Cancer 105:1054–1060. doi:10.1038/bjc.2011.312

Mehta AJ, Wang XR, Eisen EA et al (2007) Work area measurements as predictors of personal exposure to endotoxin and cotton dust in the cotton textile industry. Ann Occup Hyg 52:45–54. doi:10.1093/annhyg/mem061

Meroueh SO, Bencze KZ, Hesek D et al (2006) Three-dimensional structure of the bacterial cell wall peptidoglycan. Proc Natl Acad Sci U S A 103:4404–9. doi:10.1073/pnas.0510182103

Miller JD, Young JC (1997) The use of ergosterol to measure exposure to fungal propagules in indoor air. Am Ind Hyg Assoc J 58:39–43. doi:10.1080/15428119791013062

Milton DK, Alwis KU, Fisette L, Muilenberg M (2001) Enzyme-linked immunosorbent assay specific for (1–>6) branched, (1–>3)-beta-D-glucan detection in environmental samples. Appl Environ Microbiol 67:5420–4. doi:10.1128/AEM.67.12.5420-5424.2001

Miura NN (2005) Fate of β-glucans in vivo. In: Toxicol. 1 - 3-beta-glucans. Informa Healthcare, pp 109–126

Noss I, Doekes G, Sander I et al (2010a) Passive airborne dust sampling with the electrostatic dustfall collector: optimization of storage and extraction procedures for endotoxin and glucan measurement. Ann Occup Hyg 54:651–658. doi:10.1093/annhyg/meq026

Noss I, Wouters IM, Bezemer G et al (2010b) beta-(1,3)-Glucan exposure assessment by passive airborne dust sampling and new sensitive immunoassays. Appl Environ Microbiol 76:1158–1167. doi:10.1128/AEM.01486-09

Noss I, Wouters IM, Visser M et al (2008) Evaluation of a low-cost electrostatic dust fall collector for indoor air endotoxin exposure assessment. Appl Environ Microbiol 74:5621–7. doi:10.1128/AEM.00619-08

O'Garra A, Vieira P (2004) Regulatory T cells and mechanisms of immune system control. Nat Med 10:801–805. doi:10.1038/nm0804-801

O'Shaughnessy PT, Donham KJ, Peters TM et al (2010) A task-specific assessment of Swine worker exposure to airborne dust. J Occup Env Hyg 7:7–13. doi:10.1080/15459620903327970

OSHA (1995) Subpart Z - Toxic and hazardous substances. Standard Num: 1910.1000. In: Occupational Safety and Health Standards part 1910. Available via https://www.osha.gov/pls/oshaweb/owasrch.search_form?p_doc_type=STANDARDS&p_toc_level=1&p_keyvalue=1910. Accessed 05. Occupational Safety and Health Administration.

Park JH, Cox-Ganser JM, Kreiss K et al (2008) Hydrophilic fungi and ergosterol associated with respiratory illness in a water-damaged building. Environ Health Perspect 116:45–50. doi:10.1289/ehp.10355

Park J-H, Spiegelman DL, Burge HA et al (2000) Longitudinal study of dust and airborne endotoxin in the home. Environ Health Perspect 108:1023–1028. doi:10.1289/ehp.001081023

Paudyal P, Semple S, Niven R et al (2011) Exposure to dust and endotoxin in textile processing workers. Ann Occup Hyg 55:403–409. doi:10.1093/annhyg/meq084

Poole JA, Dooley GP, Saito R et al (2010) Muramic acid, endotoxin, 3-hydroxy fatty acids, and ergosterol content explain monocyte and epithelial cell inflammatory responses to agricultural dusts. J Toxicol Env Heal A 73:684–700. doi:10.1080/15287390903578539

Portengen L, Preller L, Tielen M et al (2005) Endotoxin exposure and atopic sensitization in adult pig farmers. J Allergy Clin Immunol 115:797–802. doi:10.1016/j.jaci.2004.11.046

Prefontaine D, Banville-Langelier A-A, Fiset P-O et al (2010) Children with atopic histories exhibit impaired lipopolysaccharide-induced Toll-like receptor-4 signalling in peripheral monocytes. Clin Exp Allergy 40:1648–1657. doi:10.1111/j.1365-2222.2010.03570.x

Radon K, Danuser B, Iversen M et al (2002) Air contaminants in different European farming environments. Ann Agric Env Med 9:41–48

Reed CE, Milton DK (2016) Endotoxin-stimulated innate immunity: a contributing factor for asthma. J Allergy Clin Immunol 108:157–166. doi:10.1067/mai.2001.116862

Renz H, Blumer N, Virna S et al (2006) The immunological basis of the hygiene hypothesis. Chem Immunol Allergy 91:30–48. doi:10.1159/000090228

Reynolds SJ, Black DW, Borin SS et al (2001) Indoor environmental quality in six commercial office buildings in the midwest United States. Appl Occup Env Hyg 16:1065–1077. doi:10.1080/104732201753214170

Rylander R (1997) Airborne (1→3)-β-d-glucan and airway disease in a day-care center before and after renovation. Arch Environ Heal An Int J 52:281–285. doi:10.1080/00039899709602199

Rylander R (2006) Endotoxin and occupational airway disease. Curr Opin Allergy Clin Immunol 6:62–66. doi:10.1097/01.all.0000202356.83509.f7

Rylander R, Bergstrom R (1993) Bronchial reactivity among cotton workers in relation to dust and endotoxin exposure. Ann Occup Hyg 37:57–63

Rylander R, Norrhall M, Engdahl U et al (1998) Airways inflammation, atopy, and (1–> 3)-beta-D-glucan exposures in two schools. Am J Respir Crit Care Med 158:1685–7. doi:10.1164/ajrccm.158.5.9712139

Rylander R, Persson K, Goto H et al (1992) Airborne beta-1,3-glucan may be related to symptoms in sick buildings. Indoor Built Environ 1:263–267. doi:10.1177/1420326X9200100502

Rylander R, Thorn J, Attefors R (1999) Airways inflammation among workers in a paper industry. Eur Respir J Off J Eur Soc Clin Respir Physiol 13:1151–1157

Sabroe I, Read RC, Whyte MKB et al (2003) Toll-like receptors in health and disease: complex questions remain. J Immunol 171:1630–1635

Saito K, Nishijima M, Ohno N et al (1992) Activation of complement and limulus coagulation systems by an alkali-soluble glucan isolated from omphalia lapidescens and its less-branched derivatives. (studies on fungal polysaccharide. XXXIX). Chem Pharm Bull (Tokyo) 40:1227–1230

Saito R, Cranmer BK, Tessari JD et al (2009) Recombinant factor C (rFC) assay and gas chromatography/mass spectrometry (GC/MS) analysis of endotoxin variability in four agricultural dusts. Ann Occup Hyg 53:713–722. doi:10.1093/annhyg/mep052

Samadi S, Heederik DJ, Krop EJ et al (2010) Allergen and endotoxin exposure in a companion animal hospital. Occup Env Med 67:486–492. doi:10.1136/oem.2009.051342

Samadi S, Rietbroek NNJ, Dwars RM et al (2011) Endotoxin and beta-(1 –> 3)-glucan exposure in poultry and ruminant clinics. J Environ Monit 13:3254–3261. doi:10.1039/c1em10566c

Samadi S, van Eerdenburg FJ, Jamshidifard AR et al (2012) The influence of bedding materials on bio-aerosol exposure in dairy barns. J Expo Sci Env Epidemiol 22:361–368. doi:10.1038/jes.2012.25

Samadi S, Wouters IM, Houben R et al (2009) Exposure to inhalable dust, endotoxins, beta(1->3)-glucans, and airborne microorganisms in horse stables. Ann Occup Hyg 53:595–603. doi:10.1093/annhyg/mep040

Sander I, Fleischer C, Borowitzki G et al (2008) Development of a two-site enzyme immunoassay based on monoclonal antibodies to measure airborne exposure to (1–>3)-beta-D-glucan. J Immunol Methods 337:55–62. doi:10.1016/j.jim.2008.05.010

Saraf A, Larsson L, Burge H, Milton D (1997) Quantification of ergosterol and 3-hydroxy fatty acids in settled house dust by gas chromatography-mass spectrometry: comparison with fungal culture and determination of endotoxin by a Limulus amebocyte lysate assay. Appl Env Microbiol 63:2554–2559

Savilahti EM, Karinen S, Salo HM et al (2010) Combined T regulatory cell and Th2 expression profile identifies children with cow's milk allergy. Clin Immunol 136:16–20. doi:10.1016/j.clim.2010.02.011

Schaub B, Lauener R, von Mutius E (2006) The many faces of the hygiene hypothesis. J Allergy Clin Immunol 117:969–77. doi:10.1016/j.jaci.2006.03.003. quiz 978

Schram D, Doekes G, Boeve M et al (2005) Bacterial and fungal components in house dust of farm children, Rudolf Steiner school children and reference children–the PARSIFAL Study. Allergy 60:611–618. doi:10.1111/j.1398-9995.2005.00748.x

Schram-Bijkerk D, Doekes G, Douwes J et al (2005) Bacterial and fungal agents in house dust and wheeze in children: the PARSIFAL study. Clin Exp Allergy 35:1272–1278. doi:10.1111/j.1365-2222.2005.02339.x

Schwartz DA, Thorne PS, Yagla SJ et al (1995) The role of endotoxin in grain dust-induced lung disease. Am J Respir Crit Care Med 152:603–608. doi:10.1164/ajrccm.152.2.7633714

Semple S, Devakumar D, Fullerton DG et al (2010) Airborne endotoxin concentrations in homes burning biomass fuel. Environ Health Perspect 118:988–91. doi:10.1289/ehp.0901605

Senthilselvan A, Beach J, Feddes J et al (2011) A prospective evaluation of air quality and workers' health in broiler and layer operations. Occup Env Med 68:102–107. doi:10.1136/oem.2008.045021

Senthilselvan A, Zhang Y, Dosman JA et al (1997) Positive human health effects of dust suppression with canola oil in swine barns. Am J Respir Crit Care Med 156:410–417

Sigsgaard T, Bonefeld-Jorgensen EC, Hoffmann HJ et al (2005) Microbial cell wall agents as an occupational hazard. Toxicol Appl Pharmacol 207:310–319. doi:10.1016/j.taap.2004.12.031

Sigsgaard T, Heederik D (2005) On the hygiene hypothesis: regulation down, up, or sideways? J Allergy Clin Immunol 115:1325–6. author reply 1326

Sigsgaard T, Jensen LD, Abell A et al (2004) Endotoxins isolated from the air of a Danish paper mill and the relation to change in lung function: an 11-year follow-up. Am J Ind Med 46:327–332. doi:10.1002/ajim.20068

Sigsgaard T, Omland Ø, Thorne PS, Parnham MJ (2010) Asthma-like diseases in agriculture. In: Sigsgaard T, Heederik D (eds) Occup. Asthma. Birkhauser Basel, Basel, Switzerland, pp 163–183

Simpson JC, Niven RM, Pickering CA et al (1999) Comparative personal exposures to organic dusts and endotoxin. Ann Occup Hyg 43:107–115. doi:S0003487898000830 [pii]

Singh U, Levin L, Grinshpun SA et al (2011) Influence of home characteristics on airborne and dustborne endotoxin and β-D-glucan. J Environ Monit 13:3246–53. doi:10.1039/c1em10446b

Smit LA, Heederik D, Doekes G et al (2008) Exposure-response analysis of allergy and respiratory symptoms in endotoxin-exposed adults. Eur Respir J 31:1241–1248. doi:10.1183/09031936.00090607

Smit LA, Wouters IM, Hobo MM et al (2006) Agricultural seed dust as a potential cause of organic dust toxic syndrome. Occup Env Med 63:59–67. doi:10.1136/oem.2005.021527

Smit LAM, Heederik D, Doekes G et al (2010) Occupational endotoxin exposure reduces the risk of atopic sensitization but increases the risk of bronchial hyperresponsiveness. Int Arch Allergy Immunol 152:151–158. doi:10.1159/000265536

Spaan S, Heederik DJ, Thorne PS, Wouters IM (2007) Optimization of airborne endotoxin exposure assessment: effects of filter type, transport conditions, extraction solutions, and storage of samples and extracts. Appl Env Microbiol 73:6134–6143. doi:10.1128/AEM.00851-07

Spaan S, Schinkel J, Wouters IM et al (2008a) Variability in endotoxin exposure levels and consequences for exposure assessment. Ann Occup Hyg 52:303–316. doi:10.1093/annhyg/men024

Spaan S, Smit LAM, Eduard W et al (2008b) Endotoxin exposure in sewage treatment workers: investigation of exposure variability and comparison of analytical techniques. Ann Agric Environ Med 15:251–261

Spaan S, Wouters IM, Oosting I et al (2006) Exposure to inhalable dust and endotoxins in agricultural industries. J Env Monit 8:63–72. doi:10.1039/b509838f

Spierenburg A, Smit L, Robbe P, et al (2016) Occupational endotoxin exposure dose-dependently protects against atopy and hay fever: Results of a longitudinal study. Eur. Respir. J. 48:

Stern DA, Riedler J, Nowak D et al (2007) Exposure to a farming environment has allergen-specific protective effects on TH2-dependent isotype switching in response to common inhalants. J Allergy Clin Immunol 119:351–358. doi:10.1016/j.jaci.2006.10.013

Szadkowska-Stańczyk I, Bródka K, Buczyńska A et al (2010) Exposure to bioaerosols among CAFO workers (swine feeding). Med Pr 61:257–69

Szepfalusi Z, Loibichler C, Pichler J et al (2000) Direct evidence for transplacental allergen transfer. Pediatr Res 48:404–407

Thilsing T, Madsen AM, Basinas I et al (2015) Dust, endotoxin, fungi, and bacteria exposure as determined by work task, season, and type of plant in a flower greenhouse. Ann Occup Hyg 59:142–57. doi:10.1093/annhyg/meu090

Thorn J, Beijer L, Rylander R (1998) Airways inflammation and glucan exposure among household waste collectors. Am J Ind Med 33:463–470. doi:10.1002/(SICI)1097-0274(199805)33:5<463::AID-AJIM5>3.0.CO;2-T [pii]

Thorn J, Rylander R (1998) Airways inflammation and glucan in a rowhouse area. Am J Respir Crit Care Med 157:1798–803. doi:10.1164/ajrccm.157.6.9706081

Thorne PS, Perry SS, Saito R et al (2010) Evaluation of the Limulus amebocyte lysate and recombinant factor C assays for assessment of airborne endotoxin. Appl Env Microbiol 76:4988–4995. doi:10.1128/AEM.00527-10

Tischer C, Gehring U, Chen CM et al (2011) Respiratory health in children, and indoor exposure to (1,3)-β-D- glucan, EPS mould components and endotoxin. Eur Respir J 37:1050–1059. doi:10.1183/09031936.00091210

Vandenbulcke L, Bachert C, Van Cauwenberge P, Claeys S (2006) The innate immune system and its role in allergic disorders. Int Arch Allergy Immunol 139:159–165. doi:10.1159/000090393

Van Strien RT, Engel R, Holst O et al (2004) Microbial exposure of rural school children, as assessed by levels of N-acetyl-muramic acid in mattress dust, and its association with respiratory health. J Allergy Clin Immunol 113:860–867. doi:10.1016/j.jaci.2004.01.783

Vogelzang PFJ, Van Der Gulden JWJ, Folgering H et al (2000) Longitudinal changes in bronchial responsiveness associated with swine confinement dust exposure. Chest 117:1488–1495. doi:10.1378/chest.117.5.1488

Vollmer W, Blanot D, De Pedro MA (2008) Peptidoglycan structure and architecture. FEMS Microbiol Rev 32:149–167. doi:10.1111/j.1574-6976.2007.00094.x

Von Mutius, Braun-Fahrländer, Schierl et al (2000) Exposure to endotoxin or other bacterial components might protect against the development of atopy. Clin Exp Allergy 30:1230–1234. doi:10.1046/j.1365-2222.2000.00959.x

Wan GH, Li CS (1999) Indoor endotoxin and glucan in association with airway inflammation and systemic symptoms. Arch Environ Health 54:172–9. doi:10.1080/00039899909602256

Wang XR, Pan LD, Zhang HX et al (2002) Follow-up study of respiratory health of newly-hired female cotton textile workers. Am J Ind Med 41:111–118. doi:10.1002/ajim.10042

Wang Z, Larsson K, Palmberg L et al (1997) Inhalation of swine dust induces cytokine release in the upper and lower airways. Eur Respir J Off J Eur Soc Clin. Respir Physiol 10:381–387

Wang Z, Malmberg P, Ek A et al (1999) Swine dust induces cytokine secretion from human epithelial cells and alveolar macrophages. Clin Exp Immunol 115:6–12

Williams DL (1997) Overview of (1→3)-β-D-glucan immunobiology. Mediators Inflamm 6:247–250. doi:10.1080/09629359791550

Williams DL, Lowman DW, Ensley HE (2005) Introduction to the chemistry and immunobiology of _ -glucans. In: Toxicol. 1 - 3-Beta-Glucans. Informa Healthcare, pp 1–34

Williams KL (2007a) Endotoxin relevance and control overview. In: Williams KL (ed) Endotoxins pyrogens, LAL test. depyrogenation. Informa Healthcare, New York, USA, p 27–45

Williams KL (2007b) Endotoxin structure, function, and activity. In: Williams KL (ed) Endotoxins pyrogens, LAL test. depyrogenation. Informa Healthcare USA, Inc, New York, USA, p 67–90

Wouters IM, Spaan S, Douwes J et al (2006) Overview of personal occupational exposure levels to inhalable dust, endotoxin, beta(1–>3)-glucan and fungal extracellular polysaccharides in the waste management chain. Ann Occup Hyg 50:39–53. doi:10.1093/annhyg/mei047

Zeković DB, Kwiatkowski S, Vrvić MM et al (2005) Natural and modified (1→3)-β-d-glucans in health promotion and disease alleviation. Crit Rev Biotechnol 25:205–230. doi:10.1080/07388550500376166

Zhang K, Petty HR (1994) Influence of polysaccharides on neutrophil function: specific antagonists suggest a model for cooperative saccharide-associated inhibition of immune complex-triggered superoxide production. J Cell Biochem 56:225–35

Zhao Z, Sebastian A, Larsson L et al (2008) Asthmatic symptoms among pupils in relation to microbial dust exposure in schools in Taiyuan, China. Pediatr Allergy Immunol 19:455–65. doi:10.1111/j.1399-3038.2007.00664.x

Chapter 9
Human Biomonitoring of Mycotoxins for the Detection of Nutritional, Environmental and Occupational Exposure

Benedikt Cramer and Hans-Ulrich Humpf

Abstract The development of methods for the detection of mycotoxins as biomarker for nutritional, environmental, and occupational exposure is reviewed in this chapter. Summaries on the current state of research on mycotoxin metabolism of aflatoxins, deoxynivalenol, fumonisins, ochratoxins, and zearalenone are given. Furthermore, the development of biomarkers from these studies is discussed and excretion rates are reported. Starting from single analyte methods such as for aflatoxin M_1, the first mycotoxin biomarker for aflatoxin exposure, the improvement of detection methods for aflatoxins, ochratoxin A, zearalenone, deoxynivalenol, fumonisins, and many other mycotoxins up to the recent HPLC-MS/MS methods is shown. Finally, the recently developed multi-mycotoxin methods for the detection of a broad spectrum of biomarkers and fungal metabolites is depicted and their current applications and future perspectives are discussed.

Keywords Mycotoxin · metabolism · biomonitoring · biomarker · urine · plasma

9.1 Introduction

Human exposure to mycotoxins has a long history. Examples are reports of St. Antony's fire in the middle ages as well as in the Assyrian Empire 600 bc, caused by consumption of sclerotia of the fungus *Claviceps pururea* containing ergot alkaloids (Schiff 2006; Guggisberg 1954). Nowadays acute intoxications with mycotoxins are rare but occur from time to time such as the aflatoxicosis outbreak in Kenya in 2004, demonstrating the acute toxic potential of these compounds (Lewis et al. 2005). Besides acute toxicity, effects of chronic exposure to lower concentration can affect larger populations, resulting for example in higher cancer incidence, reduced immune response, and many other adverse

B. Cramer (✉) · H.-U. Humpf
Westfälische Wilhelms-Universität Münster, Institute of Food Chemistry, Münster, Germany
e-mail: cramerb@wwu.de

C. Viegas et al. (eds.), *Exposure to Microbiological Agents in Indoor and Occupational Environments*, DOI 10.1007/978-3-319-61688-9_9

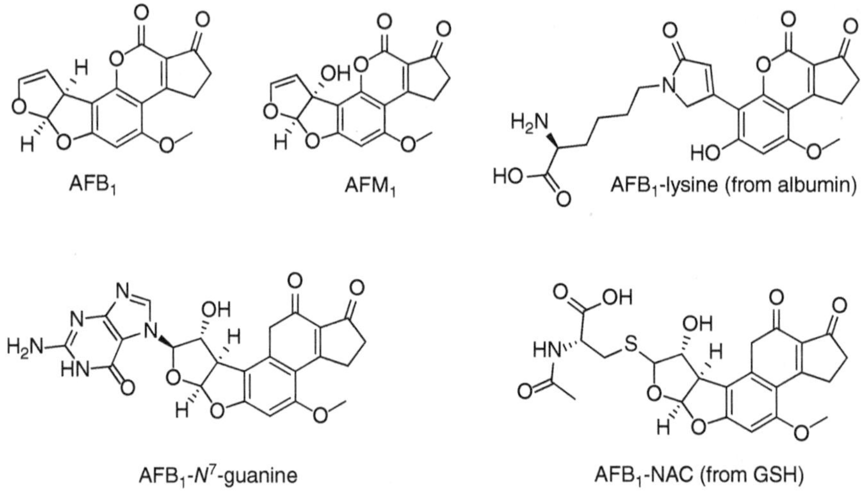

Fig. 9.1 Aflatoxin B1 (AFB_1) and metabolites applied for HBM. Aflatoxin M_1 (AFM_1), aflatoxin B_1-lysine adduct released from AFB_1 bound to albumin (AFB_1-lysine), aflatoxin B_1 bound to guanine and released from DNA (AFB_1-N^7-guanine), aflatoxin B_1-*N*-acetyl cysteine released from AFB_1 bound to glutathione (AFB_1-NAC)

effects (Wild and Gong 2010). Usually exposure to mycotoxins is caused by consumption of contaminated food including cereals, vegetables, beverages, meat, and especially fermented or (improper) stored products. However, depending on the living and working environment, housing situation or general environmental conditions, dust, bio-aerosols, or contaminated surfaces can also become major sources of mycotoxin exposure. In these cases, inhalation or dermal resorption might even be more important routes than consumption (Degen 2011). Official food control provides a good overview about average contamination of food and commodities with mycotoxins. Percentiles to describe exposure to average contaminated food for high consumers are also available, taking the diverse nutrition habits and specific risk groups into account (EFSA 2011). Nevertheless, the situation of individuals and especially non-food related exposure to humans cannot be calculated from this database. Human biomonitoring (HBM) is therefore becoming a valuable tool to eliminate these deficits via reporting of individual internal exposure data.

The first report of HBM with the focus on mycotoxins was published by Campbell et al. in 1970 (Campbell et al. 1970). In their publication, the authors investigated the relationship between the uptake of aflatoxins (Fig. 9.1) from contaminated peanut butter and excretion of the metabolite aflatoxin M_1 (AFM_1). In that study, it could be demonstrated that a direct relationship between the presence of AFM_1 in urine and aflatoxin uptake is given – although no quantitative measurement of AFM_1 was possible at that time. Subsequently new methods and trials on aflatoxin and aflatoxin metabolite analysis of serum and urine samples in the US and worldwide have been published (Wray and Hayes 1980; Onyemelukwe and Ogbadu 1981; Nelson et al. 1980). About a decade later, published in 1982, HBM of ochratoxin A started with the analysis of blood samples of patients suffering from the Balkan endemic nephropathy (BEN), an endemic kidney disease described for certain regions of The Balkans (Hult et al. 1982). In the following decades intense ochratoxin A biomonitoring in Europe and around the world was done, providing a database on ochratoxin A (OTA) exposure and changes thereof over the years (Malir et al. 2016; Märtlbauer et al. 2009). Since then, many mycotoxins and their metabolites have been detected *in vitro* and *in vivo*, resulting in a growing understanding of mycotoxin metabolism and elimination pathways. Together with the development of new, more sensitive instrumentation and its application in routine laboratories, approaches for studying the multi-mycotoxin exposure became possible. In the recent decade, LC-MS based multi-mycotoxin methods have been established as routine techniques, allowing now the detection of a broad range of compounds with less sample preparation and in shorter time. However, evaluation of the available data and mycotoxin levels detected is still challenging in this dynamic field of research as broad sets of data on large populations are currently in the process of being developed. In this chapter the current state of single-analyte as well as analyte-group methods for the most prominent mycotoxins will be discussed, followed by a comparison with the recent developments in the field of multi-mycotoxin methods.

9.2 Aflatoxins

9.2.1 Background

Aflatoxins and especially the metabolism of aflatoxins have been studied extensively, resulting in a comprehensive understanding of metabolites, intermediates and excretion pathways. Excellent reviews on this metabolism and the toxicity of aflatoxins and aflatoxin biomarkers have been published by Kensler et al. and others (Kensler et al. 2011; Turner et al. 2012). In food, the four different aflatoxins aflatoxin B_1 (AFB_1), aflatoxin B_2 (AFB_2), aflatoxin G_1 (AFG_1), and aflatoxin G_2 (AFG_2) occur in a quite constant ratio of 1.0:0.1:0.3:0.03, making AFB_1 most important regarding occurrence. As it is also the most toxic aflatoxin, metabolites of aflatoxins analyzed by HBM are usually those of AFB_1 shown in Fig. 9.1. After absorption, AFB_1 undergoes rapid phase I metabolism resulting either in hydroxylation at C-7 to aflatoxin M_1 (AFM_1) or epoxidation at C-8, C-9 either in *endo* or in *exo* configuration. AFM_1 is the first described metabolite and validated biomarker of aflatoxin exposure with an average conversion rate of about 1.5% (Campbell et al. 1970). It can be analyzed in urine as well as breast milk (Zarba et al. 1992). Epoxidized AFB_1 further reacts with DNA (*exo*-epoxide only), glutathione, or it can be hydrolyzed to AFB-diol. AFB-DNA-adducts are rapidly replaced by repair enzymes, resulting in the excretion of aflatoxin B_1-N^7-guanine (AFB_1-N^7-guanine) in urine. Aflatoxin B_1-glutathion conjugates are also excreted via urine as aflatoxin B_1-mercapturic acid (Aflatoxin B_1-*N*-acetyl-cysteine, AFB_1-NAC) conjugates. Both compounds are also applicable as urinary biomarkers for short time exposure to aflatoxins with excretion rates of approximately 0.2% for

AFB_1-N^7-guanine (Scholl et al. 1997; Groopman et al. 1993). AFB-diol on the other hand reacts as AFB-dialdehyde with amino groups such as the ε-amino function of lysine in proteins, making the aflatoxin-albumin adduct in blood the most important long-term biomarker of aflatoxin exposure. The structures of AFB_1 as well as its relevant metabolites are summarized in Fig. 9.1.

9.2.2 Methods

Biomarker detection of aflatoxins is done either in urine, breast milk or blood. In an initial step of the aflatoxin specific methods, pre-concentration via solid phase extraction (SPE) on reversed phase material is applied. Subsequently, aflatoxin fractions are eluted from the SPE-material, buffered and purified via immunoaffinity chromatography (IAC). This well-established protocol applies high performance liquid chromatography coupled with fluorescence detection (HPLC-FLD) detection and was developed in 1992 by Groopman et al. with further optimization by Tang et al. and others (Tang et al. 2008; Schwartzbord et al. 2016). Depending on the IAC material applied, besides AFM_1, also AFB_1-NAC, and with modification also AFB_1-N^7-guanine adducts can be detected. Standard HPLC systems with fluorescence detection are typically used for detection, reaching a sensitivity of 0.04 ng/mL urine, which is sufficient for detection of all three biomarkers in a moderately exposed population. Long-term exposure cannot be assessed with the above mentioned analytes as they are excreted via urine within 24 h after exposure. Instead the measurement of aflatoxin-albumin adducts is common practice. In average about 5% of AFB_1 react after activation by P450 enzymes with the ε-amino groups of lysine side chains of albumin. As albumins have an average biological half-life of 19 days, AFB_1-lysine from albumin allows the determination of aflatoxin exposure over a period of a few weeks. For analysis, aliquots of serum or plasma are treated with the enzyme mixture pronase for a period of 18 h, followed by solid phase extraction and HPLC-MS/MS (Mccoy et al. 2008). To improve accuracy stable isotope dilution analysis is applied, using AFB-lysine-d_4 as internal standard. With this approach a limit of detection (LOD) of 0.2 pg AFB_1-lysine/mg albumin can be reached (Groopman et al. 2014).

Enzyme linked immunosorbent assays (ELISA) are an alternative approach towards aflatoxin and aflatoxin-metabolite analysis, allowing fast and relatively inexpensive at least semi-quantitative detection. They can either be used for the detection of total AFB_1 in blood or for the detection of metabolites in urine. For blood analysis, a comparable treatment with pronase as for the LC-MS analysis is done to release albumin-bound AFB_1-lysine. After an additional immuno affinity chromatography step, AFB_1 and AFB_1-lysine are detected via a combined ELISA with a sensitivity of 1 ng/mL corresponding to approximately 35 pg AFB_1-lysine/mg albumin (Viegas et al. 2016). Current ELISA for AFM_1 in urine provide a sensitivity of 0.2 ng/mL urine which is sufficient for the detection in highly exposed populations (Mccoy et al. 2008). In countries with lower aflatoxin exposure

ELISA usually requires a pre-concentration step using solid phase extraction to increase sensitivity by a factor of 10–50. For example, Zhu et al. reported an ELISA based method specific for AFM_1, which applies a pre-concentration of 25 mL urine via solid phase extraction before ELISA, resulting in a limit of quantification (LOQ) of 0.03 ng/mL. Unfortunately, the most of the benefit of ELISA compared to HPLC, its speed and low costs, is mostly diminished when this additional step has to be applied (Zhu et al. 1987).

A recently published approach towards simplified AFB_1-albumin adduct detection is based on dried blood spots (DBS) (Xue et al. 2016). This technique provides the opportunity to perform minimal invasive blood sampling using a lancet as only 50–100 µL of blood are required. For DBS, blood samples are dripped on special DBS cards and dried. For subsequent analysis, disks of dried blood are punched out of the cards, and AFB_1-albumin is extracted with PBS-buffer. After incubation of the PBS-extract with pronase, AFB_1-lysine is concentrated and purified via reversed phase SPE. For analysis, both, HPLC-FLD as well as HPLC-MS/MS methods have been described. For highest accuracy, the eluate of the SPE is subjected to HPLC-MS/MS analysis, yielding a LOD of 0.2 pg/mL, which is equivalent to the sensitivity of the classical method mentioned above.

9.3 Ochratoxin A

9.3.1 Background

The nephrotoxic and carcinogenic mycotoxin ochratoxin A (OTA) is a frequent contaminant in food and feed (Malir et al. 2016). It shows strong affinity to human serum albumin in blood with 99.8% of ochratoxin A being bound to it (Chu 1971; Stojkovic et al. 1984), leading to a long biological half-life of approximately 35 days (Studer-Rohr et al. 2000). Consequently, OTA blood levels have become a primary biomarker for ochratoxin A exposure with the first measurements starting in the 1980s (Scott 2005; Hult et al. 1982). Since then, numerous studies have been conducted to determine OTA blood levels around the world. Results of these studies have been discussed in different reviews (Märtlbauer et al. 2009; Malir et al. 2016; Scott 2005). By means of the Klaassen equation, OTA blood levels have been used for the calculation of OTA intake providing a valuable and validated tool for exposure assessment (Märtlbauer et al. 2009; Duarte et al. 2011).

Determination of ochratoxin A in urine was less frequently done in the beginning of OTA HBM as only small fractions of the ingested OTA are found in urine (<3% per day) (Studer-Rohr et al. 2000; Degen 2016). Nevertheless since urine sampling is non-invasive and better accepted than traditional blood sampling, many studies rely on urine data (Muñoz et al. 2009). Additionally OTA levels found in urine seem to provide better insight into short-term exposure to ochratoxin A than changes of blood levels. Surprisingly this seems to be strongly depending on the individual situation (Castegnaro et al. 1991; Gilbert et al. 2001;

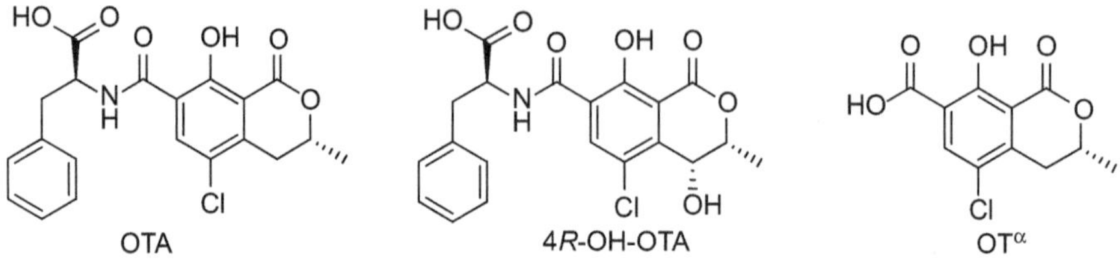

Fig. 9.2 Structures of ochratoxin A (OTA), 4R-hydroxy-ochratoxin A (4-OH-OTA), ochratoxin α (OTα)

Castegnaro et al. 2006). Treatment of urine samples with glucuronidase and sulfatase prior to extraction can increase OTA concentrations, suggesting that under certain conditions at least fractions of this toxin are conjugated to glucuronides or sulfates (Muñoz et al. 2017). These observations were supported by data from Han et al. who tentatively identified glucuronides of ochratoxin A formed *in vitro* in liver microsomes by HPLC-HRMS (Han et al. 2013). However, direct detection of ochratoxin A-glucuronides in human urine samples was so far not successful due to sensitivity issues (Muñoz et al. 2017). Further metabolites of OTA are hydroxylated ochratoxin A (4*R*-OH-OTA) and ochratoxin α (OTα) (Jonsyn-Ellis 2001; Muñoz et al., 2010; Coronel et al. 2011) (Fig. 9.2). The role of OTα as phase I metabolite of OTA is still under dispute as it might mostly be formed by the microbiota in the large intestine which has been shown to rapidly cleave OTA. Additionally only traces of OTα have been found in incubations of OTA with liver microsomes. On the other hand to a large extend conjugation of OTα with glucuronic acid was observed *in vitro* (Han et al. 2013).

9.3.2 Methods

Analysis of OTA in blood historically started from plasma, which was extracted with chloroform, re-extracted with alkaline buffer and subjected to fluorimetric measurements. For confirmation, the fluorescent fraction was treated with carboxypeptidase releasing OTα from OTA and shifting the fluorescence excitation and emission wavelength (Hult et al. 1982). Since then chloroform extraction remained popular but HPLC with fluorescence detection became the method of choice (Beker and Radic 1991). Additional purification by immuno affinity chromatography improved sensitivity and specificity. Further increase of sensitivity of the fluorescence detection (up to a factor of six) is described when post-column elevation of the pH to alkaline conditions using ammonia is applied (Petkova-Bocharova et al. 2003; Zimmerli and Dick 1995). Nowadays chloroform extraction is still a common procedure for OTA detection in plasma, but mass spectrometric instead of fluorescence detection is used more frequently (Muñoz et al., 2010). With mass spectrometric detection, also immuno affinity chromatography can be omitted, saving time and costs and providing the benefit that compounds such as OTα can be detected which

Table 9.1 Sensitivity of different methods applied for the analysis of ochratoxin A in blood, plasma, serum or urine

Method	Matrix	Limits	Literature
Chloroform extract, IAC, HPLC-FLD	Serum	LOQ: 5–10 pg/mL	(Zimmerli and Dick 1995)
Chloroform/2-propanol (97+3) extract, HPLC-MS/MS	Urine	LOD: 20 pg/mL	(Muñoz et al., 2010)
		LOQ: 50 pg/mL	
Chloroform/2-propanol (97+3) extract, HPLC-MS/MS	Plasma	LOD: 70 pg/mL	(Muñoz et al., 2010)
		LOQ 100 pg/mL	
Dried blood spot extract, HPLC-MS/MS	Blood, plasma, serum	LOD: 6–8 pg/mL	(Osteresch et al. 2016)
		LOQ: 21–26 pg/mL	
Direct analysis, HPLC-MS/MS	Urine	LOD: 10 pg/mL	(Gerding et al. 2015)
		LOQ: 75 pg/mL	
Competitive ELISA (cELISA)	Plasma	LOD: 4–20 pg/mL	(Ueno et al. 1998)

are usually not bound to IAC columns and thus not detectable. A method for the detection of OTA in DBS of coffee consumers has been published in 2015. OTA and a thermal degradation product of OTA, 2′*R*-ochratoxin A were analyzed in dried blood spots (DBS) which were either cut out completely or disks were punched from the spot. The samples were extracted with a solvent mixture of acetone, acetonitrile and water prior to HPLC-MS/MS analysis (Osteresch et al. 2016; Cramer et al. 2015). In this method, neither chloroform nor IAC were required. Data on the LOD/LOQ of the recent methods for OTA analysis are reported in Table 9.1. Competitive ELISA has also been applied for the analysis of ochratoxin A in plasma and has been widely used for screening (Pestka et al. 1981; Ueno et al. 1998). So far, the binding of the OTA phase II metabolites to the ELISA antibodies has not been studied.

A first method for the analysis of OTA in human urine was published by Orti et al. (1986). It already implemented a step of enzymatic hydrolysis to release OTA from possible glucuronic acid or sulfate conjugates followed by extraction with chloroform. The subsequent cleanup protocol appears laborious and complicated as it utilizes three subsequent SPE purification steps with cartridges of silica gel, cyano- and octyl-phases before final analysis with HPLC-FLD. Following this protocol allows the simultaneous detection of ochratoxin A as well as aflatoxin B_1 at levels of 1 ng/mL. However, with the three different columns used in the protocol and thus possible sources of error, it found only little application in publications of the following years. Castegnaro et al. replaced two of the three SPE columns by applying a preparative RP-HPLC step prior to analytical HPLC. This method allowed a LOD of 5 pg/mL and reached a recovery between 65% and 75% (Castegnaro et al. 1990). 4-OH-OTA could be detected additionally to OTA but no enzymatic treatment with glucuronidase or sulfatase was implemented in the method. In subsequent studies by Jonsyn-Ellis, the preparative HPLC step was omitted, leading to an increase of the LOD to 200 pg/mL (Jonsyn-Ellis 2001). An alternative cleanup protocol published by Wafa et al. was based on the preferred solubility of OTA in alkaline aqueous solutions (Wafa et al. 1998). Here firstly acidified urine was extracted with chloroform and the organic phase re-extracted with alkaline aqueous solution. The aqueous phase was acidified, OTA again extracted with chloroform and applied to HPLC-FLD analysis (Wafa et al. 1998). The first application of IAC for purification of OTA from human urine was reported in 2001. In this popular method, urine samples were just diluted, filtered and passed through an IAC cartridge (Pascale and Visconti 2001). After washing with water and elution with methanol, HPLC-FLD analysis reached a LOD of 5 pg/mL. Unfortunately, also this method did not use a pretreatment with glucuronidase/sulfatase to release OTA from conjugates. Alternative cleanup procedures using solid phase micro extraction of urine samples previously extracted with chloroform under alkaline conditions reached a sensitivity of 10 pg/mL have also been published but also with little further application (Vatinno et al. 2007). HPLC-MS/MS analysis of OTA and OTα in urine was first published by Munoz et al. using a liquid-liquid extraction with chloroform-2-isopropanol identical with the method described above for plasma (Muñoz et al., 2010). With MS/MS detection, a LOD of 20 pg/mL for both analytes was established. Molecular imprinted polymers are also successfully applied for OTA purification from urine with sufficient sensitivity (Xie et al. 2014). Further developments of methods for OTA and OTA metabolite detection are described in the section on multi-mycotoxin metabolite methods below.

9.4 Zearalenone

9.4.1 Background

Zearalenone (ZEN) is a *Fusarium* toxin previously named also F2 toxin that can be found in cereals and cereal products. It belongs to the group of resorcylic acid lactones which have estrogenic properties. After ingestion, ZEN is rapidly absorbed and undergoes both, phase I and phase II metabolism as reviewed by Metzler et al. (2010). In a first step, ZEN can be reduced at C-7 to α- or β-zearalenol (α-/β-ZEL) with α-ZEL being predominantly found in human and pig while β-ZEL is dominant in cows. Further hydration of the double bound between C-11 and C-12 to α-/β-zearalanol (α-/β-ZAL) is described for sheep but the levels of these compounds are more than a factor of 10 lower than those found for α-/β-ZEL (Miles et al. 1996). ZEN can also undergo oxidative metabolism but so far most metabolites have only been found *in vitro*. Only the formation of 8-hydroxy ZEN was confirmed *in vivo* in the liver and urine of rats dosed with ZEN but so far not in humans (Pfeiffer et al. 2009; Bravin et al. 2009). Glucuronic acid conjugates of ZEN, ZAN, α-/β-ZEL and α-/β-ZAL have been synthesized and

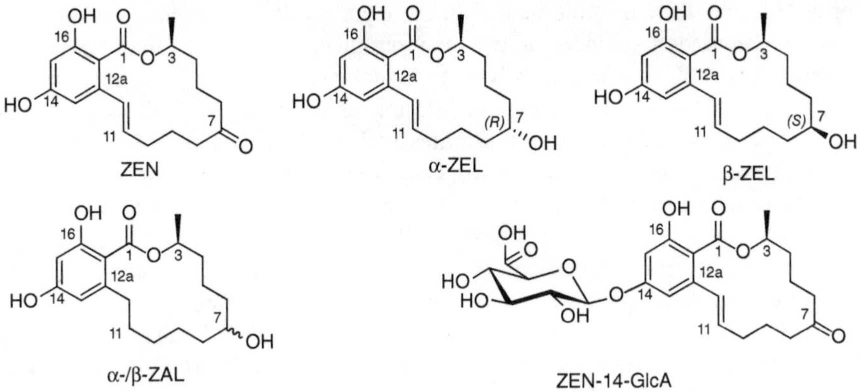

Fig. 9.3 Zearalenone (ZEN) and its metabolites applied for HBM: α-Zearalenol (α-ZEL), β-zearalenol (β-ZEL), α-/β-zearalanol (α-/β-ZAL), zearalenone-14-*O*-β-glucuronide (ZEN-14-GlcA)

applied as references but due to sensitivity issues, only ZEN-14-*O*-β-glucuronide (ZEN-14-GlcA) could be detected in human urine with a direct method (Gerding et al. 2014; Abia et al. 2013; Warth et al. 2012). The determination of the excretion profile of ZEN and ZEN-14-GlcA revealed a urinary excretion rate of approximately 7–20% total ZEN in 24 h (Mirocha et al. 1981; Warth et al. 2013). The chemical structures of ZEN and its most important metabolites are shown in Fig. 9.3.

9.4.2 Methods

Method development for the detection of ZEN and its metabolites started with animal experiments and has initially been done with TLC followed by gas chromatography (GC) (Trenholm et al. 1980). However, a first method sensitive enough for biomonitoring of ZEN and α-ZEL (LOD 0.6 ng/mL plasma) was developed for HPLC-FLD in 1981 and consists of the extraction of blood plasma with a 2-propanol diethyl-ether mixture. The extract is further purified by a cleanup via base-acid extraction utilizing the solubility of the analytes in alkaline aqueous solutions but not in acids prior to injection on a normal phase column (Trenholm et al. 1981). A subsequently developed HPLC-FLD based method for ZEN, α-ZEL and β-ZEL also covered phase II metabolites of these compounds due to an additional treatment with glucuronidase/arylsulfatase (Olsen et al. 1985). Application of HPLC-MS for analysis of ZEN in blood and urine started with thermospray ionization attached to a single quadrupole mass analyzer and covered also some trichothecenes, resulting in one of the first HPLC-MS-based multi-mycotoxin methods for biomonitoring. However due to technical limitations, method performance and sensitivity were not satisfying compared to the methods using fluorescence detection (LOD in urine 5–10 ng/mL) (Voyksner et al. 1985). GC-MS/MS used for detection of ZEN and α-ZEL reached a LOD of 0.1 ng/mL for α-ZEL and ZEN in urine making it the most sensitive technique at that time. Sample preparation includes a two-step purification with reversed phase and normal phase SPE followed by silylation to archive the required volatility for GC (Plasencia et al. 1990). Application of IAC for sample preparation allowed a sensitive detection of ZEN and its metabolites with fluorescence detection. With the development of robust atmospheric pressure ionization techniques in combination with tandem mass spectrometry, sample preparation for ZEN analysis could be reduced to the application of a simple C-18 cartridge (Jodlbauer et al. 2000). For the accurate detection of this group of compounds also stable isotope labeled standards became available and have been applied in combination with HPLC-MS/MS and GC-MS for urine analysis (Blokland et al. 2006; Launay et al. 2004). In recent years, reference compounds of ZEN-14-GlcA, ZEN-16-*O*-β-glucuronide and glucuronides of ZEL have been synthesized via enzymatic and chemical synthesis, providing references for direct detection (Stevenson et al. 2008; Pfeiffer et al. 2010; Mikula et al. 2012). A first method used for HBM of ZEN and direct detection of its metabolites was published by Warth in 2012 is already part of a multi-mycotoxin method further discussed below (Warth et al. 2012).

9.5 Deoxynivalenol

Deoxynivalenol (DON) is produced by different fungi species of the *Fusarium* genus and regarded as the main producer and source of trichothecene contamination in food. Trichothecenes can be subdivided into type A-D depending on their chemical core structure. In food the type B trichothecene deoxynivalenol together with its acetylated forms 3-acetyl-deoxynivalenol (3-AcDON), and 15-acetyl-deoxynivalenol (15-AcDON) as well as plant metabolites thereof can be seen as most important based on occurrence. However, the type A trichothecenes T-2 and HT-2 toxin as well as nivalenol (NIV) and fusarenon X (FUS) are also toxicologically relevant food contaminants. DON induces digestive disorders and oxidative damage. In cells, it is able to inhibit DNA, RNA and protein biosynthesis by interaction with ribosomes (Pestka 2010). The metabolism of trichothecenes, especially DON has extensively studied in animals but also some data on human metabolism is available. After ingestion, within 24 h DON is nearly completely excreted, to a large extend as phase II metabolites (Meky et al. 2003). Glucuronidation of DON has already been observed in animals in 1986 and was shown for humans in 2002 (Cote et al. 1986; Meky et al. 2003). The structures of the glucuronides, deoxynivalenol-3-*O*-β-glucuronide (DON-3-GlcA), deoxynivalenol-15-*O*-β-glucuronide (DON-15-GlcA) and deoxynivalenol-7-*O*-glucuronide (DON-7-GlcA) have been proposed from mass spectrometric experiments in 2011 and 2012 (Maul et al. 2012; Warth et al. 2011; Lattanzio et al. 2011). However, so far only the structures of DON-3-GlcA and DON-15-GlcA have fully been characterized while the assignment of a certain peak as DON-7-GlcA is still tentative (Fruhmann et al. 2012; Uhlig et al. 2013). A further metabolite of DON, deepoxy DON (DOM-1)

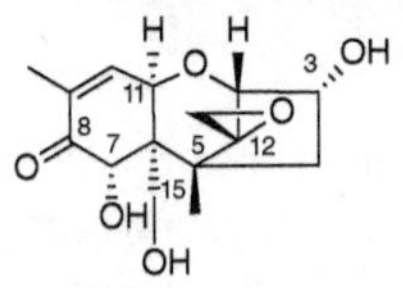

DON

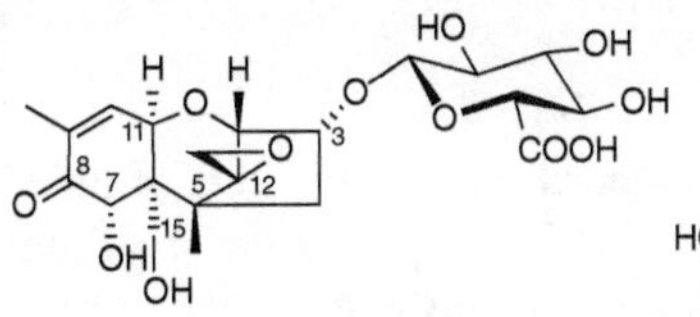

DON-3-GlcA

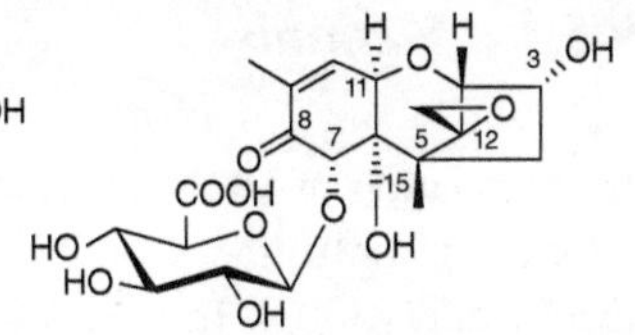

DON-7-GlcA

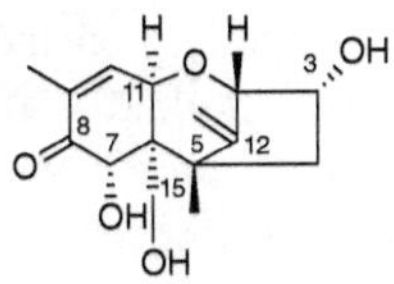

DOM-1

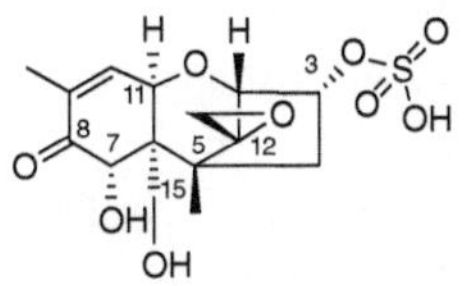

DON-3-sulfate

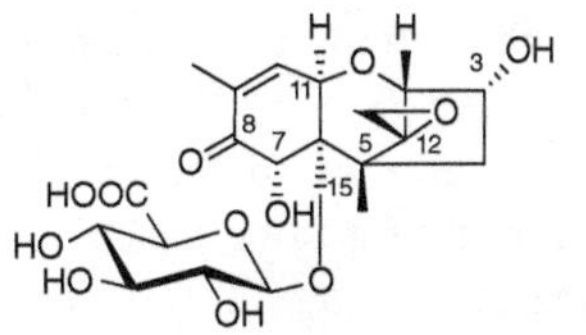

DON-15-GlcA

Fig. 9.4 Chemical structures of deoxynivalenol (DON) and its human metabolites deoxynivalenol-3-O-β-glucuronide (DON-3-GlcA), deoxynivalenol-7-O-β-glucuronide (tentative structure assignment, DON-7-GlcA,), deoxynivalenol-15-O-β-glucuronide (DON-15-GlcA), deepoxy-deoxynivalenol (DOM-1), deoxynivalenol-3-sulfate (DON-3-sulfate)

as well as its glucuronide have been detected in large quantities in animal studies but so far only in trace amounts in humans (Kouadio et al. 2014; Turner et al. 2010a; Heyndrickx et al. 2015). Recently the product of an enzymatic sulfatation of DON, DON-3-sulfate, was detected in human at levels corresponding to approximately 4% of the DON ingested (Warth et al. 2016).The structures of DON and its metabolites are shown in Fig. 9.4.

Glucuronidation rate of DON found in urine ranges roughly between 70% and 98% (Gerding et al. 2014; Turner et al. 2010b; Warth et al. 2013; Warth et al. 2012). Among the three glucuronides of DON, DON-15-GlcA was detected in the highest concentrations covering approximately 70% of all glucuronides while DON-3-GlcA accounts for approximately 20–30%. DON-7-GlcA has so far not been studied due to missing reference compounds but seems to account only for minor quantities. Heyndrickx et al. determined high levels of DOM-glucuronide especially in urine from children, exceeding even the sum of DON, DON-3GlcA and DON-15-GlcA.

A first biomarker for DON exposure was proposed by Meky et al. who studied the metabolism of radiolabeled DON in rat and compared it with urine samples from a highly exposed region in China. Based on their data from animal experiments the author conclude that approximately 30% of DON uptake is excreted via urine within 24 h and 40% of DON in urine can be recovered as the unmodified mycotoxin (Meky et al. 2003). A comprehensive survey in combination with biomonitoring allowed Turner et al. to calculate a mean transfer of DON and its glucuronides of 72% for its transfer into urine (Turner et al. 2010b). Data from a single individual also supports that urinary concentrations of DON and its glucuronides is suitable for biomonitoring with urinary excretion rate of 60–73% for DON and its glucuronides (Warth et al. 2013).

9.5.1 Methods

Determination of DON and its metabolites in blood is of little importance for biomonitoring as only traces of this polar compound can be found in this compartment (Meky et al. 2003). Instead urinary analysis of DON and its metabolites is common. Initial HBM methods for DON in human urine used IAC columns for selective binding of DON. DON glucuronides were not retained on this column and required a glucuronidase treatment before this step (Meky et al. 2003). Depending on the antibody, also DOM-1 could be retained and purified by IAC (Turner et al. 2010a). For sensitive detection, HPLC-MS/MS methods were used after purification allowing for urine a LOD of 0.1 and 0.05 ng/mL for DON and DOM-1, respectively (Turner et al. 2008). Direct detection of DON and its glucuronides was first published in 2011 (Warth et al. 2011). Due to the missing sample enrichment and matrix removal, the LOD of DON was 60 times higher than with IAC purification. For DON-3-GlcA, a LOD of 3 ng/mL was determined. Due to method optimization and increasing MS performance LOD and LOQ for DON and its glucuronides are in the recently published methods also ranging between 0.05 and 0.2 ng/mL (Huybrechts et al. 2015; Gerding et al. 2014). Availability of references for DON glucuronides is still critical for the direct detection methods. In order to overcome this shortage at least partially, in some publications chromatography is optimized in a way that DON-3-GlcA and DON-15-GlcA coelute. Due to comparable ionizability, only one standard, DON-3-GlcA, has been used for quantitation and both glucuronides reported as sum parameter (Huybrechts et al. 2015; Gerding et al. 2014).

9.6 Fumonisins

9.6.1 Background

Fumonisins are mycotoxins produced by different *Fusarium* species with *Fusarium verticillioides* being one of the most frequent producers. More than 30 different fumonisins are known among which fumonisin B_1 (FB_1) is regarded as the most important and fumonisin B_2 (FB_2) the second most important due to occurrence rate, concentration and toxic properties. Fumonisins are potent inhibitors of ceramide synthase and implicated in human diseases such as carcinogenesis, neural tube defects, and stunting in children (Stockmann-Juvala and Savolainen 2008, Wang et al. 1991). Biomonitoring of fumonisins is based on the analysis of the most important FB_1 in urine (Gong et al. 2008). Validation studies in human demonstrated that 0.12–2% of this compound is excreted via urine (Riley et al. 2012; van der Westhuizen et al. 2011). However, due to the high amounts of FB_1 found in food, often in the milligram per kilogram range, the levels of FB_1 in urine are still sufficient for quantification. The chemical structure of FB_1 is shown in Fig. 9.5. An alternative approach towards the detection of exposure to fumonisins is based on the

Fig. 9.5 Chemical structure of fumonisin B_1 (FB_1)

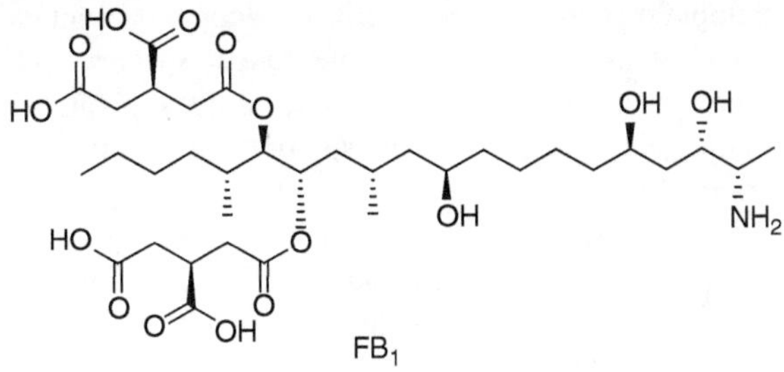

measurement of a biomarker of effect. This means that a biochemical change induced by fumonisins, in monitored. A biochemical effect of fumonisins is the inhibition of the ceramide synthase, which results in an increase of sphingoid bases and sphingoid 1-phosphates (Wang et al. 1991). Consequently Riley et al. developed a method for the detection of sphinganine 1-phosphate and sphingosine 1-phosphate in blood, based in the DBS technique. The subsequent application of this technique in a study in Guatemala revealed a correlation between increased fumonisin excretion and increased sphinganine 1-phosphate to sphingosine 1-phosphate ratio as well as sphinganine 1-phosphate concentration in blood (Riley et al 2015).

9.6.2 *Methods*

Analytical methods for the detection of FB_1 utilize the tricarballylic acid side chains of the molecule for selective enrichment on strong anion exchange resins (SAX). Thus internal standards are added to the urine samples before they are diluted with water and the pH is adjusted. After different washing steps, FB_1 is eluted with strong acids such as a solution of 2% formic acid (Gong et al. 2008). Using this set up, a LOD of 0.02 ng/mL urine could be established. Alternatively, enrichment of FB_1 on reversed phase (C18) SPE cartridges was also be applied (Riley et al. 2012). After passing the urine through the cartridge and washing, bound FB_1 is stable on the material, allowing even the shipping of the SPE from the sampling site to an analytical laboratory with HPLC-MS system. With this setup comparable LODs as reported for strong anion exchange columns can be achieved. Without previous pre-concentration or purification steps, the LOD for fumonisin B_1 is approximately 2–3 times higher and stronger influenced by matrix effects (Huybrechts et al. 2015; Gerding et al. 2014).

9.7 Multi-Mycotoxin Methods for HBM

A comprehensive overview over the nutritional, occupational or environmental exposure was the driving force behind the development of multi-mycotoxin methods for HBM. Co-exposure to several mycotoxins is obvious as for most kinds of food multiple mycotoxins are described as common contaminants. This co-contamination can be caused by just a single fungal species such as *Fusarium graminearum*, which produces zearalenone as well as deoxynivalenol or by multiple fungi growing on the field like *Fusarium* sp. or during improper storage like *Aspergillus* or *Penicillium* sp.

Healthy, diversified nutrition means multiple food sources and thus multiple sources of mycotoxin exposures. As most commodities can be contaminated with mycotoxins, it is possible that relevant sources of exposure remain unidentified. As an example, Gerding et al. analyzed and compared urine samples from Bangladesh, Germany and Haiti (Gerding et al. 2015). In their study, the authors observed comparably high concentrations of OTA in urine samples for Bangladesh with a mean level of 207 pg/mg creatinine compared to 30 pg/mg creatinine in Germany and 91 pg/mg creatinine in Haiti. Taking into account the seven most important sources of OTA in Germany: Bread, breakfast cereals, coffee, pasta, beer, chocolate, and sausages, none of these products is very popular in Bangladesh. Instead, tea is preferred over coffee, rice the main source of carbohydrates, and beer and sausages are not consumed for religious reasons (EC 2002; FAO 1999). Currently the source of OTA exposure is not clear but contaminated spices or fruits might play an important role.

Occupational exposure often also occurs as a complex mixture. Considering bakeries for instance, exposure through flour can be relevant but also through fungal growth in the humid and warm bakery environment. However, due to co-exposure through food consumption it is often difficult to attribute a certain mycotoxin exposure to environmental factors (Degen 2011).

9.7.1 Methods

The first published HBM multi-mycotoxin methods were focusing on the parallel application of multiple IAC to achieve sufficient sensitivity and purity of the physiological samples. In a first report by Ahn et al., urine samples were divided into three portions which were independently passed through three different IAC columns specific for aflatoxins, ochratoxin A and fumonisins, respectively. The purified IAC eluates were combined and analyzed in a single HPLC-MS/MS run (Ahn et al. 2010). As a further improvement, 6-in-1 IAC columns were utilized for urine analysis covering additional toxins. Rubert et al. determined T2/HT-2, DON, FB_1, FB_2, AFB_1, AFB_2, AFG_1, AFG_2, ZEN and OTA but missed the more important AFM_1 and α-ZEL. Also no glucuronidase/sulfatase treatment for deconjugation of phase II metabolites was applied (Rubert et al. 2011). Further improvement of the 6-in-1 IAC method was achieved by introduction of a deconjugation step with glucuronidase from *Helix pomatia*, incorporation of AFM_1 as well as α-ZEL and β-ZEL (Solfrizzo et al. 2011).

Besides IAC, alternative approaches applying SPE or strong anion exchange columns (SAX) for pre-concentration have been published (Ediage et al. 2012).

However with the growing number of emerging mycotoxins and mycotoxin metabolites that should be incorporated into multi-mycotoxin HBM methods require a more and more unspecific sample treatment and a method that can handle a broad range of polarities ranging from highly polar phase II metabolites such as DON-glucuronide to nonpolar enniatins, dehydrocitrinone or ZEN (Gerding et al. 2014).

In the last two decades HPLC-MS/MS-Systems for mycotoxin detection reached a 10- to 100-fold increase in sensitivity and became faster in terms of scan rate as well as in the required time for maximum selected reaction monitoring (SRM) sensitivity. While at the beginning, 25–50 ms per sensitive SRM transition were required, modern instruments need only 1–3 ms per SRM to achieve full sensitivity. Further implementations such as rapid polarity switching or the possibility to define observation windows for specific SRMs (e.g. scheduled MRM™) gave rise to the development of dilute-and-shoot multi-mycotoxin methods (Sulyok et al. 2006). Consequently, this approach was transferred to the field of mycotoxin HBM. In this case urine samples were either directly injected or diluted by a factor between 2 and 20 followed by a sensitive HPLC-MS/MS system. If required additional centrifugation or filtration steps prior to injection were applied. An overview over recent methods and their application for HBM in urine is shown in Table 9.2.

9.8 Discussion and Outlook

HBM of mycotoxins has been carried out for more than three decades. Aim of the first studies was exposure assessment in highly exposed cohorts. Often medical indications had been observed first and HBM was used to identify the source of elevated hepatitis, nephropathy or cancer rates. Based on food analysis, specific mycotoxins were in focus of the investigations and the method development optimized for a single compound or a small group of compounds. In the beginning metabolism of the toxins was rarely considered, but in some cases glucuronidase/sulfatase treatment had already been applied. Within the last decade HBM of mycotoxins shifted from a toxicity driven approach towards general approach of assessing the exposure to mycotoxins. This general approach also required a shift of the method development form single analyte techniques with specific enrichment towards multi-analyte analysis. As a first step, 6-in-1 IAC columns were implemented in sample preparation, allowing the simultaneous detection of aflatoxins, ochratoxins, fumonisins, zearalenone, deoxynivalenol, and nivalenol as a new approach towards multi analyte analysis. Benefits of this sample preparation are clearly the efficient removal of matrix as well as the possible sample enrichment. However, binding affinity of the applied antibodies are in most cases limited to the parent compounds while metabolites such as OTα and especially phase II metabolites are not retained. Thus for 6-in-1 IAC, glucuronidase/sulfatase treatment is mandatory to increase the amount of total mycotoxin which is the base of most exposure calculations. However, efficiency of the enzymatic hydrolysis is not always secure and complete, and should previously investigated with phase II

Table 9.2 HPLC-MS based multi-mycotoxin methods for urine analysis

Year	Analytes	HBM region	Laboratory/method	Literature
2011	11	Spain ($n = 27$)	University Valencia, Spain	(Rubert et al. 2011)
2011	7	Italy ($n = 10$)	ISPA Bari, Italy	(Solfrizzo et al. 2011)
2012	14	Cameroon ($n = 175$)	IFA Tulln, Austria	(Warth et al. 2012)
2012	18	Belgium ($n = 40$)	Ghent University, Belgium	(Ediage et al. 2012)
2013	15	Cameroon ($n = 175$, 83% HIV-positive)	IFA Tulln, Austria	(Abia et al. 2013)
2013	18	Cameroon ($n = 220$, children)	Ghent University, Belgium	(Ediage et al. 2013)
2013	7	Transkei, South Africa ($n = 53$, females)	ISPA Bari, Italy	(Shephard et al. 2013)
2014	7	Ivory coast ($n = 99$)	ISPA Bari, Italy	(Kouadio et al. 2014)
2014	23	Germany ($n = 101$, 57 females, 44 males)	WWU Münster, Germany	(Gerding et al. 2014)
2014	15	Thailand ($n = 60$, 33 females, 27 males)	IFA Tulln, Austria	(Warth et al., 2014)
2014	8	Italy ($n = 52$)	ISPA Bari, Italy	(Solfrizzo et al., 2014)
2014	15	Nigeria ($n = 120$, 19 children, 20 adolescence, 81 adults)	IFA Tulln, Austria	(Ezekiel et al. 2014)
2015	23	Bangladesh ($n = 95$), Germany ($n = 50$), Haiti ($n = 142$)	WWU Münster, Germany	(Gerding et al. 2015)
2015	8	Sweden ($n = 252$)	ISPA Bari, Italy	(Wallin et al., 2015)
2015	33	Belgium ($n = 394$, 155 children and 239 adults)	Ghent University, Belgium	(Heyndrickx et al. 2015)
2015	32	Belgium ($n = 32$)	CODA-CERVA Tervuren, Belgium	(Huybrechts et al. 2015)

metabolites of the parent compound. As a further disadvantage of this technique is the time-consuming sample preparation which, especially for large cohort studies, significantly reduces the number of samples that can be analyzed per day.

Following to 6-in-1 IAC, direct dilute and shoot approached for mycotoxin detection have been developed. In this case no purification nor sample enrichment but even dilution of urine is done. Also the analyte spectrum is extended as all phase II metabolites are now quantified individually. Consequently, the most sensitive HPLC-MS/MS equipment is required for this approach to allow sensitive analyte detection and quantification despite interfering urine matrix. Separate detection of the parent compound and its metabolites furthermore results in lower analyte concentrations. As an example, DON, DON-3-GlcA, DON-15-GlcA, and DON-3-sulfate have to be detected individually when a dilute and shoot approach is chosen while cleanup with glucuronidase and sulfatase treatment – in the best case – leads to one peak with a signal intensity corresponding to the sum of all compounds.

Besides the known mycotoxins discussed here, more than 300 mycotoxins and toxicologically relevant fungal secondary metabolites have been reported. For many of them only little information is available regarding occurrence rate and distribution. Dilute and shoot methods are designed to be capable to include these compounds and therefore to provide first information on the presence in biological matrices. The power of this approach can be seen in the increasing number of analytes covered by the newly developed multi-mycotoxin methods for HBM reported in Table 9.2. Among the 23 analytes detected by Gerding et al. were also dihydrocitrinone, which is a metabolite of the mycotoxin citrinin and the mycotoxins enniatin B_1, T-2 toxin, HT-2 toxin as well as HT-2-4-*O*-glucuronide (Gerding et al. 2014). Heyndrickx et al. covered even further mycotoxins and metabolites such as DOM-1 and DOM-1-glucuronide, deacetoxyscirpenol and fusarenon X (Heyndrickx et al. 2015). Detection of enniatins or DOM-1-glucuronide besides DON, ZEN and OTA with their metabolites in urine samples from Belgium or Germany confirm the important role that HBM of mycotoxins can play in the near future for exposure and risk assessment.

References

Abia WA, Warth B, Sulyok M et al. (2013) Bio-monitoring of mycotoxin exposure in Cameroon using a urinary multi-biomarker approach. Food Chem Toxicol 62:927–934

Ahn J, Kim D, Kim H et al. (2010) Quantitative determination of mycotoxins in urine by LC-MS/MS. Food Addit Contam A 27(12):1674–1682

Beker D, Radic B (1991) Fast determination of ochratoxin a in serum by liquid-chromatography – comparison with enzymatic spectrofluorimetric method. J Chromatogr B Biomed Sci Appl 570 (2):441–445

Blokland MH, Sterk SS, Stephany RW et al. (2006) Determination of resorcylic acid lactones in biological samples by GC-MS. Discrimination between illegal use and contamination with fusarium toxins. Anal Bioanal Chem 384(5):1221–1227

Bravin F, Duca RC, Balaguer P et al. (2009) In vitro cytochrome P450 formation of a mono-hydroxylated metabolite of zearalenone exhibiting estrogenic activities: possible occurrence of this metabolite in vivo. Int J Mol Sci 10(4):1824–1837

Campbell TC, Caedo JP, Bulataoj J et al. (1970) Aflatoxin-M1 in human urine. Nature 227 (5256):403–404

Castegnaro M, Canadas D, Vrabcheva T et al. (2006) Balkan endemic nephropathy: role of ochratoxins A through biomarkers. Mol Notr Food Res 50:519–529

Castegnaro M, Maru V, Maru G et al. (1990) High-performance liquid-chromatographic determination of ochratoxin a and its 4R-4-hydroxy metabolite in human urine. Analyst 115(2):129–131

Castegnaro M, Plestina R, Dirheimer G, et al. (eds)(1991) Mycotoxins, endemic nephropathy and urinary tract tumours. IARC Scientific Publications 115. IARC, Lyon.

Chu FS (1971) Interaction of ochratoxin A with bovine serum albumin. Arch Biochem Biophys 147(2):359–366

Coronel MB, Sanchis V, Ramos AJ et al. (2011) Ochratoxin A in adult population of Lleida, Spain: presence in blood plasma and consumption in different regions and seasons. Food Chem Toxicol 49(10):2697–2705

Cote LM, Dahlem AM, Yoshizawa T et al. (1986) Excretion of deoxynivalenol and its metabolite in milk, urine, and feces of lactating dairy-cows. J Dairy Sci 69(9):2416–2423

Cramer B, Osteresch B, Munoz KA et al. (2015) Biomonitoring using dried blood spots: detection of ochratoxin A and its degradation product 2′R-ochratoxin A in blood from coffee drinkers. Mol Nutr Food Res 59(9):1837–1843

Degen GH (2011) Tools for investigating workplace-related risks from mycotoxin exposure. World Mycotoxin J 4(3):315–327

Degen GH (2016) Are we ready to estimate daily ochratoxin A intake based on urinary concentrations? Environ Intl 97:254–255

Duarte SC, Pena A, Lino CM (2011) Human ochratoxin A biomarkers – from exposure to effect. Crit Rev Toxicol 41(3):187–212

EC (2002) European Community. Assessment of dietary intake of ochratoxin A by the population of EU Member States. Report of the Scientific Cooperation, Task 3.2.7 Directorate-General Health and Consumer Protection, European Commission. europaeuint/comm/food/fs/scoop/327_en.pdf

Ediage EN, Di Mavungu JD, Song SQ et al. (2013) Multimycotoxin analysis in urines to assess infant exposure: a case study in Cameroon. Environ Int 57–58:50–59

Ediage EN, Di Mavungu JD, Song SQ et al. (2012) A direct assessment of mycotoxin biomarkers in human urine samples by liquid chromatography tandem mass spectrometry. Anal Chim Acta 741:58–69

EFSA (2011) Use of the EFSA Comprehensive European Food Consumption Database in Exposure Assessment. EFSA J 9(3):2097

Ezekiel CN, Warth B, Ogara IM et al. (2014) Mycotoxin exposure in rural residents in northern Nigeria: a pilot study using multi-urinary biomarkers. Environ Int 66:138–145

FAO (1999) Bangladesh. FAO – Nutrition Country Profiles. Food and Agricultural Organization of the United Nations, Rome

Fruhmann P, Warth B, Hametner C et al. (2012) Synthesis of deoxynivalenol-3-beta-D-O-glucuronide for its use as biomarker for dietary deoxynivalenol exposure. World Mycotoxin J 5(2):127–132

Gerding J, Ali N, Schwartzbord J et al. (2015) A comparative study of the human urinary mycotoxin excretion patterns in Bangladesh, Germany, and Haiti using a rapid and sensitive LC-MS/MS approach. Mycotoxin Res 31(3):127–136

Gerding J, Cramer B, Humpf HU (2014) Determination of mycotoxin exposure in Germany using an LC-MS/MS multibiomarker approach. Mol Nutr Food Res 58(12):2358–2368

Gilbert J, Brereton P, MacDonald S (2001) Assessment of dietary exposure to ochratoxin A in the UK using a duplicate diet approach and analysis of urine and plasma samples. Food Addit Contam 18(12):1088–1093

Gong YY, Torres-Sanchez L, Lopez-Carrillo L et al. (2008) Association between tortilla consumption and human urinary fumonisin B1 levels in a Mexican population. Cancer Epidem Biomar 17(3):688–694

Groopman JD, Egner PA, Schulze KJ et al. (2014) Aflatoxin exposure during the first 1000 days of life in rural South Asia assessed by aflatoxin B-1-lysine albumin biomarkers. Food Chem Toxicol 74:184–189

Groopman JD, Wild CP, Hasler J et al. (1993) Molecular epidemiology of aflatoxin exposures – validation of aflatoxin-N7-guanine levels in urine as a biomarker in experimental rat models and humans. Environ Health Persp 99:107–113

Guggisberg H (1954) Mutterkorn – vom Gift zum Heilstoff. Karger, Basel

Han Z, Tangni EK, Di Mavungu JD et al. (2013) In vitro glucuronidation of ochratoxin A by rat liver microsomes. Toxins 5(12):2671–2685

Heyndrickx E, Sioen I, Huybrechts B et al. (2015) Human biomonitoring of multiple mycotoxins in the Belgian population: results of the BIOMYCO study. Environ Int 84:82–89

Hult K, Plestina R, Habazinnovak V et al. (1982) Ochratoxin A in human blood and Balkan endemic nephropathy. Arch Toxicol 51(4):313–321

Huybrechts B, Martins JC, Debongnie P et al. (2015) Fast and sensitive LC-MS/MS method measuring human mycotoxin exposure using biomarkers in urine. Arch Toxicol 89(11): 1993–2005

Jodlbauer J, Zollner P, Lindner W (2000) Determination of zeranol, taleranol, zearalenone, alpha- and beta-zearalenol in urine and tissue by high-performance liquid chromatography-tandem mass spectrometry. Chromatographia 51(11–12):681–687

Jonsyn-Ellis FE (2001) Seasonal variation in exposure frequency and concentration levels of aflatoxins and ochratoxins in urine samples of boys and girls. Mycopathologia 152(1):35–40

Kensler TW, Roebuck BD, Wogan GN et al. (2011) Aflatoxin: a 50-year odyssey of mechanistic and translational toxicology. Toxicol Sci 120:S28–S48

Kouadio JH, Lattanzio VMT, Ouattara D et al. (2014) Assessment of mycotoxin exposure in Côte D'ivoire (Ivory Coast) through multi-biomarker analysis and possible correlation with food consumption patterns. Toxicol Int 21(3):248–257

Lattanzio VMT, Solfrizzo M, De Girolamo A et al. (2011) LC-MS/MS characterization of the urinary excretion profile of the mycotoxin deoxynivalenol in human and rat. J Chromatogr B Analyt Technol Biomed Life Sci 879(11–12):707–715

Launay FM, Young PB, Sterk SS et al. (2004) Confirmatory assay for zeranol, taleranol and the Fusarium spp. toxins in bovine urine using liquid chromatography-tandem mass spectrometry. Food Addit Contam 21(1):52–62

Lewis L, Onsongo M, Njapau H et al. (2005) Aflatoxin contamination of commercial maize products during an outbreak of acute aflatoxicosis in eastern and central Kenya. Environ Health Persp 113(12):1763–1767

Malir F, Ostry V, Pfohl-Leszkowicz A et al. (2016) Ochratoxin A: 50 years of research. Toxins 8(7):pii: E191

Märtlbauer E, Usleber E, Dietrich R et al. (2009) Ochratoxin A in human blood serum – retrospective long-term data. Mycotoxin Res 25(4):175–186

Maul R, Warth B, Kant JS et al. (2012) Investigation of the hepatic glucuronidation pattern of the Fusarium mycotoxin deoxynivalenol in various species. Chem Res Toxicol 25(12):2715–2717

Mccoy LF, Scholl PF, Sutcliffe AE et al. (2008) Human aflatoxin albumin adducts quantitatively compared by ELISA, HPLC with fluorescence detection, and HPLC with isotope dilution mass spectrometry. Cancer Epidem Biomar 17(7):1653–1657

Meky FA, Turner PC, Ashcroft AE et al. (2003) Development of a urinary biomarker of human exposure to deoxynivalenol. Food Chem Toxicol 41(2):265–273

Metzler M, Pfeiffer E, Hildebrand AA (2010) Zearalenone and its metabolites as endocrine disrupting chemicals. World Mycotoxin J 3(4):385–401

Mikula H, Hametner C, Berthiller F et al. (2012) Fast and reproducible chemical synthesis of zearalenone-14-beta,D-glucuronide. World Mycotoxin J 5(3):289–296

Miles CO, Erasmuson AF, Wilkins AL et al. (1996) Ovine metabolism of zearalenone to α-zearalanol (Zeranol). J Agr Food Chem 44(10):3244–3250

Mirocha CJ, Pathre SV, Robison TS (1981) Comparative metabolism of zearalenone and transmission into bovine-milk. Food Cosmet Toxicol 19(1):25–30

Muñoz K, Blaszkewicz M, Degen GH (2010) Simultaneous analysis of ochratoxin A and its major metabolite ochratoxin alpha in plasma and urine for an advanced biomonitoring of the mycotoxin. J Chromatogr B 878(27):2623–2629

Muñoz K, Cramer B, Dopstadt J et al. (2017) Evidence of ochratoxin A conjugates in urine samples from infants and adults. Mycotoxin Res 33(1):39–47

Muñoz K, Degen G, Blaszkewicz M et al. (2009) Biomonitoring of ochratoxin A and its metabolite ochratoxin alpha in urine, plasma and human milk. Toxicol Lett 189:S151

Nelson DB, Kimbrough R, Landrigan PS et al. (1980) Aflatoxin and Reyes Syndrome - a case control study. Pediatrics 66(6):865–869

Olsen ME, Pettersson HI, Sandholm KA et al. (1985) Quantitative liquid-chromatographic method using fluorescence detection for determining zearalenone and its metabolites in blood-plasma and urine. J Assoc Off Anal Chem 68(4):632–635

Onyemelukwe GC, Ogbadu G (1981) Aflatoxin levels in sera of healthy 1st time rural blood-donors - preliminary-report. T Roy Soc Trop Med H 75(6):780–782

Orti DL, Hill RH, Liddle JA et al. (1986) High-performance liquid-chromatography of mycotoxin metabolites in human urine. J Anal Toxicol 10(2):41–45

Osteresch B, Cramer B, Humpf HU (2016) Analysis of ochratoxin A in dried blood spots - Correlation between venous and finger-prick blood, the influence of hematocrit and spotted volume. J Chromatogr B Analyt Technol Biomed Life Sci 1020:158–164

Pascale M, Visconti A (2001) Rapid method for the determination of ochratoxin A in urine by immunoaffinity column clean-up and high-performance liquid chromatography. Mycopathologia 152(2):91–95

Pestka JJ (2010) Deoxynivalenol: mechanisms of action, human exposure, and toxicological relevance. Arch Toxicol 84(9):663–679

Pestka JJ, Steinert BW, Chu FS (1981) Enzyme-linked immunosorbent-assay for detection of ochratoxin A. Appl Environ Microb 41(6):1472–1474

Petkova-Bocharova T, Castegnaro M, Pfohl-Leszkowicz A et al. (2003) Analysis of ochratoxin A in serum and urine of inhabitants from an area with Balkan endemic nephropathy: a one month follow up study. Facta Univ Ser Med Biol 10:62–68

Pfeiffer E, Hildebrand A, Damm G et al. (2009) Aromatic hydroxylation is a major metabolic pathway of the mycotoxin zearalenone in vitro. Mol Nutr Food Res 53(9):1123–1133

Pfeiffer E, Hildebrand A, Mikula H et al. (2010) Glucuronidation of zearalenone, zeranol and four metabolites in vitro: formation of glucuronides by various microsomes and human UDP-glucuronosyltransferase isoforms. Mol Nutr Food Res 54(10):1468–1476

Plasencia J, Mirocha CJ, Pawlosky RJ et al. (1990) Analysis of zearalenone and alpha-zearalenol in urine of ruminants using gas-chromatography tandem mass-spectrometry. J Assoc Off Anal Chem 73(6):973–980

Riley RT, Torres O, Showker JL et al. (2012) The kinetics of urinary fumonisin B1 excretion in humans consuming maize-based diets. Mol Nutr Food Res 56(9):1445–1455

Riley RT, Torres O, Matute J et al. (2015) Evidence for fumonisin inhibition of ceramide synthase in humans consuming maize-based foods and living in high exposure communities in Guatemala. Mol Nutr Food Res 59(11):2209–2224

Rubert J, Soriano JM, Manes J et al. (2011) Rapid mycotoxin analysis in human urine: A pilot study. Food Chem Toxicol 49(9):2299–2304

Schiff PL (2006) Ergot and its alkaloids. Am J Pharm Educ 70 (5):98.

Scholl PF, Musser SM, Groopman JD (1997) Synthesis and characterization of aflatoxin B1 mercapturic acids and their identification in rat urine. Chem Res Toxicol 10(10):1144–1151

Schwartzbord JR, Leroy JL, Severe L et al. (2016) Urinary aflatoxin M1 in Port-au-Prince and a rural community in north-east Haiti. Food Addit Contam A 33(6):1036–1042

Scott PM (2005) Biomarkers of human exposure to ochratoxin A. Food Addit Contam 22 (SUPPL. 1):99–107

Shephard GS, Burger HM, Gambacorta L et al. (2013) Multiple mycotoxin exposure determined by urinary biomarkers in rural subsistence farmers in the former Transkei, South Africa. Food Chem Toxicol 62:217–225

Solfrizzo M, Gambacorta L, Lattanzio VMT et al. (2011) Simultaneous LC-MS/MS determination of aflatoxin M1, ochratoxin A, deoxynivalenol, de-epoxydeoxynivalenol, alpha and beta-zearalenols and fumonisin B1 in urine as a multi-biomarker method to assess exposure to mycotoxins. Anal Bioanal Chem 401(9):2831–2841

Solfrizzo M, Gambacorta L, Visconti A (2014) Assessment of multi-mycotoxin exposure in southern Italy by urinary multi-biomarker determination. Toxins 6(2):523–538

Stevenson DE, Hansen RP, Loader JI et al. (2008) Preparative enzymatic synthesis of glucuronides of zearalenone and five of its metabolites. J Agr Food Chem 56(11):4032–4038

Stockmann-Juvala H, Savolainen K (2008) A review of the toxic effects and mechanisms of action of fumonisin B1. Hum Exp Toxicol 27(11):799–809

Stojkovic R, Hult K, Gamulin S et al. (1984) High-affinity binding of ochratoxin A to plasma constituents. Biochem Int 9(1):33–38

Studer-Rohr I, Schlatter J, Dietrich DR (2000) Kinetic parameters and intraindividual fluctuations of ochratoxin A plasma levels in humans. Arch Toxicol 74(9):499–510

Sulyok M, Berthiller F, Krska R et al. (2006) Development and validation of a liquid chromatography/tandem mass spectrometric method for the determination of 39 mycotoxins in wheat and maize. Rapid Commun Mass Sp 20(18):2649–2659

Tang LL, Tang M, Xu L et al. (2008) Modulation of aflatoxin biomarkers in human blood and urine by green tea polyphenols intervention. Carcinogenesis 29(2):411–417

Trenholm HL, Warner R, Farnworth ER (1980) Gas-chromatographic detection of the mycotoxin zearalenone in blood-serum. J Assoc Off Anal Chem 63(3):604–611

Trenholm HL, Warner RM, Farnworth ER (1981) High performance liquid chromatographic method using fluorescence detention for quantitative analysis of zearalenone and alpha-zearalenol in blood plasma. J Assoc Off Anal Chem 64(2):302–310

Turner PC, Flannery B, Isitt C et al. (2012) The role of biomarkers in evaluating human health concerns from fungal contaminants in food. Nutr Res Rev 25(1):162–179

Turner PC, Hopton RP, Lecluse Y et al. (2010a) Determinants of urinary deoxynivalenol and de-epoxy deoxynivalenol in male farmers from Normandy, France. J Agr Food Chem 58 (8):5206–5212

Turner PC, White KLM, Burley VJ et al. (2010b) A comparison of deoxynivalenol intake and urinary deoxynivalenol in UK adults. Biomarkers 15(6):553–562

Turner PC, Rothwell JA, White KLM et al. (2008) Urinary deoxynivalenol is correlated with cereal intake in individuals from the United Kingdom. Environ Health Persp 116(1):21–25

Ueno Y, Maki S, Lin J et al. (1998) A 4-year study of plasma ochratoxin A in a selected population in Tokyo by immunoassay and immunoaffinity column-linked HPLC. Food Chem Toxicol 36(5):445–449

Uhlig S, Ivanova L, Faeste CK (2013) Enzyme-assisted synthesis and structural characterization of the 3-, 8-, and 15-glucuronides of deoxynivalenol. J Agr Food Chem 61(8): 2006–2012

van der Westhuizen L, Shephard GS, Burger HM et al. (2011) Fumonisin B1 as a urinary biomarker of exposure in a maize intervention study among South African subsistence farmers. Cancer Epidem Biomar 20(3):483–489

Vatinno R, Aresta A, Zambonin CG et al. (2007) Determination of ochratoxin A in human urine by solid-phase microextraction coupled with liquid chromatography-fluorescence detection. J Pharmaceut Biomed 44(4):1014–1018

Viegas S, Veiga L, Almeida A et al. (2016) Occupational exposure to aflatoxin B1 in a Portuguese poultry slaughterhouse. Ann Occup Hyg 60(2):176–183

Voyksner RD, Hagler WM, Tyczkowska K et al. (1985) Thermospray high-performance liquid-chromatographic mass-spectrometric analysis of some Fusarium mycotoxins. J High Res Chromatog 8(3):119–125

Wafa EW, Yahya RS, Sobh MA et al. (1998) Human ochratoxicosis and nephropathy in Egypt: a preliminary study.. Hum Exp Toxicol 17(2):124–129

Wallin S, Gambacorta L, Kotova N et al. (2015) Biomonitoring of concurrent mycotoxin exposure among adults in Sweden through urinary multi-biomarker analysis. Food Chem Toxicol 83:133–139

Wang E, Norred WP, Bacon CW et al. (1991) Inhibition of sphingolipid biosynthesis by fumonisins – implications for diseases associated with *Fusarium moniliforme*. J Biol Chem 266 (22):14486–14490

Warth B, Del Favero G, Wiesenberger G et al. (2016) Identification of a novel human deoxynivalenol metabolite enhancing proliferation of intestinal and urinary bladder cells. Sci Rep 6:33854

Warth B, Petchkongkaew A, Sulyok M et al. (2014) Utilising an LC-MS/MS-based multi-biomarker approach to assess mycotoxin exposure in the Bangkok metropolitan area and surrounding provinces. Food Addit Contam A 31(12):2040–2046

Warth B, Sulyok M, Berthiller F et al. (2011) Direct quantification of deoxynivalenol glucuronide in human urine as biomarker of exposure to the Fusarium mycotoxin deoxynivalenol. Anal Bioanal Chem 401(1):195–200

Warth B, Sulyok M, Berthiller F et al. (2013) New insights into the human metabolism of the Fusarium mycotoxins deoxynivalenol and zearalenone. Toxicol Lett 220(1):88–94

Warth B, Sulyok M, Fruhmann P et al. (2012) Development and validation of a rapid multi-biomarker liquid chromatography/tandem mass spectrometry method to assess human exposure to mycotoxins. Rapid Commun Mass Sp 26(13):1533–1540

Wild CP, Gong YY (2010) Mycotoxins and human disease: a largely ignored global health issue. Carcinogenesis 31(1):71–82

Wray BB, Hayes AW (1980) Aflatoxin B1 in the serum of a patient with primary hepatic carcinoma. Environ Res 22(2):400–403

Xie LW, Zhao XS, Kong WJ et al. (2014) Determination of ochratoxin A in human urine by HPLC-FLD after cleaned-up by molecularly imprinted polymer solid phase extraction column. Yaoxue Xuebao 49(4):517–523

Xue KS, Cai W, Tang L et al. (2016) Aflatoxin B1-lysine adduct in dried blood spot samples of animals and humans. Food Chem Toxicol 98:210–219

Zarba A, Wild CP, Hall AJ et al. (1992) Aflatoxin M1 in human breast-milk from the Gambia, West Africa, quantified by combined monoclonal-antibody immunoaffinity chromatography and HPLC. Carcinogenesis 13(5):891–894

Zhu JQ, Zhang LS, Hu X et al. (1987) Correlation of dietary aflatoxin B1 levels with excretion of aflatoxin M1 in human urine. Cancer Res 47(7):1848–1852

Zimmerli B, Dick R (1995) Determination of ochratoxin A at the ppt level in human blood, serum, milk and some foodstuffs by high-performance liquid-chromatography with enhanced fluorescence detection and immunoaffinity column cleanup – methodology and Swiss data. J Chrom B 666(1):85–99

Chapter 10
Microbial Secondary Metabolites and Knowledge on Inhalation Effects

Kati Huttunen and Merja Korkalainen

Abstract Microbial secondary metabolites include compounds produced during the growth of both fungi and bacteria. These compounds are present in workplaces and indoor environments, although the concentrations of single toxins in the air are typically low. Inhalation is considered to be the most significant route of exposure for microbial secondary metabolites in indoor environments although exposure to microbial toxins may happen also via alimentary or dermal route. Inhalation effects of microbial secondary metabolites have been studied experimentally in vivo in animal models, mainly in rodents. In vitro studies with cells of respiratory system and ex vivo cultured tissues have elucidated the mechanisms of action for the most common toxins. However, there are only few epidemiological studies on health effects of mycotoxin exposure, and often the studies are limited by exposure assessment based on single compounds or surrogates of mycotoxin exposure. We summarize here studies on the inhalation effects of microbial toxins showing a wide variety of adverse health effects which are not limited to the respiratory system, and identify the knowledge gaps where future research efforts should be targeted.

Keywords Microbial secondary metabolites · mycotoxin · inhalation · toxicity

10.1 Introduction

Mycotoxins, i.e. toxic secondary metabolites of fungi have been recognized as a health concern since early 1960s, after the revelation that aflatoxin-contaminated grain was the reason for mass killings of poultry (Forgacs, 1962). The number of

K. Huttunen (✉)
University of Eastern Finland, Department of Environmental and Biological Sciences, Kuopio, Finland
e-mail: kati.huttunen@uef.fi

M. Korkalainen
National Institute for Health and Welfare, Environmental Health Unit, Kuopio, Finland
e-mail: merja.korkalainen@thl.fi

C. Viegas et al. (eds.), *Exposure to Microbiological Agents in Indoor and Occupational Environments*, DOI 10.1007/978-3-319-61688-9_10

identified mycotoxins (~400 individual compounds) is only a fraction of the total number of toxic metabolites of fungi in our environment (Council for Agricultural Science and Technology, 2003; Täubel and Hyvärinen, 2015). In addition to fungi, also bacteria are able to produce toxic compounds during their growth (Täubel et al., 2011), which increases the number of toxic microbial metabolites possibly relevant to human exposure in the order of thousands. The main exposure route for mycotoxins is alimentary, e.g. eating contaminated foods, which directed the early research efforts into studying the effects of fungi typically present in foods or feeds, and oral exposure as a point of entry. Also dermal or mucosal absorption is possible, but to a lesser extent (Boonen et al., 2012; Pinton and Oswald, 2014). Exposure through respiratory route did not appear to be as relevant, as the majority of mycotoxins are nonvolatile. However, it was soon realized that inhalation is also a major route of exposure to mycotoxins. Not only were mycotoxins found from airborne dust (Croft et al., 1986; Flannigan, 1987), but the spores and fragments of microbial growth were shown to carry mycotoxins along them to the lungs (Brasel et al., 2005).

As a portal of entry for inhaled air, the nasal airway is a primary target for many inhaled toxicants (Harkema et al., 2006). The surface epithelium lining is often the first tissue in the nose to be directly injured, for example, by spores or mycotoxins of *Stachybotrys chartarum* (Pestka et al., 2008). Also microbial volatile organic compounds (MVOCs) are known to cause eye and upper-airway irritation in experimental exposure studies both in animals and humans (Korpi et al., 2009). Even if the toxins are nonvolatile, the adverse effect of mycotoxins can be carried by spores, as was seen when studying the pulmonary toxicity of intranasal or intrathacheal exposure to spores of *S. chartarum* in mice and rats (Lichtenstein et al., 2010; Pestka et al., 2008; Yike and Dearborn, 2004; Yike et al., 2005). The effect of inhalation exposure can be further aggravated by the presence and persistence of structural components in the lungs, leading to high concentrations of toxins within spore's immediate microenvironment (Carey et al., 2012). Comparison of intranasal and oral routes of exposure to mycotoxin deoxynivalenol (DON) in mice showed clearly higher plasma and tissue concentrations, as well as higher mRNA expressions of proinflammatory cytokines in internal organs following intranasal exposure (Amuzie et al., 2008). Accordingly, acute toxicity of the trichothecene mycotoxin T-2 in mice, rats and guinea pigs has been shown to be higher after inhalation exposure than when delivered by other routes (Creasia et al., 1987; Creasia et al., 1990). Considering that the lungs are a likely entry point both for nonvolatile and volatile compounds, inhalation is arguably the most significant route of exposure for microbial secondary metabolites in indoor environments (Fig. 10.1).

First studies on inhalation exposure to mycotoxins appeared in the late 1980s, describing the acute effects of trichothecenes produced by *Stachybotrys* (Croft et al., 1986). In the early years, the research on inhalation effects focused on mycotoxins that originated from molds thriving in high water activity such as aflatoxin B1 (AFB1) produced by e.g. *Aspergillus flavus* and satratoxins from *Stachybotrys chartarum* (Bitnun and Nosal, 1999; Hodgson et al., 1998). Originally the data on mycotoxin concentrations were largely from the workplaces

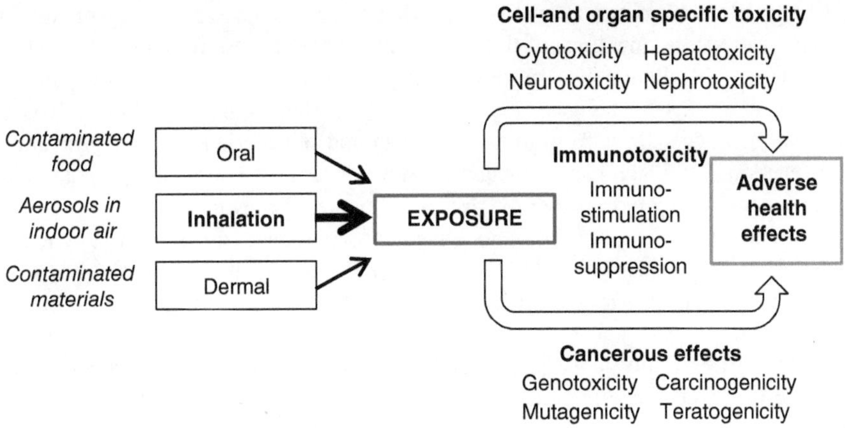

Fig. 10.1 Exposure to mycotoxins from various sources and some of the mechanisms and toxic properties leading to adverse health effects

that are routinely monitored for contamination such as factories processing grain, corn, peanut or malt. The most abundant mycotoxins found in the air of these workplaces are AFB1, DON and ochratoxin A (OTA) (Fromme et al., 2016). In addition of AFB1, DON and OTA, fumonisin and zearalenone are also agriculturally important mycotoxins with worldwide distribution (Jarvis and Miller, 2005; Miller and McMullin, 2014). Later, mycotoxins have been detected basically from all workplaces where organic material is handled including greenhouses, pigsties, cowsheds and waste treatment facilities (Degen, 2011; Mayer et al., 2008). In addition to the occupational exposure in industrial settings, the mycotoxin exposure in moisture damaged indoor environments has been a target of interest since 1980s, brought forward by the extensive media attention on cases of proposed intoxications with toxic secondary metabolites of *Stachybotrys chartarum* (Nevalainen et al., 2015). However, there is still a lack of exposure data, as the airborne mycotoxin concentrations are rarely measured, let alone associated with clinical health data. The challenge of the future studies is a better characterization of exposure to low concentrations of multiple toxins (Miller et al., 2010).

10.2 Toxicological Models of Respiratory System

Throughout the history of inhalation toxicology, small laboratory animals such as mice, rats, rabbits and guinea pigs have been used as model systems for pulmonary toxicity. In the case of a systemic effect to an inhaled toxicant, animal model is the only available approach for studying the effects in a controlled environment. In experimental conditions the most natural way of introducing mycotoxins into the lungs of laboratory animals is via inhalation (whole body exposure). However, inhalation exposure is restricted by the difficulties in assuring the actual dose in the lungs of each animal as well as the need to use expensive equipment, special expertise and large quantities of the tested materials. For these reasons, intratracheal or intranasal instillation has become widely used for respiratory toxicity studies (Driscoll et al., 2000; Hasegawa-Baba et al., 2014; Islam et al., 2006, 2007). Rodent airways, however, differ from those of human and other primates both structurally and functionally, which has motivated the development of nonhuman primate models in order to better extrapolate the results to humans (Carey et al., 2012).

With an increasing interest and legislation to reduce the use of animals in research (EU, 2003; REACH, 2006), animal models are increasingly being replaced with in vitro models using human cells or tissues. In addition to ethical issues, in vitro toxicity testing appeals with the associated low cost, high throughput and the possibility to target a specific cell type in a defined experimental setup. The function of the respiratory epithelium as a first contact point for airborne exposure agents makes it an appealing target for toxicological studies aiming to model the effects of inhaled substances. The in vitro respiratory platforms include carcinoma-derived and virus-transformed continuous cell lines, primary cells and ex vivo models. Examples of cell lines typically used in inhalation toxicology are carcinoma-derived lung epithelial or bronchial cells (e.g. A549 and Calu-3) and immortalized human epithelial or bronchial cells (e.g. BEAS-2B and 16HBE140) (Prytherch and BéruBé, 2014). The next step from monolayers of transformed cell lines would be to utilize more complicated cell models, where e.g. macrophages are cultured together with lung epithelia to mimic the components of immune system operational in the lungs. The co-culture models for alimentary exposure to mycotoxins are already in use, as was shown in a study of modulation of inflammatory response by OTA in a co-culture of intestinal and liver cells (González-Arias et al., 2015). However, for inhalation exposure to mycotoxins the studies using co-culture models of transformed cells are still lacking.

Primary cells isolated from nasal and bronchial epithelium or from the alveolar level of the lungs have been used in inhalation toxicology for producing cultures of respiratory tissue with close to realistic morphological and physiological properties (Prytherch et al., 2011). For example the impairment of motile and chemosensory functions due to aflatoxin exposure has been demonstrated with cultures of primary human sinonasal and bronchial cells (Lee et al., 2016). Cultures of primary respiratory cells have shown also the ability of airway epithelium to detoxify both AFB1 and sterigmatocystin (Cabaret et al., 2010; Van Vleet et al., 2001). Generating respiratory tissue by culturing embryonic or induced pluripotent stem cells is possible as well, but achieving accurate characteristics and homogenous bronchial epithelium still needs further development (McIntyre et al., 2015). Alternatively, complete sections of respiratory tissue can be extracted and cultured ex vivo, maintaining the complexity and function of an intact lung. Ex vivo experiments with rat tracheas have elucidated e.g. the vasoconstrictive abilities of OTA (Chatopadhyay et al., 2014) and ciliostatic abilities of several metabolites of indoor fungi (Piecková and Kunová, 2002). However, the life span of current ex vivo models is very limited, and similarly with cultures of primary cells, also cultured tissues are donor specific and thus vary in their responses.

10.3 Microbial Secondary Metabolites in Indoor Environments

Microbial secondary metabolites have been detected from water-damaged indoor environments both in private residents and public buildings, although the concentrations of single toxins are typically low and the diversity high (Bloom et al., 2009; Cai et al., 2011; Kirjavainen et al., 2015; Peitzsch et al., 2012; Polizzi et al., 2009; Täubel et al., 2011; Tuomi et al., 2000). The most widely studied mycotoxins detected indoors are aflatoxins (sterigmatocystin) (Bloom et al., 2009), OTA (Hope and Hope, 2012) and trichothecene mycotoxins (T2, DON, satratoxins, verrucarin) (Pestka et al., 2008; Straus, 2009).

Aflatoxins, especially AFB1, are known liver carcinogens, but they can also exert their effects on the lungs. The mechanism behind its carcinogenicity in lung is suggested to be oxidative DNA damage (Guindon-Kezis et al., 2014). The high local bioactivation and neuronal transport to the olfactory bulb seen after intranasal administration of AFB1 indicates a risk of tumorigenesis in nasal mucosa after high local exposure (Larsson and Tjälve, 2000). Aerosol exposures of animals to AFB1 have resulted in an increase of lymphatic leukemia in mice (Louria et al., 1974), genotoxic effects in rats (Zarba et al., 1992) and suppression of pulmonary and systemic host defenses in rats and mice (Jakab et al., 1994). Along other mycotoxins, also aflatoxins have been detected in patients with health problems related to exposure to water damaged indoor environment (Trasher et al., 2012; Brewer et al., 2013b). In the industrial setting, the inhalation exposure to aflatoxins has been linked to cancer incidence and pulmonary disorders (Dvoráčková and Pichová, 1986; Hayes et al., 1984). Elevated levels of AFB1 were also measured in bronchoalveolar lavage and serum samples of workers in food-grain handling in India, where about 50% of all food-grain workers had chronic respiratory symptoms (Malik et al., 2014). A case-control study by Lai et al., 2014 (Lai et al., 2014) showed an association between the inhalation exposure to AFB1 and the risk of hepatocellular cancer among sugar and papermaking factory workers. Thus, the inhalation of aflatoxins represents an occupational risk factor.

OTA is one of the most important and deleterious mycotoxins contaminating foodstuff, leading to exposure also via inhalation. OTA is thought to be carcinogenic, teratogenic, immunotoxic, hepatotoxic and neurotoxic, but the detailed mechanisms of its toxicity are not known (Hope and Hope, 2012; Kőszegi and Poór, 2016; Malir et al., 2016). Nevertheless, inhibition of protein synthesis and energy production, induction of oxidative stress, DNA adduct formation as well as apoptosis/necrosis and cell cycle arrest are suggested to be involved in its toxic action (Corcuera et al., 2015; Vettorazzi et al., 2013). Elevated levels of OTA have been detected in urine of patients exposed to toxins in water-damaged buildings (Hooper et al., 2009) as well as in workers exposed to airborne dust originating from the processing of contaminated foods (Iavicoli et al., 2002). Inhalation of OTA has been linked to acute renal failure and respiratory distress in workers exposed to Aspergillus-producers of OTA in a granary (Di Paolo et al., 1994). Furthermore, OTA has been found in 22% of sinonasal tissue and mucus of chronic rhinosinusitis patients, and 83% of the urine of patients suffering from chronic fatigue syndrome (CFS) (Brewer et al., 2013b).

Trichothecenes are a group of over 200 structurally related mycotoxins produced by various species of e.g. *Fusarium, Myrothecium* and *Stachybotrys*. It has been suggested that the toxicity of trichothecenes might be a reason for many of the adverse effects of *Stachybotrys chartarum*, an infamous black mold suggested to contribute to damp building related illnesses (Pestka et al., 2008). Macrocyclic trichothecenes have been detected in patients with health problems related to mold contaminated buildings in several studies (Brewer et al., 2013b; Croft et al., 2002; Dennis et al., 2009; Rea et al., 2003, Trasher et al., 2012). The toxic potencies of different inhaled mycotoxins have been compared in hamster lung cells in vitro, listing the T-2 and HT-2 toxins as the most cytotoxic mycotoxins followed by *Fusarium* toxins DON, beauvericin, and enniatin B (Behm et al., 2012). One possible mechanism for adverse health effects of trichothecenes is the impairment of host defense, as the immune system is extremely sensitive to their effects (Pestka et al., 2008; Pestka, 2010a). The effect may be bipolar, where low concentrations cause immunostimulation and high concentrations result in immunosuppression, which was shown in human immune cells in vitro (Pestka et al., 2004). Both DON and T-2 have been shown to impair host resistance and pulmonary immune responses to enteric reovirus infection in mice (Li et al., 2006; Li et al., 2007).

Ueno (Ueno, 1984) was the first to report on the inhalation toxicity of **T-2** in mice. Later, Creasia et al. (1987) demonstrated the acute effects immediately after inhaling aerosolic T-2 in mice: tremors, lethargy, stilted gait, and sometimes prostration. In rats and guinea pigs, respiratory exposure to T-2 revealed that lesions in histopathological examination were similar to those after systemic administration (Creasia et al., 1990). However, even after lethal doses, no histological evidence of acute pulmonary injury was found. The most consistent histological findings in guinea pigs after inhaled T-2 were lymphocytolysis and phagocytosis in the lymphoreticular system, and changes in the gastrointestinal tract (Marrs et al., 1986). Pang et al. (1987, 1988) reported that inhalation exposure of pigs to a sublethal dose of T-2 toxin caused morphological changes, clinical signs of toxicity and effects on hematology, serum biochemistry as well as on pulmonary and systemic immunity.

DON (vomitoxin) is a common trichothecene mycotoxin detected in grain dust. After nasal inhalation in mice, DON rapidly reaches the lungs and other organs, enhancing proinflammatory cytokine expression (Amuzie et al., 2008). Other effects of DON in multiple species are anorexia, diarrhea, growth retardation, neuroendocrine effects, and disruption of the immune system (Pestka, 2010a, 2010b). DON may be an indicator for other trichothecenes as DON, nivalenol and zearalenone have all been detected in the airborne dust from granaries. The occupational exposure to these toxins has been estimated to be below the tolerable daily intake (TDI) values, but activities such as cleaning have been shown to significantly increase the exposure of workers (Mayer et al., 2007; Niculita-Hirzel et al., 2016).

Satratoxins are acutely toxic secondary metabolites of *S. chartarum*, similarly as T-2 and HT-2 toxins. In mice models, the intranasal exposure to satratoxin G has been reported to evoke apoptosis of olfactory sensory neurons and acute inflammation in the nose and brain (Amuzie et al., 2010; Islam et al., 2006; Jia et al., 2011). Also, the nasal airways of rhesus monkeys were shown to be vulnerable to satratoxin G induced neuronal cell death and inflammation (Carey et al., 2012). Likewise, a single intranasal exposure to **roridin A** (a trichothecene mycotoxin with a similar chemical structure) has been found to induce rapid apoptosis and loss of olfactory sensory neurons in the nasal airways and the olfactory bulb in mice (Islam et al., 2007), and the effects may be even exacerbated by repeated exposure (Corps et al., 2010). These inflammatory effects have been suggested to account for nasal congestion and runny or itchy nose in people.

In addition to trichothecenes there is another mycotoxin family, **atranones**, which might contribute to the adverse health effects associated with *S. chartarum*. Pure atranones have been shown to contribute to inflammatory response by increasing the influx of immunocytes and cytokine responses in mice lungs after intratracheal instillation (Rand et al., 2006). It has been suggested that induction of inflammation by *S. chartarum* is related to atranones, while cytotoxicity to satratoxin production (Nielsen et al., 2002).

The mycotoxin research has been dominated by toxins of *Stachybotrys*, but it is good to keep in mind the wide **variety of other toxins** possibly relevant for human exposure. Less known "emerging toxins" such as sterigmatocystin and mycophenolic acid have been shown to occur frequently in agricultural products and thus could be relevant also for inhalation exposure (Gruber-Dorninger et al., 2016). There are indications that some MVOCs like 1-octen-3-ol might be a risk factor for indoor air related adverse health effects (Sahlberg et al., 2013). Toxins from *Penicillium* species common on wet building materials (brevianamide A, mycophenolic acid, and roquefortine C) have been shown to stimulate compound-specific inflammatory and cytotoxic responses in a mice model (Rand et al., 2005). Similarly, a study using doses comparable to the estimated doses of possible human exposure showed alterations in the expression of inflammation-associated genes in mice lungs after intratracheal instillation of toxins from fungi common on damp building environment (atranone C, brevianamide, cladosporin, mycophenolic acid, neoechinulin A & B, sterigmatocystin or TMC-120A) (Miller et al., 2010). A mixture of mycotoxins is also suggested to account for the ciliostatic activity of extracts of filamentous fungi isolated from mouldy buildings (Piecková and Kunová, 2002).

In addition to the reported inflammatory and cytotoxic responses, an increase of allergic immune response has also been linked with mycotoxin exposure. In a mice model of allergic asthma, the exposure to gliotoxin and patulin produced by *Aspergillus* and *Penicillium* species exacerbated the asthma-like phenotype by modulating the Th1/Th2 balance and by inducing oxidative stress (Schütze et al., 2010). However, in a cohort study measuring a broad set of microbial secondary metabolites no association between the presence of individual toxins and asthma development was observed, indicating that exposure to mycotoxins is not responsible for the increased risk of asthma in homes with moisture damage and mold (Kirjavainen et al., 2015). Overall, the in vivo findings suggest that mycotoxins may play a role in acute inflammation of airways, neurological deficits and impaired host resistance to infections, which are all symptoms associated with damp building exposure. Thus, a prolonged exposure to mycotoxins may provoke some of the adverse health effects associated with damp indoor environments (Miller et al., 2010; Miller and McMullin, 2014; Rand et al., 2011).

10.4 Exposure Assessment

Adverse health effects of inhalation of microbial toxins are better known from occupational environments such as farms and feed manufacturing plants, where exposure to massive doses of organic dust and resulting pulmonary mycotoxicoses such as organic dust toxic syndrome (ODTS) have been well documented (Seifert et al., 2003; Viegas et al., 2013a). Establishing the link between inhaled microbial toxins in moisture-damaged indoor environments has proven to be more difficult, as the levels of microbial toxins are much lower and only few studies have characterized inhalation exposure accurately. Also the lack of specificity may cause problems, e.g. hundreds of MVOCs can be measured from the air, but none of these compounds are exclusively produced by microbes (Korpi et al., 2009). The exposure assessment is often based on presence of limited number of mycotoxins in indoor dust samples or biological fluids, whereas studies describing airborne concentrations of mycotoxins are virtually nonexistent (Tables 10.1 and 10.2).

Table 10.1 Clinical studies on adverse health effects related to mycotoxin exposure

Exposure assessment	Reference	*N* cases (+controls)	Country	Study population	Findings
Mycotoxins measured in indoor air or dust	Hayes et al., 1984	71 (+67)	The Netherlands	Oilpress workers	Aflatoxin exposure associated with total and respiratory cancer incidence
	Cai et al., 2011	462	Malaysia	Random sample of schoolchildren	Verrucarol in dust swabs associated negatively with daytime breathlessness
	Zock et al., 2014	645	Finland, Spain, The Netherlands	Schoolteachers from schools with and without moisture damages	Higher mycotoxin levels in settled dust associated with higher asthma symptom score and more nasal symptoms

(continued)

Table 10.1 (continued)

Exposure assessment	Reference	*N* cases (+controls)	Country	Study population	Findings
	Kirjavainen et al., 2015	228	Finland	Random sample of children	No association between levels of mycotoxins in floor dust with risk of asthma in children
	Norbäck et al., 2016	462	Malaysia	Random sample of schoolchildren	Verrucarol in dust swabs associated positively with tiredness
Mycotoxins or biomarkers measured from biological fluids or tissues of exposed persons	Dvoráckovå and Pichová, 1986	3	Czechoslovakia	Patients with pulmonary interstitial fibrosis	AFB1 detected in lungs of patients with interstitial fibrosis
	Trout et al., 2001	6 (+2)	USA	Hotel staff exposed to mold	Specific IgG antibodies to roridin in sera not related to exposure or the one diagnosed restrictive lung disease
	Croft et al., 2002	4	USA	Mycotoxicosis patients exposed to mold contaminated buildings	MT detected from urine of mycotoxicocis patients
	Rea et al., 2003	100	USA	Patients exposed to toxic molds in their homes	MT detected in urine of patients with health problems including respiratory and neurological deficits
	Dennis et al., 2009	79	USA	CRS patients with a history of mold exposure	MT detected from 7 out of 8 analysed urine samples
	Lieberman et al., 2011	18	USA	CRS patients undergoing endoscopic sinus surgery	OT detected in 22% of sinonasal tissue and secretions of CRS patients
	Thrasher et al., 2012	5	USA	Family of two adults and three children (and a pet dog) living in a water damaged home	MT, AT and OT in detected in urine, nasal secretions and tissue samples of patients with health problems including respiratory and neurological deficits

(continued)

Table 10.1 (continued)

Exposure assessment	Reference	*N* cases (+controls)	Country	Study population	Findings
	Brewer et al., 2013b	112 (+55)	USA	Patients diagnosed with CFS	MT, AT, OTA, levels in urine associated with CFS
	Brewer et al., 2013a	1	USA	Patient diagnosed with CFS	OTA detected in urine of CFS patient
	Malik et al., 2014	46 (+44)	India	Workers handling food grains	AFB1 in serum associated with occupational exposure and respiratory symptoms
	Lai et al., 2014	181 (+203)	China	Workers handling sugar and sugarcane	AFB1 albumin adducts in serum associated with hepatocellular carcinoma

AFB1 aflatoxin B1, *AT* aflatoxin, *CRS* chronic rhinosinusitis, *CFS* chronic fatigue syndrome, *MT* macrocyclic trichothecenes, *OTA* ochratoxin A, *OT* ochratoxin

Table 10.2 Biomonitoring of mycotoxin exposure in different exposure environments

Exposure assessment	Reference	*N* cases (+controls)	Country	Study population	Findings
Mycotoxins or biomarkers measured from serum of exposed persons	Autrup et al., 1991	45 (+45)	Denmark	Workers handling raw material for animal feed production	Levels of AFB1 adducts in serum were higher after 4 weeks of work compared to levels after vacation
	Iavicoli et al., 2002	6 (+23)	Italy	Workers processing coffee, black pepper, nutmeg and cocoa beans	OTA in serum associated with occupational exposure
	Van Emon et al., 2003	5 (+5)	USA	Individuals exposed to *S. chartarum* in water damaged environments	Stachylysin in serum associated with *S.chartarum* exposure
	Brasel et al., 2004	44 (+26)	USA	Individuals exposed to indoor molds	MT in serum associated with mold exposure (ELISA-assay)

(continued)

Table 10.2 (continued)

Exposure assessment	Reference	*N* cases (+controls)	Country	Study population	Findings
	Degen et al., 2007	61	Germany	Workers handling food grain	Levels of OTA in serum similar to average values in the German population
	Hooper et al., 2009	769 (+97)	USA	Individuals with or without known exposure to toxin producing molds	MT, AF and OT in urine, nasal secretions and tissue samples associated with reported mold exposure
	Oluwafemi et al., 2012	28 (+30)	Nigeria	Workers in animal feed production	AF in serum associated with occupational exposure and ventilation of the workplace
	Viegas et al., 2012	31 (+30)	Portugal	Workers in poultry production facilities	AFB1 in serum associated with occupational exposure
	Viegas et al., 2013b	28 (+30)	Portugal	Workers in swine production facilities	AFB1 in serum associated with occupational exposure
	Viegas et al., 2013c	45 (+30)	Portugal	Workers in swine (34) and poultry production (11)	AFB1 in serum associated with occupational exposure
	Viegas et al., 2015	41 (+30)	Portugal	Workers in waste industry	AFB1 in serum associated with occupational exposure
	Viegas et al., 2016	30 (+30)	Portugal	Workers in poultry slaughterhouse	AFB1 in serum associated with occupational exposure
	Föllmann et al., 2016	17 (+13)	Germany	Workers in grain mills	Levels of DON, OTA, zearalenone and citrinin in urine not associated with occupational exposure

AF aflatoxins, *AFB1* aflatoxin B1, *MT* macrocyclic trichothecenes, *OTA* ochratoxin A, *OT* ochratoxin

Determining mycotoxins directly from air is difficult, so usually the inhalation exposure is estimated by multiplying the mycotoxin concentrations in settled dust with either amount of dust measured from the air (Mayer et al., 2008) or derived from exposure limits for dust (Brasel et al., 2005). Even more often the exposure assessment is merely based on assumption that the mycotoxin exposure is taking place because mycotoxin producing fungi were detected from the indoor environment or exposure to molds was reported by the study subjects (Hodgson et al., 1998; Kilburn, 2009; Kilburn et al., 2009). Another shortcoming of mycotoxin exposure assessment is the lack of information about the "normal" level of mycotoxin exposure in indoor environments, although mycotoxins are known to be ubiquitous in built environment and found also from outdoors (Peitzsch et al., 2012; Täubel et al., 2011).

10.5 Biomonitoring of Mycotoxins

Mycotoxins, specifically trichothecenes, aflatoxins, and ochratoxins, can be detected in human tissue and body fluids in patients who have been exposed to toxin producing molds in their environment (Table 10.2). In fact, in a study of Hooper et al. (2009) trichothecenes were found in 95% of urines, 44% of nasal secretions and 59% of tissue, whereas aflatoxins and ochratoxins were present in lesser cases. Only one study reports an association between levels of OTA in serum and occupational exposure, but AFB1 levels in serum have been consistently associated with working in waste industry, poultry farms and slaughterhouses (Oluwafemi et al., 2012; Viegas et al., 2012, 2013b, 2013c, 2015, 2016). Similarly, levels of macrocyclic trichothecenes as well as hemolysin stachylysin in serum have been associated with exposure to indoor molds (Brasel et al., 2004; Van Emon et al., 2003). Few studies report no association between occupational exposure and levels of mycotoxins in biological fluids (Degen et al., 2007; Föllman et al., 2016), suggesting that exposure levels can be controlled even in workplaces with high risk for mycotoxin contamination.

Some mycotoxins can also form metabolites or protein adducts, which could serve as biomarkers of exposure; e.g. intratracheal exposure to satratoxin G produced adducts with serum albumin in both humans and rats (Yike et al., 2006). Serum AFB1 albumin adducts have been used to assess chronic exposure via both diet (Leong et al., 2012) and inhalation (Autrup et al., 1991; Lai et al., 2014), but detectable levels of adducts were seen only after inhalation exposure to a certain level of AFB1 in an epidemiological study and in an animal model (Mo et al., 2014). Generally, several biomarkers of oral exposure exists for the agriculturally important toxins (Lee and Ryu, 2015), but for the toxins from building associated fungi there is hardly any biomarkers available. The development of biomarkers showing inhalation exposure to broader set of mycotoxins would be an important step for risk assessment.

10.6 Risk Assessment

A valid risk assessment for inhalation exposure to microbial secondary metabolites is a demanding task for several reasons. Firstly, accurate exposure assessment is needed since mycotoxins can show a wide variety of effects (e.g. immunotoxic, carcinogenic, mutagenic, toxic or teratogenic) depending on the type of toxin. The type of exposure (inhalation, oral, dermal) may also affect the adverse effects seen; the inhalation route seems to be more harmful than ingestion of certain mycotoxins. The dose and duration of exposure are important determinants as well as the toxin fraction absorbed after respiratory intake. Furthermore, the sensitivity to mycotoxins varies between exposed individuals or population groups (Fromme et al., 2016; World Health Organization, 2009). Worryingly, children appear to be at greater risk for inhaled toxicants, as the respiratory deposition of fungal fragments into to lower airways of infants was found to be four to five times higher than in adults (Cho et al., 2005).

In indoor environment toxins never exist individually, but rather are mixed with a variety of fungi and bacteria, their spores, fragments and cell wall components together with allergenic proteins, volatile organic compounds and particulates (Nevalainen et al., 2015). In experimental setups, mycotoxins have been shown to interact synergistically with each other (Vejdovszky et al., 2017), with microbes (Huttunen et al., 2004; Penttinen et al., 2005), with spores (Šegvić Klarić et al., 2015), with microbial structural components (Islam and Pestka, 2006; Islam et al., 2007; Kankkunen et al., 2009, 2014; Korkalainen et al., 2017; Zhou et al., 1999) as well as with particulate matter (PM_{10}) (Capasso et al., 2015). Similarly, there is evidence of strong synergistic or additive effects after simultaneous exposure to multiple foodborne toxins (Stoev, 2015). Synergistic effects of microbial secondary metabolites should be given due consideration, as there is a possibility of underestimating the effects of microbial exposures if the risk assessment is based on information on individual exposure agents.

Several regulatory agencies and organizations, such as World Health Organisation (WHO), Joint FAO/WHO Expert Committee on Food Additives (JECFA), EU Scientific Committee for Food (SCF) and European Food Safety Authority (EFSA) have set maximum permissible levels and produced TDIs for agriculturally important mycotoxins worldwide. International Agency for Research of Cancer (IARC) has also evaluated the carcinogenic risks of some naturally occurring toxins to humans. However, these limit values are merely directed to regulate the exposure via food. Since only limited toxicological data from inhalation exposure are available, the extended threshold of toxicological concern (TTC) concept (concentration of no toxicologic concern, CoNTC) has been introduced for safety assessment purposes (Drew and Frangos, 2007; Munro et al., 2008). The resulting airborne concentration, 30 ng/m^3, is expected not to pose hazard to humans through lifetime. Although exposures in the most extreme working conditions may exceed this concentration, measured airborne mycotoxin levels in the built indoor environment have been repeatedly found to be too low to

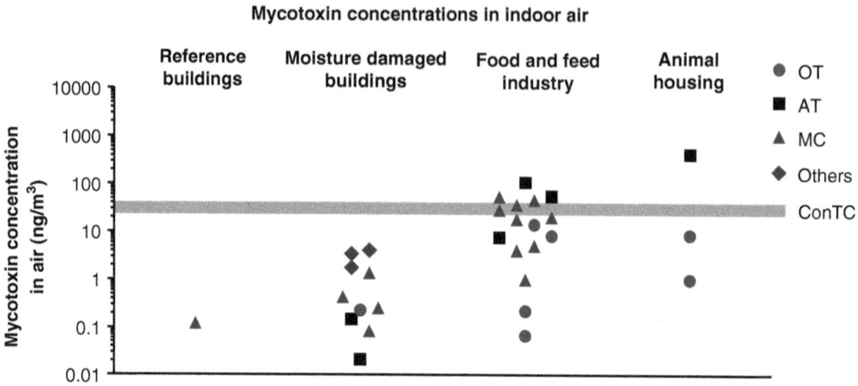

Fig. 10.2 Airborne concentrations of common mycotoxins ochratoxin (OT), Aflatoxins (AT), macrocyclic trichothecenes (MC) and other mycotoxins (Others) in different indoor environments, including reference buildings, moisture damaged buildings (Brasel et al., 2005; Gottschalk et al., 2008; Polizzi et al., 2009), food and feed industry (Burg et al., 1981, 1982; Iavicoli et al., 2002; Niculita-Hirzel et al., 2016; Sorenson et al., 1984) and animal housing (Selim et al., 1998; Skaug et al., 2001; Wang et al., 2008). The concentrations are either measured directly from air or measured from a dust sample and airborne concentration is calculated assuming a dust concentration of 15 mg/m^3. The grey line represents the concentration of no toxicologic concern (ConTC) for mycotoxins showing the level of exposure not considered to be hazardous even with lifetime exposure (Hardin et al., 2009)

pose a credible health risk, as the doses of both nonvolatile and volatile microbial metabolites are consistently below the CoNTC (Hardin et al., 2003; Hardin et al., 2009; Kelman et al., 2004; Robbins et al., 2000, Korpi et al., 2009) (Fig. 10.2). However, the presence of mycotoxins in materials or products handled in a workplace should be considered as a potential hazard to health, even though the causal connection between mycotoxin exposure and adverse health effects is not established and risk assessment requires more detailed information about the quality and quantity of the inhaled substance (Degen, 2011).

10.7 Conclusions

Microbial secondary metabolites can be found from indoor environments and particularly in workplaces where organic material is handled or moisture damage supports excessive microbial growth. Exposure to microbial toxins is possible and even likely, as the spores and fragments of microbes can act as carriers. Among microbial secondary metabolites there are numerous compounds which have been shown to be toxic to mammals. However, the concentrations of microbial toxins indoors are typically low, and same microbial secondary metabolites can be found also from outdoor air as well as microbiologically "normal" indoor environments.

In order to pose a health risk the amount of toxin in the air (or retained in the body) should exceed the threshold of toxicological concern, and at least for single toxins, this is typically not the case.

Future efforts should concentrate on the studies of multiple exposures, comorbidities of microbial toxicoses and effects of low-level chronic exposures. Epidemiological studies are needed to establish causality, but they need better measures of exposure to a wider selection of microbial secondary metabolites. There is a lack of toxicokinetic data on humans, clinical studies measuring both exposure and health outcome as well as biomonitoring studies assessing internal exposure after respiratory intake in residential settings. In addition, more experimental studies with advanced models of respiratory system are needed to elucidate the complex relationships between toxin exposure and health outcome.

References

Amuzie CJ, Harkema JR, Pestka JJ (2008) Tissue distribution and proinflammatory cytokine induction by the trichothecene deoxynivalenol in the mouse: comparison of nasal vs. oral exposure. Toxicology 248:39–44. doi:10.1016/j.tox.2008.03.005

Amuzie CJ, Islam Z, Kim JK et al. (2010) Kinetics of satratoxin G tissue distribution and excretion following intranasal exposure in the mouse. Toxicol Sci 116:433–440. doi:10.1093/toxsci/kfq142

Autrup JL, Schmidt J, Seremet T et al. (1991) Determination of exposure to aflatoxins among Danish workers in animal-feed production through the analysis of aflatoxin B1 adducts to serum albumin. Scand J Work Environ Health 17:436–440

Behm C, Föllmann W, Degen GH (2012) Cytotoxic potency of mycotoxins in cultures of V79 lung fibroblast cells. J Toxicol Environ Health A 75:1226–1231. doi:10.1080/15287394.2012.709170

Bitnun A, Nosal RM (1999) *Stachybotrys chartarum* (atra) contamination of the indoor environment: health implications. Paediatr Child Health 4:125–129

Bloom E, Nyman E, Must A et al. (2009) Molds and mycotoxins in indoor environments – a survey in water-damaged buildings. J Occup Environ Hyg 6:671–678. doi:10.1080/15459620903252053

Boonen J, Malysheva SV, Taevernier L et al. (2012) Human skin penetration of selected model mycotoxins. Toxicology 301:21–32. doi:10.1016/j.tox.2012.06.012

Brasel TL, Campbell AW, Demers RE et al. (2004) Detection of trichothecene mycotoxins in sera from individuals exposed to *Stachybotrys chartarum* in indoor environments. Arch Environ Health 59:317–323

Brasel TL, Martin JM, Carriker CG et al. (2005) Detection of airborne *Stachybotrys chartarum* macrocyclic trichothecene mycotoxins in the indoor environment. Appl Environ Microbiol 71:7376–7388

Brewer JH, Thrasher JD, Hooper D (2013a) Chronic illness associated with mold and mycotoxins: is naso-sinus fungal biofilm the culprit? Toxins (Basel) 6:66–80. doi:10.3390/toxins6010066

Brewer JH, Thrasher JD, Straus DC et al. (2013b) Detection of mycotoxins in patients with chronic fatigue syndrome. Toxins (Basel) 5:605–617. doi:10.3390/toxins5040605

Burg WA, Shotwell OL, Saltzman BE (1981) Measurements of airborne aflatoxins during the handling of contaminated corn. Am Ind Hyg Assoc J 42:1–11. doi:10.1080/15298668191419271

Burg WR, Shotwell OL, Saltzman BE (1982) Measurement of airborne aflatoxins during the handling of 1979 contaminated corn. Am Ind Hyg Assoc J 43:580–586

Cabaret O, Puel O, Botterel F et al. (2010) Metabolic detoxication pathways for sterigmatocystin in primary tracheal epithelial cells. Chem Res Toxicol 23:1673–1681. doi:10.1021/tx100127b

Cai GH, Malarstig B, Kumlin A et al. (2011) Fungal DNA and plergen levels in Swedish day care centers and associations with building characteristics. J Environ Monit 13:2018–2024. doi:10.1039/c0em00553c

Capasso L, Longhin E, Caloni F et al. (2015) Synergistic inflammatory effect of PM10 with mycotoxin deoxynivalenol on human lung epithelial cells. Toxicon 104:65–72. doi:10.1016/j.toxicon.2015.08.008

Carey SA, Plopper CG, Hyde DM et al. (2012) Satratoxin-G from the black mold *Stachybotrys chartarum* induces rhinitis and apoptosis of olfactory sensory neurons in the nasal airways of rhesus monkeys. Toxicol Pathol 40:887–898. doi:10.1177/0192623312444028

Chatopadhyay P, Tariang B, Agnihotri A et al. (2014) Synergism of ochratoxin B and calcium-channel antagonist verapamil caused mitochondrial dysfunction. Toxicol Mech Methods 24:428–432. doi:10.3109/15376516.2014.936543

Cho S, Seo S, Schmechel D et al. (2005) Aerodynamic characteristics and respiratory deposition of fungal fragments. Atmos Environ 39:5454–5465

Corcuera LA, Vettorazzi A, Arbillaga L et al. (2015) Genotoxicity of aflatoxin B1 and ochratoxin A after simultaneous application of the in vivo micronucleus and comet assay. Food Chem Toxicol 76:116–124. doi:10.1016/j.fct.2014.12.003

Corps KN, Islam Z, Pestka JJ et al. (2010) Neurotoxic, inflammatory, and mucosecretory responses in the nasal airways of mice repeatedly exposed to the macrocyclic trichothecene mycotoxin roridin A: dose-response and persistence of injury. Toxicol Pathol 38:429–451. doi:10.1177/0192623310364026

Council for Agricultural Science and Technology (2003) Mycotoxins: risk in plant, animal, and human systems. Task force report 139, Ames, Iowa

Creasia DA, Thurman JD, Jones 3rd LJ et al. (1987) Acute inhalation toxicity of T-2 mycotoxin in mice. Fundam Appl Toxicol 8:230–235

Creasia DA, Thurman JD, Wannemacher Jr RW et al. (1990) Acute inhalation toxicity of T-2 mycotoxin in the rat and guinea pig. Fundam Appl Toxicol 14:54–59

Croft WA, Jarvis BB, Yatatwara CS (1986) Airborne outbreak of trichothecene toxicosis. Atmos Environ 20:549–552

Croft WA, Jastromski BM, Croft AL et al. (2002) Clinical confirmation of trichothecene mycotoxicosis in patient urine. J Environ Biol 23:301–320

Degen G (2011) Tools for investigating workplace-related risks from mycotoxin exposure. World Mycotoxin J 4:315–327

Degen GH, Mayer S, Blaszkewicz M (2007) Biomonitoring of ochratoxin A in grain workers. Mycotoxin Res 23:88–93. doi:10.1007/BF02946032

Dennis D, Robertson D, Curtis L et al. (2009) Fungal exposure endocrinopathy in sinusitis with growth hormone deficiency: Dennis-Robertson syndrome. Toxicol Ind Health 25:669–680. doi:10.1177/0748233709348266

Di Paolo N, Guarnieri A, Garosi G et al. (1994) Inhaled mycotoxins lead to acute renal failure. Nephrol Dial Transplant 9(Suppl 4):116–120

Drew R, Frangos J (2007) The concentration of no toxicological concern (CoNTC): a risk assessment screening tool for air toxics. J Toxicol Environ Health A 70:1584–1593. 781628584 [pii]

Driscoll KE, Costa DL, Hatch G et al. (2000) Intratracheal instillation as an exposure technique for the evaluation of respiratory tract toxicity: uses and limitations. Toxicol Sci 55:24–35

Dvorácková I, Pichová V (1986) Pulmonary interstitial fibrosis with evidence of aflatoxin B1 in lung tissue. J Toxicol Environ Health 18:153–157. doi:10.1080/15287398609530856

EU (2003) EU Directive 76/768/EEC, Directive 2003/15/EC of the European Parliament and of the Council. Off J Eur Union L66:26–35

Flannigan B (1987) Mycotoxins in the air. Int Biodeterior 23:73–78

Föllmann W, Ali N, Blaszkewicz M et al. (2016) Biomonitoring of mycotoxins in urine: pilot study in mill workers. J Toxicol Environ Health A 79:1015–1025

Forgacs J (1962) Mycotoxicoses – the neglected disease. Feedstuffs 34:124–134

Fromme H, Gareis M, Völkel W et al. (2016) Overall internal exposure to mycotoxins and their occurrence in occupational and residential settings – an overview. Int J Hyg Environ Health 219:143–165. doi:10.1016/j.ijheh.2015.11.004

González-Arias CA, Crespo-Sempere A, Marin S et al. (2015) Modulation of the xenobiotic transformation system and inflammatory response by ochratoxin A exposure using a co-culture system of Caco-2 and HepG2 cells. Food Chem Toxicol 86:245–252. doi:10.1016/j.fct.2015.10.007

Gottschalk C, Bauer J, Meyer K (2008) Detection of satratoxin g and h in indoor air from a water-damaged building. Mycopathologia 166:103–107. doi:10.1007/s11046-008-9126-z

Gruber-Dorninger C, Novak B, Nagl V, et al. (2016) Emerging mycotoxins: beyond traditionally determined food contaminants. *J Agric Food Chem.* doi:10.1021/acs.jafc.6b03413

Guindon-Kezis KA, Mulder JE, Massey TE (2014) In vivo treatment with aflatoxin B1 increases DNA oxidation, base excision repair activity and 8-oxoguanine DNA glycosylase 1 levels in mouse lung. Toxicology 321:21–26. doi:10.1016/j.tox.2014.03.004

Hardin BD, Kelman BJ, Saxon A (2003) Adverse human health effects associated with molds in the indoor environment. J Occup Environ Med 45:470–478

Hardin BD, Robbins CA, Fallah P et al. (2009) The concentration of no toxicologic concern (CoNTC) and airborne mycotoxins. J Toxicol Environ Health A 72:585–598. doi:10.1080/15287390802706389

Harkema JR, Carey SA, Wagner JG (2006) The nose revisited: a brief review of the comparative structure, function, and toxicologic pathology of the nasal epithelium. Toxicol Pathol 34:252–269. H1080432VQ59X328 [pii]

Hasegawa-Baba Y, Kubota H, Takata A et al. (2014) Intratracheal instillation methods and the distribution of administered material in the lung of the rat. J Toxicol Pathol 27:197–204. doi:10.1293/tox.2014-0022

Hayes RB, van Nieuwenhuize JP, Raatgever JW et al. (1984) Aflatoxin exposures in the industrial setting: an epidemiological study of mortality. Food Chem Toxicol 22:39–43

Hodgson MJ, Morey P, Leung WY et al. (1998) Building-associated pulmonary disease from exposure to *Stachybotrys chartarum* and *Aspergillus versicolor*. J Occup Environ Med 40:241–249

Hooper DG, Bolton VE, Guilford FT et al. (2009) Mycotoxin detection in human samples from patients exposed to environmental molds. Int J Mol Sci 10:1465–1475. doi:10.3390/ijms10041465

Hope JH, Hope BE (2012) A review of the diagnosis and treatment of ochratoxin A inhalational exposure associated with human illness and kidney disease including focal segmental glomerulosclerosis. J Environ Public Health 2012:835059. doi:10.1155/2012/835059

Huttunen K, Pelkonen J, Nielsen KF et al. (2004) Synergistic interaction in simultaneous exposure to *Streptomyces californicus* and *Stachybotrys chartarum*. Environ Health Perspect 112:659–665

Iavicoli I, Brera C, Carelli G et al. (2002) External and internal dose in subjects occupationally exposed to ochratoxin A. Int Arch Occup Environ Health 75:381–386. doi:10.1007/s00420-002-0319-3

Islam Z, Amuzie CJ, Harkema JR et al. (2007) Neurotoxicity and inflammation in the nasal airways of mice exposed to the macrocyclic trichothecene mycotoxin roridin a: kinetics and potentiation by bacterial lipopolysaccharide coexposure. Toxicol Sci 98:526–541.

Islam Z, Harkema JR, Pestka JJ (2006) Satratoxin G from the black mold *Stachybotrys chartarum* evokes olfactory sensory neuron loss and inflammation in the murine nose and brain. Environ Health Perspect 114:1099–1107

Islam Z, Pestka JJ (2006) LPS priming potentiates and prolongs proinflammatory cytokine response to the trichothecene deoxynivalenol in the mouse. Toxicol Appl Pharmacol 211:53–63. S0041-008X(05)00223-1 [pii]

Jakab GJ, Hmieleski RR, Zarba A et al. (1994) Respiratory aflatoxicosis: suppression of pulmonary and systemic host defenses in rats and mice. Toxicol Appl Pharmacol 125:198–205. S0041-008X(84)71065-9 [pii]

Jarvis BB, Miller JD (2005) Mycotoxins as harmful indoor air contaminants. Appl Microbiol Biotechnol 66:367–372. doi:10.1007/s00253-004-1753-9

Jia C, Sangsiri S, Belock B et al. (2011) ATP mediates neuroprotective and neuroproliferative effects in mouse olfactory epithelium following exposure to satratoxin G in vitro and in vivo. Toxicol Sci 124:169–178. doi:10.1093/toxsci/kfr213

Kankkunen P, Rintahaka J, Aalto A et al. (2009) Trichothecene mycotoxins activate inflammatory response in human macrophages. J Immunol 182:6418–6425. doi:10.4049/jimmunol.0803309

Kankkunen P, Välimäki E, Rintahaka J et al. (2014) Trichothecene mycotoxins activate NLRP3 inflammasome through a P2X7 receptor and Src tyrosine kinase dependent pathway. Hum Immunol 75:134–140. doi:10.1016/j.humimm.2013.11.010

Kelman BJ, Robbins CA, Swenson LJ et al. (2004) Risk from inhaled mycotoxins in indoor office and residential environments. Int J Toxicol 23:3–10. doi:10.1080/10915810490265423

Kilburn KH (2009) Neurobehavioral and pulmonary impairment in 105 adults with indoor exposure to molds compared to 100 exposed to chemicals. Toxicol Ind Health 25:681–92. doi:10.1177/0748233709348390

Kilburn KH, Thrasher JD, Immers NB (2009) Do terbutaline- and mold-associated impairments of the brain and lung relate to autism? Toxicol Ind Health 25:703–10. doi:10.1177/0748233709348391

Kirjavainen PV, Täubel M, Karvonen AM et al. (2015) Microbial secondary metabolites in homes in association with moisture damage and asthma. Indoor Air 3:448–456. doi:10.1111/ina.12213

Korkalainen M, Täubel M, Naarala J et al. (2017) Synergistic proinflammatory interactions of microbial toxins and structural components characteristic to moisture-damaged buildings. Indoor Air 27:13–23. doi:10.1111/ina.12282

Korpi A, Järnberg J, Pasanen AL (2009) Microbial volatile organic compounds. Crit Rev Toxicol 39:139–93. doi:10.1080/10408440802291497

Kőszegi T, Poór M (2016) Ochratoxin A: molecular interactions, mechanisms of toxicity and prevention at the molecular level. Toxins 8:111. doi:10.3390/toxins8040111

Lai H, Mo X, Yang Y et al. (2014) Association between aflatoxin B1 occupational airway exposure and risk of hepatocellular carcinoma: a case-control study. Tumour Biol 35:9577–9584. doi:10.1007/s13277-014-2231-3

Larsson P, Tjälve H (2000) Intranasal instillation of aflatoxin B(1) in rats: bioactivation in the nasal mucosa and neuronal transport to the olfactory bulb. Toxicol Sci 55:383–391

Lee HJ, Ryu D (2015) Advances in mycotoxin research: public health perspectives. J Food Sci 80:T2970–83. doi:10.1111/1750-3841.13156

Lee RJ, Workman AD, Carey RM et al. (2016) Fungal aflatoxins reduce respiratory mucosal ciliary function. Sci Rep 6:33221. doi:10.1038/srep33221

Leong YH, Rosma A, Latiff AA et al. (2012) Associations of serum aflatoxin B1-lysine adduct level with socio-demographic factors and aflatoxins intake from nuts and related nut products in Malaysia. Int J Hyg Environ Health 215:368–372. doi:10.1016/j.ijheh.2011.12.005

Li M, Harkema JR, Cuff CF et al. (2007) Deoxynivalenol exacerbates viral bronchopneumonia induced by respiratory reovirus infection. Toxicol Sci 95:412–426

Li M, Harkema JR, Islam Z et al. (2006) T-2 toxin impairs murine immune response to respiratory reovirus and exacerbates viral bronchiolitis. Toxicol Appl Pharmacol 217:76–85. S0041-008X(06)00265-1 [pii]

Lichtenstein JH, Molina RM, Donaghey TC et al. (2010) Pulmonary responses to Stachybotrys chartarum and its toxins: mouse strain affects clearance and macrophage cytotoxicity. Toxicol Sci 116:113–121. doi:10.1093/toxsci/kfq104

Lieberman SM, Jacobs JB, Lebowitz RA et al. (2011) Measurement of mycotoxins in patients with chronic rhinosinusitis. Otolaryngol Head Neck Surg 145:327–329. doi:10.1177/0194599811403891

Louria DB, Finkel G, Smith JK et al. (1974) Aflatoxin-induced tumors in mice. Sabouraudia 12:371–375

Malik A, Ali S, Shahid M et al. (2014) Occupational exposure to Aspergillus and aflatoxins among food-grain workers in India. Int J Occup Environ Health 20:189–193. doi:10.1179/2049396714Y.0000000055

Malir F, Ostry V, Pfohl-Leszkowicz A et al. (2016) Ochratoxin A: 50 years of research. Toxins 8:191. doi:10.3390/toxins8070191

Marrs TC, Edginton JA, Price PN et al. (1986) Acute toxicity of T2 mycotoxin to the guinea-pig by inhalation and subcutaneous routes. Br J Exp Pathol 67:259–268

Mayer S, Curtui V, Usleber E et al. (2007) Airborne mycotoxins in dust from grain elevators. Mycotoxin Res 23:94–100. doi:10.1007/BF02946033

Mayer S, Engelhart S, Kolk A et al. (2008) The significance of mycotoxins in the framework of assessing workplace related risks. Mycotoxin Res 24:151–164. doi:10.1007/BF03032342

McIntyre BA, Kushwah R, Mechael R et al. (2015) Innate immune response of human pluripotent stem cell-derived airway epithelium. Innate Immun 21:504–511. doi:10.1177/1753425914551074

Miller JD, McMullin DR (2014) Fungal secondary metabolites as harmful indoor air contaminants: 10 years on. Appl Microbiol Biotechnol 98:9953–9966. doi:10.1007/s00253-014-6178-5

Miller JD, Sun M, Gilyan A et al. (2010) Inflammation-associated gene transcription and expression in mouse lungs induced by low molecular weight compounds from fungi from the built environment. Chem Biol Interact 183:113–124. doi:10.1016/j.cbi.2009.09.023

Mo X, Lai H, Yang Y et al. (2014) How does airway exposure of aflatoxin B1 affect serum albumin adduct concentrations? Evidence based on epidemiological study and animal experimentation. J Toxicol Sci 39:645–653. DN/JST.JSTAGE/jts/39.645 [pii]

Munro IC, Renwick AG, Danielewska-Nikiel B (2008) The Threshold of Toxicological Concern (TTC) in risk assessment. Toxicol Lett 180:151–156. doi:10.1016/j.toxlet.2008.05.006

Nevalainen A, Täubel M, Hyvärinen A (2015) Indoor fungi: companions and contaminants. Indoor Air 25:125–156. doi:10.1111/ina.12182

Niculita-Hirzel H, Hantier G, Storti F et al. (2016) Frequent occupational exposure to Fusarium mycotoxins of workers in the Swiss Grain Industry. *Toxins (Basel)*. 8(12). doi:10.3390/toxins8120370

Nielsen KF, Huttunen K, Hyvärinen A et al. (2002) Metabolite profiles of Stachybotrys isolates from water-damaged buildings and their induction of inflammatory mediators and cytotoxicity in macrophages. Mycopathologia 154:201–205

Norbäck D, Hashim JH, Cai GH et al. (2016) Rhinitis, ocular, throat and dermal symptoms, headache and tiredness among students in schools from Johor Bahru, Malaysia: associations with fungal DNA and mycotoxins in classroom dust. PLoS One 11:e0147996. doi:10.1371/journal.pone.0147996

Oluwafemi F, Odebiyi T, Kolapo A (2012) Occupational aflatoxin exposure among feed mill workers in Nigeria. World Mycotoxin J 5:385–389

Pang VF, Lambert RJ, Felsburg PJ et al. (1987) Experimental T-2 toxicosis in swine following inhalation exposure: effects on pulmonary and systemic immunity, and morphologic changes. Toxicol Pathol 15:308–319

Pang VF, Lambert RJ, Felsburg PJ et al. (1988) Experimental T-2 toxicosis in swine following inhalation exposure: clinical signs and effects on hematology, serum biochemistry, and immune response. Fundam Appl Toxicol 11:100–109

Peitzsch M, Sulyok M, Täubel M et al. (2012) Microbial secondary metabolites in school buildings inspected for moisture damage in Finland, The Netherlands and Spain. J Environ Monit 14:2044–2053. doi:10.1039/c2em30195d

Penttinen P, Pelkonen J, Huttunen K et al. (2005) Interactions between *Streptomyces californicus* and *Stachybotrys chartarum* can induce apoptosis and cell cycle arrest in mouse RAW264.7 macrophages. Toxicol Appl Pharmacol 202:278–288. S0041-008X(04)00347-3 [pii]

Pestka JJ (2010a) Deoxynivalenol: mechanisms of action, human exposure, and toxicological relevance. Arch Toxicol 84:663–679. doi:10.1007/s00204-010-0579-8

Pestka JJ (2010b) Deoxynivalenol-induced proinflammatory gene expression: mechanisms and pathological sequelae. Toxins 2:1300–1317. doi:10.3390/toxins2061300

Pestka JJ, Yike I, Dearborn DG et al. (2008) *Stachybotrys chartarum*, trichothecene mycotoxins, and damp building-related illness: new insights into a public health enigma. Toxicol Sci 104:4–26. doi:10.1093/toxsci/kfm284

Pestka JJ, Zhou HR, Moon Y et al. (2004) Cellular and molecular mechanisms for immune modulation by deoxynivalenol and other trichothecenes: unraveling a paradox. Toxicol Lett 153:61–73. doi:10.1016/j.toxlet.2004.04.023

Piecková E, Kunová Z (2002) Indoor fungi and their ciliostatic metabolites. Ann Agric Environ Med 9:59–63

Pinton P, Oswald IP (2014) Effect of deoxynivalenol and other Type B trichothecenes on the intestine: a review. Toxins 6:1615–1643. doi:10.3390/toxins6051615

Polizzi V, Delmulle B, Adams A et al. (2009) JEM Spotlight: fungi, mycotoxins and microbial volatile organic compounds in mouldy interiors from water-damaged buildings. J Environ Monit 11:1849–1858. doi:10.1039/b906856b

Prytherch Z, Job C, Marshall H et al. (2011) Tissue-specific stem cell differentiation in an in vitro airway model. Macromol Biosci 11:1467–1477. doi:10.1002/mabi.201100181

Prytherch ZC, BéruBé KA (2014) A normal and biotransforming model of the human bronchial epithelium for the toxicity testing of aerosols and solubilised substances. Altern Lab Anim 42:377–381

Rand TG, Dipenta J, Robbins C et al. (2011) Effects of low molecular weight fungal compounds on inflammatory gene transcription and expression in mouse alveolar macrophages. Chem Biol Interact 190:139–147. doi:10.1016/j.cbi.2011.02.017

Rand TG, Flemming J, David Miller J et al. (2006) Comparison of inflammatory responses in mouse lungs exposed to atranones A and C from *Stachybotrys chartarum*. J Toxicol Environ Health A 69:1239–1251. R5048J37649XT10X [pii]

Rand TG, Giles S, Flemming J et al. (2005) Inflammatory and cytotoxic responses in mouse lungs exposed to purified toxins from building isolated *Penicillium brevicompactum* Dierckx and *P. chrysogenum* Thom. Toxicol Sci 87:213–222. doi: kfi223 [pii]

Rea WJ, Didriksen N, Simon TR et al. (2003) Effects of toxic exposure to molds and mycotoxins in building-related illnesses. Arch Environ Health 58:399–405

REACH (2006) Regulation (EC) no 1907/2006 of the European Parliament and of the Council concerning the Registration, Evaluation, Authorisation and Restriction of Chemicals (REACH). Off J Eur Union L396:1–849

Robbins CA, Swenson LJ, Nealley ML et al. (2000) Health effects of mycotoxins in indoor air: a critical review. Appl Occup Environ Hyg 15:773–784. doi:10.1080/10473220050129419

Sahlberg B, Gunnbjörnsdottir M, Soon A et al. (2013) Airborne molds and bacteria, microbial volatile organic compounds (MVOC), plasticizers and formaldehyde in dwellings in three North European cities in relation to sick building syndrome (SBS). Sci Total Environ 444:433–40. doi:10.1016/j.scitotenv.2012.10.114

Schütze N, Lehmann I, Bönisch U et al. (2010) Exposure to mycotoxins increases the allergic immune response in a murine asthma model. Am J Respir Crit Care Med 181:1188–1199. doi:10.1164/rccm.200909-1350OC

Šegvić Klarić M, Jakšić Despot D, Kopjar N et al. (2015) Cytotoxic and genotoxic potencies of single and combined spore extracts of airborne OTA-producing and OTA-non-producing

Aspergilli in human lung A549 cells. Ecotoxicol Environ Saf 120:206–214. doi:10.1016/j.ecoenv.2015.06.002

Seifert SA, Von Essen S, Jacobitz K et al. (2003) Organic dust toxic syndrome: a review. J Toxicol Clin Toxicol 41:185–193

Selim MI, Juchems AM, Popendorf W (1998) Assessing airborne aflatoxin B1 during on-farm grain handling activities. Am Ind Hyg Assoc J 59:252–256. doi:10.1080/15428119891010514

Skaug MA, Eduard W, Stormer FC (2001) Ochratoxin A in airborne dust and fungal conidia. Mycopathologia 151:93–98

Sorenson WG, Jones W, Simpson J et al. (1984) Aflatoxin in respirable airborne peanut dust. J Toxicol Environ Health 14:525–533. doi:10.1080/15287398409530603

Stoev SD (2015) Foodborne mycotoxicoses, risk assessment and underestimated hazard of masked mycotoxins and joint mycotoxin effects or interaction. Environ Toxicol Pharmacol 39:794–809. doi:10.1016/j.etap.2015.01.022

Straus DC (2009) Molds, mycotoxins, and sick building syndrome. Toxicol Ind Health 25:617–635. doi:10.1177/0748233709348287

Täubel M, Hyvärinen A (2015) Occurrence of mycotoxins in indoor environments. In: Viegas C, Pinheiro A, Sabino R, et al (eds) Environmental mycology in public health, Amsterdam, Elsevier Academic Press pp 299–319

Täubel M, Sulyok M, Vishwanath V et al. (2011) Co-occurrence of toxic bacterial and fungal secondary metabolites in moisture-damaged indoor environments. Indoor Air 21:368–375. doi:10.1111/j.1600-0668.2011.00721.x

Thrasher JD, Gray MR, Kilburn KH et al. (2012) A water-damaged home and health of occupants: a case study. J Environ Public Health 2012:312836. doi:10.1155/2012/312836

Trout D, Bernstein J, Martinez K et al. (2001) Bioaerosol lung damage in a worker with repeated exposure to fungi in a water-damaged building. Environ Health Perspect 109:641–644. sc271_5_1835 [pii]

Tuomi T, Reijula K, Johnsson T et al. (2000) Mycotoxins in crude building materials from water-damaged buildings. Appl Environ Microbiol 66:1899–1904

Ueno Y (1984) Toxicological features of T-2 toxin and related trichothecenes. Fundam Appl Toxicol 4:S124–32

Van Emon JM, Reed AW, Yike I et al. (2003) ELISA measurement of stachylysin in serum to quantify human exposures to the indoor mold *Stachybotrys chartarum*. J Occup Environ Med 45:582–591. doi:10.1097/01.jom.0000071503.96740.65

Van Vleet TR, Klein PJ, Coulombe Jr RA (2001) Metabolism of aflatoxin B1 by normal human bronchial epithelial cells. J Toxicol Environ Health A 63:525–540. doi:10.1080/15287390152410156

Vejdovszky K, Hahn K, Braun D, et al. (2017) Synergistic estrogenic effects of *Fusarium* and Alternaria mycotoxins in vitro. Arch Toxicol 91:1147–1460. doi:10.1007/s00204-016-1795-7

Vettorazzi A, van Delft J, López de Cerain A (2013) A review on ochratoxin A transcriptomic studies. Food Chem Toxicol 59:766–783. doi:10.1016/j.fct.2013.05.043

Viegas S, Veiga L, Almeida A et al. (2016) Occupational exposure to aflatoxin B1 in a Portuguese poultry slaughterhouse. Ann Occup Hyg 60:176–183. doi:10.1093/annhyg/mev077

Viegas S, Veiga L, Figueiredo P et al. (2015) Assessment of workers' exposure to aflatoxin B1 in a Portuguese waste industry. Ann Occup Hyg 59:173–181. doi:10.1093/annhyg/meu082

Viegas S, Faisca VM, Dias H et al. (2013a) Occupational exposure to poultry dust and effects on the respiratory system in workers. J Toxicol Environ Health A 76:230–239. doi:10.1080/15287394.2013.757199

Viegas S, Veiga L, Figueredo P et al. (2013b) Occupational exposure to aflatoxin B1 in swine production and possible contamination sources. J Toxicol Environ Health A 76:944–951. doi:10.1080/15287394.2013.826569

Viegas S, Veiga L, Figueredo P et al. (2013c) Occupational exposure of aflatoxin B1: the case of poultry and swine production. World Mycotoxin J 6:309–315

Viegas S, Veiga L, Malta-Vacas J et al. (2012) Occupational exposure to aflatoxin (AFB(1)) in poultry production. J Toxicol Environ Health A 75:1330–1340. doi:10.1080/15287394.2012.721164

Wang Y, Chai T, Lu G et al. (2008) Simultaneous detection of airborne aflatoxin, ochratoxin and zearalenone in a poultry house by immunoaffinity clean-up and high-performance liquid chromatography. Environ Res 107:139–144. doi:10.1016/j.envres.2008.01.008

World Health Organization (2009) WHO guidelines for indoor air quality: dampness and mould. WHO Regional Office for Europe, Copenhagen.

Yike I, Dearborn DG (2004) Pulmonary effects of *Stachybotrys chartarum* in animal studies. Adv Appl Microbiol 55:241–273. doi:10.1016/S0065-2164(04)55009-8

Yike I, Distler AM, Ziady AG et al. (2006) Mycotoxin adducts on human serum albumin: biomarkers of exposure to *Stachybotrys chartarum*. Environ Health Perspect 114:1221–1226

Yike I, Rand TG, Dearborn DG (2005) Acute inflammatory responses to *Stachybotrys chartarum* in the lungs of infant rats: time course and possible mechanisms. Toxicol Sci 84:408–417

Zarba A, Hmieleski R, Hemenway DR et al. (1992) Aflatoxin B1 – DNA adduct formation in rat liver following exposure by aerosol inhalation. Carcinogenesis 13:1031–1033

Zhou HR, Harkema JR, Yan D et al. (1999) Amplified proinflammatory cytokine expression and toxicity in mice coexposed to lipopolysaccharide and the trichothecene vomitoxin (deoxynivalenol). J Toxicol Environ Health A 57:115–136

Zock JP, Borras-Santos A, Jacobs J, et al (2014) Dampness and microbial secondary metabolites in schools and respiratory symptoms in teachers. European Respiratory Society 2014 International Congress. Abstract 2161

Part IV
Microbial Exposures and Health Effects in Specific Indoor Settings

Chapter 11
Indoor Microbial Aerosol and Its Health Effects: Microbial Exposure in Public Buildings – Viruses, Bacteria, and Fungi

Elena Piecková

Abstract Mechanisms of aerosolization of microorganisms, composition and dynamics of microbioaerosol are characterized. As well as methods of its detection, incl. modern equipment set-ups and sampling procedures recommended are outlined. Medical impact of (indoor) air disperged viral, bacterial and fungal propagules (allergies, intoxications, infections), together with the related European legislation is summarized. An overview of real mycoaerosol conditions in our dwellings and their outdoors with different microclimate, settlement and building types, household characteristics and health state of occupants is given, too. Finally, examples of several possible health damages due to exposition to (aerosolized) fungal toxicants *in vitro* and *in vivo* are demonstrated.

Keywords Droplets · spores · hyphal fragments · aerodynamics · respiratory deposition · aeroscope

11.1 Introduction

Bioaerosols comprise microorganisms (viruses, bacteria, fungi), while other propagules may be originated from living organisms (pollen, plant seeds, wooden dust, fragments of animal hair, insects and their excrements etc.). All these components may occur in free forms or are carried by some vehicles, e. g., dust particles or droplets.

Microorganisms are released to the air from their growth sites, or colonized surfaces. Infections, allergies, intoxications, eventually even leading to pre- and canceroses can be spread around by the mean of bioaerosolization. Aerosols are discussed also because of their possible misuse (anthrax vs. bioterrorism). The forementioned notices explain the scientific and public health interest in aeromicrobiology increasing over the last years (Piecková, 2013).

During a sneeze, millions of droplets of water and mucus are expelled at about 100 m/s. The droplets initially are about 10–100 μm in diameter, but they dry rapidly

E. Piecková (✉)
Institute of Microbiology, Faculty of Medicine, Slovak Medical University, Bratislava, Slovakia
e-mail: elena.pieckova@szu.sk

C. Viegas et al. (eds.), *Exposure to Microbiological Agents in Indoor and Occupational Environments*, DOI 10.1007/978-3-319-61688-9_11

Table 11.1 Some human diseases transmitted interpersonally by inhaled airborne particles

Virosis (virus agent)	Bacterial disease (bacterial agent)
Chickenpox (Varicella)	Whooping cough (*Bordetella pertussis*)
Flu (Influenza)	Meningitis (*Neisseria* sp.)
Measles (Rubeola)	Diphtheria (*Corynebacterium diphtheriae*)
German measles (Rubella)	Pneumonia (*Mycoplasma pneumoniae, Streptococcus* sp.)
Mumps - Parotitis (Rubulavirus)	Tuberculosis (*Mycobacterium tuberculosis*)
Smallpox (Variola)	–

Open access source: http://archive.bio.ed.ac.uk

Table 11.2 Some airborne human diseases dependent on the environmental source

Disease	Source
Psittacosis (*Chlamydia psittaci*)	Dried, powdery droppings from infected birds (parrots, pigeons, etc.)
Legionnaire's disease (*Legionella pneumophila*)	Droplets from air-conditioning - HVAC systems, water storage tanks, etc., where the bacterium grows
Acute allergic alveolitis (various fungal and actinomycete spores)	Fungal or actinomycete particles from decomposing organic matter (composts, grain stores, hay, etc.)
Aspergillosis (*Aspergillus fumigatus, A. flavus, A. niger*)	Fungal particles inhaled from decomposing organic mater
Histoplasmosis (*Histoplasma capsulatum*)	Spores of the fungus, in old, weathered bat or bird droppings
Coccidioidomycosis (*Coccidioides immitis*)	Spores in air-blown dust in desert regions (Central, South and North America), where the fungus grows in the soil

Open access source: http://archive.bio.ed.ac.uk (2017)

to **droplet nuclei** of 1–4 μm, containing virus particles or bacteria. This is a major means of transmission of several diseases of humans, shown in the Table 11.1.

Several other diseases may be acquired by inhaling aerosolized microbial particles from environmental sources, not directly originated from an infected person (Table 11.2).

Psittacosis is a serious disease due to handling birds or inhaling dust from bird excrements. It is caused by the bacterium *Chlamydia psittaci*, an obligate intracellular parasite. After entering the respiratory tract, the cells are transported to the liver and spleen, multiply there and then invade the lungs, causing inflammation, haemorrhage and pneumonia (http://archive.bio.ed.ac.uk).

11.2 Viruses in Air

A virus can multiply only within a host cell. Infected cells can spread viruses directly into the surrounding air (primary aerosolization) or to fluids and surfaces, which can become sources for airborne transmission (secondary aerosolization).

Secondary aerosolization can occur for any virus, predominantly when air displacements or movements around contaminated surfaces or fluids disperse the viruses into the air. It can also occur by liquid splashes, which can aerosolize viruses in liquids or on surfaces. In fact, almost any kind of disturbance of infected organisms or materials, even the bursting of foam in seawater, can produce airborne, virus-laden particles (Aller et al., 2005).

Virus-laden particles are a complex mixture of various components (salts, proteins, and other organic and inorganic matter, including virus particles). The size of the viral particle itself does not rule the airborne particle size. The influence of viruses alone on the granulometric distribution of aerosols is likely negligible compared to that of the remainder of the aerosol. It was demonstrated that the particle size distribution of artificially produced submicrometer and ultrafine aerosols of culture media is not affected by the presence of bacteriophages (Hogan et al., 2005).

Relative humidity (RH) is the most widely studied factor that affects airborne virus infectivity. Depending on the virus, optimal preservation of infectivity may require a low RH (under 30%), an intermediate RH (30–70%), or a high RH (over 70%). Influenza virus, Japanese B encephalitis virus, Newcastle disease virus, and vesicular stomatitis virus, all of which are enveloped, are most stable at low RH, while rhinovirus, poliovirus, rhinotracheitis virus, picornavirus, and viruses of the Columbia SK group, which are nonenveloped (with the exception of the rhinotracheitis virus), are most stable at high RH. Human coronavirus 229E, pseudorabies virus, and rotavirus are most stable at intermediate RH. The first two are enveloped, while mature rotaviruses are usually nonenveloped. RH has no effect on the stability of airborne St. Louis encephalitis virus under the conditions tested (Verreault et al., 2008).

In 2001, a Norwalk-like virus outbreak in a school in the United Kingdom was believed to have been caused by airborne transmission. A similar occurrence has also been reported for a hotel restaurant (Marks et al., 2003). A retrospective cohort study conducted after a severe acute respiratory syndrome (SARS) epidemic in Hong Kong in 2003 suggested that airborne spread may have played an important role in the transmission of the disease. The same mode of transmission was also hypothesized in other studies of SARS (Li et al., 2005, Yu et al., 2005). Middle East Respiratory Syndrome (MERS) is an illness caused by a virus Middle East Respiratory Syndrome Coronavirus (MERS-CoV). Most MERS patients developed severe acute respiratory illness with symptoms of fever, cough and shortness of breath. About 3 to 4 out of every 10 patients reported with MERS have died. The disease was first reported in Saudi Arabia in September 2012. Later retrospective studies identified that the first known cases of MERS occurred in Jordan in April 2012. So far, all cases of MERS have been linked through travel to, or residence in, countries in and near the Arabian Peninsula. The largest known outbreak of MERS outside the Arabian Peninsula occurred in the Republic of Korea in 2015. The outbreak was associated with a traveler returning from the Arabian Peninsula. MERS-CoV, like other coronaviruses, is thought to spread from an infected person's respiratory secretions, such as through coughing. However, the precise ways the virus spreads are not currently well understood (https://www.cdc.gov/coronavirus/mers/, (2017)).

Seasonal variations in indoor RH have also been correlated with fluctuations in the morbidity of influenza (low RH) and poliomyelitis (high RH) viruses, with the highest morbidity occurring at the optimal RH for each virus. Seasonal variations have also been observed with measles virus and respiratory syncytial virus (Yusuf et al., 2007).

UV radiation is another factor that influences survivability of microbes. UV germicidal lamps, for instance, can be used to inactivate airborne microorganisms, including viruses, in indoor settings (First et al., 2007). However, in certain cases, RH must be taken into consideration. For example, vaccinia virus is more susceptible to UV radiation at low RH than at high RH (McDevitt et al., 2007).

The gas composition of the air can also have an influence on viruses, as ozone has been shown to inactivate airborne viruses. In fact, virus susceptibility to ozone is much higher than those of bacterial and fungal bioaerosols. However, the ozone efficacy varies from virus to virus. Ions in the air can also reduce the recovery rate of certain viruses, such as aerosolized T1 bacteriophage, with positive ions having the most detrimental effect (Tseng and Li, 2006).

While epidemiological data can help to determine the source of the contamination, direct data obtained from air samples can provide very useful information for risk assessment purposes of (viral) bioaerosols. Many types of samplers have been used over the years, including liquid impingers, solid impactors, filters, electrostatic precipitators, and many others. The efficiencies of these samplers depend on a variety of environmental and methodological factors that can affect the integrity of the virus structure. The aerodynamic size distribution of the aerosol also has a direct effect on sampler efficiency. Viral aerosols can be studied under controlled laboratory conditions, using biological or nonbiological tracers and surrogate viruses (Verreault et al., 2008).

11.3 Indoor-Air-Associated Bacteria

The majority of bacteria that are common indoors belong to the genera of *Micrococcus, Staphylococcus, Bacillus*, and *Pseudomonas*. They are distributed almost proportionaly in the indoor atmosphere as well.

Micrococcus is a sphere-shaped G+, relatively harmless bacterium. It is very common on skin, and it can also be found in soil, water, and meat products. It is generally a saprophyte and can cause spoilage of food. This organism can also be responsible for causing human sweat to smell badly. In immunocompromised patients, it can be an opportunistic pathogen. Some common species include *M. luteus, M. roseus*, and *M. varians*.

Staphylococcus is another sphere-shaped G+ bacterium. It is much more known than *Micrococcus*, especially in the context of hospitals. Nosocomial infections (hospital-acquired infections, HAI) are caused mostly by methicillin resistant *S. aureus* (MRSA) strains of this bacteria. However, staphylococci are found almost everywhere, and their presence usually does not result in infection. They are very common on skin, and can also be found in the nasal cavity, throats, and hair of 50% of healthy individuals. Food poisoning and skin infections, as well as toxic shock syndrome, are among the illnesses caused by *Staphylococcus*. *Staphylococcus* is the facultative anaerobic microbe.

Bacillus is a rod-shaped G+ bacterium. It has the ability to produce endospores – tough structures that can survive adverse environmental conditions. For the most part, bacilli are harmless saprophytes, and can be found in soil, water, dust, and sometimes within the human digestive system. Some species of *Bacillus* can cause food poisoning, and some can cause illness or infection. *B. anthracis* is the etiologic agent of anthrax — a common disease of livestock and, occasionally, of humans — and the only obligate pathogen within the genus *Bacillus*. The symptoms in anthrax depend on the type of infection and can take anywhere from 1 day to more than 2 months to appear. All types of anthrax have the potential, if untreated, to spread throughout the body and cause severe illness and even death. Four forms of human anthrax disease are recognized based on their portal of entry. Among them, inhalation, a rare but highly fatal form, is characterized by flu like symptoms, chest discomfort, diaphoresis, and body aches (Fig. 11.1). There were rather many attempts to misuse spores of *B. anthracis* in bioterroristic attacks reported so far (CDC, 2015).

Pseudomonas is another rod-shaped but G-bacterium commonly present indoors. It can be found in soil and water, and on plants. It is an opportunistic pathogen, and generally considered a nosocomial infection agent as the organism tends only to attack individuals that are immunocompromised. Along with infection, it also has the ability to produce exotoxins.

Some studies revealed also relatively high incidence of *Corynebacterium* in the indoor environment of certain public buildings (schools). These bacteria are

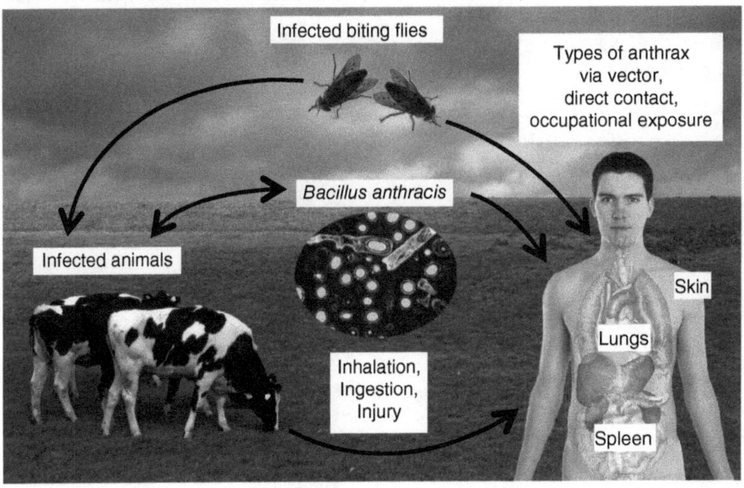

Fig. 11.1 The human exposure scenarios to Bacillus anthracis (Author: Renáta Lehotská)

rod-shaped G+ organisms omnipresent in nature. Some are useful in industrial settings, others can cause human disease, including most notably diphtheria, which is caused by *C. diphtheriae*. As with various species of a microbiota, they usually are not pathogenic but can occasionally invade into tissues (via wounds) or break through weakened host defense, that both results in certain infection process (Aydogdu et al., 2010).

While single virus or bacterium particles exist in the air, they tend to aggregate rapidly. Aggregation speed depends on the size distribution of the airborne particles, the concentration of the aerosol, and the thermodynamic conditions. It has been shown that a visually clean environment may be more contaminated by bioaerosols than a visually dirty one. This may be due to the fact that larger particles tend to settle faster than smaller particles do; the settling velocity of 0.001 μm particles is $6.75E^{-09}$ m/s, while 10 μm particles settle at $3.06E^{-03}$ m/s and 100 μm particles settle at $2.49E^{-01}$ m/s. Airborne particles in a "clean" environment are more likely to remain small and inhalable than are particles in a dirty environment, which tend to grow larger by sticking to other airborne particles (Verreault et al., 2008).

11.3.1 Legionnaire's Disease

This is a common form of bacterial pneumonia in elder or immunocompromised people (Fig. 11.2). It is seldom transmitted directly from person to person. The bacterium *Legionella pneumophila* is an aquatic rod-shaped G-species found naturally in freshwater environments, like lakes and streams. with a temperature

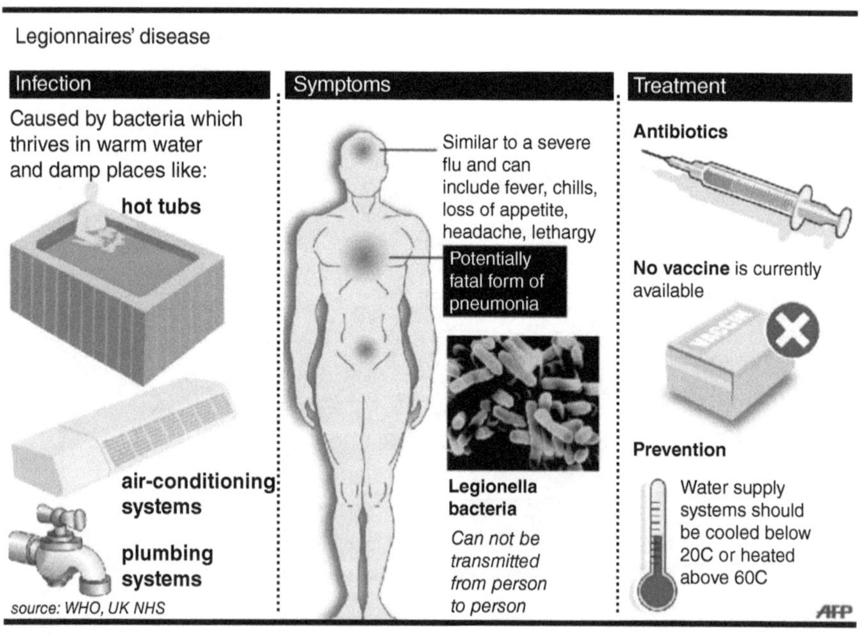

Fig. 11.2 Legionnellosis summary (Bunbury, 2015)

optimum of about 36°C, and is a common inhabitant of warm-water systems in buildings like

- hot tubs that aren't drained after each use
- hot water tanks and heaters
- large plumbing systems
- cooling towers (air-conditioning – HVAC units for large buildings)
- decorative fountains

where it can become of a health concern. Infection occurs when people inhale aerosol droplets containing the bacteria. People who get sick after being exposed to *Legionella* can develop two different illnesses: legionnaires' disease and Pontiac fever. Symptoms usually begin 2–10 days after being exposed to the bacteria, but it can take longer, for about 2 weeks after exposure.

Pontiac fever symptoms are primarily fever and muscle aches; it is a milder infection than legionnaires' disease. Symptoms begin between a few hours to 3 days after being exposed to the bacteria and usually last less than a week. Pontiac fever is different from legionnaires' disease because someone with Pontiac fever does not have pneumonia necessarily (Correia et al., 2016).

The key to preventing legionnaires' disease is to prevent *Legionella* colonization (growth) in water systems. Low water volumes combined with high temperatures and heavy bather loads make public hot tub operation challenging. The result can be low disinfectant levels that allow the growth and spread of a variety of germs (e.g. *Pseudomonas* and *Legionella*) that can cause skin and respiratory recreational water illnesses (RWI). Operators that focus on hot tub maintenance and operation to ensure continuous, high water quality are the first line of defense in preventing the spread of RWI. Guidelines for reducing the risk of *Legionella* growth and spread are available for those who maintain and manage building water systems, including systems for potable (water for drinking and showering), non-potable, and recreational water from ASHRAE (ASHRAE Guideline 2000; ASHRAE Standard, 2015).

11.4 Fungal Bioaerosol

When dealing with mouldy (organic) materials, the air concentration of viable fungal propagules can reach the level as high as 10^9 colony forming units (cfu)/m^3, incl. a broad spectrum of chemical fungal irritants (mycotoxins, volatile organic compounds). The occupants may suffer from mucous and skin irritations, but some severe acute or chronic damage of their respiratory tract (bronchitis, allergic alveolitis, "farmer lungs," pulmonary mycotoxicoses) can take place as well. The occupational hygiene must, thus, develop and apply all necessary preventive and protective measures to minimize the bioaerosol hazard in particular plants (Piecková and Jesenská, 1999).

Rather different situation is encountered in mouldy dwellings and/or public buildings like offices, schools, waiting rooms, cultural premises etc., where even the youngest children, elderly people or other vulnerable occupants are staying, even long time. From the indoor air of such dwellings, 3–450,000 fungal cfu/m^3 might be recovered. Though, the quantity of viable and cultivabe propagules of aeromycobiota depends highly on activities carried out in the houses that elevate turbulencies leading to stronger dispersion of spores and hyphal fragments from the fungal bodies (colonies) around. The physiological characteristics of particular fungal species affect their aerosolization, too. Some micromycetes form hygroscopic spores sticked in slimy heads on their aerial mycelium (acremonia, fusaria), while others produce enormous quantity of small conidia in very fragile chains and columns (penicillia, aspergilli, cladosporia). And even other ones, large macrospores (alternariae). There is also a high number of devitalized germs, which can't be entrapped onto cultivation media, but their allergenic and toxic potential remains unchanged (Piecková, 2008).

11.4.1 Methods of Indoor Aeromicrobiota Detection

The indoor air samples in buildings are collected by the mean of different volumetric apparatuses – aeroscopes: germs are immobilized on agar plates or into the liquid medium (impingers). All of them enable qualified quantitative analysis of viable micro(myco)flora present. The dilution method may be employed to characterize the fungi associated with settled dust. And swabs or adhesive tapes are used to sample microbes from the surfaces. These samples first undergo the direct microscopic observation, followed by cultivation evaluation. The modern state-of-art in sampling methods has been broaden by modifications of PCR (microarrays), showing the presence of e. g. stachybotrys directly in the affected material, incl. the air. There are also laboratory setups with propagules' optical size counters coupled to computing of the aerodynamic size of the germs and their depository potential (Sivasubramani et al., 2004).

The Anderson sampler is an example of the impactor aeroscope for selective entrapping different sizes of particles according to their size (momentum). This sampler consists of a stack of 8 metal sections that fit together with ring seals to form an air-tight cylinder. Each metal section has a perforated base, and the number of perforations is the same in each section, but the size of these perforations is progressively reduced from the top of the column to the bottom. To use this sampler, open agar plates are placed between each metal section, resting on three studs. When fully assembled with an open agar plate between each unit an electric motor sucks air from the bottom of the unit, causing spore-laden air to enter at the top and to pass down through the cylinder. The path taken by this air is shown in Fig. 11.3 (http://archive.bio.ed.ac.uk).

Figure 11.4 shows the methodology applied indoor of a building (a private gallery) with offending mouldy smell to find its causative agents. Figure 11.5 illustrates the fungal spectrum commonly recovered from the school indoors in Slovakia over last 15 years.

It is also possible to detect the aeroconcentration of fungal ergosterol or betaglucan, that measure total fungal load, i. e. vital and devitalized particles. As there are

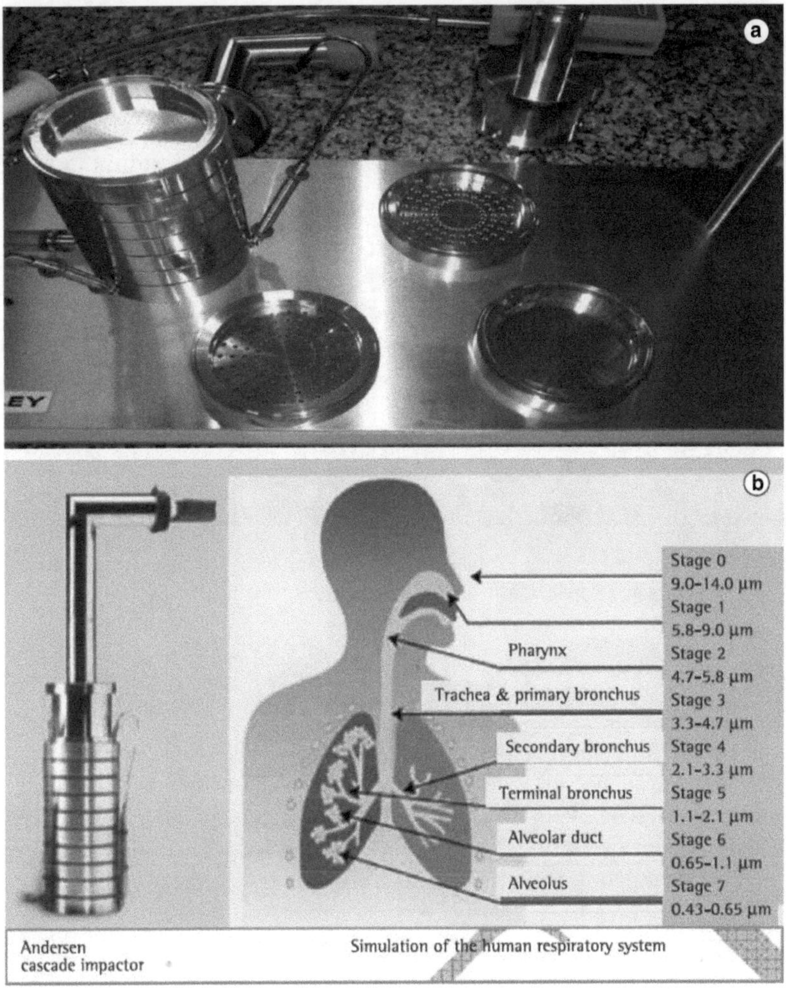

Fig. 11.3 Size distribution of airborne particles entrapped in the Anderson sampler and their airways' penetrating potential in humans (Andrade-Lima et al., 2012)

remarkable differences in the ergosterol content among fungal species (from 2 up to 14 μg/g of dry biomass), this method is recently ranked as the superficial only (Piecková, 2013).

Very recently, it has become concluded to use the fungal volatile organic compounds (VOC) as a new quantitative marker of undesired mould present in the indoor environment (Piecková, 2010).

The complex study of (fungal) bioaerosol should cover its source, properties and (health) effects.

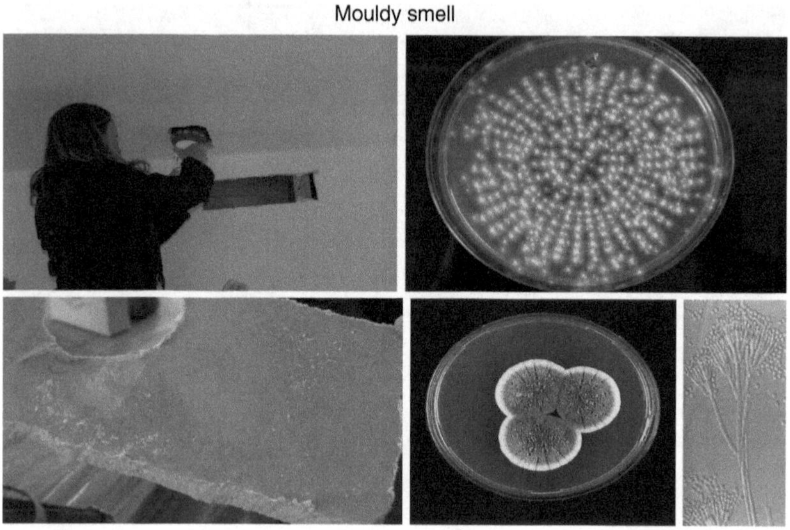

Fig. 11.4 Complex mycological evaluation of the indoor moulds in a private gallery house strongly affected by an excessive malodour

11.4.2 Bioaerosol Source Characteristics

The macro-and micromorphology of the micromycete releasing germs or toxicants into the air are usually well known. On the other hand, there is almost an absolute lack of knowledge on the biophysics of fungal growths, e. g. how strong force can cause the propagule release; what is the life dynamics of real natural fungal populations, effects of their biotop, intra- and inter-population relations etc.

11.4.3 Bioaerosol Characterization

Microscopic fungi reproduce by vegetative spores (microspores up to 10 μm, macrospores several tenths micrometers) and/or sexual spores (similar size to the microspores). The later ones are produced in fruit bodies (up to hundreds micrometers). The shapes and ornamentations of any spores are species-related. From the aerial dispersion facility point of view, the aerodynamic features of spores (do not copy with their physical size necessarily and are strongly connected with the bioaerosol concentration), their aggregation and conglomeration on its own or on any carriers (dust, pollen particles etc.; ca. 60 % of aerosol particles are present in aggregates) play the crucial role.

Macroscopic appearance of the microfungus body is represented by its colonies (thalus) of different surface structure and coloration. Hyphae of the colony grow superficially and in the substrate. It's become already possible technically to simulate mechanic irritation of the colonies (vibrations, airflow velocity) enabling

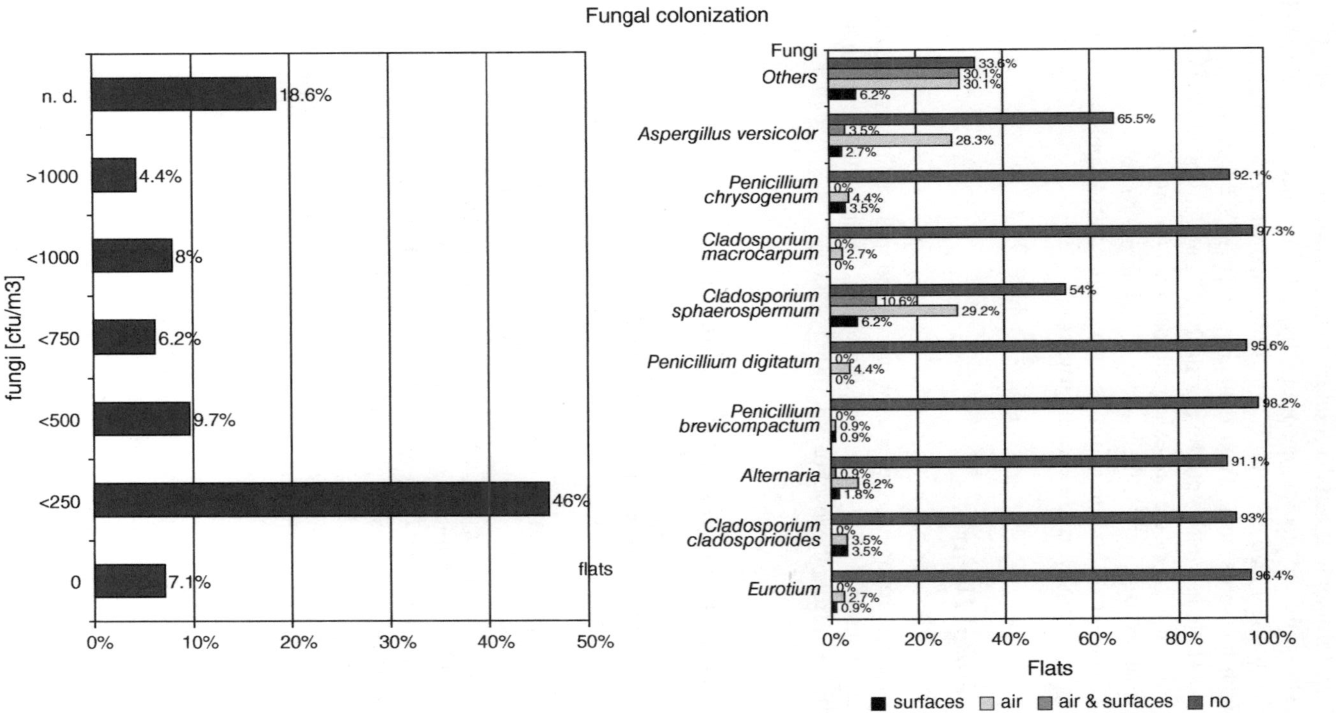

Fig. 11.5 Spectrum of indoor fungi isolated from Slovak schools over last 15 years

effective release of propagules to the air. Regarding that, e. g. the relation between colony texture and formation of aerial microturbulencies is studied recently.

Simultaneously with the spores, also the hyphal fragments are dispersed from the colonies (optical size ca. 1 μm), usually in hundred times higher counts (10^5 cfu/cm^2) than the spore ones. Their deliberation depends on the structure of the overgrown substrate as well as on the age of the culture (so called dessication stress) and the concentration of the pre-existing bioaerosol (Górny, 2004).

Settlement of aerosolized germs, their survival and ability to colonize surfaces has their specific rules, too. So, the ability to grow under hygroscopic conditions enables the fungi to colonize/invade human airways.

11.4.4 Bioaerosol Decay

The aerosolized fungal propagules are usually aggregated in the units with the diameter of the water vapour (10–20 μm), thus, they form a gentle mist. Its particles then settle in 19–5 min from the level of 3 m (Górny, 2004).

11.4.5 Inhalatory Exposition to the (Fungal) Bioaerosol

Its dynamics is not commonly characterized under normal (housing) conditions, neither at the fungal-burst situation. Depository parameters of microbial bioaerosol particles are described only very marginally. While breathing via healthy nose, 30–40% of the particles are entrapped onto the upper airways mucose and other 30–40% pass through to the lower parts of the lungs, incl. alveoli. Mouth breathing, in contrary, enables penetration of ca. 70% of the propagules to the utmost fine pulmonary chambers. The ration of bigger and smaller germs and their aggregates, remains unknown. Asexual fungal macrospores (the biggest fungal germs) finish in the nose, throat and sinuses, microspores (a- and sexual) at the trachea and bronchi, and hyphal fragments may reach alveoli easily as they belong to so called fine particular matters, their depository potential is the highest, even related to other head cavities, incl. the skull ones.

The respiratory tract is highly effective in trapping airborne particles, with sometimes serious consequences for health. The mechanisms involved depend on particle size.

1. Large particles (about 10 μm) have sufficient mass to **impact** onto surfaces, even at low air speeds. They break free from the air as it flows around obstacles. During normal breathing, the airflow in the nose and trachea is about 100 cm per second – sufficient for pollen grains and larger fungal spores, esp. macrospores (*Alternaria* sp. etc.) to be retained on the mucous, where they can cause typical **hay fever symptoms** like (allergic) rhinitis and asthma. These are the types of particle detected on the top plates of the Anderson sampler.
2. Smaller particles do not impact at these air speeds, and the air speed decreases as the respiratory system branches further down. So all the particles of 5 μm or less

are carried deep into the lungs. There they can settle out by **sedimentation** in the brief periods when the air is calm between successive breaths. Particles of 2–4 μm are optimal for alveolar deposition, and this range includes the spores of many *Aspergillus* and *Penicillium* spp. This is how some of the serious fungal infections of humans are initiated – aspergillosis, histoplasmosis, coccidioidomycosis etc.

3. Even smaller particles, such as the spores of actinomycetes along with hyphal fragments (about 1 μm) are less efficient at being deposited in the alveoli, but repeated exposure to aerosol clouds can lead to sensitisation and **extrinsic allergic alveolitis** (farmer's lungs etc.).
4. Very small particles, less than about 0.5 μm, do not impact but are moved by diffusion (**Brownian motion**) which brings them randomly into contact with surfaces in the lungs, and are able to cross other barriers in human body, e.g. meninges. This is true of the fine dusts that cause many (occupational) diseases.

Do the bacterial and viral pathogens copy this scheme as well?

The nasopharyngial viroses are associated with large sneeze droplets which impact in the upper airways. Most bacterial diseases also are initiated in the upper airways, when bacteria are carried in large droplets or on "rafts" of skin that impact onto the mucous. However, infections by *Mycobacterium* sp. (**tuberculosis**) and *Bacillus anthracis* (**anthrax**) are initiated in the lungs. These are highly virulent pathogens, and even single cells or spores (about 3 μm for *Bacillus*) can initiate infections after deposition in the alveoli.

The macroorganisms cleans itself of the bioaerosol propagules by the mean of the mucosiliary effect (self-cleaning of the upper respiratory tract involving production of mucous and epithelium cells ciliary beating), or the macrophages. The possible damaging effect occurs at the level of the contact tissues (mucouses of the airways, eyes etc.; Fig. 11.6) up to the (sub-)cellular level (pulmonary alveolar macrophages; Fig. 11.7). The mechanisms might result in respiratory tract

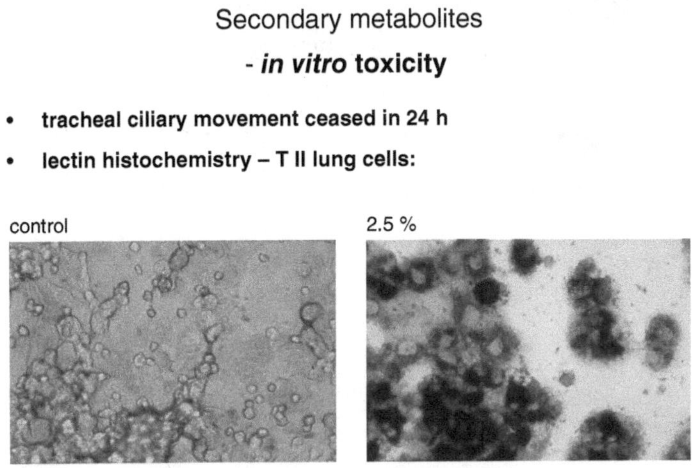

Fig. 11.6 Activity of indoor airborne *Aspergillus versicolor* toxins showing negative *in vitro* effects onto respiratory tract cells

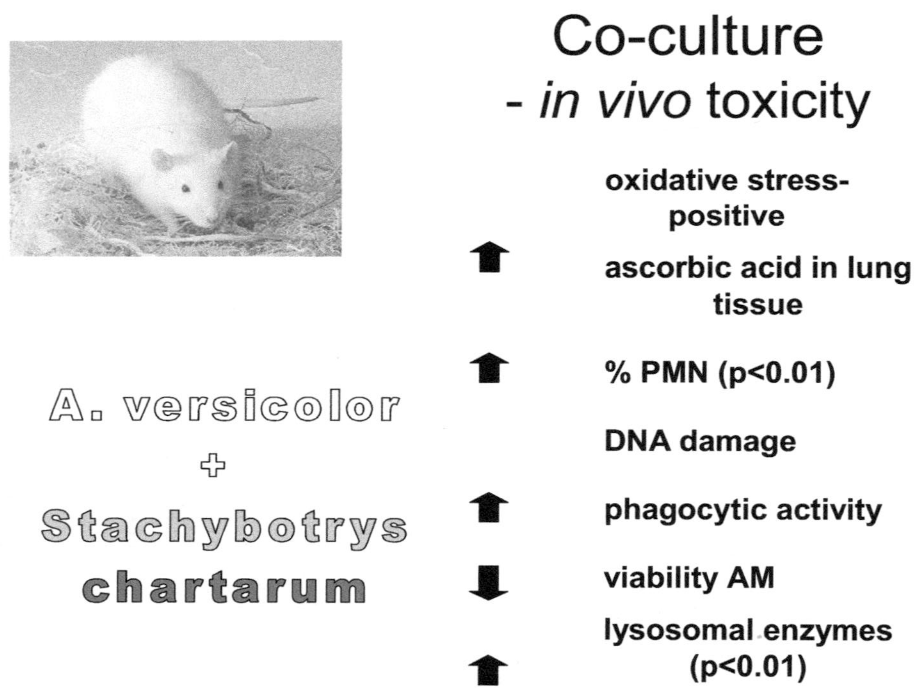

Fig. 11.7 Activity of indoor *Aspergillus versicolor* and *Stachybotrys chartarum* toxins showing negative *in vivo* effects onto respiratory tract cells (arrows indicate elevation/depression of the biochemic or cytological parameters pronouncing the cell damage)

colonization, its inflammatory or cytotoxic debilitation, decreasing of the mucosal immunity (at least) that finally leads to recurred infectious (viral, bacterial, mycotic). The infectious loads of the particular moulds necessary to cause the ill health symptoms in humans are studied by epidemiology and infectology (Piecková, 2008).

To control and sustain so called healthy indoors, there have been adopt some hygienic limits of airborne microbes, incl. fungi – EU legislation and WHO guidelines (WHO, 2009; Reg. Min. Hlth SR, 2008). The basic limit was set down at the level of 500 cfu/m^3, qualitatively copying the outdoor mycoflora, without toxic and pathogenic species.

References

Aller JY, Kuznetsova MR, Jahns CJ et al (2005) The sea surface microlayer as a source of viral and bacterial enrichment in marine aerosols. J Aerosol Sci 36:801–812

Andrade-Lima M, Ferreira Pereira LF, Godoy Fernandes AL (2012) EquivalÃªncia farmacÃªutica da formulaÃ§Ã£o combinada de budesonida e formoterol em cÃ¡psula Ãºnica com dispositivo inalador de pÃ³. Jornal Brasileiro de Pneumologia 38 (6):748–756

CDC (2015) Anthrax symptoms. Centers for Disease Control and Prevention. Retrieved 16 Nov 2015.

ASHRAE Guideline 12-2000 Minimizing the risk of legionellosis associated with building water systems. ISSN 1041-2336.

ASHRAE Standard 188-2015 Legionellosis: risk management for building water systems (ANSI approved). ISSN 1041-2336.

Aydogdu H, Asan A, Tatman Otkun M (2010) Indoor and outdoor airborne bacteria in child day-care centers in Edirne City (Turkey), seasonal distribution and influence of meteorological factors. Environ Monit Assess 164:53–66

Bunbury K (2015) Legionnaires outbreak – re your clients properly insured? Environmental Risk Managers, Moline (USA), 2017.

Correia AM, GonCalves J, Gomes JP et al (2016) Probable person-to-person transmission of legionnaires' disease. N Engl J Med 374:497–508

First M, Rudnick SN, Banahan KF et al (2007) Fundamental factors affecting upper-room ultraviolet germicidal irradiation. I. Experimental. J Occup Environ Hyg 4:321–331

Górny RL (2004) Filamentous microorganisms and their arrangements in indoor air. Ann Agric Environ Med 11:185–197

Hogan JrCJ, Kettleson EM, Lee MH et al (2005) Sampling methodologies and dosage assessment techniques for submicrometre and ultrafine virus aerosol particles. J Appl Microbiol 99:1422–1434

Li Y, Huang X, Yu IT et al (2005) Role of air distribution in SARS transmission during the largest nosocomial outbreak in Hong Kong. Indoor Air 15:83–95

Marks PJ, Vipond IB, Regan FM et al (2003) A school outbreak of Norwalk-like virus: evidence for airborne transmission. Epidemiol Infect 131:727–736

McDevitt JJ, Lai KM, Rudnick SM et al (2007) Characterization of UVC light sensitivity of vaccinia virus. Appl Environ Microbiol 73:5760–5766

Piecková E. (2008) Pseudallescheria/Scedosporium: an aerobiological approach. In: de Hoog GS (eds) Abstracts of the ECMM Workshop on Pseudallescheria/Scedosporium, Utrecht, 2008.

Piecková E. (2013) Indoor microbial aerosol and its health effects. In: Petráš D et al (eds) Proceedings of the international conference Indoor Climate 2013, Bratislava.

Piecková E, Jesenská Z (1999) Microscopic fungi in dwellings and their health implications in humans. Ann Agric Environ Med 6:1–11

Piecková E, Majorošová M, Hurbánková M et al (2010) Aspergillus versicolor [Vuill.] Tiraboschi – the most frequent indoor fungal colonizer under Slovak housing conditions. In: Petráš D et al (eds) Proceedings of the international conference Indoor Climate 2010, Bratislava.
Reg. Min. Hlth SR Nr 259/2008 Zz on details and requirements of the indoor environment and on minimal requirements on the dwellings of lower standard and the dwelling buildings for temporary staying, Bratislava, 2008.
Sivasubramani SK, Niemeier RT, Reponen T et al (2004) Fungal spore source strengh tester: laboratory evaluation of a new concept. Sci Total Environ 329:75–86
Tseng CC, Li CS (2006) Ozone for inactivation of aerosolized bacteriophages. Aerosol Sci Technol 40:683–689
Verreault D, Moineau S, Duchaine C (2008) Methods for sampling of airborne viruses. Microbiol Mol Biol Rev 72:413–444
WHO Regional Office for Europe (2009) WHO guidelines for indoor air quality: dampness and mould, Geneva, 2009.
Yu IT, Wong TW, Chiu YL et al (2005) Temporal-spatial analysis of severe acute respiratory syndrome among hospital inpatients. Clin Infect Dis 40:1237–1243
Yusuf S, Piedimonte G, Auais A et al (2007) The relationship of meteorological conditions to the epidemic activity of respiratory syncytial virus. Epidemiol Infect 135:1077–1090

Chapter 12
Microbial Exposures in Schools and Daycare Centers

Martin Täubel and Hanna K. Leppänen

Abstract The quality of indoor air – and that including the microbial quality – in schools and daycare centers is of crucial importance to pupils' and teachers' health. This is because of the extended periods of exposure daily and during lifetime, and the vulnerability of in particular pupils to air contaminants due to their maturing physiology and high intake rates. On a public health level, exposures in schools and daycares affect many individuals at once, increasing the impact of both health beneficial and adverse situations in these environments on a population level. We summarize here the scientific literature that has dealt with microbial exposure in schools and daycare centers, highlighting the key determinants that affect microbial levels in such buildings and both negative and positive health impacts of the related exposure to microbes.

Keywords Schools · daycare centers · microbes · moisture damage · respiratory health · indoor microbiome

12.1 Introduction

Schools and daycare centers are environments that contribute considerably to the everyday exposure to indoor air for children playing and learning and adults working in these spaces. In particular children spend several hours per day, most days of a week and most weeks of a year inside daycare or school buildings, making this the most important indoor environment in terms of time spent after the home environment; the situation is similar for teachers or other personnel working in schools and daycare centers. Therefore, the quality of indoor air in these environments is of crucial importance. Children are considered a vulnerable population, more susceptible to the effects of air pollution, because of their less mature physiology and relatively high intake rates and thus exposure to indoor air contaminants when compared to

M. Täubel (✉) · H.K. Leppänen
Environmental Health Unit, National Institute for Health and Welfare, Kuopio, Finland
e-mail: martin.taubel@thl.fi

C. Viegas et al. (eds.), *Exposure to Microbiological Agents in Indoor and Occupational Environments*, DOI 10.1007/978-3-319-61688-9_12

adults Fahrhat et al. (2005). Moreover, schools and daycares typically concentrate tens to hundreds of building occupants in a relatively small space so that poor indoor air quality and exposure to harmful agents will affect many individuals at once. This particular setting is similar in many parts of the world and calls health authorities for action. There are international guidelines covering certain chemical and dampness-related, biological agents that contribute to poor indoor air quality (WHO Regional Office for Europe 2009, 2010); however, availability, interpretation and implementation of guidelines and standards on national level vary widely. In addition to various indoor and outdoor sources of indoor pollutants, insufficient ventilation that fails to supply enough "clean" air and remove or dilute both chemical and biological indoor pollutants and support stable physical parameters indoors, such as temperature and relative humidity, are considered crucial issues.

Given the context of this book we restrict here to exposures of microbial origin. From studies in homes we know of the ambivalent role of microbes in health and disease, referring to the positive and negative impacts that microbial exposure may have, in particular to children's health (see Chap. 13: Microbial exposures in residential homes). Some studies indicate that schools and daycares may contribute considerably to a child's daily accumulated exposure to microbes, not only because of exposure duration, but also because of exposure levels that may clearly exceed home exposure levels, as shown for several fungal and bacterial parameters (Jacobs et al. 2013; Wan and Li 1999).

In this chapter we aim to summarize the scientific literature that reports on aspects of microbial exposure in schools and daycare environments. We attempt here to condense these reports towards presenting the key determinants of microbial levels in schools and daycare centers, as well as to summarize what has been learned so far with respect to the impact of microbial exposure in these environments on health outcomes in children and adults. We were strict with including only studies that have actually measured one or several microbial parameters directly from indoor environmental samples, utilizing some of the approaches for sampling and measuring of microbes that are described in Chapters 4, 5 and 6. Those techniques may include traditional cultivation and non-cultivation based methods, target nucleic acids, cell-wall components or secondary metabolites, or report on toxicity of particulate matter that is likely to relate to microbial components, in addition to other indoor air contaminants. In Table 12.1 we provide an overview of studies that complied with our definitions. While the authors do not claim this list to be a complete, we believe the reader will find a good overview of a major part of original scientific literature on the topic of microbial exposures in schools.

12.2 Moisture Damage in Schools and Health Effects

There is a good body of literature on exposure to moisture damage and dampness in schools and the impact on pupils' and teachers' health. However, many of those reports were assessing the impact of the moisture damaged environment on

Table 12.1 Overview of studies on microbial exposures in schools and daycare centers

Type/objective of study	Study sites and design	Microbial sampling and analyses	Main findings	Reference
Microbial exposures in schools and pupils' and teachers' health				
Fungi and mycotoxins in dust versus respiratory health in pupils	8 secondary schools in Malaysia; 32 classrooms; 462 students	Settled dust; fungal and bacterial DNA targets (qPCR); mycotoxins	Some associations between levels of microbial contaminants and respiratory health (positive for *A. versicolor* versus wheeze and daytime breathlessness and *Streptomyces* and doctor diagnosed asthma; negative (protective) for *Stachybotrys* DNA and mycotoxins versus daytime breathlessness).	Cai et al., 2011; *Pediatr Allergy Immunol* 22 (3):290–297
Association between fungal spores and children's health	School children (aged 6 to 15 years) in 44 schools in Taiwan	Burkard Personal Air Sampler; viable fungi	*Aspergillus/Penicillium* and basidiospores were significantly correlated with current asthma and with symptoms reduced on holidays or weekends.	Chen et al. 2014; *Chest* 146(1):123–134
Exposure to mold and associations with building related symptoms and lung function of teachers	522 teachers from 15 public schools, 8 water-damaged schools, and 7 non-damaged schools (107 classrooms)	Airborne dust samples, floor dust and dust from ventilation ducts; culturable fungi and LPS	Female teachers report more frequently irritation and general symptoms than male teachers. Women's symptoms reports were positively associated with mold exposure in the schools, but lung function was not associated with mold exposure, neither IL-8 and ECP.	Ebbehoj et al. 2005; *Indoor Air* 15 (Suppl 10):7–16
Management of indoor air problems in school buildings	School center with suspected mold and health problems in Finland	Air samples (six-stage impactor), surface swab-samples, material samples; viable microbes	Fungal genera found in air and surface samples contained e.g. *Aspergillus versicolor, Eurotium* and other moisture damage-related genera. Microbial results from material samples supported the conclusions made in the structural studies. Several irritative and recurrent symptoms were common, incidence of asthma was high.	Haverinen et al. 1999; *Environ Health Persp* 107:509–514

(continued)

Table 12.1 (continued)

Type/objective of study	Study sites and design	Microbial sampling and analyses	Main findings	Reference
Personal and microenvironmental microbial exposure and relation to health effects	81 elementary school teachers in Eastern Finland	Air (Button samplers), personal sampling and microenviron-mental sampling in workplaces and homes	Both personal exposure and concentrations of pollutants at home were more frequently associated with health symptoms than concentrations at work.	Haverinen-Shaughnessy et al. 2007; *J Expo Sci Environ Epidemiol* 17 (2):182–190
Health effects of moisture damage in a school, and assessing the effect of a holiday on the symptoms	A moisture damaged school, 32 employees, in Finland	Air samples (Andersen impactor), surface and material samples; viable bacteria and fungi	Presence of several microbial types indicative of moisture and mold problems. The concentrations of tumor necrosis factor alpha (TNF-alpha), interleukin-6 (IL-6), and nitric oxide (NO) in nasal lavage fluid were significantly higher in the exposed than in control subjects. This difference disappeared during summer holiday, but the production of NO and IL-6 increased again during the re-exposure in the fall term.	Hirvonen et al. 1999; *Am J Respir Crit Care Med* 160(6):1943–1946
Dampness and dampness-related determinants in schools and homes, and health effects in children	330 pupils from provincial towns in the age range 6–10 years in Denmark	Airborne dust (IOM Samplers), floor dust, settled dust; viable bacteria and fungi, endotoxin, (1-3)-β-D-glucans and NAGase	High classroom dampness, but not bedroom dampness, was negatively associated with lung function parameters and positively with wheezing. No consistent findings were reported between any individual microbial components or combination of microbial components and health outcomes.	Holst et al. 2016; *Indoor Air* 26(6):880–891
Endotoxin in primary schools and homes of children in relation to asthma and sensitization	10 schools and 169 homes in The Netherlands	Settled dust (EDCs); endotoxin	Higher endotoxin levels in classrooms compared to homes indicating the significance of exposure at school for children and teachers; indicative association of endotoxin exposure in schools with non-atopic asthmatic symptoms in pupils.	Jacobs et al. 2013; *Eur Resp J* 42(2):314–322

(continued)

Table 12.1 (continued)

Type/objective of study	Study sites and design	Microbial sampling and analyses	Main findings	Reference
Associations between school dampness, levels of microbial markers, and children's health effects	Primary schools in Finland, The Netherlands and Spain with ($N = 15$) and without ($N = 10$) moisture damage, 232 classrooms	Settled dust (EDC and cardboard boxes); endotoxin, glucans, ergosterol, qPCR	The prevalence of respiratory symptoms was higher in moisture damaged schools, being most pronounced in Finland. Levels of different microbial markers were generally higher in moisture damaged schools, lower in Finnish schools. Wheeze was inversely associated with microbial levels.	Jacobs et al. 2014; *Occup Environ Med* 71 (10):704–712
Asthma and atopy in relation to microbial and plasticizer exposure	8 primary schools in Uppsala, Sweden; measurements in 23 classrooms	Indoor and outdoor air; microbe enumeration via staining and identification via cultivation, MVOCs	At higher concentrations of total MVOC, nocturnal breathlessness and doctor diagnosed asthma were more common among pupils. Indoor MVOCs and indoor viable molds and bacteria were negatively correlated, questioning the use of MVOCs as microbial surrogates.	Kim et al. 2007; *Indoor Air* 17(2):153–163.
Effect of moisture damage and repair on microbial condition and pupils' health	A moisture damaged school (index) and a reference school	Air samples (Andersen impactor); viable bacteria and fungi, 250 + 246 samples (index + reference)	Microbial concentrations were higher in the index school than in the reference school before and during renovation. After renovation, the levels decreased to the level of the reference school and mycobiota changed. Several symptoms were more prevalent in the moisture-damaged school than in the reference school, but the differences disappeared during the renovations.	Lignell et al. 2007; *J Environ Monit* 9 (3):225–233
Effects of moisture-damage repairs on microbial exposure and symptoms in pupils	Two primary and two secondary school (2 moisture damaged, 2 reference schools) in central Finland	Air samples (Andersen impactor); viable bacteria and fungi	In one of the damaged school, the microbial levels decreased after full renovation and also the symptom prevalence decreased. No change in microbial conditions was seen after partial repairs in the other damaged school, and only slight improvement was observed in symptom prevalence.	Meklin et al. 2005; *Indoor Air* 15(Suppl 10):40–47

(continued)

Table 12.1 (continued)

Type/objective of study	Study sites and design	Microbial sampling and analyses	Main findings	Reference
Exposure to mold and association to symptoms in school children	15 Danish school with varying degrees of moisture damage; 1832 teacher and pupils	Settled dust, floor dust, ventilation duct dust, active air; viable fungi and bacteria, endotoxin	No associations between moisture damage and mold and symptoms; positive associations between symptoms and viable molds in floor dust, after adjustment for several confounding factors.	Meyer et al. 2004; *Indoor Air* 14:65–72
Testing interactions between gender and mold versus building related symptoms	15 Danish school with varying degrees of moisture damage; 1024 pupils	Settled dust, floor dust, ventilation duct dust, active air; viable fungi and bacteria, endotoxin	Associations between molds in floor dust and building related symptoms were strong among boys, but not existent among adolescent girls. Hypothesis of estrogen levels protecting from the effects of mold.	Meyer et al. 2005; *Indoor Air* 15(Suppl 10):17–24
Association between mold levels in floor dust and building-related symptoms	15 schools in Denmark, 503 school boys	Floor dust; viable fungi	Significantly increased risk for multiple symptoms (nasal congestion and dizziness in multiple logistic regression) using a variable combining high level of viable molds in floor dust and mechanical ventilation in classrooms.	Meyer et al. 2011; *Scand J Work Environ Health* 37(4):332–340
Health effects and microbial levels in displacement and mixing ventilation system	3 classrooms in Sweden (1 classroom blinded intervention, 2 control classrooms), 61 pupils	Air samples; viable bacteria and fungi	Viable bacteria were numerically lower, while total bacteria and molds were higher with displacement ventilation. Also the pupils measured health parameters were better with displacement ventilation compared with mixing ventilation system.	Norback et al. 2011; *Indoor Air* 21(5):427–437
Pupils respiratory illness versus ethnicity and microbial exposure in schools	8 secondary schools in Malaysia; 32 classrooms; 462 students	Vacuumed dust; 3-hydroxy fatty acids, muramic acid, ergosterol, fungal DNA (qPCR)	Endotoxin at school mostly protective for respiratory illness, different types may have different effects; fungal contamination (ergosterol, *Aspergillus versicolor*) can be risk factors for respiratory illness; ethnical difference in prevalence of asthma and atopy.	Norback et al. 2014; *PLoS One* 9(2):e88303

(continued)

Table 12.1 (continued)

Type/objective of study	Study sites and design	Microbial sampling and analyses	Main findings	Reference
Microbial exposure in schools vs. pupils symptoms	8 secondary schools in Malaysia; 32 classrooms; 462 students	Vacuumed dust; 3-hydroxy fatty acids, muramic acid, ergosterol, fungal DNA (qPCR)	Amount of fine dust was associated with rhinitis and ocular symptoms; negative (protective) associations between LPS and ergosterol levels and rhinitis; some positive associations for LPS subgroups and symptoms.	Norback et al. 2016; *Sci Tot Environ* 545–546:95–103
Associations between fungal DNA targets, mycotoxins and rhinitis and SBS symptoms	8 secondary schools in Malaysia; 32 classrooms; 462 students	Settled dust; fungal and bacterial DNA targets (qPCR), mycotoxins	Positive associations for total fungal DNA (rhinitis, ocular symptoms, tiredness), *Aspergillus versicolor* (ocular symptoms and tiredness) and verrucarol (tiredness); negative associations of *Streptomyces* and tiredness.	Norback et al. 2016; *PLoS One* 11(2): e0147996
Health effects of fungal exposure and its relation with the building type	2 school buildings (old and new), 301 pupils, aged 16–19 years, in Turkey	Viable fungi (Petri plate gravitational settling)	No significant difference in fungal concentrations in the two buildings. In both buildings, throughout all seasons, the most common fungal species were *Penicillium* and *Cladosporium*. The most common allergenic fungi (skin prick test = SPT) in the new building was *Epidermophyton* and in the old building *Penicillium*. The SPT results were also related with building related symptoms, except for *Alternaria*.	Orman et al. 2006; *Saudi Med J* 27(8):1146–1151
Mold exposure and mold specific IgG antibodies and their association with teachers' health effects	Teachers in three school buildings located in central Finland and follow-up after mold remediation	Air samples (Andersen impactor); viable bacteria and fungi	The total concentration of the IgG antibodies for *Tr. toseum* was lower at the end than at the beginning of the follow-up and this remained significant for the group of teachers with sinusitis. The decrease in IgG to certain molds was associated with bronchitis.	Patovirta et al. 2003; *Int J Occup Med Environ Health* 16(3):221–230

(continued)

Table 12.1 (continued)

Type/objective of study	Study sites and design	Microbial sampling and analyses	Main findings	Reference
Respiratory symptoms, production of inflammatory mediators and clinical samples of adults in index and reference school	A moisture damaged elementary school (*N* = 37 teachers) and a reference school (*N* = 23) in Finland	Air samples (Andersen impactor); viable bacteria and fungi; surface and material samples	Significantly higher airborne fungal levels in moisture damaged school with moisture damage related microbial genera. Also significant elevation of IL-1, TNF-alpha and IL-6 in nasal lavage fluid and IL-6 in induced sputum during the spring term in the subjects from the moisture-damaged school building. In addition, the exposed workers reported sore throat, phlegm, eye irritation, rhinitis, nasal obstruction and cough.	Purokivi et al. 2001; *Eur Respir J* 18(6):951–958
Association between day-to-day changes in personal endotoxin exposure and asthma severity	Schoolchildren with asthma, aged 6 to 13 years from elementary school	Particulate air samples (PM_{10} and $PM_{2,5}$), personal and stationary sampling (indoors and outdoors); endotoxin	Personal exposures did not correlate with and were significantly higher than endotoxin measurements from either indoor or outdoor stationary monitors. Increases in personal endotoxin exposures were associated with decreased lung function and increased symptoms.	Rabinovitch et al. 2005; *J Allergy Clin Immunol* 116 (5):1053–1057
Moisture-damage repairs on the exposure and on the upper airway inflammatory responses	32 school buildings in central Finland	Air samples before and after repairs; qPCR, nasal lavage samples, PM and NO	After the repairs, concentrations of the measured airborne microbes decreased. Also airborne PM caused significantly lower production of IL-6 and TNF-alpha in mouse macrophages, and IL-4 in NAL samples were significantly lower after the renovation.	Roponen et al. 2013; *Indoor Air* 23(4):295–302
Association of in vitro toxicity of indoor samples with teachers' health symptoms	15 schools (170/210 classrooms) in Southern Finland	Settled dust swabs and "fallout" agar plates; in vitro toxicity in a boar sperm motility assay	Classrooms (15) with highly toxic dust samples showed significant association with teachers' self-reported symptoms against a reference group (11) with low toxicity; similarly, classrooms with more toxic agar plate samples associated with teachers' symptoms.	Salin et al. 2017; *Environ Res* 154:234–239

(continued)

Table 12.1 (continued)

Type/objective of study	Study sites and design	Microbial sampling and analyses	Main findings	Reference
Association of fungal exposure with children's IgG response to water damage microbes, and associations with children's health effects	A moisture damaged elementary school ($N = 365$ pupils) and a reference school ($N = 176$ pupils), in Finland	Air samples (Andersen impactor), surface and material samples; viable bacteria and fungi	The mean number of positive IgG findings was significantly higher among exposed children; no correlation between the number of positive IgG antibodies and respiratory illnesses or symptoms. In the exposed group, a negative correlation was found between the number of positive IgG antibodies and the total value of allergen-specific IgE antibodies.	Savilahti et al. 2002; *Pediatr Allergy Immunol* 13(6):438–442
Indoor molds in schools and health effects in schoolchildren	21 schools (46 classrooms) in Italy, Denmark, Sweden, Norway and France; 654 school children	Fungal and bacterial DNA targets from vacuumed dust; viable molds from air samples	Viable molds and total fungal DNA higher in buildings with mold/dampness problems; Viable molds but not fungal DNA were inversely related to ventilation rate; *Aspergillus/Penicillium* DNA was positively associated with wheeze, *Aspergillus versicolor* with wheeze, rhinitis and cough; negative (protective) associations of *A. versicolor* and *Streptomyces* with lung function measures.	Simoni et al. 2011; *Pediatr Allergy Immunol* 22:843–852
Current asthma in pupils and association with school environment	11 schools, 28 classrooms, 7th grade pupils ($N = 762$) in Sweden	Vacuumed dust: endotoxin, cat, mite and dog allergens; air samples for microbes (filters and growth media)	More current asthma was associated with schools that were larger, had more open shelves, lower room temperature, higher relative air humidity, higher concentrations of formaldehyde or other volatile organic compounds, viable molds or bacteria or more cat allergen in the settled dust.	Smedje et al. 1997; *Clin Exp Allergy* 27 (11):1270–1278
Associations of airborne microbes with health effects	8 daycare centers, 8 office buildings, and 8 homes, in the Taipei area	Air samples (Andersen impactor); viable bacteria and fungi, (three-piece cassette); endotoxin and (1-3)-β-D-glucans	Viable microbial levels were highest in daycare centers, followed by homes and office buildings. Similar patterns were observed for endotoxin and (1-3)-β-D-glucans. A strong association between beta-(1-3)-β-D-glucans and lethargy/fatigue was found.	Wan and Li 1999; *Arch Environ Health* 54 (3):172–179

(continued)

Table 12.1 (continued)

Type/objective of study	Study sites and design	Microbial sampling and analyses	Main findings	Reference
Microbial components in relation to SBS symptoms in pupils	10 junior high schools (33 classrooms) in Taiyuan, China; 1143 pupils	Vacuumed dust: 3-hydroxy fatty acids, muramic acid, ergosterol; settled dust: cat and dog allergens, fungal DNA	New onset of mucosal symptoms was negatively associated with concentration of muramic acid, total LPS, and shorter lengths of 3-hydroxy fatty acids. Onset of general symptoms was negatively associated with C18 LPS. Onset of school-related symptoms was negatively associated with C16 LPS, but positively associated with total fungal DNA.	Zhang et al. 2011; *Sci Tot Environ* 409:5253–5259
Associations of microbial exposure with asthmatic symptoms in pupils	10 junior high schools in Taiyuan, China; 39 classrooms; 1993 pupils	Vacuumed dust; 3-hydroxy fatty acids, muramic acid, ergosterol	Negative association of muramic acid and wheeze and daytime breathlessness; total LPS positively associated with daytime breathlessness, but negative associations of lower chain length LPS with wheezing (i.e. effect of LPS varies with type of LPS exposure).	Zhao et al. 2008; *Pediatr Allergy Immunol* 19:455–465
Determinants of microbial exposures in schools				
Determinants of airborne bacterial concentrations in schools	39 elementary schools in Canada	Air samples (Andersen impactor); viable bacterial counts	Naturally ventilated rooms had higher bacterial levels than mechanically ventilated ones. Bacterial counts were negatively correlated with supply and exhaust flow rates, air exchange rates, and occupant activity. Bacterial groups indicated various sources of the bacterial aerosol, including occupancy and site variables.	Bartlett et al. 2004a; *J Occup Environ Hyg* 1 (10):639–647
Evaluation of airborne fungal concentrations and determinants	39 elementary schools in Canada	Air samples (Andersen impactor); viable fungi	Season affected significantly fungal concentrations and composition. Mechanically ventilated rooms had significantly lower concentrations than naturally ventilated ones. Environmental determinants accounted for 58% of the variation in fungal concentrations.	Bartlett et al. 2004b; *Ann Occup Hyg* 48 (6):547–554

(continued)

Table 12.1 (continued)

Type/objective of study	Study sites and design	Microbial sampling and analyses	Main findings	Reference
Examine the diversity and concentration of molds in inner-city schools	12 schools, 180 classrooms in Boston, USA	Active air collection with Burkard Indoor Recording, mold spore counts	Classrooms within the same school had differing mold levels and diversity scores; classroom accounted for the majority of variance in the total mold count and mold diversity score (62% and 56%). Visible mold may be a predictor for higher mold spore counts. Mold concentrations were significantly higher in fall versus spring.	Baxi et al. 2013; *Pediatr Allergy Immunol* 24 (7):697–703.
Fungal DNA and associations with building characteristics and allergen levels	Allergen avoidance (*N* = 11) and ordinary day care centers (*N* = 11) in Sweden, 70 rooms monitored	Swab-samples, floor dust and settled dust; total fungal DNA, cat, dog, dust- mite and horse allergens	Linoleum floor material, carpets, reported dampness/mold and odor were increasing the level of total fungi significantly. Levels were decreased by wooden construction. Total fungal DNA was related to cat, dog, horse and total allergen levels.	Cai et al. 2009; *Indoor Air* 19(5):392–400
Relationship between building characteristics and IAQ	Day care centers in Sweden (*N* = 21), 103 rooms monitored	Settled dust; fungal DNA targets (qPCR)	Total fungal DNA levels were higher in risk construction buildings with increased risk of high humidity within the building fabric, in rooms with linoleum flooring and rotating heat exchangers; associations between total fungal DNA and some animal allergens.	Cai et al. 2011; *J Environ Monit* 13:2018–2024
Establish indoor pollution levels for IAQ benchmarking	18 classrooms from 5 primary schools and one nursery in the Greater London area.	Vacuumed settled dust from above floor level; qPCR targeting several fungal and bacterial groups	No differences in fungal and bacterial levels between urban and suburban schools, except for Mycobacterium spp.Underfloor heating with the presence of carpets increase levels of *Penicillium/ Aspergillus* spp. and *Aspergillus versicolor*. Higher levels of some microbes in nurseries compared to primary school classrooms.	Chatzidiakou et al. 2015; *Intell Build Int* 7 (2–3):130–146.

(continued)

Table 12.1 (continued)

Type/objective of study	Study sites and design	Microbial sampling and analyses	Main findings	Reference
Associations between observational dampness scores and measurements of microbial agents	Primary, middle and high schools ($N = 3$) in northeastern USA, 121 rooms monitored	Floor dust; total viable fungi, viable bacteria, ergosterol, glucans, muramic acid, endotoxin	Dampness scores associated significantly with culturable bacteria and mixed microbial exposure indices (MMEI). Rooms with higher dampness scores had significantly higher levels of most microbial levels, MMEIs, relative moisture content and recent water leaks.	Cho et al. 2016; *Indoor Air* 26(2):168–178
Association of carpeting with biocontaminant levels	1 tiled school, 1 carpeted school in North Carolina, USA	Air samples, floor dust; viable fungi, spores, dust mass (PM2.5), endotoxin, β-1,3 glucans	The carpeted surfaces generally had higher surface loadings of the biocontaminants, while the airborne levels were significantly higher over tiled floors, suggesting that carpet flooring was not the major contributor to airborne levels of biocontaminants.	Foarde and Berry 2004; *J Expo Anal Environ Epidemiol* 14(Suppl 1):41–48
Effect of occupancy on microbial levels	Occupied and unoccupied classrooms	Air samples; muramic acid and 3-hydroxy fatty acids	Levels of muramic acid and 3-hydroxyfatty acids were higher (range 5- to 50-fold) in occupied rooms than in unoccupied school rooms.	Fox et al. 2003; *J Environ Monit* 5(2):246–252
Monitoring of air quality in German schools.	92 and 75 classrooms in Bavaria, Germany in two seasons	Settled dust samples; endotoxin determinations (LAL assay)	Levels of endotoxin varied significantly between the sampling periods, but were independent of room surface area, room volume and flooring. Levels were generally considered low exposure levels.	Fromme et al. 2008; *Das Gesundheitswesen* 70 (2):88–97
Characterize bacterial and fungal microbiome in indoor environments and outdoor air samples	1 school/classroom; several house dust and outdoor air particulate matter samples	Vacuumed floor and settled dust sample from multiple surfaces in the classroom; bacterial 16S and fungal ITS sequencing	Indoor dust and outdoor air particulate matter harbor remarkably different microbial profiles (eg. Gram positive Firmicutes and Actinobacteria and Ascomycota indoors versus Gram negative Proteobacteria and Basidiomycota outdoors). Similar bacterial profiles but different fungal profiles in school vs home dust.	Hanson et al. 2016; *Environ Sci Process Impacts* 18(6):713–724

(continued)

Table 12.1 (continued)

Type/objective of study	Study sites and design	Microbial sampling and analyses	Main findings	Reference
Effect of human occupancy on indoor airborne bacteria	A university classroom while occupied and when vacant in the northeastern USA	Ventilation filter dust, floor dust, total PM; qPCR and sequencing of bacterial 16S rDNA	Resuspended floor dust is an important contributor to bacterial aerosol populations during occupancy. Direct human shedding may also significantly impact the concentration of indoor airborne PM.	Hospodsky et al. 2012; *PLoS One* 7:e34867
Association of emission rates with fungal and bacterial concentrations	6 classroom in 4 different countries	Air samples (Andersen impactor); viable bacteria and fungi, qPCR	In heavily occupied classrooms, human occupancy was significantly increasing the concentrations of both airborne bacteria and fungi. Occupied indoor/outdoor (I/O) ratios consistently exceeded vacant I/O ratios.	Hospodsky et al. 2015; *Indoor Air* 25 (6):5098–5106
Association between microbial components and inflammatory responses in moisture-damaged vs. non-damaged schools	14 moisture-damaged schools and 11 reference schools in Spain, The Netherlands, Finland	Settled dust samples; inflammatory markers, selected microbial groups, fungal and bacterial cell wall markers	Dust from moisture-damaged schools in Sp and Nl, but not Fi induced stronger immunotoxicological responses compared to samples from reference schools. IL-6 and apoptosis responses were most strongly associated with moisture status. Muramic acid concentration predicted immunotoxicological potential of dust.	Huttunen et al. 2016; *Indoor Air* 26(3):380–390
Endotoxin levels and associations with potential determinants	23 schools in Spain, The Netherlands and Finland, 237 classrooms	Settled dust (EDC) (645 samples); endotoxin	Endotoxin levels were higher in lower grades and in classrooms with higher occupancy. There was geographical and seasonal variation in endotoxin levels.	Jacobs et al. 2014; *Indoor Air* 24(2):148–157
Microbial exposures in different microenvironments	44 classrooms at 11 elementary schools, 42 bars, 41 Internet cafes, and 20 homes	Air samples (Andersen impactor); viable bacteria and fungi	Individual fungi occurred in this order: *Cladosporium*, *Penicillium*, *Aspergillus*, and *Alternaria*. Indoor and outdoor microbial levels were affected by: microenvironment type, time of sampling in elementary school classrooms, seasonal variation and facility location.	Jo and Seo 2005; *Chemosphere* 61 (11):1570–1579

(continued)

Table 12.1 (continued)

Type/objective of study	Study sites and design	Microbial sampling and analyses	Main findings	Reference
Effect of architectural design on indoor bacterial communities	Four-story classroom and office building (155 spaces) in Oregon, USA	Settled dust; sequencing of the bacterial 16S gene	Most communities were dominated by *Proteobacteria*, *Firmicutes*, and *Deinococci*. Architectural design characteristics related to space type, building arrangement, human use and movement, and ventilation source had a large influence on the structure of bacterial communities.	Kembel et al. 2014; *PLoS One* 9:e87093
Determination of allergen and glucan levels in schools and homes	19 schools and 169 homes (of respective pupils) in The Netherlands	Airborne settled dust collected with EDCs; immunological assay for determination of beta-1,3-glucans	Glucan levels were highest in classrooms compared to other locations in the schools, impacted by season, socioeconomic status, moisture damage and pet ownership of the pupils. Repeated measurements showed significant variation of levels over time. Exposure to glucans was about twofold higher in schools compared to homes.	Krop et al. 2014; *PLoS One* 9(2):e88871
Monitoring of indoor air parameters in Hong Kong schools	5 classrooms in different schools in Hong Kong	Various environmental determinations, including airborne viable bacteria	Average bacterial counts were below Hong Kong IAQ guidelines using 1000 CFU/m^3. Indoor samples had lower bacterial counts compared to outdoor samples.	Lee and Chang 2000; *Chemosphere* 41:109–113
Viability and diversity of bacteria in a child-care center	1 day care center (4 classrooms)	Surface swabs from toys and furniture; cultivation for bacteria and 16S rRNA gene sequencing	Combination of culture and culture-independent methods is powerful for determining viability/diversity. *Bacillus* spp., *Staphylococcus* spp. and *Microbacterium* spp. were most common in culture; sequencing predicted 190 bacterial species, dominated by *Pseudomonas* spp.	Lee et al. 2007; *BMC Microbiology* 7:27
Measurement of viable bacteria and cell wall components in indoor air of school buildings	2 elementary schools in the USA; 18 classrooms	Vacuumed surface dust and active air samples in- and outdoors; viable bacteria, muramic acid, 3-hydroxy fatty acids	Levels of bacteria and their cell wall components are correlated with total surface dust and total airborne particles (TSP) in classrooms. High correlations between levels of CO_2 and levels of bacteria and their cell wall constituents indicate correlation of bacterial measures with occupancy.	Liu et al. 2000; *J Air Waste Manag Assoc* 50 (11):1957–1967

(continued)

Table 12.1 (continued)

Type/objective of study	Study sites and design	Microbial sampling and analyses	Main findings	Reference
Size distributions of viable fungi and bacteria in schools and impact of moisture damage and building frame	32 school buildings in Finland	Active air sampling for viable bacteria and fungi using a cascade impactor (Andersen impactor)	Highest fungal levels in the size range of 1.1–4.7 μm; concentrations of fungi in all size classes higher in wooden as compared to concrete schools. Moisture damage associated differences were observed only in concrete schools in the size range of 1.1–2.1 μm. Mean diameter of total viable fungi was smaller in wooden compared to concrete and moisture damaged compared to reference schools.	Meklin et al. 2002b; *Atmos Environ* 36: 6031–6039
Association of microbial IAQ with children's health in moisture damaged versus reference schools.	32 school buildings in central Finland	Airborne sampling (Andersen impactor); viable fungi and bacteria	Clear effect of moisture damage on levels of viable fungi in concrete/brick constructions, but not in wooden schools. Occurrence of *Cladosporium*, *Aspergillus versicolor*, *Stachybotrys* and actinobacteria showed some indicator value for moisture damage.	Meklin et al. 2002a; *Indoor Air* 12:175–183
Effect of building frame material and moisture damage on microbes	17 wooden and 15 concrete/brick school buildings, Finland	Air samples (Andersen impactor); viable bacteria and fungi	Mean concentrations of viable airborne fungi were significantly higher in wooden schools than in concrete schools. Clear effect of moisture damage on levels of viable fungi in concrete/brick constructions, but not in wooden schools.	Meklin et al. 2003; *AIHA J (Fairfax, Va)* 64 (1):108–116
Effect of ventilation on microbial levels in child day care centers	9 day care centers, 52 classrooms in Portugal	Air samples, floor dust samples; total bacteria and fungi, chemicals, PM, house dust mites	Total bacterial concentrations were >50-fold higher in the nursery and kindergarten than outdoors. Gram-negative bacteria were associated with low airflow rates.	Mendes et al. 2014; *J Toxicol Environ Health A* 77(14–16):931–943
Sources, seasonal and diurnal variations of bioaerosol levels	2 houses, 2 offices, 1 kindergarten and 1 primary school; 14 sampling sites, Turkey	Air samples; viable bacteria and fungi (Andersen impactor), volatile organic compounds and PM	The highest bacterial levels were observed in classrooms. Bioaerosol levels were higher in winter than in summer. The levels showed remarkable daily and diurnal variations. Bacterial levels were higher than fungal levels.	Mentese et al. 2012; *J Air Waste Manag Assoc* 62 (12):1380–1393

(continued)

Table 12.1 (continued)

Type/objective of study	Study sites and design	Microbial sampling and analyses	Main findings	Reference
Prevalence of microbial secondary metabolites in school buildings in three European countries	97 school buildings in The Netherlands, Finland and Spain; 445 classrooms, in total 741 samples	Settled dust and surface swabs; 186 microbial secondary metabolites determined with HPLC-Mass spectrometry	More than 50% of samples were positive for at least one of the targeted metabolites; prevalence of individual metabolites was typically low; 30 different metabolites detected. Differences in metabolite profiles were observed for different countries and also when comparing school buildings with and without moisture damages.	Peitzsch et al. 2012; *J Environ Monit* 14:2044–2053
Effect of occupancy on size-resolved emission rates of airborne microbes	A university classroom while occupied and when vacant in the USA	Air samples (8-stage impactor); qPCR and 16SrDNA sequencing	Significant increases of total particle mass and bacterial genome concentrations were observed during the occupied period compared to the vacant case. About 18% of the bacterial emissions were from taxa that are closely associated with the human skin microbiome.	Qian et al. 2012; *Indoor Air* 22(4):339–351
Effect of children's occupancy and performance of cleaning procedures on endotoxin and allergen levels	60 sites in Brazil: 15 day care centers, 15 preschools, 15 kindergartens, 15 elementary schools	Floor dust, mattress dust, vacuumed (300 samples); allergens (157 samples); endotoxin	Levels of endotoxin in day care centers and preschools were three times higher than in elementary schools. Highest levels of group 1 mite allergens were detected in day care centers and preschool floors; levels of the cockroach allergen were highest in elementary schools.	Rullo et al. 2002; *J Allergy Clin Immunol* 110 (4):582–588
Explore levels of submicron particles in elementary schools and determinants	70 classrooms from 8 elementary schools, in- and outdoor measurements	(1,3)-β-D-glucan in submicron fragments, airborne mold and bacteria, PM10	Indoor/outdoor ratios of glucan in submicron fragments were >1 in all schools. Post rainy season, PM10, airborne mold and bacteria and glucan concentrations decreased. Relative humidity was negatively correlated with concentration of submicron fungal fragments.	Seo et al. 2015; *Environ Sci Process Impacts* 17 (6):1164–1172.

(continued)

Table 12.1 (continued)

Type/objective of study	Study sites and design	Microbial sampling and analyses	Main findings	Reference
Exposure to bioaerosols in childcare facilities	5 day care centers and five elementary schools in Seoul, Korea	Air samples (filter cassettes, 50 samples); fungal and bacterial pyrosequencing	The bacterial community in indoor air appeared to contain diverse bacteria associated with both humans and the outside environment. In contrast, the fungal community was largely derived from the surrounding outdoor environment and not from human activity.	Shin et al. 2015; *PLoS One* 10:e0126960
Study the influence of furnishings and cleaning on the indoor air quality of schools.	181 classrooms in 48 schools in Sweden	Various environmental determinations, including airborne viable molds and bacteria	Classrooms cleaned with wet mopping had more viable airborne bacteria, but less settled dust. Furnishings and textiles in the classroom act as reservoirs of irritants and allergens.	Smedje and Norback 2001; *Indoor Air* 11 (2):127–133
Microbial exposure and association with geography, rural location and season	Schools in Jordan, Sweden and Poland ($N = 6$), urban and rural environments	Settled dust (85 samples); ergosterol, 3-hydroxy fatty acids	Levels of ergosterol and 3-hydroxy fatty acids were differing between countries, between seasons, and between urban and rural environments.	Wady et al. 2004; *J Expo Sci Environ Epidemiol* 14 (4):293–299
Association between allergenic fungal aerosol particles and indoor emissions	7 classrooms in four different countries	Air samples (allergenic fungal aerosol particles) indoors and outdoors; qPCR and multiplexed DNA sequencing	76% of indoor fungal aerosol particles and 80% of airborne allergenic fungal taxa were associated with indoor emissions, mostly relating to occupant-generated emissions. Indoor emissions affect more allergenic fungal exposures in classrooms than outdoor contributions. Geographical variations in fungal communities were found.	Yamamoto et al. 2015; *Environ Sci Technol* 49 (8):5098–5106
Pathogens, viruses, hygiene				
Evaluation of keratinophilic fungi in 24 kindergartens in Jordan.	72 classrooms in 24 kindergarten schools	Floor dust; cultivation of viable keratinophilic fungi	33 species of keratinophilic fungi were isolated form the classrooms, including many known human and animal mycotic agents and potential pathogens.	Ali-Shtayeh and Al-Sheikh 1988; *Mycopathologia* 103 (2):69–73

(continued)

Table 12.1 (continued)

Type/objective of study	Study sites and design	Microbial sampling and analyses	Main findings	Reference
Bacteria and viruses in hygiene intervention and control classrooms	6 elementary classrooms, 3 with disinfecting wipes intervention	Swabs; heterotrophic bacterial plate count (cultivation), norovirus and influenza A (RT-PCR)	Frequently contacted surfaces were the most contaminated. Children in the control classrooms were more likely to report absenteeism and for extended periods. Improved classroom hygiene may reduce the incidence of infection and student absenteeism.	Bright et al. 2010; *J School Nurs* 1:33–41
Determine the prevalence of pathogenic bacteria in daycare centers	Daycare centers, 15 locations in each daycare center	Swabs; bacteria, respiratory viruses, gastrointenstinal viruses	Coliform bacteria were found in the toilet and kitchen; nasopharyngeal bacteria on toys and fabric surfaces; respiratory viruses were omnipresent in day care centers, especially on toys.	Ibfelt et al. 2015; *J Environ Health* 78 (3):24–29
Other studies on microbial exposure in school environments				
Evaluation of toxicity of actinomycete isolates from indoor air from schools and day cares	Several actinomycete strains isolated from air samples	Evaluation of the toxicity of actinomycete strains using boar spermatozoa assay	Toxicity of several indoor isolates was established and valinomycin production of a *Streptomyces griseus* strain was described.	Andersson et al. 1998; *Appl Environ Microbiol* 64(12):4767–4773.
Study the cultivable dust and airborne bacterial diversity in schools and day care centers	9 schools and day care centers in Southern Finland with history of water damage	Settled dust; cultivation and identification of bacterial strains	Bacterial flora cultivated from settled dust in day care centers and schools were dominated by Gram positive bacilli and actinomycetes.	Andersson et al. 1999; *J Appl Microbiol* 86:622–634
Evaluate presence, quantity and origin of potentially toxic airborne substances	25 buildings, including some schools and daycare centers (10 rooms)	Various indoor aerosols and dusts; analysis of toxicity with boar spermatozoa assay	"Toxic" aerosols were found from various indoor locations with reported or suspected building related ill health, including daycares and classrooms.	Andersson et al. 2010; *Toxicol In Vitro* 24 (7):2041–2052.

(continued)

Table 12.1 (continued)

Type/objective of study	Study sites and design	Microbial sampling and analyses	Main findings	Reference
Characterize fluorescent biological aerosol particle (FBAP) levels in a classroom	1 university classroom monitored during occupied and non-occupied periods	Real-time monitoring of FBAPs (1–15 μm) with a UV aerodynamic particle sizer	Between days and seasonal changes in FBAP number concentrations were small compared to within day variation (attributable to occupancy, occupant behavior, and ventilation system operation). Emissions from human occupants were determined as a strong source of FBAPs peaking at 3–4 μm, characteristic for airborne particles with associated microbes.	Bhangar et al. 2014; *Indoor Air* 24:604–617
Diurnal variations in indoor microbes in child day care, effect of disinfection	2 day care centers in Tainan City, Taiwan; 2 classrooms	Burkard air sampling for viable bacteria and fungi	Some diurnal variation in indoor microbes, with highest bacterial levels at noon and highest fungal levels in the morning and at noon; variation between weekdays. Partly significant reduction in bacterial and fungal levels post disinfection activities.	Chen et al. 2012; *J Environ Monit* 14:2692–2697
Associations between microbial exposure and respiratory symptoms using individual exposure versus group means	4 cross-sectional studies on occupants of water damaged buildings	Floor dust; culturable fungi, ergosterol, endotoxin	Using group-average exposure estimates provides less attenuated associations between microbial exposures and health symptoms in damp buildings as compared to individual exposure measurements. In particular buildings, floor was found to be useful in constructing exposure groups for microbial agents in epidemiological analyses.	Cho et al. 2013; *J Exp Sci Environ Epidemiol* 1–7
Follow-up aerosol monitoring by particle size in school air – effect of occupancy	3 schools	Follow-up aerosol monitoring by particle size	The particle size distribution was shown to be quite different in occupied versus unoccupied schoolrooms. Within individual classrooms, concentrations of airborne particles ≥0.8 μm in diameter, and CO_2 were correlated, suggesting that the increased levels of larger particles are responsible for elevation of bacterial markers during occupation.	Fox et al. 2005; *J Environ Monit* 7(5):450–456

(continued)

Table 12.1 (continued)

Type/objective of study	Study sites and design	Microbial sampling and analyses	Main findings	Reference
Investigate how microbial conditions of kitchen facilities differ from those in other school facilities	8 schools in Finland, measurements in kitchen and other facilities	Airborne sampling with Andersen impactor; viable bacteria	Total concentrations of airborne microbes were lower in kitchens than in other facilities. Moisture damage increased microbial concentrations in kitchen and other locations. Gram negative bacteria dominated in kitchen, but also thermophilic and mycobacteria.	Lignell et al. 2005; *Int J Occup Med Environ Health* 18(2):139–150
Assessing indoor fungal levels	20 primary schools (73 classrooms) in Portugal	AirIdeal portable air sampler; viable fungi	The most prevalent fungi found indoors were *Penicillium* spp. (>70%) and *Cladosporium* spp. Indoor fungal concentrations were higher than outdoors.	Madureira et al. 2014; *J Toxicol Environ Health A* 77:816–826
Pupils symptoms versus microbial exposure and other indoor air pollutants in schools and homes in Portugal	20 primary schools, 1600 children; homes of a subsample of 68 children (asthmatics and non-asthmatics)	Air samples indoors and outdoors; viable bacteria and fungi. Monitoring of VOC, aldehydes, CO, PM, CO_2	Bacterial concentrations did not differ significantly between case and control homes. Fungal concentrations were higher in control homes. Bacterial levels were higher in schools compared to homes, while highest fungal levels were observed in homes.	Madureira et al. 2015; *J Toxicol Environ Health A* 78(13–14):915–930
Analyses of bacterial microbiota from different surfaces in a classroom	12 university classroom; swabs from four different surfaces (desks, chairs, floors, walls; $n = 70$)	Surface swab samples using nylon flocked swab; bacterial 16S amplicon sequencing	Human-associated microbial communities can be transferred to indoor surfaces following contact; proximity to other surfaces in the classroom does not influence community composition.	Meadow et al. 2014; *Microbiome* 2:7
Comparison of sampling methods	Day care center (5 rooms) and office building (8 rooms) in France	Electrostatic dust collector (EDC), air sampling; culturable fungi	Significant correlation in fungal levels using EDC sampling and repeated air-impaction. Similarities between the fungal species isolated from EDCs and multiple air-impaction samples. Poor reliability of a single air-impaction measurement.	Normand et al. 2016; *Ann Occup Hyg* 60 (2):161–175

(continued)

Table 12.1 (continued)

Type/objective of study	Study sites and design	Microbial sampling and analyses	Main findings	Reference
Indoor air mold distribution in elementary schools in Turkey	18 classrooms from schools in Denizli, Turkey	Active air samples; cultivation and identification of fungi	Molds belonging to ten different genera, most prominently *Penicillium*, *Aspergillus Cladosporium* and *Alternaria*, were isolated from classrooms.	Ovet et al. 2012; *Mirobiyol Bul* 46 (2):266–275
Measurement of total airborne concentrations of virus-like and bacteria-like particles	Several locations in Virginia, USA, including a classroom, daycare center and outdoors	Active air sampling onto filter membrane; staining of DNA and RNA and count of virus-like and bacteria-like particles	No significant differences in total counts of virus-like (VLP) and bacteria-like particles (BLP) between different indoor environments; virus to bacteria ratio was 0.9. VLP and BLP concentrations were 2.6 and 1.6 times higher outdoors than in indoor air.	Prussin et al. 2015; *Environ Sci Technol Lett* 2(4):84–88
Case report of a school with an unusual combination of indoor air contaminants	One school in Finland	Various microbial and chemical exposure measurements	Case report describing the combined exposure to chemicals and microbes in a school building. Results indicate that chemical contaminants were the source of indoor air complaints and that remediation improved IAQ and health status of children.	Putus et al. 2004; *Arch Environ Health* 59 (4):194–201
Measurements of biological contaminants in two elementary schools.	2 schools (Minneapolis, USA) that serve inner city minority populations	Airborne and carpet viable fungi	The school environment can be an important exposure source of fungi for children.	Ramachandran et al. 2005; *J Occup Environ Hyg* 2(11):553–566.
Comparing inner-city endotoxin exposure to exposure in homes in children with asthma	12 urban elementary schools, homes of students with asthma	Settled and airborne dust samples from schools ($N = 229$) and homes ($N = 118$); endotoxin	Endotoxin levels were almost twice as high in schools than in homes.	Sheehan et al. 2012; *Ann Allergy Asthma Immunol* 108(6):418–422

(continued)

Table 12.1 (continued)

Type/objective of study	Study sites and design	Microbial sampling and analyses	Main findings	Reference
Comparing the personal exposure to particles and bioaerosols measured by stationary samplers in the main microenvironments	Elementary school teachers (N=81) from two cities in Eastern Finland	Personal air sampling and micro-environmental air sampling in homes and workplaces; PM, viable and total microbes	Personal microbial exposure is higher than exposure in home and workplace. Concentrations usually lower in homes compared to workplaces. The fungi and bacteria results showed high variation, which emphasizes the importance of quality assurance (duplicates and field blanks).	Toivola et al. 2002; *J Environ Monit* 4 (1):166–174
Compare personal exposure to particles and microbes in schools and homes.	38 schools and 81 corresponding homes of teachers in Eastern Finland	Active air sampling with Button samplers, including personal air, stationary air in the home and in the school; PM, viable bacteria and fungi	Bacterial and fungal levels were higher at the workplace compared to homes. Significant correlations were found for microbial measurements between work and home environments. Stationary samples are only surrogates of personal exposure. Time-weighted micro-environmental model is useful to assess exposure to some microbes, but may underestimate personal exposure to others.	Toivola et al. 2004; *Indoor Air* 14(5):351–359
Evaluation of a new passive settled dust sampling device in a Danish school.	85 classrooms in 15 Danish primary schools (with and without moisture damage)	Air samples (IOM sampler), airborne settled dust, floor dust; viable molds; (1-3)-β-D-glucan	Measurements of culturable fungi from air samples correlated with airborne settled dust in mechanically ventilated classrooms. Floor dust and settled dust measurements did not correlate well. Mass concentration of culturable fungi from airborne settled dust did correlate with the degree of dampness in the schools.	Wurtz et al. (2005); *Indoor Air* 15(9):33–40

health, using observational measures such as visible mold, dampness indicators, observations of moisture damage or mold odor, without including microbial measurements in these assessments. Following our definitions for this chapter, i.e. focusing on studies that have included actual measurement of microbes in schools or daycares, this literature is not included in the overview of Table 12.1. However, the issue of moisture damage in buildings is a highly relevant one, and the related, health relevant exposures are assumed to be mostly or at least in part of microbial origin. Therefore we want to briefly summarize in the following the current evidence on moisture damage in schools and related health effects.

Adverse health outcomes associated with moisture damage, dampness and mold in buildings have been extensively reported, reviewed and evaluated (for example WHO Regional Office in Europe 2009, Institute of Medicine 2004, Mendell et al. 2011; Kanchongkittiphon et al. 2015). The evidence is consistent for an association between moisture damage and dampness in buildings and various respiratory symptoms, respiratory infections, exacerbation and new onset of asthma. While it is well established that living or working in moisture damaged buildings increases the occupant's risk for ill (respiratory) health, it is also acknowledged that our understanding of the underlying mechanisms and knowledge of causal agents is still very limited. Moisture damage and dampness in buildings can cause microbial proliferation and subsequent increased exposure of building occupants to various microbial and chemical agents involved in these processes, including fungal and bacterial spores, cell fragments, metabolites and mycotoxins, and chemical emissions from building materials (WHO Regional Office in Europe 2009). These exposures are widely considered to be key in contributing to the observed adverse health outcomes. Moisture damage, dampness and mold are prevalent in the building stock across countries, with prevalence ranging from 10% to 50% as estimated in two of the reviews above (WHO Regional Office in Europe 2009, IOM Institute of Medicine 2004). A recent study carried out in schools in three European countries – Spain, Finland and The Netherlands – estimated prevalence of moisture problems with 41%, 24% and 20%, respectively (Haverinen-Shaughnessy et al. 2012).

Observations of moisture damage, dampness and mold have been associated with respiratory symptoms and asthma and atopy (one study) among school children in several studies (Meklin et al. 2002a, 2005; Savilahti et al. 2000, 2001; Sahakian et al. 2008; Mi et al. 2006; Borrás-Santos et al. 2013). Building frame, i.e. wooden versus concrete construction, and country were found to effect the associations between mold observations and respiratory symptoms in two of these studies. Interventions, that is here remediation of the moisture and mold damage, has shown an improvement in symptom prevalence as compared to the pre-remediation situation in a few studies (Savilahti et al. 2000; Meklin et al. 2005; Haverinen-Shaughnessy et al. 2004). One of these studies indicated that part of the health effects of exposure to moisture damage and dampness may be short-term and reversible, once the exposure is ceased. A systematic review of Sauni et al. (2015) on the impact of mold remediation on reducing respiratory symptoms, infections and asthma concluded, however, that the evidence of a positive effect of a mold remediation in school environment on respiratory symptoms of children is inconsistent and that only prevalence of respiratory infections might decrease. This review also calls for more high quality studies with sufficient statistical power to investigate the effectiveness of moisture damage remediation on pupils' and teachers' health.

12.3 Studies on Microbial Exposures in Schools and Daycare Centers

Table 12.1 provides an extensive overview of studies that have considered indoor microbial exposures in schools and this table should serve as a source for further references for the interested reader. We have structured the table according to studies that analyzed health outcomes against microbial exposures, studies that have more specifically explored determinants of microbes in schools, and other studies that do not fit strictly into either of these two main categories. In the text below, we summarize some of the key aspects emerging from these reports.

12.3.1 Determinants

Moisture damage in buildings and its associations with microbial levels has been explored in numerous studies. Overall, these studies rather consistently find that moisture damage increases levels of some microbes or microbial markers in schools. There are few studies assessing the effect of renovation of moisture damaged schools on exposures that report significantly decreased microbial levels after intervention. Also outdoor environmental factors affect microbial levels in schools. There are studies showing geographical and seasonal effects on indoor microbial levels. Ventilation has an important role in filtering indoor emissions, transferring outdoor derived microbes indoors and also in resuspension of settled particles. As building factors, flooring and construction materials can affect microbial levels to a great extent. Human occupancy and activity are key factors when it comes to indoor sources of microbes. Not only human shedding, but also resuspension of floor dust during human activity have been shown to affect microbial levels. Pet ownership and farming activities can have an effect on microbes even in schools, as microbes can be transferred into school environments through e.g. clothes, shoes and hair.

12.3.1.1 Moisture Damage

Microbial levels in schools that are affected by dampness, mold and moisture damages have been assessed in many studies. The levels of different microbial agents - e.g. viable bacteria and fungi, endotoxin, muramic acid, ergosterol, (1–3)-β-D-glucans, bacterial and fungal DNA - have been found to be higher in both air and dust samples from the damaged schools and classrooms compared to non-damaged ones (Purokivi et al. 2001; Lignell et al. 2005; Cai et al. 2009; Simoni et al. 2011; Jacobs et al. 2014; Krop et al. 2014; Cho et al. 2016). Some of these studies in moisture damaged buildings report the presence of species, genera or microbial groups indicative for moisture and mold problems. Occurrence of e.g. *Cladosporium, Aspergillus versicolor, Eurotium, Stachybotrys* and actinobacteria have been shown to associate with moisture damage, when viable microbes from air samples have been assessed (Haverinen et al. 1999; Meklin et al. 2002a). In addition, the microbial secondary metabolite profiles, composed of mycotoxins and bacterial secondary metabolites, of affected school buildings are different from non-damaged buildings (Peitzsch et al. 2012). In an effort to assess and quantify moisture damage severity in classrooms, Cho et al. (2016) suggested the use of a dampness score, applying a 4-point-scale scores (0–3) of water damage, water stains, visible mold, moldy odor, and wetness for each of eight room components, such as ceiling, walls, windows, floor, etc. Also this study reports that rooms with higher dampness scores had significantly higher levels of most microbial levels in floor dust.

Moisture damages in buildings are caused by several factors, including failures in structures such as leakage on roofs or water pipes. Cai et al. (2011) found some risk constructions buildings with increased risk of high humidity within the building fabric, which affected the levels of total fungal DNA determined by qPCR in day care centers in Sweden. The type of building material seems to also have an effect on how microbes behave and proliferate, when dampness is available. Meklin et al. (2002a) saw clear effect of moisture damage on levels of airborne viable fungi in schools with concrete/brick construction, but failed to see so in wooden schools. The mean diameter of total viable fungi was smaller in moisture damaged compared to reference schools (Meklin et al. 2002b), possibly indicating differences in the mycoflora.

There are only a few studies published that have assessed the effect of repairing the moisture and mold damages in schools, and most of those studies suffer from small building and sample numbers. Lignell et al. (2007) studied viable microbial levels in air samples from two schools before the renovation, during the renovation and afterwards. Before the renovation, the microbial concentrations were higher in the index school than in the reference school. During the renovation the situation stayed similar. Afterwards, the microbial levels of the repaired index school decreased to the level of the reference school; the study also reports a change in the mycobiota in the damaged school following moisture thorough moisture damage repairs. Similarly, Roponen et al. (2013) found decreased concentrations of airborne microbial levels after repairs, when studying 32 moisture damaged school buildings in central Finland before and after intervention.

12.3.1.2 Other Indoor Sources

A recent study (Yamamoto et al. 2015) showed that indoor emissions affect more allergenic fungal exposures in classrooms than do outdoor contributions from ventilation, when assessed from air samples with qPCR and multiplexed DNA sequencing.

This research also concluded that, on average, 81% of allergenic fungi from indoor sources originate from occupant-generated emissions. For bacteria, Qian et al. (2012) observed that about 18% of the bacterial emissions originated from taxa that were closely associated with the human skin microbiome. Hospodsky et al. (2012) showed that not only human shedding, but also resuspension of floor dust during human occupancy affects the airborne bacterial levels considerably. The effect of occupancy has been evaluated also in other studies that found higher microbial levels in lower grades where pupils have supposedly higher activity levels and in classrooms with higher occupancy (Fox et al. 2003; Bartlett et al. 2004a; Jacobs et al. 2013; Hospodsky et al. 2015). Kembel et al. (2014) assessed bacterial communities with 16S rDNA sequencing and found that several of the bacterial taxa most strongly associated with restrooms as well as with high occupant density space types, such as classrooms. Bacterial communities in lower occupant density areas such as offices and mechanical support spaces in schools were more related to the outdoor environment, reflecting for example soil bacterial communities. According to Shin et al. (2015), the sources of bacterial and fungal communities in indoor air vary, as bacterial community contains diverse bacteria associated with both humans and the outside environment, while the fungal community seems to be largely derived from the surrounding outdoor environment and less from human activity. Pet ownership affects the indoor microbial and allergen levels even in schools, since the microbes and allergens are carried from homes to schools via e.g. clothes, shoes and hair. Krop et al. (2014) reported that fungal glucan levels in settled dust appeared to be impacted by pet ownership of the pupils. Similarly, farming can affect the microbial levels in schools and daycares. Wady et al. (2004) found that ergosterol and 3-hydroxy fatty acid levels differed between rural and urban environments. Chatzidiakou et al. (2015) did not find significant differences in microbial levels in settled dust determined by qPCR of schools located in urban and suburban areas, except for *Mycobacterium* spp, which showed higher levels in suburban schools.

The type of flooring has been shown to have an effect on microbial concentrations in school environments: linoleum flooring and carpets appear to increase microbial levels in classrooms (Foarde and Berry 2004; Cai et al. 2009, 2011). Fleecy and porous materials are great sinks for particles to settle and get attached to, which explains generally higher microbial levels in dust samples from carpet flooring. However, when matching floor dust and active air sampling in classrooms, Foarde and Berry (2004) found that while the carpeted surfaces generally had higher surface loadings of the biocontaminants, the airborne levels were significantly higher over tiled floors, suggesting that carpet flooring was not the major contributor to airborne levels of biocontaminants. Underfloor heating combined with the presence of carpets has been noticed to increase levels of *Penicillium/Aspergillus* spp. and *Aspergillus versicolor* determined with qPCR Chatzidiakou et al. (2015). Also the building structure has been noticed to affect microbial levels in school buildings in some studies, though with contradictory results. Meklin et al. (2003) found significantly higher concentrations of viable airborne fungi in wooden compared to concrete schools, whereas Cai et al. (2009) observed decreased fungal levels in dust samples collected from wooden school buildings.

12.3.1.3 Outdoor Environmental Factors

The outdoor air is an important determinant affecting indoor microbial levels in schools. There are several studies that have pointed out the effect of geographical variation and also seasonality on the variation of microbial levels in indoor dust and air samples (Bartlett et al. 2004b; Wady et al. 2004; Jo and Seo (2005); Mentese et al. 2012; Baxi et al. 2013; Jacobs et al. 2013, 2014, Krop et al. 2014; Yamamoto et al. 2015). A multi-center study evaluated endotoxin levels of settled dust from schools in Spain, The Netherlands and Finland (Jacobs et al. 2013). Lowest endotoxin levels were found in Finnish schools, where the snow covers the ground during winter time, and hence outdoor microbes do not affect much on indoor microbial levels during winter months. Wady et al. (2004) studied levels of different microbial agents of settled dust in Jordan, Sweden and Poland and found that the different agents behave differently when analyzing seasonal variation in different geographical regions. For example, the concentrations of fungal ergosterol in indoor dust were higher in summer than in winter for both Poland and Sweden, whereas the opposite was true for Jordan. For muramic acid, a marker of Gram-positive bacteria that are thought to be largely contributed by human presence in indoor spaces, the concentrations were higher in summer than in winter in all countries. In a study carried out in Turkey, the viable bacterial and fungal levels in air samples were higher in winter than in summer (Mentese et al. 2012), which could be due to the higher outdoor relative humidity and moderate temperatures during the winter months in the Mediterranean climate region, supporting the presence of airborne fungi outdoors. In a study carried out in Boston, US, the concentrations of major mold groups in active air samples in schools were significantly higher in fall versus spring (Baxi et al. 2013). During the fall season, the outdoor molds are often at a peak concentration in the Northeast United States. Yamamoto et al. (2015) made principal coordinate analysis, which revealed geographical variations in fungal communities among sites in China, Europe, and North America, demonstrating that geography may affect personal exposures to allergenic fungi.

The main route for the transfer of outdoor air to indoors is usually ventilation, either via mechanical ventilation systems or through open windows and doors, but air also infiltrates through leakages in the building envelope. Naturally ventilated schools have been noticed to contain higher microbial levels than schools with mechanical ventilation, which seems to be a due to more efficient filtering of microbes from incoming air and transfer of biocontaminants from indoors to outdoors in mechanically ventilated spaces (Bartlett et al. 2004a, 2004b). Ventilation type can also affect the *composition* of the microbiome of indoor air (Kembel et al. 2014). This study showed that specific bacterial taxa were associated with mechanically ventilated offices, while other taxa were more closely associated with window ventilated offices. Mendes et al. (2014) studied the effect of ventilation on airborne microbial levels in child care centers and schools and found that higher Gram-negative bacterial levels were associated with low airflow rates. Simoni et al. (2011) showed that viable mold concentrations in air samples were inversely related to ventilation rates. Insufficient ventilation poses a risk for poor indoor air quality, including biocontaminants, and hence can cause undesired health effects. Well planned and maintained ventilation decreases indoor pollutant levels effectively.

12.3.2 Health Effects

There is a good body of literature that has included microbial measurements in assessments of indoor exposures in schools and daycare centers on pupils' and/or teachers' health (Table 12.1). The microbial determinations in these studies rely on various different techniques, such as cultivation, determination of biomass by quantification of cell wall components, DNA based analyses, and on different sampling approaches, including samples from air, previously airborne dust that settles on elevated surfaces and floor dust. With respect to health assessments, symptom reports are most frequently used; only few studies include clinical measurements, such as lung function measurements or determination of inflammatory markers from clinical samples.

Generally speaking, more studies report a positive association of higher microbial levels with adverse health outcomes, mostly self-reports of respiratory symptoms; however, also negative (that is protective) associations of higher microbial levels in schools with pupils' health are being reported repeatedly. In some cases, where multiple microbial agents or groups are measured, both positive and negative associations are reported in the same study, depending on the microbial agent or group in question. Thus, it is obvious that no general conclusions such as "high microbial exposure is adverse to pupils' and teachers' health" or "microbes in classrooms are good" can be made, but that the nature and composition of the exposure and its sources appear to be relevant in determining positive or negative health outcomes.

In a series of publications relating to a study carried out in eight secondary schools in Malaysia, the authors find both positive and negative associations of various microbial exposures with respiratory and other symptoms that were reported by the students (Norback et al. 2016, 2014; Cai et al. 2011). As a general trend from these publications, different types of bacterial lipopolysaccharides (endotoxins) with different chain lengths – and assumingly derived from different bacteria – had differential effects on respiratory symptoms and illness, though mostly negative (protective) associations were reported. Fungal contamination, measured as ergosterol, total fungal DNA and *Aspergillus versicolor* DNA, was more often identified as a risk factor for respiratory health and general symptoms, such as fatigue. The variable effect of different types of bacterial endotoxin on pupils health was also reported in other studies (Zhao et al. 2008), where also negative associations between muramic acid and respiratory health of pupils were observed. Generally, protective effects of bacterial cell wall components (muramic acid and lipopolysaccharide) on mucosal and general symptoms, and adverse effects of total fungal DNA were reported in a study from China (Zhang et al. 2011). While these reports - in summary - point towards a more positive potential of bacterial exposures and a negative quality of fungal exposures in school environments, a study conducted in a vulnerable population of school children with asthma found that increases in personal endotoxin exposures were associated with decreased lung function and increased symptom reports (Rabinovitch et al. 2005).

Some studies have not found conclusive results with respect to the impact of microbial exposures in schools and pupils' and teachers' health reports. In a large and longitudinal study in Europe, wheeze was found to be generally inversely related to microbial levels (Jacobs et al. 2014), while otherwise no consistent associations between multiple microbial factors and pupils' health were observed. A study conducted in Danish school children similarly did not find consistent associations between any individual microbial components or combination of microbial components and health outcomes (Holst et al. 2016).

As mentioned earlier, there are more reports on adverse effects of fungal exposure in the school environment. A study conducted in 46 classrooms in five European countries concluded on associations of *Aspergillus/Penicillium* fungi and *Aspergillus versicolor* with respiratory symptoms such as wheeze, rhinitis and cough, but no such associations and in some cases negative associations were found between microbial groups and lung function measures (Simoni et al. 2011). In a Swedish study, more current asthma was associated with schools that had higher viable fungal and bacterial levels in dust (Smedje et al. 1997). Another study conducted in Taiwan found viable molds to be significantly associated with current asthma (Chen et al. 2014). Two Danish studies reported associations between viable molds in floor dust and building related symptoms in adolescent school children and teachers (Meyer et al. 2004; Ebbehoj et al. 2005). No associations of mold exposure and lung function in teachers were found in this study. A longitudinal study in Finnish schools reported higher airborne fungal levels in a moisture damaged school, as well as elevation of inflammatory markers in nasal lavage and induced sputum and higher prevalence of respiratory symptoms in adults working in the moisture damaged school (Purokivi et al. 2001). Another Finnish study found higher inflammatory marker levels in nasal lavage fluid of workers in a moisture damaged school compared to control subjects, and a reduction and re-elevation of inflammatory markers during and after holidays, respectively (Hirvonen et al. 1999).

A multi-center research implementing a school study in identical set-up in three different countries in Europe found differences in the associations between moisture damage observations and pupils' health by country and also between microbial exposures in the study schools and health (Jacobs et al. 2014; Borras-Santos et al. 2013). This could indicate that differences in outdoor climatic conditions and/or national differences or characteristics in the school building stock - and potentially related to that, differences in moisture damage and microbial contamination inside buildings and their structures - may be effect modifiers for the interaction between microbial exposure and occupants' health in schools. In line with this hypothesis, a study carried out in Denmark found an increase in the prevalence of a variety of symptoms in schools with high viable mold levels in floor dust, specifically in mechanically ventilated classrooms, as compared to naturally ventilated classrooms (Meyer et al. 2011).

The impact of moisture damage intervention on microbial exposures and school building occupants' health reports have not been very widely researched. A systematic review of Sauni et al. (2015) concluded that the evidence of a mold repair in school environment on respiratory symptoms of children is inconsistent and that only respiratory infections might decrease after the intervention. Concerning adults, the review shows that there is very low-quality evidence that asthma-related symptoms and other respiratory symptoms in a mold-damaged school are similar to those working in a non-damaged school, both before and after mold repairs. Studies carried out in Finnish school buildings report in part a reduction of microbial exposure levels upon moisture damage renovations, correlating with an improvement of health symptoms and inflammatory markers (Roponen et al. 2013), However, all of these studies were rather limited with respect to the numbers of school buildings and study participants.

12.4 Conclusions

A large number of studies have investigated determinants of microbial levels and composition and related impacts on pupils' and teachers' health in schools, less so in daycare centers. The outdoor environment, human occupancy and activity, and building factors including most notably conditions of moisture damage and dampness and ventilation, are the key determinants affecting indoor microbes quantitatively and qualitatively. As also found for residential home environments, both negative and positive impacts of exposure to microbes in schools on respiratory and general health symptoms have been reported. Based on the studies conducted so far there is a notion that bacterial exposure may also link to beneficial effects, whereas fungal exposures are more often linked to adverse health effects. However, for example differential effects of different types of bacterial endotoxins have been reported repeatedly, supporting the general conclusion that not only levels but also the quality and type of the bacterial and fungal exposures are important when evaluating health impacts on children and adults in school and daycare environments.

There is support that certain microbial exposures in schools may exceed home exposure levels. This, together with the fact that children and teachers spend many hours daily in schools and daycare centers, suggests that microbial exposure in these environments can contribute considerably to the total daily exposure for these population groups, and that for many years. Good indoor air quality in schools and daycares, including its microbial aspects, is of major concern to public health and must be a goal pursued with great effort and dedication.

References

Ali-Shtayeh MS, Al-Sheikh BS (1988) Isolation of keratinophilic fungi from the floor dust of Arab kindergarten schools in the West Bank of Jordan. Mycopathologia 103:69–73

Andersson A, Weiss N, Rainey F et al. (1999) Dust-borne bacteria in animal sheds, schools and children's day care centres. J Appl Microbiol 86:622–634

Andersson MA, Mikkola R, Rasimus S et al. (2010) Boar spermatozoa as a biosensor for detecting toxic substances in indoor dust and aerosols. Toxicol In Vitro 24:2041–2052

Andersson M, Mikkola R, Kroppenstedt R et al. (1998) The mitochondrial toxin produced by Streptomyces griseus strains isolated from an indoor environment is valinomycin. Appl Environ Microbiol 64:4767–4773

Bartlett K, Kennedy S, Brauer M et al. (2004a) Evaluation and a predictive model of airborne fungal concentrations in school classrooms. Ann Occup Hyg 48:547–554

Bartlett K, Kennedy S, Brauer M et al. (2004b) Evaluation and determinants of airborne bacterial concentrations in school classrooms. J Occup Environ Hyg 1:639–647

Baxi SN, Muilenberg ML, Rogers CA et al. (2013) Exposures to molds in school classrooms of children with asthma. Pediatr Allergy Immunol 24:697–703

Bhangar S, Huffman JA, Nazaroff WW (2014) Size-resolved fluorescent biological aerosol particle concentrations and occupant emissions in a university classroom. Indoor Air 24:604–617

Borràs-Santos A, Jacobs JH, Täubel M et al. (2013) Dampness and mould in schools and respiratory symptoms in children: the HITEA study. Occup Environ Med 70(10):681–687

Bright KR, Boone SA, Gerba CP (2010) Occurrence of bacteria and viruses on elementary classroom surfaces and the potential role of classroom hygiene in the spread of infectious diseases. J School Nurs 26:33–41

Cai G, Broms K, Malarstig B et al. (2009) Quantitative PCR analysis of fungal DNA in Swedish day care centers and comparison with building characteristics and allergen levels. Indoor Air 19:392–400

Cai G, Hashim JH, Hashim Z et al. (2011) Fungal DNA, allergens, mycotoxins and associations with asthmatic symptoms among pupils in schools from Johor Bahru, Malaysia. Pediatr Allergy Immunol 22:290–297

Cai G, Malarstig B, Kumlin A et al. (2011) Fungal DNA and pet allergen levels in Swedish day care centers and associations with building characteristics. J Environ Monit 13:2018–2024

Chatzidiakou L, Mumovic D, Summerfield A et al. (2015) Indoor air quality in London schools. Part 2: long-term integrated assessment. Intell Build Int 7:130–146

Chen C, Chao J, Chan C et al. (2014) Current asthma in schoolchildren is related to fungal spores in classrooms. Chest 146:123–134

Chen NT, Su YM, Hsu NY et al. (2012) Airborne fungi and bacteria in child daycare centers and the effectiveness of weak acid hypochlorous water on controlling microbes. J Environ Monit 14:2692–2697

Cho SJ, Cox-Ganser JM, Kreiss K et al. (2013) Evaluation of individual-based and group-based exposure estimation of microbial agents in health effects associated with a damp building. J Expo Sci Environ Epidemiol 23:409–415

Cho SJ, Cox-Ganser JM, Park J- (2016) Observational scores of dampness and mold associated with measurements of microbial agents and moisture in three public schools. Indoor Air 26:168–178

Ebbehoj N, Meyer H, Wurtz H et al. (2005) Molds in floor dust, building-related symptoms, and lung function among male and female schoolteachers. Indoor Air 15(Suppl. 10):7–16

Fahrhat SCL, Paulo RPL, Shimoda TM et al. (2005) Effect of air pollution on pediatric respiratory emergency room visits and hospital admissions. Braz J Med Biol Res 38:227–235

Foarde K, Berry M (2004) Comparison of biocontaminant levels associated with hard vs. carpet floors in nonproblem schools: results of a year long study. J Expo Anal Environ Epidemiol 14:S41–S48

Fox A, Harley W, Feigley C et al. (2003) Increased levels of bacterial markers and CO2 in occupied school rooms. J Environ Monit 5:246–252
Fox A, Harley W, Feigley C et al. (2005) Large particles are responsible for elevated bacterial marker levels in school air upon occupation. J Environ Monit 7:450–456
Fromme H, Heitmann D, Dietrich S et al. (2008) Air quality in schools – classroom levels of carbon dioxide(CO(2)), volatile organic compounds (VOC), aldehydes, endotoxins and cat allergen. Gesundheitswesen 70:88–97
Hanson B, Zhou Y, Bautista EJ et al. (2016) Characterization of the bacterial and fungal microbiome in indoor dust and outdoor air samples: a pilot study. Environ Sci 18:713–724
Haverinen U, Husman T, Toivola M et al. (1999) An approach to management of critical indoor air problems in school buildings. Environ Health Perspect 107:509–514
Haverinen-Shaughnessy U, Pekkanen J, Nevalainen A et al. (2004) Estimating effects of moisture damage repairs on students' health-a long-term intervention study. J Expo Anal Environ Epidemiol 14(Suppl 1):S58–64
Haverinen-Shaughnessy U, Borras-Santos A, Turunen M et al. (2012) Occurrence of moisture problems in schools in three countries from different climatic regions of Europe based on questionnaires and building inspections – the HITEA study. Indoor Air 22:457–466
Haverinen-Shaughnessy U, Toivola M, Alm S et al. (2007) Personal and microenvironmental concentrations of particles and microbial aerosol in relation to health symptoms among teachers. J Exp Sci Environ Epidemiol 17:182–190
Hirvonen M, Ruotsalainen M, Roponen M et al. (1999) Nitric oxide and proinflammatory cytokines in nasal lavage fluid associated with symptoms and exposure to moldy building microbes. Am J Resp Crit Care Med 160:1943–1946
Holst GJ, Host A, Doekes G et al. (2016) Allergy and respiratory health effects of dampness and dampness-related agents in schools and homes: a cross-sectional study in Danish pupils. Indoor Air 26:880–891
Hospodsky D, Yamamoto N, Nazaroff WW et al. (2015) Characterizing airborne fungal and bacterial concentrations and emission rates in six occupied children's classrooms. Indoor Air 25:641–652
Hospodsky D, Qian J, Nazaroff WW et al. (2012) Human occupancy as a source of indoor airborne bacteria. PLoS One 7:e34867
Huttunen K, Tirkkonen J, Täubel M et al. (2016) Inflammatory potential in relation to the microbial content of settled dust samples collected from moisture-damaged and reference schools: results of HITEA study. Indoor Air 26:380–390
Ibfelt T, Engelund EH, Permin A et al. (2015) Presence of pathogenic bacteria and viruses in the daycare environment. J Environ Health 78:24–29
IOM (Institute of Medicine) (2004) Damp indoor spaces and health. the National Academic Press, Washington DC
Jacobs JH, Krop EJ, Borras-Santos A et al. (2014) Endotoxin levels in settled airborne dust in European schools: the HITEA school study. Indoor Air 24:148–157
Jacobs JH, Krop EJM, de Wind S et al. (2013) Endotoxin levels in homes and classrooms of Dutch school children and respiratory health. Eur Resp J 42:314–322
Jacobs J, Borras-Santos A, Krop E et al. (2014) Dampness, bacterial and fungal components in dust in primary schools and respiratory health in schoolchildren across Europe. Occup Environ Med 71:704–712
Jo W, Seo Y (2005) Indoor and outdoor bioaerosol levels at recreation facilities, elementary schools, and homes. Chemosphere 61:1570–1579
Kanchongkittiphon W, Mendell MJ, Gaffin JM et al. (2015) Indoor environmental exposures and exacerbation of asthma: an update to the 2000 review by the Institute of Medicine. Environ Health Perspect 123(1):6–20
Kembel SW, Meadow JF, O'Connor TK et al. (2014) Architectural design drives the biogeography of indoor bacterial communities. PLoS One 9:e87093
Kim JL, Elfman L, Norbaeck D (2007) Respiratory symptoms, asthma and allergen levels in schools – comparison between Korea and Sweden. Indoor Air 17:122–129

Krop EJM, Jacobs JH, Sander I et al. (2014) Allergens and beta-glucans in Dutch homes and schools: characterizing airborne levels. PLoS One 9:e88871

Lee L, Tin S, Kelley ST (2007) Culture-independent analysis of bacterial diversity in a child-care facility. BMC Microbiol 7:27

Lee S, Chang M (2000) Indoor and outdoor air quality investigation at schools in Hong Kong. Chemosphere 41:109–113

Lignell U, Meklin T, Putus T et al. (2005) Microbial exposure, symptoms and inflammatory mediators in nasal lavage fluid of kitchen and clerical personnel in schools. Int J Occup Med Environ Health 18:139–150

Lignell U, Meklin T, Putus T et al. (2007) Effects of moisture damage and renovation on microbial conditions and pupils' health in two schools - a longitudinal analysis of five years. J Environ Monit 9:225–233

Liu LJ, Krahmer M, Fox A et al. (2000) Investigation of the concentration of bacteria and their cell envelope components in indoor air in two elementary schools. J Air Waste Manag Assoc 50:1957–1967

Madureira J, Paciencia I, Ramos E et al. (2015) Children's health and indoor air quality in primary schools and homes in Portugal – study design. J Toxicol Environ Health 78:915–930

Madureira J, Pereira C, Paciencia I et al. (2014) Identification and levels of airborne fungi in Portuguese primary schools. J Toxicol Environ Health 77:816–826

Meadow JF, Altrichter AE, Kembel SW et al. (2014) Bacterial communities on classroom surfaces vary with human contact. Microbiome 2:7

Meklin T, Husman T, Vepsalainen A et al. (2002a) Indoor air microbes and respiratory symptoms of children in moisture damaged and reference schools. Indoor Air 12:175–183

Meklin T, Hyvarinen A, Toivola M et al. (2003) Effect of building frame and moisture damage on microbiological indoor air quality in school buildings. AIHA J 64:108–116

Meklin T, Potus T, Pekkanen J et al. (2005) Effects of moisture-damage repairs on microbial exposure and symptoms in schoolchildren. Indoor Air 15(Suppl. 10):40–47

Meklin T, Reponen T, Toivola M et al. (2002b) Size distributions of airborne microbes in moisture-damaged and reference school buildings of two construction types. Atmos Environ 36:6031–6039

Mendell MJ, Mirer AG, Cheung K et al. (2011) Respiratory and allergic health effects of dampness, mold, and dampness-related agents: a review of the epidemiologic evidence. Environ Health Perspect 119(6):748–756

Mendes A, Aelenei D, Papoila AL et al. (2014) Environmental and ventilation assessment in child day care centers in Porto: the Envirh Project. J Toxicol Environ Health 77:931–943

Mentese S, Rad AY, Arisoy M et al. (2012) Multiple comparisons of organic, microbial, and fine particulate pollutants in typical indoor environments: diurnal and seasonal variations. J Air Waste Manage Assoc 62:1380–1393

Meyer HW, Suadicani P, Nielsen PA et al. (2011) Moulds in floor dust – a particular problem in mechanically ventilated rooms? A study of adolescent schoolboys under the Danish moulds in buildings program. Scand J Work Environ Health 37:332–340

Meyer HW, Wurtz H, Suadicani P et al. (2004) Molds in floor dust and building-related symptoms in adolescent school children. Indoor Air 14(1):65–72

Meyer HW, Wurtz H, Suadicani P et al. (2005) Molds in floor dust and building-related symptoms among adolescent school children: a problem for boys only? Indoor Air 15(s10):17–24

Mi YH, Norbäck D, Tao J et al. (2006) Current asthma and respiratory symptoms among pupils in Shanghai, China: influence of building ventilation, nitrogen dioxide, ozone, and formaldehyde in classrooms. Indoor Air 16:454–464

Norback D, Markowicz P, Cai GH et al. (2014) Endotoxin, ergosterol, fungal DNA and allergens in dust from schools in Johor Bahru, Malaysia – associations with asthma and respiratory infections in pupils. PLoS One 9:e88303

Norback D, Wieslander G, Zhang X et al. (2011) Respiratory symptoms, perceived air quality and physiological signs in elementary school pupils in relation to displacement and mixing ventilation system: an intervention study. Indoor Air 21:427–437

Norback D, Hashim JH, Cai G et al. (2016) Rhinitis, ocular, throat and dermal symptoms, headache and tiredness among students in schools from Johor Bahru, Malaysia: associations with fungal DNA and mycotoxins in classroom dust. PLoS One 11:e0147996

Norback D, Hashim JH, Markowicz P et al. (2016) Endotoxin, ergosterol, muramic acid and fungal DNA in dust from schools in Johor Bahru, Malaysia – Associations with rhinitis and sick building syndrome (SBS) in junior high school students. Sci Total Environ 545:95–103

Normand A, Ranque S, Cassagne C et al. (2016) Comparison of air impaction and electrostatic dust collector sampling methods to assess airborne fungal contamination in public buildings. Ann Occup Hyg 60:161–175

Orman A, Korcan E, Konuk M et al. (2006) Determination of fungal frequency and comparison of allergic symptoms related with buildings and fungi in Afyon, Turkey. Saudi Med J 27:1146–1151

Ovet H, Ergin C, Kaleli I (2012) Investigation of mold fungi in air samples of elementary schools and evaluation of allergen – specific IgE levels in students' sera. Mikrobiyol Bul 46:266–275

Patovirta RL, Reiman M, Husman T et al. (2003) Mould specific IgG antibodies connected with sinusitis in teachers of mould damaged school: a two-year follow-up study. Int J Occup Med Environ Health 16:221–230

Peitzsch M, Sulyok M, Täubel M et al. (2012) Microbial secondary metabolites in school buildings inspected for moisture damage in Finland, The Netherlands and Spain. J Environ Monit 14:2044–2053

Prussin 2nd AJ, Garcia EB , Marr LC (2015) Total virus and bacteria concentrations in indoor and outdoor air. Environ Sci Technol Lett 2:84–88

Purokivi M, Hirvonen M, Randell J et al. (2001) Changes in pro-inflammatory cytokines in association with exposure to moisture-damaged building microbes. Eur Resp J 18:951–958

Putus T, Tuomainen A, Rautiala S (2004) Chemical and microbial exposures in a school building: adverse health effects in children. Arch Environ Health 59:194–201

Qian J, Hospodsky D, Yamamoto N et al. (2012) Size-resolved emission rates of airborne bacteria and fungi in an occupied classroom. Indoor Air 22:339–351

Rabinovitch N, Liu A, Zhang L et al. (2005) Importance of the personal endotoxin cloud in school-age children with asthma. J Allergy Clin Immunol 116:1053–1057

Ramachandran G, Adgate J, Banerjee S *et al.* (2005) Indoor air quality in two urban elementary schools-measurements of airborne fungi, carpet allergens, CO2, temperature, and relative humidity. J Occup Environ Hyg 2:553–566

Roponen M, Meklin T, Rintala H et al. (2013) Effect of moisture-damage intervention on the immunotoxic potential and microbial content of airborne particles and on occupants' upper airway inflammatory responses. Indoor Air 23:295–302

Rullo V, Rizzo M, Arruda L et al. (2002) Daycare centers and schools as sources of exposure to mites, cockroach, and endotoxin in the city of Sao Paulo, Brazil. J Allergy Clin Immunol 110:582–588

Sahakian NM, White SK, Park JH et al. (2008) Identification of mold and dampness-associated respiratory morbidity in 2 schools: comparison of questionnaire survey responses to national data. J Sch Health 78(1):32–37

Salin JT, Salkinoja-Salonen M, Salin PJ et al. (2017) Building-related symptoms are linked to the in vitro toxicity of indoor dust and airborne microbial propagules in schools: a cross-sectional study. Environ Res 154:234–239

Sauni R, Verbeek JH, Uitti J et al. (2015) Remediating buildings damaged by dampness and mould for preventing or reducing respiratory tract symptoms, infections and asthma. Coch Datab Syst Rev 19:4. CD011621

Savilahti R, Uitti J, Laippala P et al. (2000) Respiratory morbidity among children following renovation of a water-damaged school. Arch Environ Health 55(6):405–410

Savilahti R, Uitti J, Roto P et al. (2001) Increased prevalence of atopy among children exposed to mold in a school building. Allergy 56(2):175–179

Savilahti R, Uitti J, Laippala P et al. (2002) Immunoglobulin G antibodies of children exposed to microorganisms in a water-damaged school. Pediatr Allergy Immunol 13:438–442
Seo S, Ji YG, Yoo Y et al. (2015) Submicron fungal fragments as another indoor biocontaminant in elementary schools. Environ Sci Process Impacts 17:1164–1172
Sheehan WJ, Hoffman EB, Fu C et al. (2012) Endotoxin exposure in inner-city schools and homes of children with asthma. Ann Allergy Asthma Immunol 108:418–422
Shin SK, Kim J, Ha SM et al. (2015) Metagenomic insights into the bioaerosols in the indoor and outdoor environments of childcare facilities. PLoS One 10:e0126960
Simoni M, Cai G, Norback D et al. (2011) Total viable molds and fungal DNA in classrooms and association with respiratory health and pulmonary function of European schoolchildren. Pediatr Allergy Immunol 22:843–852
Smedje G, Norback D (2001) Irritants and allergens at school in relation to furnishings and cleaning. Indoor Air 11:127–133
Smedje G, Norback D, Edling C (1997) Asthma among secondary schoolchildren in relation to the school environment. Clin Exp Allergy 27:1270–1278
Toivola M, Alm S, Reponen T et al. (2002) Personal exposures and microenvironmental concentrations of particles and bioaerosols. J Environ Monit 4:166–174
Toivola M, Nevalainen A, Alm S (2004) Personal exposures to particles and microbes in relation to microenvironmental concentrations. Indoor Air 14:351–359
Wady L, Shehabi A, Szponar B et al. (2004) Heterogeneity in microbial exposure in schools in Sweden, Poland and Jordan revealed by analysis of chemical markers. J Expo Anal Environ Epidemiol 14:293–299
Wan GH, Li CS (1999) Indoor endotoxin and glucan in association with airway inflammation and systemic symptoms. Arch Environ Health 54:172–179
WHO Regional Office for Europe (2009) WHO Guidelines for indoor air quality: dampness and mould. WHO Regional Office for Europe, Copenhagen
WHO Regional Office for Europe (2010) WHO Guidelines for indoor air quality: selected pollutants. WHO Regional Office for Europe, Copenhagen
Wurtz H, Sigsgaard T, Valbjorn O et al. (2005) The dustfall collector – a simple passive tool for long-term collection of airborne dust: a project under the Danish Mould in Buildings program (DAMIB). Indoor Air 15(Suppl 9):33–40
Yamamoto N, Hospodsky D, Dannemiller KC et al. (2015) Indoor emissions as a primary source of airborne allergenic fungal particles in classrooms. Environ Sci Technol 49:5098–5106
Zhang X, Zhao Z, Nordquist T et al. (2011) A longitudinal study of sick building syndrome among pupils in relation to microbial components in dust in schools in China. Sci Total Environ 409:5253–5259
Zhao Z, Sebastian A, Larsson L et al. (2008) Asthmatic symptoms among pupils in relation to microbial dust exposure in schools in Taiyuan, China. Pediatr Allergy Immunol 19:455–465

Chapter 13
Microbial Exposures in Residential Homes

Lidia Casas

Abstract Since we spend most of our time indoors, the characteristics of the indoor environment, and in particular, the characteristics of the microbial environment, have an impact in our health and well-being. The composition of the indoor microbial environment in homes is determined by regional factors like climate, building factors (e.g. ventilation or presence of mould or dampness) or the occupants and their indoor activities (e.g. crowdedness, pet ownership, indoor smoking, cleanliness or use of gas for cooking). The "westernization" process has modified some of these factors leading to changes in the characteristics of the indoor microbial environment and a consequent impact in our health. To date, most studies have investigated the beneficial and negative effects of the indoor microbial environment on respiratory and allergic health, mainly among children. Nevertheless, some recent studies point to a potential impact on mental health. Among children and adults, the exposure to mould or damp at home seems to be a risk factor for the development of asthma and allergies, and may have a negative effect on neuropsychological development and mood disorders. On the other hand, growing up in a farm protects against allergies and asthma. Nevertheless, for children and adults, the health effects of the indoor microbial environment are only consistent in observation/report of indoor moisture damage or when comparing farming vs non-farming homes. When actual measures targeting indoor microbial exposure in (sub-) urban areas are included, the picture becomes less clear and the associations appear inconsistent. Similarly, the presence of mould or dampness at home has been associated with lower cognitive function and with behavioural problems in children. Among adults, living in damp homes may be related with mood disorders. However, the inclusion of actual measures of the indoor microbial environment in human studies does not confirm these

L. Casas (✉)
Centre for Environment and Health, Department of Public Health and Primary Care, KU Leuven, Herestraat 49,3000 Leuven, Belgium
e-mail: lidia.casasruiz@kuleuven.be

C. Viegas et al. (eds.), *Exposure to Microbiological Agents in Indoor and Occupational Environments*, DOI 10.1007/978-3-319-61688-9_13

associations. Further epidemiological studies are needed to obtain sufficient evidence of the effects of the indoor microbial environment on mental health.

Keywords Residential microbial environment · mould · dampness · microbial diversity · respiratory health · allergy · mental health

13.1 Introduction

Nowadays, we spend most of our time indoors and the built environment, particularly in high income countries, has become our natural ecosystem. Therefore, our home microbial environment is a key factor in relation with our health. We are born mostly sterile (Aagaard et al., 2014) and are subsequently colonized by microbes that are present in our closest environment (Lax et al., 2014). The composition of the human microbiome has recently emerged as a relevant factor in human health. For example, it has shown to impact on inflammatory and allergic diseases or on the neuropsychological development (West et al., 2015; West, 2014; Borre et al., 2014). Nevertheless, the mechanisms involved in the observed associations are largely unclear (Human Microbiome Project Consortium, 2012). In addition to the human microbiome, microbial agents present in indoor air such as endotoxins, muramic acid, extracellular polysaccharides or glucans, are known to have strong immune stimulatory and pro-inflammatory properties. The exposure to these agents during infancy or childhood may be beneficial for the development of the immune system towards protecting against the development of allergies and related respiratory disorders like asthma (Douwes et al., 2002). However, the compositional characteristics of the indoor microbial environment appear to determine the direction of the effects on health, from being protective against diseases to being risk factors for the development and exacerbation of certain disorders. While a highly diverse microbial environment such as the one present in farms may protect from the development of asthma and allergies (Ege et al., 2011), the characteristics of the microbial environment in damp homes may have opposite effects on respiratory health and allergies among children and adults, and may also have an impact on their mental health (Mendell et al., 2011; Shenassa et al., 2007; Lidia Casas, Tiesler, et al., 2013). This chapter focuses on the determinants of the indoor microbial environment in homes and its health effects in children and adults.

13.2 Determinants of the Characteristics of the Indoor Environment in Homes

Several epidemiology studies have investigated the potential determinants of the indoor microbial environment in homes. Largely, these studies rely on the analyses of microbes from house dust samples, such as floor and mattress dust, or airborne settled dust. Home factors such as the presence of mould or dampness indoors, cat or dog ownership, building age, indoor smoking, ventilation, type of heating, frequency of cleaning, type of flooring, number of rooms, crowdedness (i.e. number of people living in the home), or specific (abnormal) microbial sources such as firewood or waste storages have been considered as potential determinants. The early studies aiming to characterize the determinants of the indoor microbial environment mainly focused on bacterial endotoxin levels in indoor dust (Thorne et al., 2009; Chen et al., 2012; Giovannangelo et al., 2007; Gehring et al., 2004; Waser et al., 2004; L Casas et al., 2013). They found increased indoor endotoxin concentrations or loads in homes with cats or dogs, homes with dampness, homes were occupants (or visitors) regularly smoked indoors, and in crowded homes. Moreover, some studies evaluated the potential determinants of the home levels of fungal exopolysaccharides (EPS) or glucans in indoor dust (L Casas et al., 2013; Schram et al., 2005; Sordillo et al., 2011; Gehring et al., 2001). EPS levels were determined by the frequency of cleaning, room ventilation, age of the home, indoor smoking, and gas cooking, while glucan levels in indoor dust were largely determined by the type of flooring. When studying the indoor microbial communities, researchers observe that the composition of bacterial communities is mainly dependent on the ventilation types and the occupants (Barberán et al., 2015; Meadow et al., 2014; Lax et al., 2014; Täubel et al., 2009; Hospodsky et al., 2012; Kembel et al., 2012; Weikl et al., 2016), while the composition of indoor fungal communities may depend more on the presence of mould or dampness at home, the tightness of the windows, and the heating (Weikl et al., 2016). However, for fungal communities, the strongest determinants of their composition are outdoor factors such as season or specific characteristics of the outdoor environment (Barberán et al., 2015; Weikl et al., 2016; Adams et al., 2013; Amend et al., 2010).

The outdoor factors are mainly determined by the geographical location (or region) of the home which explains most of the variation reported in the characteristics of the indoor microbial environment (L Casas et al., 2013; Chen et al., 2012). Among the studied outdoor factors, the most consistent determinant has been local climate and seasons (L Casas et al., 2013; Chen et al., 2012). However, recent studies suggest that the characteristics of the outdoor environment in terms of degree of urbanity, farming and greenness may be also important determinants of the composition of the indoor microbial communities (Weikl et al., 2016). Farming homes as compared to non-farming homes have been rather consistently linked with higher quantity and diversity of the indoor microbial content determined from house dust.

13.3 The Home Indoor Microbial Environment and Child Health

13.3.1 Asthma and Allergies

Asthma is a heterogeneous condition with different clinical expressions, characterized by a chronic inflammation of the respiratory tract, which is mediated by increased expression of multiple inflammatory proteins (Chipps et al., 2011). When the intensity of the inflammation increases, acute episodes or asthma exacerbations occur. Allergies are overreactions of the immune system to an allergen. Asthma and allergies are common and widespread conditions (Asher et al., 2006) that have increased in their prevalence worldwide during the last years (Asher et al., 2006; Pearce et al., 2007). Among other factors, the changes in the characteristics of the indoor microbial environment during infancy and childhood due to western lifestyles or the so called "westernization", which tends to create hygienic environments, have been suggested to partly explain the worldwide rise in the prevalence of asthma and allergies.

In this regard, the indoor microbial environment may play an either positive or negative role in the development and exacerbation of asthma and allergies depending on its characteristics. Spending time in moisture damaged homes facilitates the development of asthma (Mendell et al., 2011; Pekkanen et al., 2007). In addition, excessively hygienic environments with low quantity and variety of indoor microbes may impair the development of the immune system and contribute, as well, to increase the prevalence of asthma and allergies. In contrast, growing up on a farm, with exposure to high quantity and variety of microbes in the environment, may protect from asthma and allergies (von Mutius and Vercelli, 2010). When studying the effects of damp indoor environments or the farming environment effect as such (i.e. asking study participants questions like "is there mould or dampness in your home?" or "do you live in a farm?"), the results obtained are consistent throughout studies. However, including actual measures of indoor microbes that are thought to be involved in the relationships between damp or farming environments and health does not provide a clear picture. Furthermore the mechanisms involved in such associations are largely unclear.

13.3.1.1 The Negative Effects of the Indoor Microbial Environment: Moisture Damage and Dampness

According to the most recent reviews of the existing studies on the health effects of moisture damage and asthma, excessive indoor dampness is a public health problem and there is sufficient evidence to take preventive measures against dampness in buildings (IOM, 2004; Bornehag et al., 2001; Mendell et al., 2011; Fisk et al., 2007). Most child studies performed to date have a cross-sectional design and assess the exposure to home dampness and mould through parental report, not including any actual measures of the indoor microbial environment. They show that the prevalence of asthma and/or asthma symptoms like wheezing is higher among children living in homes with (reported) visible mould (C. Tischer, Chen, et al., 2011). Cross-sectional studies are observational studies (i.e. no intervention is performed) in which information on both exposure (e.g. home dampness) and outcome (e.g. asthma) are collected at the same point in time. The main advantage of these studies is that they are less time consuming to be carried out in large populations compared to, for example, longitudinal studies, and their main limitation is that they cannot test causality. Here, reverse causation (e.g. asthmatics tend to report dampness because they are aware of the effect of dampness on their health condition) must always be considered as a potential study limitation, in particular when both exposure and outcome are reported and not measured. Longitudinal studies may solve the issue of reverse causation present in the former cross-sectional studies. Like the cross-sectional studies, longitudinal studies are observational. However, in longitudinal studies the information on the exposure is collected when the outcome is not (yet) present, and the information on the health outcome is collected some time later. Cohort studies are longitudinal studies in which the health of the studied participants (and often the exposure) is followed for a long period of time. To study the effects of mould or dampness on child health avoiding the potential for reverse causation, *birth* cohort studies are interesting. These studies are longitudinal studies that follow up children and their exposures since birth or even pregnancy. So far, several birth cohort studies have investigated the health effects of early life exposure to mould or dampness at home. Their results confirmed the observations reported in cross-sectional studies, showing that the presence of mould or dampness at home during infancy had a negative effect on childhood asthma (Karvonen et al., 2015, 2009; C. G. Tischer et al., 2011; Pekkanen et al., 2007).

The previously mentioned cross-sectional and longitudinal studies did not, however, include measured exposures or investigate the mechanisms underlying the observed associations. So far, specific microbes, microbial cell wall agents, secondary metabolites produced by fungi and bacteria, volatile organic compounds, and allergens, that are more common or occur at higher levels in moisture damaged buildings, have been proposed as potential agents contributing to the health effects of moisture damage and dampness (Schlink et al., 2010; Claeson et al., 2009; Benndorf et al., 2008; Peitzsch et al., 2012; Reponen et al., 2012). Nevertheless, results of epidemiological studies including measures of microbial agents, species and products are inconclusive (Cai et al., 2011; Kirjavainen et al., 2015; Mendell et al., 2011; Tischer, Casas, et al., 2015). The Institute of Medicine reviewed the epidemiology studies that included measurements of indoor microbials or its products. The authors of this review concluded that there is limited evidence of a causal relationship between asthma exacerbation and exposure to indoor culturable *Penicillium* in sensitized children, and to total culturable fungal exposure regardless sensitization (Kanchongkittiphon et al., 2015). Another recent review suggests that, within the limited number of longitudinal studies, higher exposure levels to indoor *Penicillium, Aspergillus*, and *Cladosporium* species is associated with new cases of asthma in children (Sharpe et al., 2015).

Yet, these reviews include studies that rely only on microscopic observations and cultivation-based techniques, therefore drawbacks related to the measurement techniques should be considered. For example, the presence of non-culturable species, the need for varying growth conditions among different species and the occurrence of overgrowing of some species may impair the determination of indoor fungi exposure by cultivation (Nevalainen et al., 2014). In addition, fungal exposure never occurs in isolation, it is embedded in a broader fungal and bacterial community (Nevalainen et al., 2014, 2005). To overcome such limitations, USA researchers have created the Environmental Relative Moldiness Index

(ERMI) that includes 36 fungal species frequently present in indoor dust (Vesper et al., 2007; Méheust et al., 2012): 26 species associated with moisture damage (group 1), and 10 species not associated with moisture damage (group 2). The ERMI is calculated by subtracting the sum of the log-transformed concentrations of group 2 species from the sum of the log-transformed concentrations of group 1 species. Using the ERMI as exposure, cross-sectional studies in the US have observed that children living in homes with high ERMI values had reduced lung function (i.e. lung volumes) as compared with children living in homes with low ERMI values (Vesper et al., 2013). Longitudinal studies assessing the ERMI during early life did not find differences with infant wheezing (Rosenbaum et al., 2015), but higher prevalence of asthma at the age of 7 years among children living in homes with high ERMI values. In particular, their findings were more evident for *Aspergillus ochraceus*, *Aspergillus unguis*, and *Penicillium variabile* species (Reponen et al., 2011). However, the applicability of the ERMI in different countries with differences in outdoor, building and moisture damage characteristics is questionable and has been challenged (Täubel et al., 2015).

13.3.1.2 The Positive Effects of the Indoor Microbial Environment

In 1989, Strachan (Strachan, 1989) proposed an explanation for the worldwide rise in the prevalence of asthma and allergic disorders, the "hygiene hypothesis." He suggested that living in more hygienic conditions could lead to the development of allergic diseases, hypothesizing that the exposure to microbes plays a role in the maturation of the immune system. In hygienic environments, the number and diversity of microbes may be reduced, leading to Th2-oriented reactivity and Th2-type over-reactions to environmental allergens, and further to allergic manifestations (von Mutius and Vercelli, 2010). In this regard, the westernization tends to create hygienic living environments and studies comparing farming and non-farming environments, as well as those comparing Eastern and Western Germany shortly after unification or the Finnish and the Russian Karelia, initially serve to test this hypothesis. Such studies showed a strong protective effect for asthma and allergies of living in farming environments (Genuneit, 2012) and higher prevalence of asthma and allergies in the Western Germany and in the Finnish Karelia (Krämer et al., 2015; Hugg et al., 2008).

After the publication of the first farm studies in the early 2000s, studies in (sub-) urban environments aimed to replicate the findings including measurements of the indoor microbial environment. In particular, most studies performed in (sub-) urban areas focused in measuring the concentrations or loads of indoor microbial cell wall agents being indicative of microbial biomass such as endotoxin, EPS or glucans in home dust. However, the results obtained were not as clear as those observed in the farm studies. Cross-sectional studies including infants and children aged 1–6 years old reported protective effects on asthma and asthma-like symptoms such as wheezing of high concentrations of microbial agents in indoor dust (Iossifova et al., 2007, 2009; Gehring et al., 2007;

C. Tischer, Gehring, et al., 2011). However, the large number of longitudinal studies that included microbial agents measured in indoor dust during early life (i.e. during the first year of life) and asthma or allergies up to the age of 10 years presented an inconclusive picture (Tischer, Casas, et al., 2015; Douwes et al., 2006; Campo et al., 2006; Iossifova et al., 2007; Lidia Casas, Tischer, et al., 2013; C. Tischer, Gehring, et al., 2011; Rosenbaum et al., 2010; Perzanowski et al., 2006; Horick et al., 2006; Gillespie et al., 2006).

It is not yet clear whether the overall composition or rather the functionality of few species of the microbial profile in dust is accountable for the associations between the indoor microbial environment and health. Currently, investigations in urban areas have just begun to explore how the indoor dust microbiome may affect the pathogenesis of asthma and allergic diseases. Next generation amplicon sequencing has been used in a birth cohort study in the USA to describe the fungal microbiome in settled house dust collected during early life. The results of this study showed that low fungal diversity in dust was associated with increased risk of asthma at 7 years of age. In contrast, elevated moisture in these homes was associated with an increase in fungal diversity, and thus the interrelation of dampness, fungal diversity, and asthma could not be fully explained by this study (Dannemiller et al., 2014, 2016a, 2016b).

In a German birth cohort, fungal and bacterial diversity was measured in home dust collected during early life and related to the later development of asthma and allergic sensitization. Consistently with the results reported in the American study, the German researchers showed that higher fungal (but not bacterial) diversity protected from allergen sensitization at the age of 6 and from wheezing up to the age of 10 (Tischer et al., 2016). The effects of combined early life exposure to allergens and bacteria on asthma symptoms and allergy among children at high risk for asthma (i.e. children of parents with asthma or allergies) living in urban environments were investigated in another American birth cohort. The researchers of this study concluded that exposure to high levels of allergen together with a certain subset of bacterial taxa (*Firmicutes* and *Bacteriodetes*) during early life may protect against allergic sensitization and wheezing outcomes at the age of 3 years (Lynch et al., 2014).

13.3.2 Neuropsychological Development

So far, the potential mental health effects of the indoor microbial environment have not been extensively studied. Results of experimental studies in animal models suggest that pre-natal and early life immune activation through the exposure to endotoxins can inhibit neurogenesis, (Cui et al., 2009; Schwarz and Bilbo, 2011). In addition, the exposure to endotoxins later in life leads to immune-activation related mood disorders and cognitive disturbances (DellaGioia and Hannestad, 2010; Eisenberger et al., 2010; Reichenberg et al., 2001; Yirmiya, 1996). Also, experimental studies show that mycotoxins from mould have inflammatory and a neurotoxic effects that are potentiated by the co-exposure to endotoxins (Corps et al., 2010; Doi and Uetsuka, 2011; Karunasena et al., 2010; Pestka et al., 2008; Islam et al., 2007; Korkalainen et al., 2016).

To date, the number of studies evaluating the neuropsychological effects of the indoor microbial environment in children is scarce. Three European birth cohorts have published results that are in line with the results presented in animal studies. The studies performed in a Spanish and a Polish birth cohort showed that children living in homes with visible mould or dampness during early life presented lower cognitive function and social competences at the ages of 4 and 6 years, while occasional farm animal contact showed beneficial effects among the Spanish children (Lidia Casas, Torrent, et al., 2013; Jedrychowski et al., 2011). In a German birth cohort, indoor visible mould, dampness and pet ownership were considered as surrogates of the indoor microbial environment. The researchers reported higher prevalence of behavioural problems among children living in damp homes and having pets at home (Lidia Casas, Tiesler, et al., 2013). Although the results presented in human studies are consistent with those shown in animal studies, we still cannot draw a strong conclusion. The number of human studies is limited and their findings could not be confirmed by objective measures of the indoor microbial environment.

13.4 The Indoor Microbial Environment and Adult Health

13.4.1 Asthma and Allergies

The first study focusing on the effect of the home microbial environment on adult's respiratory and allergic health was conducted in Belgium in the 90s (Michel et al., 1996). This study included 69 asthmatic adults and floor and mattress dust samples of their bedrooms were analysed. They observed that individuals with higher concentrations of endotoxin in dust tended to have worse lung function, higher need for asthma medication and higher frequency of asthma and allergy symptoms. One decade later, a larger study including dust samples from more than 2500 homes in the United States was published (Thorne et al., 2005). This study evaluated the association of the concentrations of endotoxin in indoor dust with asthma symptoms and medication among a representative nationwide sample of general population. In line with the results reported in the Belgian study, the findings of the American study suggested that household endotoxin exposure could be a risk factor for asthma.

In Europe, a large ongoing cohort study called the European Community Respiratory Health Survey (ECRHS) has allowed to further investigate the effects of the home microbial environment on adult's respiratory health, including diverse markers of the indoor microbial environment and respiratory health outcomes. The ECRHS (Burney et al., 1994) is a multicentre cohort study that began in 1990 aiming to estimate the variation in the prevalence of asthma, its risk factors and treatment in Europe. Since the 90s, this study follows up more than 10,000 adults from 14 countries, mostly European. Mattress dust samples of the participating individuals were collected in 1999–2001 and information on the home characteristics (e.g. presence of mould or dampness at home) and respiratory and allergic health from each participant was collected every 10 years, approximately.

The first studies published using the information provided by the ECRHS study evaluated the association between the participants' report of home mould or dampness and respiratory and allergic health outcomes. Gunnbjörnsdóttir et al (Gunnbjörnsdóttir et al., 2006) included only the ECRHS north European countries in their study. They reported that subjects living in damp housing had a higher prevalence of respiratory symptoms and asthma. In addition, the onset of respiratory symptoms was more common and the remission of nocturnal respiratory symptoms was less common among subjects living in damp homes. Later, two longitudinal studies including all the ECRHS participant countries showed that the presence of dampness and/or mould at home increased the incidence of asthma (i.e. new cases of asthma) (Norbäck et al., 2013) and that aspects related to building dampness (water leakage or damp spots) were associated with an accelerated lung function decline only among women (Norbäck et al., 2011).

The relationship between the concentrations of several microbial agents in mattress dust and respiratory and allergic health outcomes in the ECRHS study were investigated by Bakolis et al (Bakolis et al., 2012) and Tischer et al (Tischer, Zock, et al., 2015). The first investigated the associations between respiratory health and endotoxin exposure. Although they could not prove an association with respiratory symptoms or allergy (IgE sensitization), they report an effect modification by a polymorphism in the CD14 gene (CD14/-260 genotype) in the association between endotoxin levels in mattress dust and lung function, meaning that the individual's specific genotype may determine the effect of endotoxin exposure on lung function (no effect vs reductions in lung function). Tischer et al (Tischer, Zock, et al., 2015) investigated the association of the quantity of several microbial agents measured in mattress dust samples (endotoxin, (1,3)-ß-D-glucan, fungal and bacterial qPCR and muramic acid) with the household characteristics (dampness and mould, exposure to pets, current smoking and whether the bed is placed in the living room), and asthma and lung function. They observed that current signs of dampness in the bedroom or home environment and having a cat or dog were significant predictors for higher fungal and bacterial agent concentrations in mattress dust. In addition, high concentrations of muramic acid were associated with higher incidence of asthma.

13.4.2 Mental Health

As previously mentioned, experimental studies in animals models suggest that endotoxin and mycotoxin from moulds may contribute to immune-activation related mood disorders and cognitive disturbances (DellaGioia and Hannestad, 2010; Eisenberger et al., 2010; Reichenberg et al., 2001; Yirmiya, 1996; Corps et al., 2010; Doi and Uetsuka, 2011; Karunasena et al., 2010; Pestka et al., 2008; Islam et al., 2007). However, as for children, the potential mental health effects of the "real-life" indoor microbial environment among human adults have been barely studied. So far, only three epidemiological studies looking at the associations between the indoor microbial environment and adult's mental health have been performed. All of them focused on water-damaged homes and none included measures of the indoor microbial environment. The first studies were conducted during the late 1980s in the United Kingdom. They observed increased prevalence of mood disorders among individuals living in homes with mould or dampness and awarded a social explanation to their findings (Martin et al., 1987; Platt et al., 1989). More recently, the Large Analysis and Review of European housing and Health Status Study, a multicentre study including eight European cities, reported similar results after considering the effects of several socio-economic factors and physical illnesses of the study participants (Shenassa et al., 2007).

13.5 Conclusions

There is sufficient evidence that the characteristics of the indoor microbial environment have an impact on human respiratory and allergic health. Studies in both children and adult population have shown clear negative effects of the presence of mould or dampness at home and protective effects of growing up in a farm on asthma and allergies. However, when actual measures of the indoor microbial environment and (sub-) urban areas are considered, the picture is not yet clear.

Regarding mental health, the evidence of adverse effects of the indoor microbial environment is still limited due to the small number of studies performed to date. Nevertheless, the results shown in the few published studies point to a negative impact on the cognitive development, behaviour and mood of persons who have the presence of mould and dampness at home.

Studies using next generation sequencing techniques to estimate not only the quantity but also the composition of the microbial environment indoors may be key to understand the relationship between health and the indoor environment.

References

Aagaard K et al. (2014) The placenta harbors a unique microbiome. Sci Transl Med 6(237): p.237ra65

Adams RI et al. (2013) Dispersal in microbes: fungi in indoor air are dominated by outdoor air and show dispersal limitation at short distances. ISME J 7(7):1262–73

Amend AS et al. (2010) Indoor fungal composition is geographically patterned and more diverse in temperate zones than in the tropics. Proc Natl Acad Sci U S A 107(31):13748–53

Asher MI et al. (2006) Worldwide time trends in the prevalence of symptoms of asthma, allergic rhinoconjunctivitis, and eczema in childhood: ISAAC Phases One and Three repeat multicountry cross-sectional surveys. Lancet 368(9537):733–743

Bakolis I et al. (2012) Respiratory health and endotoxin: associations and modification by CD14/-260 genotype. Eur Resp J39(3):573–81
Barberán A et al. (2015) The ecology of microscopic life in household dust. Proc R Soc B 282:1814
Benndorf D et al. (2008) Identification of spore allergens from the indoor mould Aspergillus versicolor. Allergy 63(4):454–460
Bornehag CG et al. (2001) Dampness in buildings and health. Nordic interdisciplinary review of the scientific evidence on associations between exposure to "dampness" in buildings and health effects (NORDDAMP). Indoor Air 11(2):72–86
Borre YE et al. (2014) Microbiota and neurodevelopmental windows: implications for brain disorders. Trends Mol Med 20(9):509–518
Burney PG et al. (1994) The European Community Respiratory Health Survey. Eur Res J 7(5):954–60
Cai G-H et al. (2011) Fungal DNA, allergens, mycotoxins and associations with asthmatic symptoms among pupils in schools from Johor Bahru, Malaysia. Pediatr Allergy Immunol 22(3):290–7
Campo P et al. (2006) Influence of dog ownership and high endotoxin on wheezing and atopy during infancy. J Allergy Clin Immunol 118(6):1271–1278
Casas L et al. (2013) Endotoxin, extracellular polysaccharides, and β(1-3)-glucan concentrations in dust and their determinants in four European birth cohorts: results from the HITEA project. Indoor Air 23(3):208–18
Casas L, Tiesler C et al. (2013) Indoor factors and behavioural problems in children: the GINIplus and LISAplus birth cohort studies. Int J Hyg Environ Health 216(2):146–54
Casas L, Tischer C et al. (2013) Early life microbial exposure and fractional exhaled nitric oxide in school-age children: a prospective birth cohort study. Environ Health 12(1):103
Casas L, Torrent M et al. (2013) Early life exposures to home dampness, pet ownership and farm animal contact and neuropsychological development in 4 year old children: a prospective birth cohort study. Int J Hyg Environ Health 216(6):690–7
Chen C-M et al. (2012) Geographical variation and the determinants of domestic endotoxin levels in mattress dust in Europe. Indoor Air 22(1):24–32
Chipps BE, Bacharier LB, Harder JM (2011) Phenotypic expressions of childhood wheezing and asthma: implications for therapy. J Pediatrics 158(6):878–884. e1
Claeson A-S, Nordin S, Sunesson A-L (2009) Effects on perceived air quality and symptoms of exposure to microbially produced metabolites and compounds emitted from damp building materials. Indoor Air 19(2):102–112
Corps KN et al. (2010) Neurotoxic, inflammatory, and mucosecretory responses in the nasal airways of mice repeatedly exposed to the macrocyclic trichothecene mycotoxin roridin A: dose-response and persistence of injury. Toxicol Pathol 38(3):429–451
Cui K et al. (2009) Effects of prenatal immune activation on hippocampal neurogenesis in the rat. Schizophren Res 113(2-3):288–297
Dannemiller KC et al. (2014) Next-generation DNA sequencing reveals that low fungal diversity in house dust is associated with childhood asthma development. Indoor Air 24(3):236–47
Dannemiller KC et al. (2016a) Indoor microbial communities: influence on asthma severity in atopic and nonatopic children. J Allergy Clin Immunol 138(1):76–83. e1
Dannemiller KC et al. (2016b) Influence of housing characteristics on bacterial and fungal communities in homes of asthmatic children. Indoor Air 26(2):179–192
DellaGioia N, Hannestad J (2010) A critical review of human endotoxin administration as an experimental paradigm of depression. Neurosci Biobehav Rev 34(1):130–143
Doi K, Uetsuka K (2011) Mechanisms of mycotoxin-induced neurotoxicity through oxidative stress-associated pathways. Int J Mol Sci 12(8):5213–5237
Douwes J, Pearce N, Heederik D (2002) Does environmental endotoxin exposure prevent asthma? Thorax 57(1):86–90
Douwes J et al. (2006) Does early indoor microbial exposure reduce the risk of asthma? The Prevention and Incidence of Asthma and Mite Allergy birth cohort study. J Allergy Clin Immunol 117(5):1067–1073

Ege MJ et al. (2011) Exposure to environmental microorganisms and childhood asthma. New Engl J Med 364(8):701–9
Eisenberger NI et al. (2010) Inflammation-induced anhedonia: endotoxin reduces ventral striatum responses to reward. Biol Psychiatry 68(8):748–754
Fisk WJ, Lei-Gomez Q, Mendell MJ (2007) Meta-analyses of the associations of respiratory health effects with dampness and mold in homes. Indoor Air 17(4):284–296
Gehring U et al. (2001) Beta(1–>3)-glucan in house dust of German homes: housing characteristics, occupant behavior, and relations with endotoxins, allergens, and molds. Environ Health Perspect 109(2):139–144
Gehring U et al. (2004) Levels and predictors of endotoxin in mattress dust samples from East and West German homes. Indoor Air 14(4):284–92
Gehring U et al. (2007) Bacteria and mould components in house dust and children's allergic sensitisation. Eur Resp J 29(6):1144–1153
Genuneit J (2012) Exposure to farming environments in childhood and asthma and wheeze in rural populations: a systematic review with meta-analysis. Pediatr Allergy Immunol 23(6):509–18
Gillespie J et al. (2006) Endotoxin exposure, wheezing, and rash in infancy in a New Zealand birth cohort. J Allergy Clin Immunol 118(6):1265–70
Giovannangelo M et al. (2007) Determinants of house dust endotoxin in three European countries – the AIRALLERG study. Indoor Air 17(1):70–79
Gunnbjörnsdóttir MI et al. (2006) Prevalence and incidence of respiratory symptoms in relation to indoor dampness: the RHINE study. Thorax 61(3):221–5
Horick N et al. (2006) Home endotoxin exposure and wheeze in infants: correction for bias due to exposure measurement error. Environ Health Perspect 114(1):135–40
Hospodsky D et al. (2012) Human occupancy as a source of indoor airborne bacteria. PLoS One 7(4):e34867
Hugg TT et al. (2008) Exposure to animals and the risk of allergic asthma: a population-based cross-sectional study in Finnish and Russian children. Environ Health 7:28
Human Microbiome Project Consortium, T.H.M.P (2012) Structure, function and diversity of the healthy human microbiome. Nature 486(7402):207–14
Institute of Medicine (2004) Damp Indoor Spaces and Health. The National Academies Press, Washington, DC, https://doi.org/10.17226/11011
Iossifova YY et al. (2007) House dust (1-3)-beta-D-glucan and wheezing in infants. Allergy 62(5):504–513
Iossifova YY et al. (2009) Mold exposure during infancy as a predictor of potential asthma development. Ann Allergy Asthma Immunol 102(2):131–137
Islam Z et al. (2007) Neurotoxicity and inflammation in the nasal airways of mice exposed to the macrocyclic trichothecene mycotoxin roridin a: kinetics and potentiation by bacterial lipopolysaccharide coexposure. Toxicol Sci 98(2):526–541
Jedrychowski W et al. (2011) Cognitive function of 6-year old children exposed to mold-contaminated homes in early postnatal period. Prospective birth cohort study in Poland. Physiol Behav 104(5):989–995
Kanchongkittiphon W et al. (2015) Indoor environmental exposures and exacerbation of asthma: an update to the 2000 review by the Institute of Medicine. Environ Health Perspect 123(1):6–20
Karunasena E et al. (2010) Building-associated neurological damage modeled in human cells: a mechanism of neurotoxic effects by exposure to mycotoxins in the indoor environment. Mycopathologia 170(6):377–390
Karvonen AM et al. (2009) Confirmed moisture damage at home, respiratory symptoms and atopy in early life: a birth-cohort study. Pediatrics 124(2):e329–338
Karvonen AM et al. (2015) Moisture damage and asthma: a birth cohort study. Pediatrics 135(3): e598–606
Kembel SW et al. (2012) Architectural design influences the diversity and structure of the built environment microbiome. ISME J 6(8):1469–79

Kirjavainen PV et al. (2015) Microbial secondary metabolites in homes in association with moisture damage and asthma. Indoor Air 26:448–456

Korkalainen M et al (2016) Synergistic proinflammatory interactions of microbial toxins and structural components characteristic to moisture-damaged buildings. Indoor Air 27:13–23

Krämer U et al. (2015) What can reunification of East and West Germany tell us about the cause of the allergy epidemic? Clin Exp Allergy 45(1):94–107

Lax S et al. (2014) Longitudinal analysis of microbial interaction between humans and the indoor environment. Sci (New York, N.Y.) 345(6200):1048–52

Lynch SV et al. (2014) Effects of early-life exposure to allergens and bacteria on recurrent wheeze and atopy in urban children. J Allergy Clin Immunol 134(3):593–601. e12

Martin CJ, Platt SD, Hunt SM (1987) Housing conditions and ill health. Br Med J (Clin Res Ed.) 294(6580):1125–1127

Meadow JF et al. (2014) Indoor airborne bacterial communities are influenced by ventilation, occupancy, and outdoor air source. Indoor Air 24(1):41–8

Méheust D et al. (2012) Correlation between Environmental Relative Moldiness Index (ERMI) values in French dwellings and other measures of fungal contamination. Sci Total environ 438:319–24

Mendell MJ et al. (2011) Respiratory and allergic health effects of dampness, mold, and dampness-related agents: a review of the epidemiologic evidence. Environ Health Perspect 119(6):748–756

Michel O et al. (1996) Severity of asthma is related to endotoxin in house dust. Am J Resp Crit Care Med 154(6 Pt 1):1641–1646

Nevalainen A et al. (2005) Of microbes and men. Indoor Air 15(Suppl 9):58–64

Nevalainen A, Täubel M, Hyvärinen A (2014) Indoor fungi: companions and contaminants. Indoor Air 25:125–126

Norbäck D et al. (2011) Lung function decline in relation to mould and dampness in the home: the longitudinal European Community Respiratory Health Survey ECRHS II. Thorax 66(5):396–401

Norbäck D et al. (2013) Mould and dampness in dwelling places, and onset of asthma: the population-based cohort ECRHS. Occup Environ Med 70(5):325–31

Pearce N et al. (2007) Worldwide trends in the prevalence of asthma symptoms: phase III of the International Study of Asthma and Allergies in Childhood (ISAAC). Thorax 62(9):758–766

Peitzsch M et al. (2012) Microbial secondary metabolites in school buildings inspected for moisture damage in Finland, The Netherlands and Spain. J Environ Monitor 14(8):2044–2053

Pekkanen J et al. (2007) Moisture damage and childhood asthma: a population-based incident case-control study. Eur Resp J 29(3):509–515

Perzanowski MS et al. (2006) Endotoxin in inner-city homes: associations with wheeze and eczema in early childhood. J Allergy Clin Immunol 117(5):1082–9

Pestka JJ et al. (2008) Stachybotrys chartarum, trichothecene mycotoxins, and damp building-related illness: new insights into a public health enigma. Toxicol Sci 104(1):4–26

Platt SD et al. (1989) Damp housing, mould growth, and symptomatic health state. BMJ 298 (6689):1673–1678

Reichenberg A et al. (2001) Cytokine-associated emotional and cognitive disturbances in humans. Arch Gen Psychiatry 58(5):445–452

Reponen T et al. (2011) High environmental relative moldiness index during infancy as a predictor of asthma at 7 years of age. Ann Allergy Asthma Immunol 107(2):120–126

Reponen T et al. (2012) Infant origins of childhood asthma associated with specific molds. J Allergy Clin Immunol130(3):639–644. e5

Rosenbaum PF et al. (2010) Indoor airborne fungi and wheeze in the first year of life among a cohort of infants at risk for asthma. J Exp Sci Environ Epidemiol 20(6):503–15

Rosenbaum PF et al. (2015) Environmental relative moldiness index and associations with home characteristics and infant wheeze. J Occup Environ Hyg 12(1):29–36

Schlink U et al. (2010) Quantile regression of indoor air concentrations of volatile organic compounds (VOC). Sci Total Environ 408(18):3840–3851

Schram D et al. (2005) Bacterial and fungal components in house dust of farm children, Rudolf Steiner school children and reference children–the PARSIFAL Study. Allergy 60(5):611–618

Schwarz JM, Bilbo SD (2011) LPS elicits a much larger and broader inflammatory response than Escherichia coli infection within the hippocampus of neonatal rats. Neurosci Lett 497(2):110–115

Sharpe RA et al. (2015) Indoor fungal diversity and asthma: a meta-analysis and systematic review of risk factors. J Allergy Clin Immunol 135(1):110–22

Shenassa ED et al. (2007) Dampness and mold in the home and depression: an examination of mold-related illness and perceived control of one's home as possible depression pathways. Am J Public Health 97(10):1893–1899

Sordillo JE et al. (2011) Home characteristics as predictors of bacterial and fungal microbial biomarkers in house dust. Environ Health Perspect 119(2):189–195

Strachan DP (1989) Hay fever, hygiene, and household size. BMJ 299(6710):1259–1260

Täubel M et al. (2009) The occupant as a source of house dust bacteria. J Allergy Clin Immunol 124(4):834–40. e47

Täubel M et al. (2015) Application of the Environmental Relative Moldiness Index in Finland. Appl Environ Microbiol 82(2):578–84

Thorne PS et al. (2005) Endotoxin exposure is a risk factor for asthma: the national survey of endotoxin in United States housing. Am J Resp Crit Care Med 172(11):1371–1377

Thorne PS et al. (2009) Predictors of endotoxin levels in U.S. housing. Environ Health Perspect 117(5):763–771

Tischer C et al (2016) Urban dust microbiome: impact on later atopy and wheezing. Environ Health Perspect 124:1919–1923

Tischer C, Casas L et al. (2015) Early exposure to bio-contaminants and asthma up to 10 years of age: results of the HITEA study. Eur Resp J 45(2):328–37

Tischer C, Chen C-M, Heinrich J (2011) Association between domestic mould and mould components, and asthma and allergy in children: a systematic review. Eur Resp J 38(4):812–824

Tischer C, Gehring U et al. (2011) Respiratory health in children, and indoor exposure to (1,3)-β-D-glucan, EPS mould components and endotoxin. Eur Resp J 37(5):1050–9

Tischer C, Zock J-P et al. (2015) Predictors of microbial agents in dust and respiratory health in the Ecrhs. BMC Pulm Med 15:48

Tischer CG et al. (2011) Meta-analysis of mould and dampness exposure on asthma and allergy in eight European birth cohorts: an ENRIECO initiative. Allergy 66(12):1570–1579

Vesper S et al. (2007) Development of an Environmental Relative Moldiness index for US homes. J Occup Environ Med 49(8):829–33

Vesper SJ et al. (2013) Decreased pulmonary function measured in children exposed to high environmental relative moldiness index homes. Open Resp Med J 7:83–6

von Mutius E, Vercelli D (2010) Farm living: effects on childhood asthma and allergy. Nat Rev Immunol 10(12):861–868

Waser M et al. (2004) Determinants of endotoxin levels in living environments of farmers' children and their peers from rural areas. Clin Exp Allergy 34(3):389–397

Weikl F et al. (2016) Fungal and Bacterial Communities in Indoor Dust Follow Different Environmental Determinants. PLoS One 11(4):e0154131

West CE (2014) Gut microbiota and allergic disease: new findings. Cur Opin Clin Nutr Metab Care 17(3):261–6

West CE et al. (2015) The gut microbiota and inflammatory noncommunicable diseases: associations and potentials for gut microbiota therapies. J Allergy Clin Immunol 135(1):3–13

Yirmiya R (1996) Endotoxin produces a depressive-like episode in rats. Brain Res 711(1-2):163–174

Chapter 14
Environmental Fungal Risk in Health Facilities

Jean-Pierre Gangneux

Abstract Indoor fungal contamination is associated with a wide range of adverse health effects, including infectious diseases, allergies and toxic effects. Various indoor environments are concerned by fungal health risk, particularly hospitals and critical care environments where immunosuppressed patients are hosted. Prevention of nosocomial invasive fungal infections in health-care facilities is mainly based on air treatment with monitoring environmental fungal contamination and epidemiological follow-up of invasive fungal infections. After reminding the reservoir of fungi at hospitals and the routes of transmission, we review what measures of fungal risk control should be implemented and tools that are relevant for an efficient environmental monitoring.

Keywords Hospital · nosocomial fungal infections · air treatment · prevention · monitoring

14.1 Definitions

14.1.1 The Environmental Fungal Risk

Micromycetes, yeast-like, filamentous or dimorphic, are ubiquitous in our environment: soil, air, water, food etc. Thus, humans and animals are in constant contact with these eukaryotes and have developed relationships commensalism or parasitism.

Exposure of an individual to an environmental fungal reservoir may in some situations be at the origin of very diverse diseases: infections (since benign

J.-P. Gangneux (✉)
Mycology and Parasitology Laboratory, Rennes Teaching Hospital, CHU de Rennes, 2 rue Henri le Guilloux, Rennes, France

Institut de Recherche en Santé Environnement et Travail, INSERM UMR U1085, Université Rennes 1, Rennes, France
e-mail: Jean-pierre.gangneux@univ-rennes1.fr

C. Viegas et al. (eds.), *Exposure to Microbiological Agents in Indoor and Occupational Environments*, DOI 10.1007/978-3-319-61688-9_14

superficial infections to life-threatening invasive multisystem infection), allergies, or poisoning via mycotoxins.

Here we will discuss only hospital environmental fungal infection risk because it represents a major risk in terms of mortality. It is the other best defined and identified, allowing to implement prevention strategies and monitoring in health facilities.

14.1.2 The Hospital Fungal Infection Risk

This is the risk of contracting and developing a fungal disease during hospitalization in a health institution. The most common and worrying invasive fungal infection (IFI) is invasive aspergillosis (IA) due to *Aspergillus fumigatus* in 80% of cases. It must be the subject of preventive measures (Patterson et al., 2016). Preventive measures should be emphasized during construction and renovation works. The biodiversity of potentially pathogenic fungi is then much higher.

Potentially pathogenic fungi disseminated during construction/renovation works

from the most frequent to the most uncommon according to the litterature

- *Aspergillus fumigatus in most cases*
- *A. non fumigatus (A. flavus, A. niger, A. terreus, A. nidulans, and others)*
- *Fusarium sp. (F. solani, F. oxysporum, F. moniliform)*
- *Zygomycetes (Rhizopus sp., Mucor sp., Absidia sp. Cunninghamella sp. and others)*
- *Scedosporium (S. apiospermium, S. prolificans)*
- *Dematiaceous (Alternaria sp., Exophiala sp. Ulocladium sp. Scopulariopsis sp., Curvularia sp.)*
- *Acremonium sp.*
- *Paecilomyces sp.*
- *Trichoderma sp.*

Two major difficulties exist in evaluating the hospital fungal infection risk:

- identify risk populations, thus eligible for prevention. The risk of IA is mainly known during hematological malignancies: particularly hematopoietic stem cell allograft or acute leukemia, neutropenia or graft against the host disease being the most at risk episodes (Patterson et al., 2016; Perfect et al., 2014; Lortholary et al., 2011). However, epidemiological data have been refined and revealed other contributing factors: corticosteroid therapy, solid organ transplantation, chronic lung diseases, systemic diseases and long-term immunosuppressive biotherapies (Cornillet et al., 2006),
- prove hospital acquired infection. This is a difficulty because the incubation period of IA is still not definitively known. On the other hand, biodiversity of

genotypes in the environment and patients rarely allows to affirm the character of nosocomial transmission.

14.2 Reservoir of Fungi at Hospital

14.2.1 Reservoir and Vectors of Micromycetes

Molds can be found in various places, sometimes unexpected, in the rooms or common areas of a hospital. Plants and cut flowers, boxes and packaging, engines and equipment vents fitted with engines, televisions and electric ramps, air humidifiers and oxygen bubblers etc… are examples of mold reservoirs. It is therefore necessary to maintain low or no biocontamination by banning and/or maintaining an effective biocleaning of such reservoirs in premises hosting high-risk patients fungal.

Yeasts and a number of filamentous fungi such as *Fusarium* sp. are particularly found in wetland persistent ways: bung sink or tub, nozzles and filter valves, among others. Water-related risk varies with the source of water (drawing water or surface water). The percentage of contaminated samples also varies from one study to another depending on the geographical location. In France, the combination of mostly drawing water and effective filtration systems provides in most cities assessed a percentage of samples contaminated very low, apart from rare exceptions (Gangneux et al., 2002). In the US, studies show sometimes much higher percentages of contamination, up to 21% of the positive samples (Anaissie et al., 2003).

The meal tray is also a source of both filamentous and yeast-like fungi: because of their presence in food themselves, but also spores on packaging, food trays, etc… (Vermorel-Faure et al., 1993. Bouakline et al., 2000).

Finally, the man may be carrying, but also vector when passing a contaminated locale to a clean ward with air treatment. Mold spores can thus be found on the hair and clothes essentially and yeasts on the skin, then causing a risk of dissemination and hand transmission.

14.2.2 Risk in Areas Not Receiving Air Treatment

The presence of small quantities of micromycetes is usual in these premises, and results from the accumulation of spores or yeast essentially *via* air or water. Their presence in higher amounts is often the result of excess moisture (water leak, for example) and/or deficiency in biocleaning. Thus, moldy areas, dust laden *Aspergillus* spores, sink plugs contaminated with yeast or *Fusarium* represent potential reservoirs of fungi. This is analogous situations in other premises, such as homes, but that could be partially mastered in everyday practice at the hospital, through technical building maintenance and cleaning measures.

14.2.3 Risk in Areas Receiving Air Treatment

The air treatment is an important strategy to prevent the accumulation of fungal spores in the environment of a patient at risk. Associated with a bio-cleaning and specific accompanying measures, it aims to lack of fungal reservoir. Its effectiveness varies with the level of filtration or purification in place. Accidental risk may be considered in case of failure or maintenance of the air handling system, or in case of introduction of spores through other vectors than air (people, equipment etc …).

14.3 Primary Routes of Transmission

14.3.1 Air Transmission

Air is the essential way of spore transmission to one or more patients. Conidia of *A. fumigatus* for example, with a small size between 2 and 5 microns in diameter are suspended in the contaminated air and therefore can reach the alveoli or the sinuses after inhalation (Latgé, 1999). Only the filamentous form is present in the lung tissue of patients and human transmission of filamentous fungi hardly ever occurs.

14.3.2 Transmission by Digestive Route

The digestive tract is a rare but possible gateway for molds. Several cases of gastrointestinal aspergillosis primary infection have been described in the literature (Kazan et al., 2011). It is the same with mucorales (Cheng et al., 2009). This gateway is more natural for the yeast in the diet or in water, which can translocate and therefore be responsible for fungemia. Fungemia cases after treatment with *Saccharomyces boulardii* probiotic yeast illustrate this risk of systemic infection after gastrointestinal translocation (Piechno et al., 2007).

14.3.3 Hand Transmission

The risk of transmission from patient to patient or caregiver to patient is well documented by several grouped cases of infections with *Candida parapsilosis*, a commensal yeast of the skin, but also by other species (Asmundsdóttir et al., 2008, Brillowska-Dabrowska et al., 2009). It usually results from a breach of basic rules of hygiene. Theoretically, such risk of transmission for filamentous fungi is possible, but exceptionally described.

14.4 Fungal Risk Control

Fungal risk prevention strategies are based on two approaches:

- chemoprophylaxis drug, anti-yeast or anti-filamentous referred,
- and control of the environmental fungal risk.

These two strategies are not mutually exclusive and must be viewed as complementary. Here we will only address methods for fungal environmental risk mastery.

14.4.1 Air Treatment

14.4.1.1 Air Filtration

This air treatment strategy is considered the gold standard. Optimally, the passage of air over HEPA (High Efficiency Particulate Air) filters, can retain 99.97% of particles larger than 0.3 microns. These systems must be subject to regular technical maintenance below ISO 14644 standards (Table 14.3).

It is primarily used in hematology wards, cell therapy sectors, but also in operating theaters. In hematology wards, technologies based either on unidirectional laminar air flow or on air ceiling filtration with the ultimate goal to achieve ISO classification 5. In operating theaters, air distribution types associated with filtration can rely on laminar or turbulent flow achieving from ISO 8 to ISO 5 classification, depending on the medical activity within the operating theater.

Clinical studies have underlined the benefit of air filtration in the prevention of invasive pulmonary aspergillosis during renovation works in hematology wards (Cornet et al., 1999). This technique is also recognized as a gold standard by the CDC (Centers for Disease Control and Prevention, 1997) or various national standards (Chang et al., 2014; French Consensus Conference, 2000).

14.4.1.2 Electromagnetic Epuration

This technology was developed for the MIR space station and used in the International Space Station Alpha. It is based on the treatment of all biological and particulate material via a plasma reactor without the need for filters. Therefore, the system eliminates a heavy maintenance filter change. This reactor is now developed either as a wireless system that can be installed as a plenum on top of a bed in a conventional room or at the terminal level of an air plant (Poirot et al., 2007). Furthermore mobile units are also available to clean the air of rooms and have proved effective in decreasing the aeropollution (Fernandez-Gerlinger et al., 2016; Sixt et al., 2007). These systems available in Europe have also received approval from the Federal Drug Administration in the United States of America.

14.4.1.3 Purification by Photocatalysis

Air treatment systems based on photocatalytic reactors using the properties of titanium dioxide has the ability, under the effect of light irradiation (UVA, B or C) to absorb pollutants and degrade into CO2, H2O and O2. Several devices have recently been put on the market by manufacturers.

14.4.1.4 Cyclonic Purification

This is a technology recently arrived in the area of health. The operating principle is the following: the ambient air is sucked by a centrifugal fan. Under the effect of the rotation, air enters tangentially in the upper part of the box and is driven in a swirling downward movement.

In cyclonic movement, the air is cooled and condenses on concentric layers of coils supplied with chilled water. A very large part of the dust is trapped by this swirling, which is achieved through the combined action of turbulent air movement in the cyclone exchanger and condensate runoff, which then flow into lower part of the cyclone.

Arriving at the lowest level, the supply air then rises in the output stack. Depending on the humidity requirements, it can be dried and warmed by passage through a coil module traversed by hot water.

14.4.2 Additional Measures

Areas benefiting from optimum air control, such as protected areas of hematology wards or operating theaters, are defined by the existence of an air handling system associated with various additional measures: overpressure of the locale compared to common areas, entry procedures for equipments and people (requiring the presence of a lock), operating procedures in routine use and during maintenance. All these procedures must be drafted and adhered to.

14.4.3 Food Decontamination

Many foods may be contaminated with fungal spores, sometimes in large quantities (Vermorel-Faure et al., 1993 Bouakline et al., 2000). Spices including pepper, are particularly rich in spores (De Bock et al., 1989), as well as cereals, due to the contamination of silos, cheeses and unpasteurized dairy products, vegetables and fruits, corn, coconut, peanuts, cashew nuts, coffee, tea, herbal teas of all kinds, the lyophilized products (including packet soups). Several working groups and teams have been asked to define:

- a "decontaminated food" category for heavy hematology units practicing auto- and allogeneic stem cell transplantation or implementing very haematotoxic chemotherapies as induction therapy of acute leukemia,
- and a "safe food" category when the risk is more limited (Vermorel-Faure et al., 1993 ; Gangneux et al., 2004).

Unfortunately, compliance with these rules is hardly compatible with the presentation of appetizing meal trays and organoleptic qualities.

14.4.4 Prevention in Case of Hospital Works

Construction and renovation works performed within a healthcare establishment are responsible for dust and re-suspended "clouds" of fungal spores (Barreiros et al., 2015). The level of fungal environmental risk must first be classified by type of work (Table 14.1). Then, the quantitative relationship between environmental contamination and infection risk for aspergillosis in high-risk patients must be analyzed using several works of literature (Arnow et al., 1991, Alberti et al., 2001, Vonberg et al., 2006, SF2H-SFMM working group, 2011).

Prevention measures must meet two major objectives:

- containing bioaerosols within the construction site
- and avoiding their dissemination into areas occupied by patients with a risk of fungal infection.

Several studies in the literature have shown the efficacy of such measures applied on the fungal contamination of the environment or reducing the incidence of aspergillosis (Arnow et al., 1991, Loo et al., 1996, Araujo et al., 2008). The technical measures reported as effective include the following:

- The central processing air with HEPA filtration in care units (Sherertz et al., 1987, Cornet et al., 1999, Bénet et al., 2007). The benefit of HEPA filtration on mortality and the incidence of fungal infections is however disputed in a meta-analysis of 16 studies (Eckmanns et al., 2006).
- The air treatment by mobile units (Mahieu et al., 2000, Sautour et al., 2007, Bergeron et al., 2007).

Table 14.1 Classification of construction works according to the volume of dust they produce, as defined by [Anonymous Canada, 2001; Anonymous Ireland, 2001; Haiduven 2009]

Types of construction work (*non exhaustive lists*)	
TYPE A. Non-invasive control work/ internal work with minimum production of dust – removal of suspended ceiling panels for inspection, limited to 1 plate/m^2 – painting without sanding – paperhanging – minor electrical work – minor plumbing with water cutoff in the room lasting <15 min – other inspection work requiring neither recesses in the walls, nor more extensive interventions on suspended ceilings	**TYPE B. Short-duration, minor construction work producing small quantities of dust** – wire recesses in the walls or ceilings, with controlled production of dust for minor electrical installations or repairs on ventilation components, telephone or computer cabling – removal of floor covering (limited area) – minor construction work on suspended ceilings – sanding/grinding of the walls for paint removal or wallpapering involving the repair of only a small area – plumbing work with water cutoff affecting ≥2 rooms for less than 30 min – any construction work that can be performed by a single building trade
TYPE D. Any construction work producing moderate to high levels of dust, or requiring the demolition or removal of any fixed item (e.g. sinks, boards…) – sand blasting/sanding of walls for painting or wallpapering; any construction work with plaster elements – minor demolition – removal of floor coverings and suspended ceilings – construction of new walls; installation of new partitions – minor construction – minor piping or electrical wiring work in the ceilings – minor excavation – major wiring activities – any activity that requires several building trades – any plumbing work with water cutoff affecting >2 rooms for >30 min, but <1 h	**TYPE E. Major demolition, renovation, construction work/major external construction work with significant dust production** – demolition or renovation of an entire wiring system – new construction involving several building trades – plumbing with water cutoff affecting >two rooms, for >1 h – major excavations

Several additional points still questioning, include:

- the homogenization of procedures for prevention of an institution to another,
- the validation of measurement tools for highly polluted environments,
- and the determination of acceptable pollution threshold values.

14.5 Validation of Effectiveness of Preventive Measures for Environmental Monitoring

14.5.1 When Implementing a Fungal Environmental Monitoring in the Hospital?

It is not reasonable to propose a non-targeted fungal environmental monitoring for two major reasons: cost and performance limits in air are not mastered. The "indications" of fungal environmental monitoring currently recommended or recognized are essentially:

- monitoring the effectiveness of measures to suppress fungal pollution, that is to say monitoring areas benefiting from controlled air;
- checking for the effectiveness of preventive measures implemented during construction and renovation works;
- and investigating on the origin of clusters of airborne nosocomial fungal infections.

14.5.2 What Measurement Methods Using for Environmental Monitoring?

Spores from the asexual reproduction of filamentous fungi are currently the most used marker for environmental monitoring. Thus, biocontamination is monitored via air and surface samples then cultivated on growth medium. This combination of tests provides informations on the air quality at the time of sampling and reflects also the history of the deposition of any fungal spores suspended in the air (Gangneux et al., 2002). Monitoring of other possible vectors of spores such as water or food is not systematic but can occasionally meet specific investigations.

Other fungal elements can also be found in the environment and made searchable by non culture methods: DNA and fungal fragments, the membrane constituents and the cell wall such as ergosterol or beta-1-3-D-glucans, volatile organic compounds or mycotoxins. However, these detection methods are more readily used for environmental monitoring in homes and workplaces where identifying the allergenic burden is more relevant than in hospitals where the risk evaluation mainly concerns nosocomial infection.

Note that the particulate control is not a substitute for biological control because it does not identify the exact nature of the recorded particles. Several studies have also shown possible discrepancies between particulate counts and biological controls (Landrin et al., 2005).

14.5.3 What Indicators Relevant to Consider?

Looking for *Aspergillus* in the hospital environment is a priority. But whereas this fungal genus is responsible for over 80% of pulmonary invasive fungal infections in humans, it is usually less than 10% of the fungal flora found in the environment. Therefore the calculation of the total fungal flora after culture on specific medium is a broader indicator and is correlated with the presence of *Aspergillus* (Alberti et al., 2001). In addition, identification of the main genera of human infections such as *Aspergillus* sp., *Rhizopus* sp. and other mucorales, *Fusarium* sp., *Scedosporium* sp., *Paecilomyces* sp., or *Acremonium* sp. are currently performed.

14.5.4 Proposal of a Sampling Plan in Areas with Air Treatment

14.5.4.1 Patient Room

As an example, a French working group have proposed a sampling plan (Gangneux et al., 2002). In the case of laminar air flow, two air samples could be of interest: one in the flow output area, and one off the flow. In the absence of unidirectional laminar flow, one air sample will be enough. Surface samples could be taken in parallel: at least five per room, collected systematically at each control, in places near the patient (bedside table) but also among sites favorable to developing mold (electric ramp, windowsill, baseboards, air extraction grids, television, telephone, etc.).

14.5.4.2 Common Areas of Wards Receiving Air Treatment

The number of air sampling and surface and locations will be determined based on the architecture of the ward: nursing stations, rest area, storage area, office, corridors …. A monthly rate is usually proposed but can be weekly in some hospitals (Reboux et al., 2014).

14.5.4.3 Operating Theaters

In the absence of specific microbiological standards, the sampling plan will be adapted to that of bacteriology. The check will be carried out activity.

14.5.5 Plan for Environmental Monitoring During Construction/ Renovation Works

This surveillance is not systematic, should be discussed with local authorities, depending on the existence of at-risk patients, the type of work and the level of

Table 14.2 Proposed frequency of environmental monitoring to be implemented, and responsibilities

Overall quantification of risk	Monitoring				
	Frequency and persons in charge				
	Visual	Pressure	Particulates	Airborne contamination	Surfaces
	Healthcare Unit	Technical Staff*	ICT**	ICT/laboratories	ICT/ laboratories
High «Protected areas»	Once daily	Once daily	At the end of work	Once weekly and at the end of work	Once weekly and at the end of work
High Other areas	Once daily	Once daily	–	To be defined by the ICT*** et at the end of work	At the end of work
Average	Once daily	–	–	–	At the end of work
Low	Once weekly	–	–	–	–

*Technical Department or Biomedical Department.
**ICT: Infection Control Team.
***For information and according to the duration of construction work, once or twice monthly.

protection expected. The purpose of sampling is to assess the effectiveness of preventive measures in place and to evaluate the fungal environmental quality prior to occupancy by patients at risk.

Monitoring should combine various methods:

- visual check,
- control of the depression of the work area if it is put in place,
- and particulate control and biological monitoring in areas with air treatment.

Frequency of environmental monitoring to be implemented are proposed in Table 14.2.

14.6 Tools for The Environmental Fungal Monitoring in Health Facilities

Culture methods are most often used in the environmental fungal monitoring in health care facilities. These specialized techniques require trained personnel (bio-hygienist technician or equivalent), with material decontaminated and packaged before entering a protected area, and culture and fungal identification performed by a trained staff. More recently, non-cultural methods have been developed but are more often used in the field of home surveillance as explained below (Méheust et al., 2014).

14.6.1 *Environmental Samples*

14.6.1.1 Air Samples

The air samples are taken with various air samplers which can rely on several types of technology. Bio-impaction is the most commonly used and responds to different standardized criteria: the speed of impingement and the air flow. Several bio-impactors with a convenience of use are on the market, and have been evaluated in various works of literature (Gangneux et al., 2006; Méheust et al., 2013, 2014).

It is important to always use the same device, the same sampling sites, and to respect volumes defined. In a controlled air where the level of aerobiocontamination is assumed zero or very low, the volume of a cubic meter of air is appropriate. In uncontrolled area, volume is often less so as not to saturate the agar (usually between 100 and 500 L, depending on the assumed level).

14.6.1.2 Surface Sampling

They can be performed with contact agar or wet swab. Contact agar are more sensitive but can only be used on plane surfaces. When some areas need to be monitored such as windowsill, air extraction grids, baseboards …, then swabs are more practical. Whatever the technique, it must be done in a standardized way.

14.6.2 *Mycological Analysis*

14.6.2.1 Culture Media

Two culture media are currently recommended for hospital surveillance: Sabouraud Agar and Malt extract Agar. Malt extract Agar is at least as sensitive as the Sabouraud medium and further less favorable for bacterial growth, promotes sporulating certain fungal species and therefore their identification (Méheust et al., 2014).

14.6.2.2 Incubation Temperatures

An incubation temperature of 25°C to 30°C is suitable for the cultivation of most fungi in the air (Méheust et al., 2014). The overall level of fungal contamination is a marker of the risk of *Aspergillus* environmental contamination (Alberti et al., 2001). Culturing at 37°C only allows to select specifically thermophilic species *Aspergillus*.

14.6.2.3 Duration of Incubation and Identification Methods

A first enumeration of the number of colonies is performed on days 2–3 of culture, which will again be counted on days 5–7. It is at this point that it becomes possible to identify colonies macro- and microscopically. The main genera, i.e. *Aspergillus* should be identified precisely at the species level. New identification tools such as molecular biology or MALDI-TOF mass spectrometry are increasingly accessible in routine.

14.6.3 Epidemiological Surveillance

In health facilities, monitoring of environmental risk must be coupled with epidemiological surveillance of patient infection risk. A rigorous and comprehensive monitoring allows:

- an accurate picture of local epidemiology,
- to detect clusters of cases and/or outbreaks,
- validate/correct preventive measures in place.

Several recommendations stressed the value of a local structure of epidemiological surveillance of invasive aspergillosis during periods of work, better still lasting for a prospective analysis of cases (Centers for Disease Control and Prevention, 1997). This task force must have all the skills directly involved in the prevention and include the following actors: hygienists, mycologists, representatives of medical wards, technical engineer in case of work, and administration officials.

A sustainable organization allows a prospective epidemiological surveillance of cases, but can also be involved as a crisis unit with a rapid implementation in the event of epidemic alert, i.e. a significant increase in the usual incidence of IFI or after reporting a nosocomial IFI. The procedure to be proposed then includes various steps (Davoudi et al., 2015): confirm the presence of outbreak, establish case definition, exhaustive search of all cases, determine environmental involvement using various measures, description of the spatial location of the event, implement control and prevention measures, and alert key partners.

14.7 Regulations and Standards Associated with Environmental Fungal Risk in Health Facilities

Several areas are affected by regulations or standardizations to be respected as part of the accreditation procedures.

14.7.1 Definition of Risk Areas

Four risk areas are defined by the ISO 14698 (Table 14.3). For each of them, the particle class, the kinetic particle decontamination class at 0.5 microns, the kinetics of class bio-decontamination, and the ventilation rate are defined. Other physical indicators, such as temperature, humidity of air or pressure are also defined. Two risk areas are of particular interest in health facilities, namely zone 3 defined as high risk area and zone 4 defined as very high risk of infection corresponding to particulate class ISO 7 and ISO 5, respectively.

14.7.2 Types of Controls to Set Up

The standard EN ISO 14644 for clean rooms of health institutions indicates the need to combine particulate and microbiological controls for qualifying the installation of a system or a system in use (Table 14.3).

Table 14.3 Standards related to health environmental management from the International organization for standardization

ISO 14644 standards	– ISO 14644-1: Classification of air cleanliness – ISO/DIS 14644-1.2(2014): Classification of air cleanliness by particle concentration – ISO 14644-2: Specifications for testing and monitoring to prove continued compliance with ISO 14644-1 – ISO/DIS 14644-2.2(2014):Monitoring to provide evidence of cleanroom performance related to air cleanliness by particle concentration – ISO 14644-3: Test Methods – ISO 14644-4: Design, Construction, and Start-up – ISO 14644-5: Operations – ISO 14644-6: Vocabulary – ISO 14644-7: Separative devices (clean air hoods, gloveboxes, isolators and minienvironments) – ISO 14644-8: Classification of airborne molecular contamination – ISO 14644-9: Classification of surface particle cleanliness – ISO 14644-10: Classification of Surface Cleanliness by Chemical Concentration – ISO 14644-12: Classification of Air Cleanliness by Nanoscale Particle Concentration
ISO 14698 standards	– ISO 14698-1, Cleanrooms and associated controlled environments – Biocontamination control, Part 1: General principles and methods – ISO 14698-2, Cleanrooms and associated controlled environments – Biocontamination control, Part 2: Evaluation and interpretation of biocontamination data

14.7.3 Achievement of Air Samples by Bioimpaction

The recent ISO 14698-1 defines a number of parameters for the samplers based on impact and trapping. Bio-impactors shall respect speed impaction less than 20 m/s and an air flow of 100 L/min.

14.8 Challenges Ahead in Environmental Fungal Risk in Hospitals

The major challenge is the prevention of nosocomial infection risk, i.e. the prevention of invasive fungal infection in a patient exposed to an environmental reservoir in a health facility. This prevention is part of the hospital quality initiative in the fight against nosocomial infections related to air or water and usually lacks of homogeneisation (Lequilliec et al., 2017).

It is currently based on two complementary strategies : drug chemoprophylaxis and prevention of the environmental risk.

A number of recommendations have been published in recent years and are summarized in this document. However, they need to be evaluated in routine practice, in terms of their feasibility and effectiveness. They should also be evaluated in terms of cost and benefit/risk ratio. Indeed, the increasing incidence of fungal infections and additional costs resulting from the increase in length of hospital stay and antifungal usage encourage to optimize prevention of these infections. In the absence or in addition to effective chemoprophylaxis, environmental preventive measures are known to be at the forefront of the fight against fungal nosocomial infections. The long-standing monitoring of bacterial infections has now to be associated with the monitoring of fungal infections but also viral with regards to the high-risk populations of patients.

References

Alberti C, Bouakline A, Ribaud P et al (2001) Relationship between environmental fungal contamination and the incidence of invasive aspergillosis in hematology patients. J Hosp Infect 48:198–206

Anonymous Canada (2001) Canadian Communicable Disease Report. Construction-related nosocomial infections in patients in health care facilities. Volume 27S2, July 2001. http://www.phac-aspc.gc.ca/publicat/ccdr-rmtc/01vol27/27s2/index.html

Anonymous Ireland (2001) National Disease Surveillance Centre Ireland. National guidelines for the prevention of nosocomial invasive aspergillosis during construction/renovation activities. http://www.ndsc.ie/hpsc/A-Z/Respiratory/Aspergillosis/Guidance/File,896,en.pdf. 2001

Arnow PM, Sadigh M, Costas C et al (1991) Endemic and epidemic aspergillosis associated with in-hospital réplication of *Aspergillus* organisms. J Infect Dis 164:998–1002

Araujo R, Carneiro A, Costa-Oliveira S et al (2008) Fungal infections after haematology unit renovation: evidence of clinical environmental and economical impact. Eur J Haematol 80:436–443

Asmundsdóttir LR, Erlendsdóttir H, Haraldsson G et al (2008) Molecular epidemiology of candidemia: evidence of clusters of smoldering nosocomial infections. Clin Infect Dis 47: e17–24

Anaissie EJ, Stratton SL, Dignani MC et al (2003) Pathogenic molds (including *Aspergillus* species) in hospital water distribution systems: a 3-year prospective study and clinical implications for patients with hematologic malignancies. Blood 101:2542–2546

Barreiros G, Akiti T, Magalhães AC et al (2015) Effect of the implosion and demolition of a hospital building on the concentration of fungi in the air. Mycoses 58(12):707–13

Bénet T, Nicolle MC, Thiebaut A et al (2007) Reduction of invasive aspergillosis incidence among immunocompromised patients after control of environmental exposure. Clin Infect Dis 45:682–686

Bergeron V, Reboux G, Poirot JL et al (2007) Decreasing airborne contamination levels in high-risk hospital areas using a novel mobile air-treatment unit. Infect Control Hosp Epidemiol 28:1181–1186

Bouakline A, Lacroix C, Roux N et al (2000) Fungal contamination of food in hematology units. J Clin Microbiol 38:4272–4273

Brillowska-Dabrowska A, Schön T, Pannanusorn S et al (2009) A nosocomial outbreak of *Candida parapsilosis* in southern Sweden verified by genotyping. Scand J Infect Dis 41: 135–142

Centers for Disease Control and Prevention (CDC) (1997) CDC guidelines for prevention of nosocomial pneumonia. MMWR 39:1192–1236

Chang CC, Ananda-Rajah M, Belcastro A et al (2014) Consensus guidelines for implementation of quality processes to prevent invasive fungal disease and enhanced surveillance measures during hospital building works, 2014. Intern Med J 44(12b):1389–97

Cheng VC, Chan JF, Ngan AH et al (2009) Outbreak of intestinal infection due to *Rhizopus microsporus*. J Clin Microbiol 47:2834–2843

Cornet M, Levy V, Fleury L et al (1999) Efficacy of prevention by high-efficiency particulaite air filtration or laminar airflow against *Aspergillus* airborne contamination during hospital renovation. Infect Control Hosp Epidemiol 20:508–513

Cornillet A, Camus C, Nimubona S et al (2006) Comparison of epidemiological, clinical, and biological features of invasive aspergillosis in neutropenic and nonneutropenic patients: a 6-year survey. Clin Infect Dis 43:577–584

Davoudi S, Graviss LS, Kontoyiannis DP (2015) Healthcare-associated outbreaks due to Mucorales and other uncommon fungi. Eur J Clin Invest 45(7):767–73

De Bock R, Gyssens I, Peetermans M et al (1989) *Aspergillus* in pepper. Lancet 2:331–332

Eckmanns T, Rüden H, Gastmeier P (2006) The influence of high-efficiency particulate air filtration on mortality and fungal infection among highly immunosuppressed patients: a systematic review. J Infect Dis 193:1408–1418

Fernandez-Gerlinger MP, Jannot AS, Rigaudeau S et al (2016) The plasmair decontamination system is protective against invasive Aspergillosis in neutropenic patients. Infect Control Hosp Epidemiol 37(7):845–51

French Consensus Conference (2000) [Prevention of invasive aspergillosis in immunosuppressed at risk population]. Hygiènes VII(6)

Gangneux JP, Bretagne S, Cordonnier C et al (2002) Prevention of nosocomial fungal infection: the French approach. Clin Infect Dis 35:343–345

Gangneux JP, Noussair L, Bouakline A (2004) Experimental assessment of disinfection procedures for eradication of *Aspergillus fumigatus* in food. Blood 104:2000–2002

Gangneux JP, Robert-Gangneux F, Gicquel G et al (2006) Bacterial and fungal counts in hospital air: comparative yields for 4 sieve impactor air samplers with 2 culture media. Infect Control Hosp Epidemiol 27:1405–1408

Haiduven D (2009) Nosocomial aspergillosis and building construction. Med Mycol 47(Suppl 1): S210–6

Kazan E, Maertens J, Herbrecht R et al (2011) A retrospective series of gut aspergillosis in haematology patients. Clin Microbiol Infect 17:588–594

Landrin A, Bissery A, Kac G (2005) Monitoring air sampling in operating theatres: can particle counting replace microbiological sampling? J Hosp Infect 61:27–9

Latgé JP (1999) *Aspergillus fumigatus* and aspergillosis. Clin Microbiol Rev 12:310–350

Lequilliec N, Raymond R, Vanjak D, Baghdadi N, Boulestreau H, Zahar JR, Gangneux JP, (2017) Practices of infectious control management during neutropenia: A survey from 149 French hospitals. Journal de Mycologie MÃ©dicale / Journal of Medical Mycology 27 (2): 227–231

Loo VG, Bertrand C, Dixon C et al (1996) Control of construction-associated nosocomial aspergillosis in an antiquated hematology unit. Infect Control Hosp Epidemiol 17:360–364

Lortholary O, Gangneux JP, Sitbon K et al (2011) Epidemiological trends in invasive aspergillosis in France: the SAIF network (2005-2007). Clin Microbiol Infect 17:1882–1889

Mahieu LM, De Dooy JJ, Van Laer FA et al (2000) A prospective study on factors influencing *Aspergillus* spore load in the air during renovation works in a neonatal intensive care unit. J Hosp Infect 45:191–197

Méheust D, Gangneux JP, Cann PL (2013) Comparative evaluation of three impactor samplers for measuring airborne bacteria and fungi concentrations. J Occup Environ Hyg 10(8):455–9

Méheust D, Le Cann P, Reboux G et al (2014) Indoor fungal contamination: health risks and measurement methods in hospitals, homes and workplaces. Crit Rev Microbiol 40(3):248–60

Patterson TF, Thompson 3rd GR, Denning DW et al (2016) Practice guidelines for the diagnosis and management of Aspergillosis: 2016 Update by the Infectious Diseases Society of America. Clin Infect Dis 63(4):e1–e60

Perfect JR, Hachem R, Wingard JR (2014) Update on epidemiology of and preventive strategies for invasive fungal infections in cancer patients. Clin Infect Dis 59(Suppl 5):S352–5

Piechno S, Seguin P, Gangneux JP (2007) Septiciémie due à *Saccharomyces boulardii*: attention aux levures. Can J Anaesth 54:245–246

Poirot JL, Gangneux JP, Fischer A et al (2007) Evaluation of a new mobile system for protecting immune-suppressed patients against airborne contamination. Am J Infect Control 35:460–466

Reboux G, Gbaguidi-Haore H, Bellanger AP et al (2014) A 10-year survey of fungal aerocontamination in hospital corridors: a reliable sentinel to predict fungal exposure risk? J Hosp Infect 87(1):34–40

Sautour M, Sixt N, Dalle F (2007) Prospective survey of indoor fungal contamination in hospital during a period of building construction. J Hosp Infect 67:367–373

SF2H-SFMM Working Group (2011) Risk of fungal infections and construction works in hospitals. Hygienes 19:1-52.

Sherertz RJ, Belani A, Kramer BS et al (1987) Impact of air filtration on nosocomial *Aspergillus* infections. Unique risk of bone marrow transplant recipients. Am J Med 83:709–718

Sixt N, Dalle F, Lafon I (2007) Reduced fungal contamination of the indoor environment with the Plasmair system (Airinspace). J Hosp Infect 65:156–162

Vermorel-Faure O, Lebeau B, Mallaret MR et al (1993) [Food-related fungal infection risk in agranulocytosis. Mycological control of 273 food items offered to patients hospitalized in sterile units]. Presse Med 22:157–160

Vonberg RP, Gastmeier P (2006) Nosocomial aspergillosis in outbreak settings. J Hosp Infect 63:246–254

Chapter 15
Bacterial Bioburden in Hospital Environment

Ana Monteiro and Sandra Cabo Verde

Abstract Hospital buildings may be regarded as dynamic environments affected by season, weather conditions, indoor ventilation system design and operation, intrusion of moisture, outdoor microbial load and the number of occupants, visitors and human activities. Airborne microorganisms, such as bacteria, can originate not only from humans (including patients), but can also be spawned by various indoor hospital characteristics and outdoor environmental sources. These factors may be associated with conditions for microbial growth, leading to bacterial air pollution. Poor hospital indoor air quality may cause hospital-acquired infections, sick hospital syndrome and various occupational hazards. Therefore, ensuring effective disinfection procedures and regular monitoring measures of hospital environment is essential for microbial control by detecting the irregular introduction of airborne particles via clothing of visitors and medical staff or carriage by personal and medical materials. Achieving a satisfactory microbial control in hospitals environment is thus a multidisciplinary challenge for health care professionals, hospital managers and occupational health.

Keywords Bacterial bioburden · hospital environment · indoor air · occupational health

15.1 Introduction

The complex hospital environment requires special attention to protect patients and healthcare workers against hospital-acquired (nosocomial) infections and occupational diseases (Leung and Chan, 2006), once the risk of nosocomial infection, in

A. Monteiro (✉)
Research Group Environment & Health, Lisbon School of Health Technology/Polytechnic Institute of Lisbon, Lisbon, Portugal
e-mail: ana.monteiro@estesl.ipl.pt

S. Cabo Verde
Centro de Ciências e Tecnologias Nucleares - C2TN, Instituto Superior Técnico, Universidade de Lisboa, Estrada Nacional, Bobadela LRS, Portugal
e-mail: sandracv@ctn.tecnico.ulisboa.pt

C. Viegas et al. (eds.), *Exposure to Microbiological Agents in Indoor and Occupational Environments*, DOI 10.1007/978-3-319-61688-9_15

hospital environments is higher than in other environments (Klevens et al., 2007). Intensive Care Units are considered potential reservoirs for pathogenic (opportunistic) microbial strains (Russotto et al., 2015). The microbiological quality of the air in hospital environments is increasingly a worldwide concern with regard to occupant and patient health. Surveys in hospitals revealed that certain bacterial infections in patients are suspected to have been acquired in a hospital setting, but also the patients may serve as source and vehicle of pathogenic bacteria to staff and hospital visitors (Obbard and Fang, 2003). The transmission by the patients, contaminated instruments, equipment's or surfaces and the hospital environments are the main risk of infection for workers (Luksamijarulkul et al., 2014).

Most bacteria do not cause adverse health effects, some of which are even essential for man and the environment, but there are health risks when the concentrations of some species are extremely high (Goye et al., 2001). Some of the bacterial species have pathogenic properties, allergic, infectious or toxic to humans (Douwes et al., 2003) and exposure to these microorganisms may have a number of health effects (Rintala et al., 2008). The aerosol particles of microbiological origin e.g., viruses, bacteria and fungal spores, have been associated with allergies, asthma and respiratory diseases (Luksamijarulkul et al., 2014). The severity of the effects depends, among other factors, on their toxicity, microbial load and exposure time, as well as on the age and nutritional status of exposed individuals (Huff et al., 1998).

15.2 Reservoirs and Routes of Transmission of Bacteria in Hospital Environment

Bacteria are found in the air, their main route of transmission is by aerosolization (Goyer et al., 2001), and they exist in higher concentration in indoor than in outdoor (Goyer et al., 2001; Kim and Kim, 2007; Madureira et al., 2015; Mentese et al., 2009). In the outdoor environment, bacteria mainly originate from water, soil and plants and are also associated with the presence of humans and animals (Goyer et al., 2001).

In hospitals, the sources of infectious agents are the normal endogenous microbial flora of patients (Luksamijarulkul et al., 2014). The common sources of aerial transmission include respiratory ejection from the mouth and nose, skin exudates, infected lesions, as well as respiratory diseases (Bartlett et al., 2004; Tang et al., 2009; Wan and Chung, 2011). Even if patients do not present disease symptoms or carriers of chronic disease may be potential sources and hosts of infectious agents (Luksamijarulkul et al., 2014). Also the workers and visitors (Luksamijarulkul et al., 2014; Tang, 2013), air conditioning system, ventilation rates and human factors such as overcrowding in restricted spaces (Bartlett et al., 2004; Tang, Chung and Lin, 2009; Wan and Chung, 2011) are considered important sources of microbial contamination by air (Tang, 2013).

The facilities of the health units can be considered as dynamic environments, which are affected by numerous factors that can boost microbial growth, namely, the time of year, weather, design and operation of the ventilation system, humidity, microbial load outdoor, number of occupants, visitors and human activities (Klánova and Hollerová, 2003; Park et al., 2013). Also an inadequate maintenance of HVAC systems can be a source of particles and microorganisms (Charkowska, 2001; Pejtersen, 1996).

The higher concentration of bacterial may be due to temperature, humidity, presence of unhygienic attached toilets, poor waste management system, poor or improper ventilation system (Gizaw et al., 2016; Napoli et al., 2012), wear and tear of buildings, incomplete disinfection of wards and, the high number of patients and visitors (Eslami et al., 2016).

15.3 Bacterial Load in Hospital Settings

Some studies were carried out with the objective of determining the bacterial load in air samples. High bacterial load (>103 CFU/m^3) has been reported in a ward in Gondar University teaching hospital in Northwest Ethiopia (Gizaw et al., 2016), and similar results have been found in a study realized in 37 hospitals in Taiwan (Jung et al., 2015). Nevertheless, other hospital services were described to present high airborne bacterial concentration. For example, the eye operating room (22 CFU/plate/h) of the teaching hospitals in Ghazvin University of Medical Sciences in Iran (Eslami et al., 2016); the hospital rooms (102 CFU/m^3) at a hospital in Murcia, Spain (Ortiz et al., 2009); and the emergency service (102 CFU/m^3) at a hospital in Setubal, Portugal (Cabo Verde et al., 2015).

Determining the incidence of microflora in hospitals may be more important than their amount determination in the air, not only for an understanding of the types of infections and allergies that may arise from them, as well as to play a role in preventing cross infection. Considering the nature of the activities performed in hospital environments, a large spectra of occupants can be found, namely immunocompromised people, patients who use respiratory therapy and catheters, submitted to surgery among other invasive medical process (Nunes et al., 2005; Ekhaise et al., 2008).

15.4 Most Frequent Bacteria Found in Indoor Hospital Environment

The predominant gram-positive bacteria isolated in the air belong to the genera *Arthrobacter*, *Micrococcus*, *Bacillus*, *Staphylococcus*, *Kocuria* and *Streptococcus*. The gram-negative bacteria commonly found are: *Escherichia*, *Pseudomonas*, *Acinetobacter*, *Klebsiella*, *Citrobacter*, *Enterobacter*, *Moraxella*, *Flavobacterium*, *Xanthomonas* and *Legionella* (water) (Goyer et al., 2001). Specifically, *Pseudomonas aeruginosa* is frequently detected and abundant in the hospital environment (Huang et al., 2013). Other bacterial species belonging to the genus

Staphylococcus are normally found in the indoor environment, and the pathogen, *Staphylococcus aureus*, is recognized as the most important nosocomial agent in the hospital setting (Badri, 2014). Bacterial isolates identified in some studies (Table 15.1) included coagulase-negative *Staphylococcus* spp., a non-pathogenic group of bacteria which belong to the normal skin microflora and mucous membranes of the respiratory tract. Other bacterial genera identified in some hospital locations, also normal in microflora of the skin, as well as the respiratory, gastro-intestinal and urinary tracts were *Corynebacterium*, *Acinetobacter*, *Pseudomonas* and *Moraxella*. The opportunistic bacteria identified i.e., *Acinetobacter* spp. and *Flavobacterium* spp. can be transmitted via respiratory droplets, causing infection particularly in susceptible individuals such as immunocompromised patients (Obbard and Fang, 2003).

In Intensive Care Unit, objects and surfaces near patients, indicated to have more human gut, hair and skin associated bacteria, such as *Staphylococcus*, *Propionibacteria*, *Corynebacteria*, *Lactobacillus*, *Micrococcus* and *Streptococcus* (Handorean et al., 2015). The typical bacteria on the surfaces of Neonatal Intensive Care Units were recognized to belong to the genera *Staphylococcus*, *Enterococcus*, *Acinetobacter*, Bacteroides, *Burkholderia*, *Clostridia*, *Pseudomonas* and *Streptococcus* (Hewitt et al., 2013; Brooks et al., 2014). In turn, the bacterial diversity found on the surfaces of the Operating Rooms is more restricted, being *S. aureus* the most commonly isolated microorganisms as well as atypical skin-associated microbe (Shin et al., 2015)

15.5 Measures to Control Bacterial Contamination in Hospital Environment

Hospital air quality is considered an important risk factor for the health of staff and patients. The monitoring of bioaerosols present in hospitals' indoor air, particularly in intensive care units and operation rooms, seems essential for the epidemiological study of nosocomial infections, characterization and control of bioaerosols (Eslami et al., 2016). Although, it must be defined the methodology to be used for the sampling of bioaerosols, since there is still no consensus among experts (Nunes et al., 2005).

Ensuring hospital environmental control procedures can play an important role in the prevention of nosocomial infections, hospital syndrome and other occupational hazards, especially in immunosuppressed and immunocompromised patients, which are highly susceptible to the adverse effects of various chemical and aerial microbes (Eslami et al., 2016).

Although microbial counts in indoor air depend on different variables, such as velocity, humidity, temperature, ventilation, number of occupants, people's activities, a concentration of particulate matter or dust, and the outdoor air quality (Luksamijarulkul et al., 2014), it is through air sampling that is possible to evaluate the microbial contamination in environments with high risk of infection.

Table 15.1 Bacterial genera or species reported in hospital indoor air samples

Setting and number of studied samples	Studied period	Bacteria	Reference
Hospital (operating room) air $n = 50$	2005	*Staphylococcus aureu*, *Enterococcus* spp. *Streptococcus* z gr. B, *Pseudomonas aeruginosa* *Acinetobacter lwoffii* *Alcaligenes faecalis*	(Fleischer et al., 2006)
Hospital (male, female and children ward, theater and bacteriological laboratory) air	2008	*Staphylococcus aureus* *Staphylococcus epidermidis* *Escherichia coli* *Pseudomonas aeruginosa*	(Ekhaise et al., 2008)
Hospital air $n = 37$	2 months	*Staphylococcus epidermidis* *Staphylococcus saprophyticus*	(Eslami et al., 2016)
Hospital air $n = 28$	2015	*Staphylococcus aureus* *Streptococcus pyogenes*	(Gizaw et al., 2016)
Pediatric hospital air	12 months (2014)	*Staphylococcus* *Bacillus* *Corynebacterium* *Micrococcus*	(Okten and Asan, 2012)
Hospital ward of the pneumonological department $n = 240$	1 Jun 1981–28 May 1982	*Staphylococcus epidermidis*	(Augustowska and Dutkiewicz, 2006)
Hospital air $n = 114$	Jun 2013–February 2014	*Staphylococcus aureus* *Staphylococcus capitis* *Staphylococcus hominis* *Staphylococcus epidermidis* *Staphylococcus warneri* *Micrococcus luteus* *Micrococcus lylae*	(Cabo Verde et al., 2015)
Hospital (lobby, ward, pharmacy) air $n = 15$	1 month	*Staphylococcus* spp. *Corynebacterium* *Acinetobacter* *Pseudomonas* *Moraxella* spp. *Acinetobacter* spp. *Flavobacterium* spp.	(Obbard and Fang, 2003)

Air control measures are crucial for reducing dissemination of airborne biological particles in hospitals (Okten and Asan, 2012). To prevent the microbial growth several actions should be followed such as, avoidance of wet surfaces, keeping relative humidity levels below 70%, implement team's hygiene procedures (Napoli et al., 2012), proper HVAC system operation and maintenance and effective filtration of particulates (Luksamijarulkul et al., 2014). Strict cleaning protocols, monitoring of microbial colonization and educational interventions of the cleaning procedures and results, it's an important factors for preventing infections in Intensive Care Units (Mora et al., 2016). The importance of the monitoring is not only because can significantly reduce the contamination of surfaces near patients, but because can point out the weaknesses of current protocols (Carling, 2013).

In healthcare facility, mechanical ventilation is essential for diluting indoor air pollutants by exhausting the contaminated indoor air and introducing clean outdoor air into an air-conditioned building (Leung and Chan, 2006). Air-ventilation has been referred to one of the principle techniques in reducing and controlling the spread of infectious agents through air in hospitals (Eslami et al., 2016). In some areas, such as operating theatres and delivery rooms, ventilation must be only by outdoor air, a suitable appropriate filtration system should be adopted to capture and remove particulate contaminants, and maintenance of air pressure between a hospital facility should be controlled to ensure clean-to-less-clean air flows (Leung and Chan, 2006).

Another measure to prevent microbial contamination and to provide patients and hospital staff with maximum protection is the implementation of a management program designed for controlling the emissions sources, coordinating preventive activities and promoting staff awareness (Leung and Chan, 2006).

References

Augustowska M, Dutkiewicz J (2006) Variability of airborne microflora in a hospital ward within a period of one year. Ann Agric Environ Med 13(1):99–106. http://doi.org/13099 [pii]

Badri RM (2014) Identification and characterization of bacteria air pathogens from homes in some areas of the Baghdad City. *Material Method*. 2(6): 384–388

Bartlett KH, Martinez M, Bert J (2004) Modeling of occupant-generated CO2 dynamics in naturally ventilated classrooms. J Occup Environ Hyg 1(3):139–148

Brooks B, Firek BA, Miller CS et al. (2014) Microbes in the neonatal intensive care unit resemble those found in the gut of premature infants. Microbiome 2(1):1. doi:10.1186/2049-2618-2-1

Cabo Verde S, Almeida SM, Matos J et al. (2015) Microbiological assessment of indoor air quality at different hospital sites. Res Microbiol 166(7):557–563. doi:10.1016/j.resmic.2015.03.004

Carling P (2013) Methods for assessing the adequacy of practice and improving room disinfection. Am J Infect Control 41(5 Suppl.):S20–S25. doi:10.1016/j.ajic.2013.01.003

Charkowska A. (2001) Concentration of micro-organisms in hospital air-conditioning systems – comparison between winter and summer conditions, In Proceedings 7th REHVA World Congress, Clima 2000/Napoli 2001 World Congress, Naples 15–18 September

Douwes J, Thorne P, Pearce N et al. (2003) Bioaerosol health effects and exposure assessment: progress and prospects. Ann Occup Hyg 47(3):187–200. doi:10.1093/annhyg/meg032

Ekhaise FO, Ighosewe OU, Ajakpovi OD (2008) Hospital indoor airborne microflora in private and government owned hospitals in Sagar city, India. World J Med Sci 3(1):19–23

Eslami A, Karimi F, Karimi Z et al. (2016) A survey of the quantity and type of biological aerosols in selected wards of a teaching hospital in Ghazvin. Electron Physician 8(4):2281–2285. doi:10.19082/2281

Fleischer M, Bober-Gheek B, Bortkiewicz O et al. (2006) Microbiological control of airborne contamination in hospitals. Indoor Built Environ 15(1):53–56. doi:10.1177/1420326X06062230

Gizaw Z, Gebrehiwot M, Yenew C (2016) High bacterial load of indoor air in hospital wards: the case of University of Gondar teaching hospital, Northwest Ethiopia. *Multidiscip Respir Med* 1–7. doi:10.1186/s40248-016-0061-4

Goyer N, Lavoie J, Lazure et al. (2001) Bioaerosols in the workplace: evolution, control and prevention guide. Institut de Recherche en Santé et en Sécurité du Travail du Québec. Québec

Handorean A, Robertson CE, Harris JK et al. (2015) Microbial aerosol liberation from soiled textiles isolated during routine residuals handling in a modern health care setting. Microbiome 3(1):72. doi:10.1186/s40168-015-0132-3

Hewitt KM, Mannino FL, Gonzalez A, et al. (2013) Bacterial diversity in two neonatal intensive care units (NICUs). *PLoS One*, *8*(1):1–7. doi:10.1371/journal.pone.0054703

Huang PY, Shi ZY, Chen CH et al. (2013) Airborne and surface-bound microbial contamination in two intensive care units of a medical center in central Taiwan. Aerosol Air Qual Res 13(3):1060–1069. doi:10.4209/aaqr.2012.08.0217

Huff W, Kubena L, Harvey R et al. (1998) Mycotoxin interactions in poultry and swine. J Anim Sci 66(9):2351–5

Jung CC, Wu PC, Tseng CH et al. (2015) Indoor air quality varies with ventilation types and working areas in hospitals. Build Environ 85:190–195. doi:10.1016/j.buildenv.2014.11.026

Kim KY, Kim CN (2007) Airborne microbiological characteristics in public buildings of Korea. Build Environ 42:2188–2196

Klánova K, Hollerová J (2003) Hospital indoor environment: screening for microorganisms and particulate matter. Indoor Built Environ 12(1–2):61–67

Klevens RM, Edwards JR, Richards Jr. CLet al. (2007) Estimating health care-associated infections and deaths in U.S. hospitals, 2002. Public Health Rep 122(2):160–166. doi:10.2307/20057111

Leung M, Chan AHS (2006) Control and management of hospital indoor air quality. Med Sci Monit 12(3):SR17–R23

Luksamijarulkul P, Aiempradit N, Vatanasomboon P (2014) Microbial contamination on used surgical masks among hospital personnel and microbial air quality in their working wards: a hospital in Bangkok. Oman Med J 29(5):346–350. doi:10.5001/omj.2014.92

Madureira J, Paciência I, Rufo J et al. (2015) Assessment and determinants of airborne bacterial and fungal concentrations in different indoor environments: homes, child day-care centres, primary schools and elderly care centres. Atmos Environ 109:139–146

Mentese S, Arisoy M, Rad AY et al. (2009) Bacteria and fungi levels in various indoor and outdoor environments in Ankara, Turkey. Clean-Soil Air Water 37:487–493

Mora M, Mahnert A, Koskinen K et al. (2016) Microorganisms in confined habitats: microbial monitoring and control of intensive care units, operating rooms, cleanrooms and the international space station. Front Microbiol 7(October):1–20. doi:10.3389/fmicb.2016.01573

Napoli C, Marcotrigiano V, Montagna M (2012) Air sampling procedures to evaluate microbial contamination: a comparison between active and passive methods in operating theatres. BMC Public Health 12(1):594. doi:10.1186/1471-2458-12-594

Nunes ZG, Martins AS, Altoe ALF et al. (2005) Indoor air microbiological evaluation of offices, hospitals, industries, and shopping centers. Mem Inst Oswaldo Cruz 100(4):351–357. http://doi.org//S0074-02762005000400003

Obbard JP, Fang LS (2003) Airborne concentrations of bacteria in a hospital environment in Singapore. Water Air Soil Pollut 144(1–4):333–341. doi:10.1023/A:1022973402453

Okten S, Asan A (2012) Airborne fungi and bacteria in indoor and outdoor environment of the Pediatric Unit of Edirne Government Hospital. Environ Monit Assess 184(3):1739–1751. doi:10.1007/s10661-011-2075-x

Ortiz G, Yagüe G, Segovia M et al. (2009) A study of air microbe levels in different areas of a hospital. Curr Microbiol 59(1):53–58. doi:10.1007/s00284-009-9398-7

Park DU, Yeom JK, Lee WJ et al. (2013) Assessment of the levels of airborne bacteria, gram-negative bacteria, and fungi in hospital lobbies. Int J Environ Res Public Health 10(2): 541–555. doi:10.3390/ijerph10020541

Pejtersen J (1996) Sensory pollution an microbial contamination of ventilation filters. Indoor Air 6:239–248

Rintala H, Pitkäranta M, Toivola M et al. (2008) Diversity and seasonal dynamics of bacterial community in indoor environment. BMC Microbiol 8:56. doi:10.1186/1471-2180-8-56

Russotto V, Cortegiani A, Raineri SM et al. (2015) Bacterial contamination of inanimate surfaces and equipment in the intensive care unit. J Intensive Care 3(1):54. doi:10.1186/s40560-015-0120-5

Shin H, Pei Z, Martinez KA et al. (2015) The first microbial environment of infants born by C-section: the operating room microbes. Microbiome 3(1):59. doi:10.1186/s40168-015-0126-1

Tang CS, Chung FF, Lin MC et al. (2009) Impact of patient visiting activities on indoor climate in a medical intensive care unit: a 1-year longitudinal study. Am J Infect Control 37(3):183–188

Tang CS, Wan G (2013) Air quality monitoring of the post-operative recovery room and locations surrounding operating theaters in a medical center in Taiwan. PLoS One 8(4):e610. doi:10.1371/journal.pone.0061093

Wan GH, Chung FF, Tang CS (2011) Long-term surveillance of air quality in medical center operating rooms. Am J Infect Control 39(4):302–308

Chapter 16
Virus Bioburden in Hospital Environment

Edna Ribeiro

Abstract For the past years, major concerns have emerged regarding the prevalence of pathogenic viruses, potentially transmitted by airborne and droplet, in health facilities particularly in the context of immunocompromised patients exposure. The ability of numerous viruses to persist in inanimate environments may endorse viral nosocomial outbreaks on health care workers and patients with significant associated mobility and mortality. Thus, considering the complexity of virus transmission in indoor environment, the development of accurate viral bioburden assessments in health care contexts is unquestionably imperative in order to implement proper infection control strategies.

Keywords Viral nosocomial infections · environmental assessment

16.1 Introduction

In hospital environment, occupationally exposed individuals are frequently exposed to several viruses through diverse routes such as injuries, direct patient care and contaminated environment. The prevalence of pathogenic viruses potentially transmitted by airborne and droplet represents a significant treat for both workers and patient health and the ability of numerous viruses to persist in inanimate environments (e.g. surfaces) is also particularly concerning. It is acknowledged that virus transmission in indoor environment is complex and multifarious. Thus effective measures focused in identification and isolation of infectious patients, health care personnel formation (e.g. hand hygiene), preventive vaccination and immediate clinical evaluation after unprotected exposure, are fundamental in order to prevent

E. Ribeiro (✉)
Research Center LEAF - Linking Landscape, Environment, Agriculture and Food - Instituto Superior de Agronomia, Lisbon University, Portugal

Research Group Environment & Health, Lisbon School of Health Technology/Polytechnic Institute of Lisbon, Lisbon, Portugal
e-mail: edna.ribeiro@estesl.ipl.pt

C. Viegas et al. (eds.), *Exposure to Microbiological Agents in Indoor and Occupational Environments*, DOI 10.1007/978-3-319-61688-9_16

nosocomial and occupationally associated infections (Weber and Rutala, 2016). Protocols regarding transmission based precautions in both hospital personnel and patients are continuously updated at the international level (La Rosa et al. 2013).

In order to develop proper and effective preventive measures the assessment of viral bioburden in hospital environment is crucial as it enables the identification of critical contaminated settings and can direct the selection of effective infection control measures in order to decrease viral associated nosocomial infections.

16.2 Main Etiological Agents of Viral Nosocomial Infections

Outbreaks of viral infection are particularly concerning in pediatric hospitals since children are more vulnerable to contagious diseases and are regularly admitted with diagnostics of community associated respiratory or gastrointestinal viral infections (Posfay-Barbe, 2012). Among young children common viral infections, rotavirus is still of particular concern as estimated over 570.000 children die per year from dehydration caused by associated severe diarrhea. Rotavirus outbreaks arise promptly with substantial contamination of the environment due to elevated virions numbers (100–1000) shed per milliliter of stool during diarrhea (Dennehy, 2012). Rotavirus are frequently associated with healthcare setting infections spread by air, hand and surface contamination and are one of the most assessed virus in hospital settings as demonstrated in Table 16.1.

Moreover, incidences of infection may significantly increase environmental contamination as some virus, such as Norovirus, in addition to be extremely stable in the environment are also quite resistant to frequently used disinfectants (Wu et al. 2005). Published studies demonstrate that door handles and hands were found to be one of the critical contamination sites (Akhter et al.1995; D'Arcy et al. 2014) which indicates that proper and effective disinfection and handwashing programs may have key roles in decrease virus transmission.

16.3 Virus Detection in Environmental Samples

Currently clinical virology applies booth culture-dependent (e.g. Cell cultures) and culture-independent approaches, particularly based in nucleic acid amplification techniques, for the assessment of viral infections and viral bioburdens, as discussed in the chapter Analyses approaches for virus. The development of precise and effective markers for viral contamination, such as torque-teno virus which is ubiquitously spread, is also extremely important (D'Arcy et al. 2014). Torque teno virus can be detected in air and surface samples and are identified with molecular tools such as nested RT-PCR (Carducci et al. 2011) or qRT-PCR (D'Arcy et al. 2014).

Studies performed with the aim to assess viral contamination of hospital surfaces and air, summarized in Table 16.1, reveal the widespread presence of viral

Table 16.1 Virus contamination in hospital indoor environment samples

Setting and samples analyzed	Studied period	Virus	References
Pediatric hospital Air and Surfaces n = 78	3 months	Norovirus, adenovirus, respiratory syncytial virus, Rotavirus, human metapneumovirus, Cytomegalovirus and Torque teno virus	(D'Arcy et al. 2014)
Intensive care unit Surfaces n = 504	6 months	Rotavirus	(Ganime et al. 2012)
Hospital Air n = 62 Surfaces n = 144	15 months	Adenovirus; Norovirus; Torque teno virus	(Carducci et al. 2011)
Pediatric primary immunodeficiency unit Surfaces n = 23 General pediatric ward Surfaces n = 24	Winter 2005/2006	Norovirus, Astrovirus, and Rotavirus	(Gallimore et al. 2008)
Pediatric primary immunodeficiency unit Surfaces n = 132	6 months	Norovirus, Astrovirus, and Rotavirus	(Gallimore et al. 2006)
General pediatric ward Surfaces n = 155	6 months	Rotavirus	(Akhter et al. 1995)

DNA, on both disinfected surfaces and filtered air with no association between viral, biochemical and bacterial indicators. These studies validate the importance of implementation of viruses in routine infection control monitoring.

However, due to virus biological characteristics, resistance markers, epidemiology and ecology, which are significantly different from bacteria, the regularly utilized indicators for microbial environmental monitoring do not truly represent viral contamination. Currently, there is still an obvious gap of scientific knowledge regarding indoor virome particularly associated with health care facilities. This gap may be due to the lack of a general identification target gene (e.g. bacteria and Archaea 16S rRNA), limited databases, and identification of virus with small genomes (e.g. RNA) and relatively low viral prevalence in indoor environment (D'Arcy et al., 2014).

Considering the major impact of viral nosocomial outbreaks on health care workers and patients, the development of accurate viral bioburden assessments in hospital environment are indisputable imperative in order to implement effective infection control strategies.

References

Akhter J, al-Hajjar S, Myint S,QS (1995) Viral contamination of environmental surfaces on a general paediatric ward and playroom in a major referral centre in Riyadh. Eur J Epidemiol 11(5):587–90

Carducci A, Verani M, Lombardi R et al. (2011) Environmental survey to assess viral contamination of air and surfaces in hospital settings. J Hosp Infect 77(3):242–247. doi:10.1016/j.jhin.2010.10.010

D'Arcy N, Cloutman-Green E, Klein N et al. (2014) Environmental viral contamination in a pediatric hospital outpatient waiting area: Implications for infection control. Am J Infect Control 42(8):856–860. doi:10.1016/j.ajic.2014.04.014

Dennehy PH (2012) Rotavirus infection. an update on management and prevention. Adv Pediatrics 59(1):47–74. doi:10.1016/j.yapd.2012.04.002

Gallimore CI, Taylor C, Gennery AR et al. (2006) Environmental monitoring for gastroenteric viruses in a pediatric primary immunodeficiency unit. Society 44(2):395–399. doi:10.1128/JCM.44.2.395

Gallimore CI, Taylor C, Gennery AR et al. (2008) Contamination of the hospital environment with gastroenteric viruses: Comparison of two pediatric wards over a winter season. J Clin Microbiol 46(9):3112–3115. doi:10.1128/JCM.00400-08

Ganime AC, Carvalho-Costa FA, Mendonça MCL et al. (2012) Group A rotavirus detection on environmental surfaces in a hospital intensive care unit. Am J Infect Control 40(6):544–547. doi:10.1016/j.ajic.2011.07.017

La Rosa G, Fratini M, Della Libera S et al. (2013) Viral infections acquired indoors through airborne, droplet or contact transmission. Ann Ist Super Sanita 49(2):124–32. doi:10.4415/ANN_13_02_03

Posfay-Barbe KM (2012) Infections in paediatrics: old and new diseases. Swiss Med Wkly 142 (JULY):1–10. doi:10.4414/smw.2012.13654

Weber DJ, Rutala WA (2016) Occupational health update: focus on preventing the acquisition of infections with pre-exposure prophylaxis and postexposure prophylaxis. Infect Dis Clin North Am 30(3):729–57. doi:10.1016/j.idc.2016.04.008

Wu HM, Fornek M, Schwab KJ et al. (2005) A norovirus outbreak at a long-term-care facility: the role of environmental surface contamination. Infect Control Hosp Epidemiol. 26(10): 802–10. doi:10.1086/502497

Part V
Occupational Settings with Potential High Microbial Exposure

Chapter 17
Bioburden Exposure in Highly Contaminated Occupational Environments

Carla Viegas, Lenka Smajdova, Tiago Faria, Anita Quintal Gomes and Susana Viegas

Abstract The presence of high levels of bioaerosols is frequently the result of the natural colonization of an organic substrate present in the workplace. Therefore bioaerossol composition depends of the type of setting and materials that are handled/used. Each of the components have normally health-based recommended exposure limit but workers are normally exposed to the mixture present in the bioaerosols and this can implicate different and more severe health effects than being exposed to a singular component. The purpose of this chapter is to review the microbiota and metabolites concentrations found in occupational environments with high probability to contain high microbiota load. Additionally, It also addresses the most frequently encountered fungal and bacteria species, the sampling strategy selected and the measured metabolites and tasks involving high exposure. The review focuses on 43 articles that were considered relevant to this topic (scientific studies published between 1991 and 2017). It was possible to obtain relevant data regarding the settings and tasks that involve higher exposure to bioaerosols and highlighting also the future challenges to ensure a suitable exposure assessment of the microbiota burden.

Keywords Bioburden · occupational exposure · highly contaminated · exposure assessment

C. Viegas · S. Viegas (✉)
GIAS, ESTeSL - School of Health Technology of Lisbon, Polytechnic Institute of Lisbon, Lisbon, Portugal

Center for Research in Public Health, National School of Public Health, University of Lisbon, Lisbon, Portugal
e-mail: susana.viegas@estesl.ipl.pt

C. Viegas
e-mail: carla.viegas@estesl.ipl.pt

L. Smajdova · T. Faria · A.Q. Gomes
GIAS, ESTeSL - School of Health Technology of Lisbon, Polytechnic Institute of Lisbon, Lisbon, Portugal

C. Viegas et al. (eds.), *Exposure to Microbiological Agents in Indoor and Occupational Environments*, DOI 10.1007/978-3-319-61688-9_17

17.1 Introduction

The term "organic dust" is often used synonymously with "bioaerosol" in the occupational health context and can be defined as "particles with biological origin suspended in air" (Eduard and Halstensen 2009). A broader definition can be applied, with bioaerosols being defined as airborne particles that include fungal spores and hyphae, bacteria, endotoxin, β(1→3)-glucans, mycotoxin or high-molecular-weight allergens, and organic dusts in general that are composed of or derived from biological matter (Oppliger 2014; Walser et al. 2015). The presence of high levels of bioaerosols is frequently the result of the natural colonization of an organic substrate present in the workplace but may also be integral to the processes at work and are deliberately added (for instance: breweries, wineries, and biotechnology production) (Oppliger 2014). Bioaerosols can be very complex; for example, grain dust may contain fragments from grain, husk and straw, soil particles, pollen, bacterial spores and cells, fungal spores and hyphae, fragments and faeces of storage mites and insects, and microbial components such as endotoxins, glucans, peptidoglycans, mycotoxins, antigens, and allergens (Eduard and Halstensen 2009).

The purpose of this chapter is to review the microbiota and metabolites concentrations found in occupational environments with high probability to contain high microbiota load. The review focuses on scientific studies published between 1991 and 2017. It also addresses the most frequently encountered fungal and bacteria species, the sampling strategy selected and the measured metabolites and tasks involving longer and higher exposure in each setting. In addition, this chapter also intends to highlight the future challenges to ensure a suitable exposure assessment of the microbiota burden.

An exhaustive search was made for papers available in scientific databases reporting microbiota burden exposure in high microbial load occupational environments. The articles considered were obtained using different scientific databases such as PubMed, Scopus and Google Scholar with the following keywords related with (1) types of industry: animal confinement buildings (swine, poultry and dairy farms) feed production (grain/plant handling), waste management (collection, sorting, composting and incineration), sawmills, food industries (cheese factories, bakeries, among others); (2) types of microorganisms and/or their metabolism: fungi, bacteria, organic dust, mycotoxins, ergosterol, endotoxins, microbial volatile organic compounds (MVOCs) and (3) microbial occupational exposure. The articles that included any combination of these three issues were chosen for further analyses. At the end of the selection process, 43 articles were considered relevant to this review (Table 17.1).

17.2 Sampling Strategy

Air sampling collection is the most common method to access occupational exposure to bioburden in highly contaminated settings. This approach was the chosen in most of the settings reported in the analyzed papers (42 out of 43 from the

Table 17.1 Overview of studies selected for further analyses

Setting	Samples (air/surfaces/dust)	Parameters (fungi/bacteria/metabolite) assessed	Analytical methods applied	Range/Maximum load	References
Dairy farms	Silage, Hay, Grain, Bedding samples	Mesophilic bacteria, Xerophilic fungi, Thermotolerant fungi, Thermophilic actinomycetes	Culture based-methods	Amount of microbes [CFU/g] in autumn, winter, and spring Autumn G mean Range / Winter G mean Range / Spring G mean Range / ***P*-value** Material **Silage** 89 48–210 25 8–63 97 65–560 *P*<0.001 **Grain treated with acid** 980 270–12,000 1000 180–8200 990 160–8800 NS **Grain dried with wann air** 1600 650–7700 3200 920–29,000 2000 1400–2900 NS **Grain dried with cool air** 3100 270–9100 3400 1500–15,000 3100 1200–15,000 NS **Hay dried in storage** 2400 1500–3200 1200 1100–1500 700 670–750 NS **Hay dried on poles** 1500 280–38,000 1100 580–17,000 3900 1700–15,000 NS **Baled hay** 18,000 1400–47,000 6300 480–52,000 13,000 1500–77,000 NS **Wood shavings** 69 35–390 480 160–30,000 1500 380–21,000 *P*<0.05 **Straw** 8700 2400–34,000 2800 790–39,000 17,000 6600–100,000 NS Percentage of particles<5 urn **Group of Microbes** / 500–9999 CFU/m³ Arithmetic mean / 10,000–99,999 CFU/m³ Arithmetic mean / ≥100,000 CFU/m³ Arithmetic mean Mesophilic bacteria 42.9 64.2 82.9 Xerophilic fungi 63.7 77.0 90.01 Mesophilic fungi 63.7 67.7 85.5 Thermotolerant fungi 58.6 81.2 89.4 Thermophilic actinomycetes 43.5 66.9 91.4	Kotimaa et al. 1991
Animal Feed Industry	Air	Fungi, dust and endotoxins	Culture based-methods LAL method (endotoxins)	**Endotoxin Fraction** Total (ng/m³) 2.6 6.7 0.1–1850 Inspirable (ng/m³) 1.9 4.9 0.2–160 Thoraeie (ng/m³) 0.8 5.0 0.2–610 Respirable (ng/m³) 0.1 4.0 0.1–17 Fungi (CFU/m³ × 1000) 2.3 2.6 0.13–15.3	Smid et al. 1992

(continued)

Table 17.1 (continued)

Setting	Samples (air/surfaces/dust)	Parameters (fungi/bacteria/metabolite) assessed	Analytical methods applied	Range/Maximum load	References
Feed industry	Air	Airbone dust and endotoxin	Airborne grain dust was monitored by gravimetric personal air sampling	**Endotoxin (ng/m^3)** Unloaders Average – 14.2 Range – (0.3–48.9) Cranedrivers Average – 0.41 Range: (0.1–0.7) Facility operators Average: 1.4 Range:(0.2–0.8) Press operators Average 21.7 Range:(0.9–93.1) Expedition workers Average:7.3 Range: (0.1–56.9) Truckdrivers Average: 83.9; Range: (0.2–7.2) Jacksofalltrades Average: 29.1 Range: (0.3–317.2)	Jorna et al. 1994
Grain dust farms	Air samples	Fungi and Bacteria	Culture based-methods (Fungi and bacteria) Sporophore gross morphology and a range of biodegradation tests (Actinomycetes)	**During harvesting of barley**. Airborne fungal spore concentrations ranged from 8.3×10^4 to 4.5×10^5 CFU/m^3. Concentrations of airborne bacteria ranged from 1.2×10^5 to 1.3×10^7 CFU/m^3 **During harvesting of wheat**. Concentrations of airborne fungal spores ranged from 1.8×10^3 to 1.3×10^7CFU/m^3. Concentrations of airborne bacteria ranged from 5.8×10^4 to 1.0×10^9CFU/m^3. Airborne actinomycetes ranging from none detected to 2.3×10^3 CFU/m^3. **During handling of stored grain**. Airborne fungal spore concentrations ranged from 6.8×10^3 to 1.1×10^6 CFU/m^3. Airborne bacterial concentration ranged from 1.3×10^4 to 2.1×10^7 CFU/m^3. Airborne actinomycetes were present in small numbers from none detected to 9.3×10^3 CFU/m^3 **Docks**. Concentrations of airborne fungal spores ranged from 2.5×10^4 to 6.5×10^9 CFU/m^3. Airborne bacterial concentrations ranged from 8.1×10^3 to 1.4×10^{11} CFU/m^3. Airborne actinomycetes were present in very small numbers ranging from none detected to 3.9×10^4 CFU/m^3.	Swan et al. 1998
Swine confinement buildings	Air	Microbiological pollutants (Bacteria, Thermophilic Actinomycetes and Fungi) Endotoxins	Culture based-methods LAL method (endotoxins)	Bacteria: Average > 4.3×10^5 CFU/m^3 Thermophilic Actinomycetes: very low counts (data not shown) Fungi: From 2.82×10^2 to 3.82×10^3 CFU/m^3 Endotoxins: up to 10,000 EU/m^3	Duchaine et al. 2000

(continued)

Table 17.1 (continued)

Setting	Samples (air/surfaces/dust)	Parameters (fungi/bacteria/metabolite) assessed	Analytical methods applied	Range/Maximum load	References
Grain Production	Air and settled dust	Fungi and ochratoxin	Culture based-methods Quantification of ochratoxin by competitive direct ELISA.	The grain dust contained median 40 CFU of *Penicillium* spp./mg dust (range 0–32,000 CFU/mg dust) and median 0 (range 0–5,300 CFU/mg dust) *Aspergillus* spp. Culturable *Penicillium* spp. and *Aspergillus* spp. were detected in 52% (n = 51) and 20% (n = 20) of the samples, All samples contained OTA. The median OTA concentration was 4 μg/kg (range 2–128 μg/kg, Tab. 3). Thirteen percent of the samples had 3–30 times higher OTA levels than the median (range 13–128 μg/kg).	Halstensen et al. 2004
Swine and poultry sheds	Air	Total viable bacteria and fungi	Culture based-methods	Swine: GM / Max.; Poultry: GM / Max. **Total viable Bacteria** Winter: Swine 133.887 / >55.7173 summer: Swine 32.931 / 163.760; Poultry 279.559 / >55.7173 **Total viable fungi** Winte: Swine 454 / 2.587 Summer: Swine 7.111 / 22.728; Poultry 7.423 / 32.212	Wan-KuenJo et al. 2005
Plant Products in small agricultural facilities	Air	Airborne microorganisms dust and endotoxins	Culture based methods (Fungi and bacteria) LAL method	Total mesophilic bacteria – Median 17.6; Max: 25.5, Min: 0.75 Gram-negative bacteria – Median: 0, Max: 307.5, Min: 0 Fungi – Median: 3.0, Max: 26.25, Concentration of dust (mg/m^3) – Median: 62.5, Max: 257.5 Min:2.5 Concentration of endotoxin (μg/m^3). Median: 18.75, Max: 125, Min: 0.0625	Krysińska-Traczyk et al. 2005
Valerian root on farms	Air	Fungi and Bacteria	Culture based-methods	Wide range from 0.95–7966.6 × 10^3 CFU/m^3. Though median was relatively low (10.75 × 10^3 CFU/m^3), on 4 farms, the concentrations exceeded the level of 10^{5}CFU/m^3 and on 1 farm at the level of 10^6 CFU/m^3 The median amounts of mesophilic bacteria and fungi were similar, while thermophilic actinomycetes formed only small portion of total microflora.	Skorska et al. 2005
Grain Farming	Air	Dust, fungal spores, bacteria, endotoxins, b-(1/3)-glucans	Scanning electron microscopy (SEM) for fungi and bacteria LAL method endotoxins Inhibition enzyme immunoassay (EIA) for glucans	**Range** Fungal spores (10^{6}m^3) – 5200 Actinomycetes (10^{6}m^3) – 3000 Hyphae (10^{6}AU m^3) – 200 Bacteria (10^{6}m^3) – 420 Endotoxins (10^{3}EU m^3) – 700	Halstensen et al. 2007

(continued)

Table 17.1 (continued)

Setting	Samples (air/surfaces/dust)	Parameters (fungi/bacteria/metabolite) assessed	Analytical methods applied	Range/Maximum load	References
Pig-confinement building	Airborne pollutants	Total airborne bacteria Total airborne gram negative bacteria Total airborne fingi	Culture based-methods	**Mean (Summer)** **Mean (Winter)** total airborne bacteria log(CFU/m^3) 6.13 5.67 total airborne fungi log(CFU/m^3) 4.26 4.37 total airborne gram-negative bacteria log(CFU/m^3) 4.32 3.98	Kim et al. 2007
Grain harvest	Sample of grain and Sample of settled grain dust	Fungi	Culture based-methods	Fungi and mycotoxins in rye grain and rye grain dust. The median concentration of the total microfungi able to grow on malt agar was insignifi cantly greater in the samples of rye dust compared to rye grain (401.1 vs. 5.0 × 103 CFU/g, $0.05<p<0.1$) The median concentration of the total microfungi able to grow on malt agar was signifi cantly greater in the samples of barley dust compared to barley grain (632.1 vs. 5.0×10^3 CFU/g, $p<0.01$)	Krysińska-Traczyk et al. 2007
Swine farms	Air	Airborne endotoxin	LAL method	**Airborne endotoxin** <166.660 EU/m^3	Mc Donnell et al. 2008
Poultry confinement buildings	Air	Bacteria and endotoxin	Quantitative polymerase chain reaction - quantity of total airborne bacteria and total airborne *Staphylococcus* Species). The most frequent Staphylococcus bacterial taxa isolated and identified by enzymatic test kits LAL method (endotoxins)	Stage of fattening (1–2 days old) (24–27 days old) (40–53 days old) Staphylococci as percentage of total bacteria 0.75–1.2 9.6–2.4 11–1.5 Endotoxin concentration 6198 EU m^{-3} air	Oppliger et al. 2008
Review - Multiple locations	Air	Fungi	Culture based-methods	Workplaces in which fungi and actinomycetes have been associated with hypersensitivity pneumonitis and asthma; summarised from Lacey and Crook (1988). Exposure levels of culturable and countable fungi and actinomycetes during farm work (Adapted from Eduard, 1997).	Eduard 2009

(continued)

Table 17.1 (continued)

Setting	Samples (air/ surfaces/dust)	Parameters (fungi/bacteria/ metabolite) assessed	Analytical methods applied	Range/Maximum load	References
Cowsheds	Air and Dust	Bacteria and Fungi	Culture based-methods	The most frequent fungal genera identified in the dusts were *Eurotium* (mainly *Eurotium amstelodami* and *Eurotium umbrosum* [46.6% of the total amount of fungi in the air and 34.8% of the total in the dust]), *Wallemia* (25.1% and 19.8%), *Aspergillus* (10.5% and 10.4%), and *Penicillium* (7.0% and 10.4%). Bacterial prevalence 57.4% of gram-positive cocci in the air and 68.2% in the dust, 32.2% of gram-positive bacilli in the air and 31.6% in the dust, and 10.3% of gram-negative bacilli in the air and 29.4% in the dust. No gram-negative cocci were identified.	Normand et al. 2009
Poultry Houses	Air	Bacteria and Fungi	Culture based-methods	Number of microorganisms (as CFU/m^3) in poultry houses ranged within: 1.7×10^3 to 8.8×10^3 for mesophilic bacteria; 3.5×10^1 to 8.3×10^2 for hemolytic bacteria; 1.5×10^3 to 4.6×10^4 for staphylococci; 5.0×10 to 2.0×10^2 for coli-group bacteria 1.7×10^2 to 2.4×10^4 for fungi.	Lonc and Plewa 2010
Handling of corn silage and oilseed cakes in agricultural environment	Air	Fungi	Culture based-methods	Range: 4.3×10^2 to 6.2×10^5 CFU m^{-3}. In the personal bioaerosols, the values for fungi per cubic meter of air varied from 3.3×10^3 to 1.7×10^6 CFU m^{-3}	Lanier et al. 2010
Poultry farms	Air	Moulds Endotoxin	Culture based-methods LAL method (endotoxins)	**Range** **Total fungi** (CFU/m^3) 4.9 103–6.89104 **Endotoxin levels** (EU/m^3) 230–284	Rimac et al. 2010
Poultry farms	Air	Mesophilic bacteria, staphylococci, coli group bacteria, fungi, *Salmonella* sp.	Culture based-methods	**Microorganisms in the atmospheric air of poultry houses**: **range (CFU/m^3)** mesophilic bacteria: 4×10^1 to 7.2×10^3, staphylococci: 0 to 1.3×10^4, coli group bacteria: 0 to 7×10^1, fungi 2×10^1 to 1.3×10^4 Salmonella sp. : not found **Microorganisms inside the poultry houses:** **range (CFU/m^3)** mesophilic bacteria: 1.3×10^5 to 5.2×10^5, staphylococci : 1.4×10^5 to 2.6×10^5, coli group bacteria : 2.0×10^2 to 1.5×10^4 Fungi 3.6×10^4 to 1.1×10^5 Salmonella sp not found	Lonc and Plewa 2011

(continued)

Table 17.1 (continued)

Setting	Samples (air/surfaces/dust)	Parameters (fungi/bacteria/metabolite) assessed	Analytical methods applied	Range/Maximum load	References
Poultry farms	Air	Fungi	Culture based-methods	Total fungi concentration × 10^2[CFU/m^3] in the hatchery Spring / Summer / Autumn **Indoor** Arithmetic mean: 920.99 / 810.16 / 4121.85 Min.: 68.35 / 263.93 / 2182.08 Max.: 2391.30 / 1709.40 / 5865.92 **Outdoor** Artithmetic mean: 0.35 / 150.94 / 650.68	Sowiak et al. 2012
Cotton and soybean mills	Air	Fungi	Culture based methods	Airborne fungal counts ranged between 6 × 102 to 1.66 × 104 CFU m^{-3} and 7 × 101 to 2.4 × 103 CFU m^{-3} with median values of 2.2 × 103 CFU m^{-3} and 2 × 103 CFU m^{-3}	Hameed et al. 2012
Cattle Shed	Air	Fungi	Culture based-methods	Fungi per cubic meter of air varied from 2.5–25 CFU/m^3 (median: 6.48), and from 19 to 5.9 × 102 CFU/m^3 (median: 76.4), during the first and the second period, respectively.	Lanier et al. 2012
Waste sorting plant.	Air	Fungi	Culture based-methods	Plates overloaded with *Penicillium* sp. and *Rhizopus* sp.	Malta-Vacas et al. 2012
Handling of biomass for power production purposes	Air	Fungi and Bacteria	Culture based-methods	Bacterial and fungal concentrations at workplaces ranged between 5.1×10^2 CFU/m^3 and 2.0×10^4 CFU/m^3, and between 2.2×10^2 CFU/m^3 and 2.3×10^4 CFU/m^3, respectively.	Ławniczek-Wałczyk et al. 2012
Poultry and swine farms	Air and surfaces	*Aspergillus* spp.	Culture based-methods	Range / Max. **Poultries:** Air [CFU/m^3]: 0–2000 / *A. flavus* 2000 Surface [CFU/m^2]: 0–3 × 10^{-2} / *A. versicolor* 3 × 10^{-2} Coverage of the floor [CFU/g]: 0–12.5 × 10^3 / *A. flavus* 12.5 × 10^3 **Swine** Air [CFU/m^3]: 0–>2000 / A. versicolor 2000 Surface[CFU/m^2]: 0–3 × 10^{-2} / A. versicolor 3 × 10^{-2} Coverage of the floor [CFU/g]: 0–5 × 10^3 / A. versicolor 5 × 10^3	Sabino et al. 2012

(continued)

Table 17.1 (continued)

Setting	Samples (air/ surfaces/dust)	Parameters (fungi/bacteria/ metabolite) assessed	Analytical methods applied	Range/Maximum load	References
Poultry houses	Air	Airborne microorganisms endotoxins, β-glucans	Culture based-methods LAL method (endotoxins) (1″3)–β-D-glucan analysis- quantitative kinetic Glucatell assay	**Range** **Bacterial aerosols** [CFU/m^3] $2.5 \times 10^2 - 2.9 \times 10^6$ **Fungal aerosols** [CFU/m^3] $1.8 \times 10^2 - 1.8 \times 10^5$ **Max. level** Endotoxins ng/m^3 8364 0–8364 β-glucans ng/m^3 6886 0.8–6886	Lawniczek-Walczyk et al. 2013
Feed Industry	Air	Grain dust, fungal spores, β-1→3-glucans, total bacteria, and endotoxins	Quantification of bacteria by epi-fluorescence microscopy. Quantification of fungal spores by scanning electron microscopy (SEM). LAL method endotoxins Inhibition enzyme immunoassay (EIA) for glucans	Geometric means (GMs), Geometric standard deviations (GSDs). Endotoxin (Eu m^{-3}) GM: 628; GSD: 5.9; Min: 1; Max: 64250 Bacteria (×10^4 m^{-3}) GM: 21; GSD: 7.9; Min: <DL; Max: 4000 Spores (×10^4 m^{-3}) GM: 3.6; GSD: 3.4; Min: <DL; MAX: 640	Halstensen et al. 2013
Swine	Air Surface	Fungi	Culture based-methods	**Air Frequency (CFU/m^3) (n*; %)** *Aspergillus versicolor*. 3210; 20.9 *Socpulariopsis brevicaulis* 2620; 17.0 *Penicillium* sp. 2160; 14.1 Others 7380; 48.0 **Surfaces Frequency (CFU/cm^2) (n; %)** *Aspergillus versicolor* 712; 26.6 *Cladosporium* sp. 600; 22.4 *Scopulariopsis brevicaulis* 469; 17.5 Others 899; 33.5	Viegas et al. 2013
Pig confinement building	Air, Surfaces and floor cover	Fungi	Culture based methods	Most frequent species **Air:** CFU/m^3 % *Aspergillus versicolor* 3210 21% *Scopulariopsis brevicaulis* 2620 17% *Penicillium* sp. 2160 14% **Surfaces:** *Aspergillus versicolor* 712 26.6% *Cladosporium* sp. 600 22.4% *Scopulariopsis brevicaulis* 469 17.5%	Viegas et al. 2013

(continued)

Table 17.1 (continued)

Setting	Samples (air/surfaces/dust)	Parameters (fungi/bacteria/metabolite) assessed	Analytical methods applied	Range/Maximum load	References
				New floor coverage: *Chrysosporium* sp. 13,750 38.5% *Trichoderma* sp. 10,000 28% *Aspergillus versicolor* 5000 14% *Acremonium* sp. 5000 14% **Used floor coverage:** *Mucor* sp. 42,500 25.1% *Geotrichum* sp. 35,000 20.6% *Trichoderma* sp. 31,000 18.3%	
Poultries	Air, Surfaces and Litter	Fungi Focused (*A. flavus* and *A. fumigatus* complexes)	Culture based-methods	Air (CFU/m^3) Surfaces (CFU/m^2) New litter (CFU/g) Aged litter (CFU/g) ***A. flavus* species complex** Air: 5840 CFU/m^3; Surfaces: 0.3 CFU/m^2; New litter: 0 CFU/g; Aged litter: 12,500 CFU/g ***A. fumigatus* species complex** Air: 80 CFU/m^3; Surfaces: 0.0001 CFU/m^2; New litter: 15,000 CFU/g; Aged litter: 10,000 CFU/g	Viegas et al. 2014
Waste application facilities	Air Liquid impinger sampler	Fungi	Culture based-methods	The mesophilic and thermophilic fungal concentrations ranged between 0–1,300 CFU/m^3 upwind and 0–7,280 CFU/m^3	Hameed et al. 2015
Wastewater treatment plant	Air and surfaces	Fungi	Culture based methods Molecular tools (RT-PCR)	Completely undergrounded WWTP – 20 – 460 CFU/m^3 –Air 10,000 – 440,000 CFU/m^3 – surface Partial undergrounded WWTP 20 – 600 CFU/m^3 – Air 20,000 – 100,000 CFU/m^3 – surface *Aspergillus flavus* (toxigenic strians were not detected	Viegas et al. 2014

(continued)

Table 17.1 (continued)

Setting	Samples (air/surfaces/dust)	Parameters (fungi/bacteria/metabolite) assessed	Analytical methods applied	Range/Maximum load	References
Composting plants	Air, surfaces and dust	Bacteria and fungi Endotoxin in dust samples	**All bacteria and yeasts isolates:** – Colony morphologies and selected biochemical tests (gram staining, catalase test, and oxidase test). – API tests – Nucleotide sequences of the 16S rRNA gene **Isolated molds:** – Macroscopic and microscopic observations following culture on CYA and YES media, using taxonomic keys, – ITS1/2sequence of the rDNA region Endotoxins: gas chromatography–mass spectrometry (GC-MS)	Bacteria surface: Range: 1×10^1 CFU/100 cm^2 to 3.1×10^3 CFU/100 cm^2 Bacteria air: Range: 9.4×10^2 CFU/100 cm^2 to 3.9×10^5 CFU/100 cm^2 Fungi surface: Range: 5.1×10 CFU/100 cm^2 to 2.0×10^3 CFU/100 cm^2 Fungi air: Range: 2.5×10^1 CFU/100 cm^2 to 4.2×10^4 CFU/100 cm^2	Gutarowska et al. 2015
Cork industry	Air, surfaces	Fungi	Culture based methods Molecular tools (RT-PCR)	The conventional identification of fungal species in air confirmed the presence of countless colonies of *C. sitophila* in Plant A. In Plant B C. sitophila was also prevalent, whereas Plant C presented a larger diversity of fungal species, among which the most prevalent were *Penicillium* genus (76.5 %) and *Geotrichum* (11.8 %). The distribution of fungal species in the surface samples of Plants A and B was similar, with isolates from the *A. fumigatus* complex being the only ones found in addition to *C. sitophila*. In Plant C the most prevalent genera were *Trichoderma* and *Penicillium* (52.9 %; 29.4 %). *Aureobasidum* sp. and species belonging to *A. fumigatus* complex were also isolated. Real-time PCR identified the *P. glabrum* complex in 10 out of the 12 air samples, that is, in six more sampling sites than the conventional method did	Viegas et al. 2015b

(continued)

Table 17.1 (continued)

Setting	Samples (air/surfaces/dust)	Parameters (fungi/bacteria/metabolite) assessed	Analytical methods applied	Range/Maximum load	References
Feed Industry	Air	Endotoxins, β-1→3-glucans, bacteria, and fungal spores	Bacteria quantification by epifluorescenceicroscopy Fungal spores quantification by scanning electron microscopy (SEM) Endotoxins by LAL method Inhibition enzyme immunoassay (EIA) for glucans	The overall exposure was 662 EU m^{-3} (5.9) of endotoxins, 7.4 μg m^{-3} (5.6) of β-1→3-glucans, 22 × 10^4 m^{-3} (7.9) of bacteria, and 4 × 10^4 m^{-3} (3.4) of fungal spores.	Straumfors et al. 2015
Culatello manufacturing plants	Air and surfaces	Fungi	Culture based-methods	The contamination level detected in Plant 1 was 4.15 and 4.26 Log CFU/m^3 air/room in winter and summer, respectively, and a total of 684 fungal isolates were obtained. The contamination level detected in Plant 2 was 5.15 and 5.11 Log CFU/m^3 air/room in winter and summer, respectively, and a total of 908 fungal isolates were obtained The contamination level detected in Plant 3 was 4.15 and 4.20 Log CFU/m^3 air/room in winter and summer, respectively, and a total of 438 fungal isolates were obtained	Scaramuzza et al. 2015
Waste industry	Air and surfaces	Fungi	Culture based methods Molecular tools (RT-PCR)	The air fungal load ranged from 180 to 5,280 CFU m^{-3} before cleaning and from 220 to 2,460 CFU m^{-3} after cleaning. The surfaces present results that ranged from 29 × 10^4 to 109 × 10^4 CFU m^{-2} before cleaning and from 11 × 10^4 to 89 × 10^4 CFU m^{-2} after cleaning Eight species/genera of filamentous fungi were identified in air samples by PCR	Viegas et al. 2015a
Feed industry	Air and surfaces	Fungi	Culture based methods Molecular tools (RT-PCR)	The fungal load in the air ranged 0–144 CFU × m^{-3}. The surfaces produced results that ranged 0–18 × 10^4 CFU × m^{-2} Through molecular methods was checked for the presence of toxic strains from *A. flavus* complex, *A. fumigatus* complex and *S. chartarum* by real-time PCR	Viegas et al. 2016b

(continued)

Table 17.1 (continued)

Setting	Samples (air/ surfaces/dust)	Parameters (fungi/bacteria/ metabolite) assessed	Analytical methods applied	Range/Maximum load	References
Slaughterhouse	Nasal swab	Methicillin-resistant *Staphylococcus aureus* (MRSA), Methicillin-sensitive *S.aureus* (MSSA)	SA Nasal Complete Assay, PCR, Catalase,Coagulase test, Gram Staining, Mec A, SCCmec and staphylococal protein A screening, Scn gene presence, Kirby-Bawer disk diffusion	Prevalence: *S.aureus* 27.0% MSSA -5 workers (3.6%) MRSA- 0%	Leibler et al. 2016
Slaughterhouses	Air and surfaces	Fungi	Culture based methods Molecular tools (RT-PCR)	Air fungal load ranged from 16 CFU/m^3 to 970 CFU/m^3 in the poultry slaughterhouse, 20 CFU/m^3 to 440 CFU/m^3 in the swine/bovine slaughterhouse and 10 CFU/m^3 to 36 CFU/m^3 in the large animal slaughterhouse. The surfaces present results that ranged from 0 CFU/m^2 to 10,000 CFU/m^2 in the poultry slaughterhouse and 0 CFU/m^2 to 90,000 CFU/m^2 in the large animal slaughterhouse. No fungal isolates were found in swine/bovine slaughterhouse surfaces. *Aspergillus fumigatus* was detected in additional sampling sites where was not found with culture based-method	Viegas et al. 2015a
Various	Air	Fungi	Culture based methods Molecular tools (RT-PCR)	Range coriolis: 49 – 100,000 CFU × m^{-3}. Range filter: 19 – 100,000 CFU × m^{-3}.	Viegas et al. 2017a
Farms	Air	Inhalable dust, endotoxin, and total volatile organic compounds (TVOCs)	Endotoxins by LAL method	GM 128 GSD 2.5 Min-Max 26–900	Basinas et al. 2017

selected papers). However, the air sampling can be done in different ways and the choice of the method is dependent on whether a quantitative or qualitative analysis is desired (Viegas et al. 2015a).

Different methods can be applied to assess air contamination and these are divided into passive, such as gravity plates (Sudharsanam et al. 2009), swabs (Viegas et al. 2016b) or dust sampling methods (Normand et al. 2009) or active methods (impact, impingers and filters) (Viegas et al. 2015c). This information was covered extensively in a previous chapter (Chap. 4 – Sampling for Microbial Determinations).

The results obtained with passive methods are very dependent on the microorganism dispersion ability and the different conditions of the surrounding air (turbulent versus calm atmosphere for instance depending of the type of activities being performed) (Jürgensen and Madsen 2016). In the case of gravity plates, sampling can be done by exposing media-containing plates for 30 minutes (Sudharsanam et al. 2009) and may be used especially for settings like healthcare facilities with less contamination (Sudharsanam et al. 2012). However, this method is not suitable for highly contaminated settings.

The use of swabs for occupational exposure assessments has been reported as a method to complement air sampling results from fungal burden since a wider mycobiota diversity is obtained due to species isolation only in surfaces (Viegas et al. 2012-2016), but unfortunately it is not widely applied. The results can be expressed in CFU/m^2 and can be quantitatively compared with different areas from the same setting and, importantly, to identify different species from the ones detected in air samples, thus providing a more complete and detailed picture from the bioburden occupational exposure scenario (Viegas et al. 2016a).

Concerning dust sampling methods, although widely applied in indoor air quality studies (Madsen et al. 2012; Adams et al. 2015), only one study was found using this method in highly contaminated settings (Normand et al. 2009). Passive methods, like the ones described, do not allow air quantitative results (results are expressed in CFU/plate or CFU/m^2), as it is almost impossible to have information about the volume of air sampled (Viegas et al. 2015a). As such, each result obtained from passive methods can be compared between different settings with the awareness that working and environment conditions cannot be replicated easily (Viegas et al. 2015a). However, it should be highlighted that the main advantage of passive methods is that they can collect contaminants during a larger period of time, whereas air samples can only reflect the load from a shorter period of time, correspondent to the sampling period (Viegas et al. 2017a, 2017b).

Concerning active methods, these involve sampling a known volume of air per unit time and the samplers, named volumetric, are equipped with a suction device and allow bioburden quantification (results expressed in colony forming units (CFU/m^3) when collected volumes are previously defined. These methods permits comparison with outdoor data, the main source of indoor contamination, allowing understand if the contamination is coming from outdoor sources (De Nuntiis et al. 2003). In addition, results obtained can be compared with relevant guidelines or legal requirements/limits suggested by scientific and/or technical organizations and the strictest reference levels should be applied for an occupational health approach (Viegas et al. 2016b, 2016c). Impact methodology presents the best results when compared with other sampling methods. This method, also called "Slit-to-Agar" method, is the selected one in several studies (Blomquist et al. 1984; Smid et al. 1989; Brenier-Pinchart et al. 2009; Vacova et al. 2006) and recommended by different organizations such as, ACGIH (Verhoeff et al. 1992) and Canadian Health Organization (Health Canada 1993). Also in highly contaminated occupational environments, this is the most commonly chosen method (20 out of 43 papers).

(Liquid) Impingers (formerly known as scrubbers) are good technical options when fungal spore's mortality is an issue (De Nuntiis et al. 2003). This method was used in 7 out of 43 papers and was particularly used when molecular identification was part of the study as it is the most common method to collect samples for DNA isolation (Normand et al. 2009; Viegas et al 2012–2016). A limitation of impingers is that they cannot operate for long periods due to the inherent liquid evaporation, which can hamper microorganisms' viability (De Nuntiis et al. 2003). However, this approach is useful to overcome bioburden assessment restrictions in higher contaminated settings, since this approach allows dilution of the sample prior to plate incubation (Thorne and Heederik 1999).

Concerning the filters-based method (used in 13 out 43 papers), this can provide both direct and indirect analyses. The air passing through a membrane filter, allows for particulate deposition and the filter can be removed and directly placed under a microscope for analysis (De Nuntiis et al. 2003). The filter membrane can also be placed on a culture media and incubated to allow microbial growth or digested with a buffered solution prior to inoculation in the selected media (Sudharsanam et al. 2012). Due to the risk of dehydration this method is only suited for resistant microorganisms or fungal spores (De Nuntiis et al. 2003) (9 out of 13 publications were used for fungal assessment).

17.3 Bacteria and Fungal Loads Reported

The range of concentrations varies greatly between the different settings under study for fungi and for bacteria. Air fungal load analysed in the several settings varied between 2.5 CFU/m^3 (Lanier et al. 2012) and 6.5×10^9 CFU/m^3 (Halstensen et al. 2007), whereas air bacterial load ranged from 460 CFU/m^3 (Kim et al. 2007) to 1.4×10^{11} CFU/m^3 (Swan and Crook 1998). The setting with the highest air load of both bacteria and fungi was the group of grain dust farms reported by Halstensen et al. (2007) with values varying between 1.8×10^3 CFU/m^3 to 6.5×10^9 CFU/m^3 for fungi and 8.1×10^3 CFU/m^3 to 1.4×10^{11} CFU/m^3 for bacteria. The lowest concentrations of air fungal load were found in the slaughterhouses, with values between 10 and 970 CFU/m^3 (Viegas et al. 2016b; Scaramuzza et al. 2015). The lowest bacteria air load was reported in the swine industry with values ranging between 460 CFU/m^3 and 1.34 × 10^5 CFU/m^3 (Kim et al. 2007; Wan-Kuen et al. 2005). Probably related with the extensive use of antibiotics on those farms and, unfortunately, still being the common practice in animal production (Aarestrup et al. 2008).

Other places under study such as feed Industries, cotton and soybean mills, waste water treatment plants and composting plants showed variations in fungal concentrations of 0.4 to 640 × 10^4 CFU/m^3 (Halstensen et al. 2013; Jorna et al. 1994; Smid et al. 1992; Straumfors et al. 2015), 7.0 × 10^1 to 1.66 × 10^4 CFU/m^3 (Hameed et al. 2012), 20 to 600 CFU/m^3 (Viegas et al. 2014a, 2014b, 2014c) or 2.5 × 10^1 to 4.2 × 10^4 CFU/m^3, respectively. On the other hand, composting plants and feed industry showed variations in bacteria concentrations of 9.4 × 10^2 to 3.9 × 10^5 CFU/m^3 (Gutarowska et al. 2015) and 4 x10^7 CFU/m^3 (Halstensen et al. 2013; Straumfors et al. 2015).

In striking contrast to the air sampling studies, the number of studies with bacteria and fungi assessed both from surface swabs and air sampling is much smaller. Only 8 out of 42 studies address this approach. In those studies, surfaces fungal load varied between 5.1 × 10 CFU/m^2 (Gutarowska et al. 2015) and 2.9 × 10^6 CFU/m^2 (Lawniczek-Walczyk et al. 2013). The setting where the highest concentrations of fungi was found on surfaces were the swine farms (2.5 × 10^2 to 2.9 × 10^6 CFU/m^2) (Lawniczek-Walczyk et al. 2013) and the lowest levels were observed in composting plants (5.1 × 10 to 2.0 × 10^3 CFU/m^2; Gutarowska et al. 2015).

The bacterial load in surfaces ranged between 1 × 10^1 CFU/m^2 and 3.1 × 10^3 CFU/m^2 in composting plants, with only one study assessing this issue (Gutarowska et al. 2015).

17.4 Metabolites More Commonly Assessed

Environment often contains various microorganisms producing metabolites with direct impact on the human organism. Endotoxin is the most common metabolite detected in agricultural facilities such as feed industry, grain production houses, poultry, swine farms and dairy farms (Table 17.2). The presence of endotoxins in ambient air is related to the presence of gram negative bacteria or cell wall fragments from such bacteria in airborne organic dust (Health Council of the Netherlands, 2010). This fact also explains why endotoxins are one of the major markers of microbiological contamination (Duquenne et al. 2012). Endotoxin penetration and behavior of biological particles in the human respiratory system is strongly influenced by their size, shape, density, chemical composition and reactivity (Vincent 2005; Liao et al. 2009). Occupational exposure to endotoxins is a risk factor for the development of bronchial sensitivity and dyspnea. Furthermore, asthma-related conditions are more common in endotoxin-exposed workers (Health Council of the Netherlands 2010). Despite the reality of occupational exposure to endotoxins and the acknowledgement of its effects on the health of

Table 17.2 Studies reporting metabolites assessment

Setting	Endotoxin	*β*-glucans	Mycotoxins	References
Animal Feed Industry	**(ng/m^3)** GM 2.6; GSD 6.7; Range 0.1–1850	NOB	NOB	Smid et al. (1992)
Feed industry	**(ng/m^3)** Range: 0.1–317.2	NOB	NOB	Jorna et al. (1994)
Grain production	NOB	NOB	Ochratoxin A Median: 4 **μg/kg**; range: 2–128 **μg/kg**	Halstensen et al. (2004)
Plant products in small agricultural facilities	**(μg/m^3)** Median: 18.75; Max: 125; Min: 0.0625	NOB	NOB	Krysińska-Traczyk et al. (2005)
Grain farming	**(10^3 EU m^3)** Max: 700	NOB	NOB	Halstensen et al. (2007)
Swine farms	**EU/m^3** Range: <166,660	NOB	NOB	Mc Donnell et al. (2008)
Poultry farms	**(EU/m^3)** Range 230–284	NOB	NOB	Rimac et al. (2010)
Poultry houses	**ng/m^3** max. 8364; range 0–8364	**ng/m^3** max. 6886; range 0.8–6886	NOB	Lawniczek-Walczyk et al. (2013)
Feed industry	**(EU m^{-3})** GM: 628; GSD: 5.9; Min: 1; Max: 64,250	NOB	NOB	Halstensen et al. (2013)
Feed Industry	**EU m^{-3}** GM 662; GSD 5.9	GM 7.4 **μg m^{-3}**; GSD 5.6 **μg m^{-3}**	NOB	Straumfors et al. (2015)
Dairy farms	**EU m^{-3}** GM 128 GSD 2.5	NOB	NOB	Basinas et al. (2017)

NOB, Not observed.

workers, an occupational health limit has not yet been established (Duquenne et al. 2012).

However, more recently, the Dutch Expert Committee on Occupational Safety recognised that a health-based recommended exposure limit of 90 EU/m^3

(8-hour time-weighted average) affords adequate protection against the effects of both acute and chronic exposure to endotoxins (Health Council of the Netherlands 2010).

There is relatively little information regarding the measurement strategy and consequently different methods are used from one laboratory to another, although the most reported is the LAL assay (9 out of 10 papers). Therefore it is recommended to use the same units for the expression of results and standardization is needed (Duquenne et al. 2012).

Regarding fungal contamination, β-glucans are the best indicator of fungal presence in the environment (Duquenne et al. 2012). β-glucans are water-insoluble structural cell-wall components of most of the filamentous fungi (and yeasts), some bacteria, most higher and many lower plants (Lawniczek-Walczyk et al. 2013). Inhalation of (1–>3)-beta-D-glucan in humans causes symptoms of the upper respiratory tract and induces cytokine production in blood monocytes. The relationship between the amount of (1–>3)-beta-D-glucan and the extent of symptoms as well as lung function changes and inflammatory markers have been described (Rylander 1999).

Two of the studied settings, namely feed industry (Straumfors et al. 2015) and poultry houses (Lawniczek-Walczyk et al. 2013) contained β-1/3 glucan data. The results obtained from the analyzed articles indicate the presence of β-glucans in the poultry houses and feed industry, with values ranging from 0.8 to 6886 ng/m^3, with a geometric mean of 7.4 (μg/m^3) in each of the settings, respectively.

The existence of hygienic standards for β-glucans is still quite "grey". There are neither exposure standards nor proposals of threshold limit values for this parameter; however, β-glucans have remained within the range of interest of many research studies worldwide. Among the sparse data are the results obtained by Rylander, who showed that the airborne level of β-glucans in indoor environments within the range of 0.1–5.2 ng/m^3 may determine the frequency of adverse health effects occurrence in the exposed population (Rylander 1999; Lawniczek-Walczyk et al. 2013).

Mycotoxins are also secondary metabolites produced by fungi that are capable of causing disease and death in humans and other animals. As major and more reported mycotoxins we can mention aflatoxins, citrinin, ergot alkaloids, fumonisins, ochratoxin A, patulin, trichothecenes and zearalenone (Bennett and Klich 2003). From the reviewed articles, only one setting, with grain production showed Ochratoxin A (OTA) presence ranging between 2 and 128 μg/kg with a median value of 4 μg/kg (Halstensen et al. 2004). OTA is a mycotoxin produced by several species of *Aspergillus* and *Penicillium* genus and is commonly detected worldwide in various food commodities and feed sources. Studies have shown that this mycotoxin can have several toxicological effects such as nephrotoxicity, hepatotoxicity, neurotoxicity, immunotoxicity and also acts as a teratogenic agent (Khoury and Atoui 2010). In 2002, IARC classified OTA as a human carcinogen of 2B group, supported by a wide evidence of carcinogenicity reported in several animal studies (IARC 2002; Viegas et al. 2013).

17.5 Critical Tasks for Bioburden Exposure

The knowledge of the determinants of exposure in each setting is an important requirement for the identification of the preventive measures to apply and for determining how to prioritize the intervention (Thilsing et al. 2015). Nevertheless, exposure assessment and the respective sampling campaigns are often complicated by large exposure variability due to frequent changes in tasks, activities, work processes, and locations over time (Burdorf 2005).

Previous published work has already shown that a valid registration of these changes during work days contributes to a more rigorous evaluation of the exposure determinants (Viegas 2012; Thilsing et al. 2015; Viegas et al. 2016d). Considering this need of detailed information, some researchers have developed exposure assessments based on work tasks. Identifying all the tasks developed in each workplace and conducting a task-based exposure assessment helps to refine the exposure characterization and to reduce assessment errors (Wijnand and Bakke 1999). Additionally, a task-based exposure assessment can provide a better evaluation of exposure variability, instead of assessing personal exposures using continuous 8-hour time weighted average measurements (Bello et al. 2010). Another important advantage of task-based exposure assessment is the lower probability of covering exposure peaks when comparing with the time weighted average exposure over long processes or over work shifts (Wijnand and Bakke 1999; Bello et al. 2010).

In some working environments, this detailed information has already resulted in improvements on health quality and allowed to reduce bioaerosol exposure. Ventilation or an exhaust system has been applied successfully in several settings (Madsen et al. 2014). Some of those settings are, for example, waste management (Breum et al. 1996; Viegas et al. 2014a, 2014b, 2014c), research animal laboratories (Thulin et al. 2002) and biofuel plants (Madsen 2011).

In the grain dust setting, Health Safety Executive (2010) has presented a report with estimations of current exposure levels, both long-term average and task-specific exposures. The grain industry involves several sectors, all of which have exposure to organic dust. The sectors are farming, grain importation, commercial storage, barley malting, flour milling, animal feed milling and grain exporting (HSE 2010). In the HSE report (HSE 2010), they concluded that workers are most highly exposed to grain dust during manual tasks such as cleaning plant and premises and maintenance. These tasks were of short duration but had potentially high exposures if respiratory protection or other controls were not used. Therefore, one of the report conclusions was that measures should be taken in order to ensure that respirators are used during tasks with significant exposure levels. However, the high level of automation in the import/export, commercial grain storage, milling and malting sectors has led to an important reduction of exposure levels and consequently in the number of workers involved on tasks with potential high exposure (HSE 2010).

Straumfors et al. (2015) found the same trend in the results obtained in a study developed in the feed industry. The most important determinants for elevated dust exposure were cleaning and process controlling. Cleaning increased the dust exposure level by a factor of 2.44 of the reference, from 0.65 to 1.58 mg/m^{-3}, whereas process controlling increased the dust exposure level by a factor of 2.97, from 0.65 to 1.93 mg/m^{-3} (Straumfors et al. 2015).

Regarding the waste management setting, a previous work published by Viegas et al. (2016d) showed that the task involving higher exposure to particles was the waste discharge, with statistical difference when comparing with the other tasks developed in a waste sorting unit. Probably, the open space where the discharge was done promoted particles dispersion to all unit.

Regarding poultry production, Lawniczek-Walczyk and colleagues (2013) found that workers' exposure to airborne microorganisms increased with consecutive stages of the chicken production cycle.

17.6 Challenges for the Bioburden Exposure Assessment

Culture-dependent methods were by far the most widely used procedures for assessing the microbiological content of bioaerosols in the selected papers (41 out of 43).

Although the viability of a microorganism is a critical factor, as it determines infectivity, no common analysis method can be considered perfectly accurate for detecting only live cells. With culture based methods, there is the well-described phenomenon of viable but non-culturable cells (Meheust et al. 2014). Moreover, dead airborne bacteria or fungi retain their allergenic or toxic properties and are therefore also relevant to any occupational health assessment (Oppliger 2014). However, it is now widely accepted that such methods underestimate the total quantity of microorganisms present since the vast majority of them cannot be cultivated (Oppliger et al. 2008). This situation can be exacerbated in highly contaminated occupational environments, since there is increased probability that fungi with higher growth rate impair and inhibit other fungi to growth (Viegas et al. 2015c). However, cultivation based-methods can give important insights regarding bioburden exposure assessments, since it's possible to identify and quantify organisms posing a higher occupational risk from inhalation and compare their levels with legal and scientific guidelines (Viegas et al. 2015b). Importantly, culture based-methods limitations can be overcome by the use of molecular tools in parallel as shown in eight papers (Normand et al. 2009; Viegas et al. 2014–2017). However, although qPCR provides a faster, more sensitive method than culture-based techniques for testing environmental samples, it does not differentiate between viable and non-viable cells without a pre-treatment step (Meheust et al. 2014). Quantitative PCR could be considered as a benchmark method for fungal exposure assessment (Meheust et al. 2014), but for occupational exposure assessments molecular tools should be combined with culture based-methods (Viegas et al. 2017a) due to the above mentioned reasons.

Increasing the throughput of analysis and improving innovative approaches, such as sequencing and mass spectrometry technologies, are important measures in microbiological exposure assessment. The recent introduction of DNA sequencing promises thus to revolutionize the global understanding of aerosol science (Peccia et al. 2011; Meheust et al. 2014).

The indirect measurement of microorganism levels by measuring their components (traditionally endotoxins for Gram-negative bacteria and β-d-glucans for fungi) is another very frequently used approach that allows researchers to correlate the concentration of biological components with health effects since these two components have inflammatory properties. However, measuring only endotoxins or β-glucans could be limiting because they are specific to certain microorganisms. Thus, it will be necessary to develop new methods and in parallel measure the most relevant indicators for each occupational scenario (Oppliger 2014; Viegas et al. 2017a, 2017b).

Additionally, for risk assessment purposes, it is important to consider that workers are normally exposed to the mixture present in the bioaerosols and this can implicate different and more severe health effects than being exposed to a singular component. New tools have to be developed and applied in order to consider the possible health effects arising from the complex mixture present in the air of these high microbial contaminated settings.

References

Aarestrup F, Oliver Duran C, Burch D (2008) Antimicrobial resistance in swine production. Anim Health Res Rev 9(2):135–148. doi:10.1017/S1466252308001503

Adams RI, Tian Y, Taylor JW et al. (2015) Passive dust collectors for assessing airborne microbial material. Microbiome 3:46

Basinas J, Cronin G, Hogan V et al. (2017) Exposure to inhalable dust, endotoxin, and total volatile organic carbons on dairy farms using manual and automated feeding system. Ann Work Expo Health 61(3):344–355

Bello A, Quinn MM, Perry MJ et al. (2010) Quanti-tative assessment of airborne exposures generated during common cleaning tasks: a pilot study. Environ Health 9:76, http://www.ehjournal.net/content/9/1/76

Bennett JW, Klich M (2003) Mycotoxins. Clin Microbiol Rev 16(3):497–516

Blomquist G, Palmgren U, Ström G (1984) Improved techniques for sampling airborne fungal particles in highly contaminated environments. Scand J Work Environ Health 10(4):253–258

Brenier-Pinchart MP, Lebeau BB, Quesada JL et al. (2009) Influence of internal and outdoor factors on filamentous fungal flora in hematology wards. Am J Infect Control 37:631–637

Breum NO, Nielsen BH, Nielsen EM et al. (1996) Bio-aerosol exposure during collection of mixed domestic waste – an intervention study on compactor truck design. Waste Manag Res 14:527–536

Burdorf A (2005) Identification of determinants of exposure: consequences for measurement and control strategies. Occup Environ Med 62:344–350

De Nuntiis P, Maggi O, Mandrioli P et al. (2003) Cultural heritage and aerobiology – methods and measurement techniques for biodeterioration monitoring. Kluwer Academic Publishers, Dordrecht, p 107–144

Duchaine C, Grimard Y, Cormier Y (2000) Influence of building maintenance, environmental factors, and seasons on airborne contaminants of swine confinement buildings. AIHAJ – Am Ind Hyg Assoc 61(1): 56–63

Duquenne P, Marchand G, Duchain C (2012) Measurement of endotoxins in bioaerosols at workplace: a critical review of literature and a standardization issue. Ann Occup Hyg 57(2):137–172. doi:10.1093/annhyg/mes051

Eduard W (2009) Fungal spores: a critical review of the toxicological and epidemiological evidence as a basis for occupational exposure limit setting. Crit Rev Toxicol 39(10):799–864

Eduard W, Halstensen A (2009) Quantitative exposure assessment of organic dust. Scand J Work Environ Health Suppl 7:30–35

Gutarowska B, Skóra J, Stępień Ł et al. (2015) Assessment of microbial contamination within working environments of different types of composting plants. J Air Waste Manag Assoc 65(4):466–478

Halstensen AS, Heldal KK, Wouters IM et al. (2013) Exposure to grain dust and microbial components in the Norwegian Grain and Compound Feed Industry. Ann Occup Hyg 57(9):1105–1114

Halstensen AS, Nordby KC, Wouters IM et al. (2007) Determinants of microbial exposure in grain farming. Ann Occup Hyg 51(7):581–592

Halstensen AS, Nordby KCH, Elen O et al. (2004) Ochratoxin A in grain dust – estimated exposure and relations to agricultural practices in grain production. Ann Agric Environ Med 11:245–254

Hameed AAA, Ayesh AM, Mohamed MAR et al. (2012) Fungi and some mycotoxins producing species in the air of soybean and cotton mills: a case study. Atmos Pollut Res 3:126–131

Hameed AAA, Habeebuallah T, Mashat B et al. (2015) Airborne fungal pollution at waste application facilities. Aerobiologia 31(3):283–293.

Health Canada – Indoor air quality in office buildings: a technical guide (1993) Vancouver: Health Canada. http://www.hc-sc.gc.ca/ewh-semt/alt_formats/hecs-sesc/pdf/pubs/air/office_buildingimmeubles_bureaux/93ehd-dhm166-eng.pdf

Health Council of the Netherlands (2010) Endotoxins – Health-based recommended occupational exposure limit. Health Council of the Netherlands, The Hague, https://www.gezondheidsraad.nl/en/task-and-procedure/areas-of-activity/healthy-working-conditions/endotoxins-health-based-recommended

Health Safety Executive (2010) Current control standards for tasks with high exposure to grain dust. Prepared by the Institute of Occupational Medicine. Research Report RR829

IARC (2002) Some traditional herbal medicines, some mycotoxins, naphthalene and styrene. IARC Monogr Eval Carcinog Risks Hum 82:1–556. PMID:12687954

Jorna THJM, Borm PJA, Valks J et al. (1994) Respiratory symptoms and lung function in emiel animal feed workers. Chest 106(4):1050–1055

Jürgensen CW, Madsen AM (2016) Influence of everyday activities and presence of people in common indoor environments on exposure to airborne fungi. AIMS Environ Sci 3(1):77–95

Khoury A, Atoui A (2010) Ochratoxin A: general overview and actual molecular status. Toxins 2(4):461–493

Kim KY, Ko HJ, Kim HT et al. (2007) Influence of extreme seasons on airborne pollutant levels in a pig-confinement building. Arch Environ Occup Health 62(1):27–32

Kotimaa MH, Oksanen L, Koskela P et al. (1991) Feeding and bedding materials as sources of microbial exposure on dairy farms. Scand J Work Environ Health 17(2): 117–122

Krysińska-Traczyk E, Pande BN, Skórska C et al. (2005) Exposure of Indian agricultural workers to airborne microorganisms, dust and endotoxin during handling of various plant products. Ann Agric Environ Med 12:269–275

Krysińska-Traczyk E, Perkowski J, Dutkiewicz J et al. (2007) Levels of fungi and mycotoxins in the samples of grain and grain dust collected from five various cereal crops in eastern Poland. Ann Agric Environ Med 14:159–167

Lanier C, André V, Séguin V et al. (2012) Recurrence of Stachybotrys chartarum during mycological and toxicological study of bioaerosols collected in a dairy cattle shed. Ann Agric Environ Med 19:61–67

Lanier C, Richard E, Heutte N et al. (2010) Airborne molds and mycotoxins associated with handling of corn silage and oilseed cakes in agricultural environment. Atmos Environ 44:1980–1986

Ławniczek-Wałczyk A, Gołofit-Szymczak M, Cyprowski M et al (2012) Exposure to harmful microbiological agents during the handling of biomass for power production purposes. Med Pr 63:395–407

Ławniczek-Wałczyk A, Gorny RL, Gołofit-Szymczak M et al. (2013) Occupational exposure to airborne microorganisms, endotoxins and β-glucans in poultry houses at different stages of the production cycle. Ann Agric Environ Med 20(2):259–268

Leibler JH, Jordan JA, Brownstein K et al (2016) Staphylococcus aureus Nasal Carriage among Beefpacking workers in a Midwestern United States Slaughterhouse. PLoS One 11(2): e0148789

Liao VH, Chou WC, Chio CP et al. (2009) A probabilistic approach to quantitatively assess the inhalation risk for airborne endotoxin in cotton textile workers. J Hazard Mater 177:103–108

Lonc E, Plewa K (2010) Microbiological air contamination in poultry houses. Polish J Environ Stud 19(1):15–19

Lonc E, Plewa K (2011) Comparison of indoor and outdoor bioaerosols in poultry farming. In: Prof. Anca Moldoveanu (ed) Advanced topics in environmental health and air pollution case studies. InTech, doi: 10.5772/20096

Madsen AM (2011) Identification of Work Tasks Causing High Occupational Exposure to Bioaerosols at Biofuel Plants Converting Straw or Wood Chips, Environmental Impact of Biofuels. In: Dr. Dos Santos Bernardes MA (ed), InTech, DOI: 10.5772/17859. Available from: https://www.intechopen.com/books/environmental-impact-of-biofuels/identification-of-work-tasks-causing-high-occupational-exposure-to-bioaerosols-at-biofuel-plants-con

Madsen AM, Matthiesen CB, Frederiksen MW et al. (2012) Sampling, extraction and measurement of bacteria, endotoxin, fungi and inflammatory potential of settling indoor dust. J Environ Monit 14:323

Madsen AM, Tendal K, Frederiksen MW (2014) Attempts to reduce exposure to fungi, β-glucan, bacteria, endotoxin and dust in vegetable greenhouses and a packaging unit. Sci Total Environ 468–469:1112–1121

Malta-Vacas J, Viegas J, Sabino R et al (2012) Fungal and microbial volatile organic compounds exposure assessment in a waste sorting plant. J Toxicol Environ Health 75:1410–1417

Mc Donnell PE, Coggins MA, Hogan VJ et al. (2008) Exposure assessment of airborne contaminants in the indoor environment of Irish swine farms. Ann Agric Environ Med 15:323–326

Meheust D, Le Cann P, Reboux G et al. (2014) Indoor fungal contamination: health risks and measurement methods in hospitals, homes and workplaces. Crit Rev Microbiol 40(3):248–260

Normand AC, Vacheyrou M, Sudre B et al. (2009) Assessment of dust sampling methods for the study of cultivable-microorganism exposure in stables. Appl Environ Microbiol 74(24):7617–7623

Oppliger A (2014) Advancing the science of bioaerosol exposure assessment. Ann Occup Hyg 58(6):661–663

Oppliger A, Charrière N, Droz PO et al. (2008) Exposure to bioaerosols in poultry houses at different stages of fattening; use of real-time PCR for airborne bacterial quantification. Ann Occup Hyg 52(5):405–412

Peccia J, Hospodsky D, Bibby K (2011) New directions: a revolution in DNA sequencing now allows for the meaningful integration of biology with aerosol science. Atmos Environ 45:1896–1897

Rimac D, Macan J, Varnai VM et al. (2010) Exposure to poultry dust and health effects in poultry workers: impact of mould and mite allergens. Int Arch Occup Environ Health 83(1):9–19

Rylander R (1999) Indoor air-related effects and airborne (1→3)-β-D-glucan. Environ Health Perspect 107:501–503

Sabino R, Faísca VM, Carolino E et al (2012) Occupational exposure to aspergillus by swine and poultry farm workers in Portugal. J Toxicol Environ Health 75(22–23): 1381–1391

Scaramuzza N, Diaferia C, Berni E (2015) Monitoring the mycobiota of three plants manufacturing Culatello (a typical Italian meat product). Int J Food Microbiol 203:78–85. doi:10.1016/j.ijfoodmicro.2015.02.034

Skorska C, Sitkowska J, Krysińska-Traczyk E et al (2005) Exposure to airborne microorganisms, dust and endotoxins during processing of valerian root on farms. Ann Agric Environ Med 12:119–126

Smid T, Heederik D, Mensink G et al. (1992) Exposure to dust, endotoxins, and fungi in the animal feed industry. Am Ind Hyg Assoc J 53(6):362–368

Smid T, Schokkin E, Boleij JS (1989) Enumeration of viable fungi in occupational environments: a comparison of samplers and media. Am Ind Hyg Assoc J 50(5):235–239

Sowiak M, Bródka K, Kozajda A et al (2012) Fungal aerosol in the process of poultry breeding – Quantitative and qualitative analysis. Med Pr 63(1):1–10

Straumfors A, Heldal KK, Wouters IM et al. (2015) Work tasks as determinants of grain dust and microbial exposure in the Norwegian grain and compound feed industry. Ann Occup Hyg 59(2):142–157. doi:10.1093/annhyg/meu090

Sudharsanam S, Srikanth P, Krishnamurthy S et al. (2009) Microorganisms in bioaerosols in indoor air of hospital and non-hospital settings. Sri Ramachandra J Med 2:52

Sudharsanam S, Swaminathan S, Ramalingam A et al. (2012) Characterization of indoor bioaerosols from a hospital ward in a tropical setting. Afr Health Sci 12(2):217–225

Swan JR, Crook B (1998) Airborne microorganisms associated with grain handling. Ann Agric Environ Med 5:7–15

Thilsing T, Madsen AM, Basinas I, et al. (2015) Dust, endotoxin, fungi, and bacteria exposure as determined by work task, season, and type of plant in a flower greenhouse. Ann Occup Hyg, 59(2):142–157

Thorne PS, Heederik D (1999) Assessment methods for bioaerosols. In: Salthammer T (ed.) Organic indoor air pollutants – occurrence, measurement, evaluation. Wiley/VCH, Weinheim, p 85–103

Thulin H, Björkdahl M, Karlsson AS et al. (2002) Reduction of exposure to laboratory animal allergens in a research laboratory. Ann Occup Hyg 46:61–68

Vacova M, Buchta V, Prymula R et al. (2006) The occurrence of microscopic fungi in air samples from a transplant intensive care unit. Indoor Built Environ 15:115–118

Verhoeff AP, Van Wijnen JH, Brunekreef B (1992) Presence of viable mould propagules in indoor air in relation to house damp and outdoor air. Allergy 47(2):83–91

Viegas S (2012) Estudo da exposição profissional a formaldeído laboratórios hospitalares de anatomia patológica. Caminhos do Conhecimento. http://www.edi-colibri.pt/Detalhes.aspx?ItemID=1576

Viegas S, Veiga L, Figueiredo P et al. (2013) Occupational exposure to aflatoxin B1: the case of poultry and swine production. World Mycotoxin J 6(3):309–315

Viegas S, Almeida-Silva M, Viegas C (2014a) Occupational exposure to particulate matter in 2 Portuguese waste-sorting units. Int J Occup Med Environ Health 27(5):854–862. doi:10.2478/s13382-014-0310-8

Viegas C, Dias R, Gomes AQ et al. (2014b) Aspergillus flavus contamination in two portuguese wastewater treatment plants. J Toxicol Environ Health A 77:796–805. doi:10.1080/15287394.2014.909300

Viegas S, Veiga L, Figueiredo P et al. (2014c) Assessment of workers' exposure to aflatoxin B1 in a Portuguese waste industry. Ann Occup Hyg 59(2):173–181 doi:10.1093/annhyg/meu082.

Viegas S, Veiga L, Almeida A et al. (2015a) Occupational exposure to aflatoxin B1 in a Portuguese poultry slaughterhouse. Ann Occup Hyg 60(2):176–183 doi:10.1093/annhyg/mev077

Viegas C, Sabino R, Botelho D et al. (2015b) Assessment of exposure to the Penicillium glabrum complex in cork industry using complementing methods. Arch Ind Hyg Toxicol 66(3):203–207. doi:10.1515/aiht-2015-66-2614

Viegas C, Pinheiro C, Sabino R, et al. (2015c) Environmental mycology in public health: fungi and mycotoxins risk assessment and management. Academic Press, London

Viegas C, Faria T, Meneses M et al. (2016a) Analysis of surfaces for characterization of fungal burden – does it matter? Int J Occup Med Environ Health 29(4):623–632. doi:10.13075/ijomeh.1896.00562

Viegas C, Faria T, Carolino E et al. (2016b) Occupational exposure to fungi and particles in animal feed industry. Medycyna Pracy 67(2):143–154. doi:10.13075/mp.5893.00289

Viegas C, Faria T, dos Santos M et al. (2016c) Slaughterhouses fungal burden assessment: a contribution for the pursuit of a better assessment strategy. Int J Environ Res Public Health 13:297. doi:10.3390/ijerph13030297

Viegas S, Almeida-Silva M, Faria T et al. (2016d) Occupational exposure assessment to particles with task-based approach. In: Arezes IV (eds) Occupational safety and hygiene. Taylor and Francis Group, London, p 1–6. ISBN: 978-1- 138-02942-2

Viegas C, Faria T, Caetano LA, et al. (2017a) Aspergillus spp. prevalence in different Portuguese occupational environments: what is the real scenario in high load settings? J Occup Environ Hyg 13:0. doi:10.1080/15459624.2017.1334901

Viegas S, Aranha Caetano L, Korkalainen L et al. (2017b) Cytotoxic and inflammatory potential of air samples from occupational settings with exposure to organic dust. Toxics 5(8):1–16

Viegas C, Malta-Vacas J, Sabino R et al. (2014) Accessing indoor fungal contamination using conventional and molecular methods in Portuguese poultries. Environ Monit Assess 186(3): 1951–1959.

Vincent JH (2005) Health-related aerosol measurement: a review of existing sampling criteria and proposals for new ones. J Environ Monit 7(11):1037–1053

Walser SM, Gerstner DG, Brenner B et al. (2015) Evaluation of exposure–response relationships for health effects of microbial bioaerosols – a systematic review. Int J Hyg Environ Health 218:577–589

Wan-Kuen J, Jung-Hwan K (2005) Exposure levels of airborne bacteria and fungi in Korean swine and poultry sheds. Arch Environ Occup Health 60(3):140–146

Wijnand E, Bakke B (1999) Experiences with task-based exposure assessment in studies of farmers and tunnel workers. Norsk Epidemiologi 9(1):65–70

Chapter 18
What Is Known About Zoonotic Bacteria in Cattle and Pig Farms, and What Should Be Done to Prevent Occupational Infections

Julia G. Kraemer and Anne Oppliger

Abstract About 60% of all human infectious diseases originate and spread from animals. These zoonotic infections can be caused by bacteria, viruses, fungi or parasites, and they can be transmitted by inhalation, direct or indirect contact, ingestion or different vectors (insects or mites). Some of them are very dangerous and impose great economic burdens, particularly in tropical regions. Today, due to global warming, several microorganisms are expanding across new geographical ranges and emerging in zones where they had previously been absent. Farmers are among the people who interact the most with animals. Indeed, animal farmers work in close contact with animals and are therefore exposed to many potential pathogenic microorganisms every day. The present chapter will review the zoonotic bacteria encountered most frequently in animal farms over the last 10 years. The emphasis will be on explaining which types of risks are present in which types of animal farms. Recommendations on preventing and managing those risks will be discussed. This chapter will not treat the issues of virus and parasite transmission, nor the risks associated with poultry farming.

Keywords Farming · biosecurity · zoonosis · biological risks · occupational exposure · cattle · pigs

18.1 Introduction

Wild, domestic and pet animals are reservoirs for many pathogenic microorganisms. These microorganisms often have a broad range of hosts and can be transmitted from one animal species to another, including humans. There are four main routes of

J.G. Kraemer
Institute for Work and Health, Universities of Lausanne and Geneva, Lausanne, Switzerland

Institute for Infectious Diseases, University of Bern, Bern, Switzerland

A. Oppliger (✉)
Institute for Work and Health, Universities of Lausanne and Geneva, Lausanne, Switzerland
e-mail: Anne.Oppliger@hospvd.ch

C. Viegas et al. (eds.), *Exposure to Microbiological Agents in Indoor and Occupational Environments*, DOI 10.1007/978-3-319-61688-9_18

transmission: (1) inhalation, via a bioaerosol; (2) direct contact with a contaminated animal (including bites and scratches) or indirect contact after contact with a surface, an object, soil or water contaminated with biological liquids from a contaminated animal; (3) ingestion of contaminated meat or milk, or ingestion of other contaminated food; and (4) other vectors, usually insects or arthropods.

Besides transmitting the pathogenic microorganisms which can be responsible for infectious human diseases, livestock also generates huge amounts of non-pathogenic or opportunistic pathogenic bacteria. Indeed, in concentrated animal feeding operations, bacteria present in the litter or the feed and bacteria colonising animals' skin, hair or faeces are all easily aerosolised. Numerous studies have observed that exposure to high levels of such bioaerosols is associated with a variety of different symptoms in animal farmers, including chronic coughing, rhinitis, irritation, lung inflammation, hypersensitivity pneumonitis and decline of lung function. Concerns about these issues are very well documented (see the review by Heederik et al. 2007) and it is now widely accepted that technical measures to reduce the level of airborne organic dust in animal farms must be implemented and that wearing personal protective equipment (PPE) is recommended for specific tasks that generate a lot of organic dust. Thus the present chapter will not address the issue of non-infectious risks, but will focus exclusively on infectious risks.

In order to assess the infectious risks in animal farms, we need to know which infectious agents are present, what their routes of transmission are and what their host range is. Other parameters, such as the immune status of the people exposed and whether or not they wear PPE, also need to be taken into account. The micro-organisms present on a farm will greatly depend on the animal species raised and the local climate; the risks will greatly depend on working conditions.

With regard to occupational exposure in animal farms, we will focus solely on the most important pathogens that are transmitted by inhalation or direct contact; the risks associated with food-borne pathogens are not specific to farmers and are not usually considered occupational risks. Table 18.1 summarises the essential points of each zoonotic disease described.

18.2 Zoonotic Diseases Infecting Different Animal Species

18.2.1 Q Fever

18.2.1.1 General Information

According to the USA's Centers for Diseases Control and Prevention (CDC), Q fever is a disease found worldwide. It is primarily an occupational hazard (Dorko et al. 2012), and although it has acute and chronic clinical manifestations, half of infected humans show no symptoms. The causative agent, *Coxiella burnetii*, is a small, obligate, intracellular Gram-negative bacterium. Cattle, sheep and goats are the primary reservoirs although a variety of species may be infected, including pigs. Bacteria are excreted in the milk, urine and faeces of infected

Table 18.1 Summary information on the most frequent zoonoses in cattle, small ruminant and pig farming

Disease	Farm animals	Location	Main route of occupational transmission	Symptoms in animal	Pathogenicity for humans	Vaccine for animals	Vaccine for humans
Q fever		Worldwide except New Zealand	Inhalation	Bone or abortion	-/++	Yes	Yes
Campylobacteriosis		Worldwide	Animal contact	None	-/+	No	No
Salmonellosis		Worldwide	Animal contact	Sometimes	-/+	Yes	No

(continued)

Table 18.1 (continued)

Disease	Farm animals	Location	Main route of occupational transmission	Symptoms in animal	Pathogenicity for humans	Vaccine for animals	Vaccine for humans
Brucellosis		Middle East, Mediterranean basin, Africa, China, India, Peru, Mexico, south-eastern and central Asia	Animal contact (abortion discharge)	Abortion	-/++	Yes	No
Leptospirosis		Worldwide	Animal contact	None	+/++	Yes	Yes
Bovine tuberculosis		Worldwide, but most present in Africa, Asia and South America	Inhalation	Yes, cough	+++	Yes, but not allowed in all countries	Yes

(continued)

Table 18.1 (continued)

Disease	Farm animals	Location	Main route of occupational transmission	Symptoms in animal	Pathogenicity for humans	Vaccine for animals	Vaccine for humans
Listeriosis		Worldwide	Animal contact and inhalation	None or abortion	+/++	No	No
Infection with *S. suis*		Worldwide	Skin contact	Yes	++	Yes	No
Yersiniosis		Worldwide	Animal contact	None	++	No	No
Erysipeloid		Worldwide	Skin contact	Yes	+	Yes	No

animals. During birthing, the organisms are shed in high numbers in the amniotic fluid and placenta. The microorganism is very resistant to heat, drying and many common disinfectants; this enables the bacteria to survive in the environment for long periods. Infection of humans usually occurs via the inhalation of dust from the dried placental material, birth fluid or excreta of infected animals (CDC). Transmission by the ingestion of unpasteurised milk or dairy products is rare (Angelakis and Raoult 2010). Humans are often very susceptible to the disease, and very few organisms may be required to cause infection. Symptoms are non-specific and can include high fever, severe headache, general malaise, myalgia, chills and/or sweats, non-productive coughing, vomiting, diarrhoea, and abdominal and chest pain (CDC). Most people with an acute Q fever infection recover, although the main symptom of chronic infections, endocarditis, often has a high mortality rate (see details on clinical manifestations in Angelakis and Raoult 2010). Infected pregnant women may risk pre-term delivery or miscarriage.

Infection in animals, however, is in most cases asymptomatic. In the USA, it was estimated that agricultural workers were six times more likely to have antibody evidence of a *C. burnetii* infection than those employed in other occupations (Walsh 2012).

18.2.1.2 Reported Infections in Farmers

From 2007–2010, Q fever was a major public health concern in the Netherlands with >4000 reported human cases (Dijkstra et al. 2012) and probably more than 40,000 infected individuals (Schneeberger et al. 2014). Large-scale interventions targeting small ruminants were carried out to control the epidemic (Schimmer et al. 2014). From 2009 to 2010, seroprevalence in dairy goat farmers was very high (73.5%). Moreover, the seroprevalence of their spouses and children was also high (66.7% and 57.1%, respectively) (Schimmer et al. 2012). From 2010 to 2011, the same research group investigated seroprevalence in cow and sheep farmers and observed rates of 87.2% and 66.7%, respectively (Schimmer et al. 2014, De Lange et al. 2014).

In Italy, it was reported that 50% of animal breeders were seropositive, whereas only 32.2% of agricultural workers not involved in animal breeding activities were seropositive (Tabibi et al. 2013). In southern Italy, a study found that 73.4% of the subjects exposed to farm animals were positive for *C. burnetii* IgG (titer > 20 IU/mL). Among them, 84% of workers involved in animal breeding were seropositive compared to only 13.6% of the control subjects (Monno et al. 2009). Another study carried out in southern Italy showed that about half of all farmers were seropositive (54.5%) (Fenga et al. 2015). Furthermore, in western Sicily, a seroprevalence of 25% was reported among workers operating outdoors, but not in contact with animals (Verso et al. 2016). The seroprevalence among farm workers in Poland was estimated to be from 31.1% to 39.1% (Szymanska et al. 2014).

In Ireland, a seroprevalence study among agriculture workers showed that 7.2% (6/83) of participants were positive. However, this rate was no different from the prevalence observed in groups considered to have a moderate or low exposure risk (laboratory technicians and office workers) (Reid and Malone 2004).

In Denmark, a seroepidemiological survey of farmers and other people with occupational contact with dairy cattle (Bosnjak et al. 2010) showed that farmers were less exposed than veterinarians (3% vs 36%, respectively) and that none of the seropositive subjects had the symptoms of acute Q fever infection. However, these results differed from those from a previous Swedish study showing that 28% of sheep farmers and only 13% of veterinarians were seropositive (Macellaro et al. 1993).

In Australia, two states have some of the highest rates of reported human Q fever in the world, with 50–110 cases/100,000 population per year (Gidding et al. 2009). Three outbreaks were reported: among goat farmers in 2003 (Miller et al. 2005), goat and sheep farmers in 2013 (Bond et al. 2016) and cow farmers in 2013 (Gullan et al. 2012). Interestingly, in the latter case, the authors reported that the outbreak was related to changed farming practices. This change consisted of placing placental tissue from cattle on an open composting pile, to be used in the farm's gardens as part of its organic gardening process. This composting pile was situated on top of a hill overlooking the main farm buildings and gardens. Until this change was implemented, it had been usual for placental tissues to be buried or left out for foxes to consume (Gullan et al. 2012).

A systematic review revealed that Q fever is a neglected zoonosis in Kenya, however most of the reported outbreaks do not concern farmers (Njeru et al. 2016).

18.2.1.3 Specific Recommendations

In addition to general preventive measures (Table 18.2), and according the CDC, the following specific measures should be used to prevent and control of Q fever:

- The public should be educated about the sources of infection.
- Ensure that holding facilities for sheep are located away from populated areas. Animals should be routinely tested for antibodies to *C. burnetii*.
- Individuals at the highest risk for developing chronic Q fever should be warned and advised, especially those with existing heart valve disease or vascular grafts.
- Animals in infected flocks should be vaccinated with an efficient vaccine, as should animals in uninfected flocks close to them. This can prevent abortions and the shedding of bacteria (Angelakis and Raoult 2010). A vaccine for Q fever has been developed and has successfully protected humans in occupational settings in Australia. However, its use is difficult because it must not be administrated to individuals who have already been in contact with *C. burnetii*. Indeed, this may lead to serious adverse reactions such as sterile abscesses or systemic inflammatory syndrome (Schneeberger et al. 2014).
- A study in the Netherlands showed that wearing boots and protective clothing and keeping pets out of goat stables decreased the risk of farm worker contamination (Schimmer et al. 2012).

Table 18.2 Preventive measures to avoid zoonoses in livestock farming: points in bold are measures easy to implement and essential

Animal management
• **Closed herd, if not possible purchase from known healthy flocks; 4 weeks quarantine for all new animals;** • Depending on the causal agent (brucellosis, bovine tuberculosis, q-fever, salmonellosis, leptospirosis), test and/or vaccine all animals; • **Identify and isolate adequately all ill animals or animals having aborted and call immediately the veterinarian;** • Avoid contact with neighbouring herds with an optimal grazing agenda. If not possible use fencing and even double fencing; • Wait 3 weeks before grazing on lands after spreading manure; • **Dispose all birth products and discharges (placenta, fetal membrane, aborted foetuses) in a dedicated disposal.**
Farm management
• Use vehicle deeply disinfected to transport animals; • Prohibit that person having occupational or leisure contact with other farms drive or park their vehicles in areas accessible to the animals, if not possible, require that they clean and disinfect their vehicle; • **Use separate equipment for healthy and ill animals and routinely clean and disinfect feeding and cattle handling equipment**; • Avoid the presence of cats and dogs from the feed and bedding stores and from the animal buildings in general; • **Control rodents, birds and insects by having an effective pest control plan;** • **Limit access of visitors. Require that all visitors respect the sanitary measures, i.e: disinfection of shoe/boots, use of protective clothing and hand disinfection**.
Workers behaviour
• **Observe the basic hygiene rules: use of foot baths; wash hands and take a shower after working in barn areas, use dedicated clothing washed separately from other clothing;** • **Wear protective gloves and facial protection when handling sick animal or their body fluids;** • **Do not eat, drink or smoke in the barn environment;** • **Provide adequate information and training to all employees concerning the biological risk**.

18.2.2 Campylobacteriosis

18.2.2.1 General Information

Campylobacteriosis is the most reported zoonosis in the EU (European Food Safety Authority). It is caused by bacteria of the *Campylobacter* genus, a spiral-shaped Gram-negative bacterium (CDC). The disease is usually caused by *C. jejuni*, normally found in cattle, swine, birds and in the environment, where it is non-pathogenic. However, the illness can also be caused by *C. coli* (also found in cattle, swine and birds). The gastroenteritis caused by the disease affects over 1.3 million people every year. However, accounting for under-reporting led one study to estimate 9.2 million cases in the EU alone (Havelaar et al. 2009). Symptoms are diarrhoea, abdominal cramps or pain and fever within 2–5 days of exposure to the organism. Diarrhoea may be bloody and can be accompanied by nausea and vomiting. The illness typically lasts about 1 week. Some infected people show no symptoms. In individuals with weakened immune systems, *Campylobacter* occasionally spreads to the bloodstream and causes a serious, life-threatening infection. Although *Campylobacter* infection does not commonly cause death, it has been estimated that approximately 76 people with *Campylobacter* infections die each year. Ingesting contaminated food (generally unpasteurised milk or undercooked or poorly handled poultry) or drinking contaminated water are the main routes of infection. However, contact with infected poultry, ruminants, pigs or household pets, especially puppies, can also cause the disease. It has also been shown (Inglis et al. 2010) that the *Campylobacter* shed in cattle faeces persist for long periods (around 10 months) in manure and subsequently in compost, thus enhancing the risk of environmental or human transmission.

18.2.2.2 Reported Infections in Farmers

Most occupational campylobacteriosis is due to direct contact with poultry, but also, although to a lesser extent, to direct contact with ruminants; pig farming is less of a risk. Indeed, a study carried out in France (Denis et al. 2009) showed that *Campylobacter* isolates from pigs were not genetically related to those from humans; 46% of those human Campylobacter isolates were genetically related to genotypes found in poultry. Furthermore, a significant part of these campylobacterioses were not due to the ingestion of contaminated food, but rather to occupational contact with poultry. Two other studies, from the Luxembourg and Switzerland (Mossong et al. 2016; Jonas et al. 2015), showed the same trends, with a majority of human cases of campylobacteriosis attributed to poultry (61.2% and 44%, respectively) and ruminants (33.3% and 36%, respectively). Environmental water (4.9%) and pigs (0.6% and 20%, respectively) made up the remainder of cases. Unfortunately, these studies did not take occupation into account to assess the occupational risk factors for human campylobacteriosis. In the Netherlands (Mughini-Gras et al. 2012), a type of 980 *Campylobacter* isolate from human infections also showed that the majority of human infections originated from chickens (66.2%), followed by cattle (20.7%), the environment (10.1%), sheep (2.5%) and finally pigs (0.3%). Among non-food risk factors, this study showed that occupational exposure to animals was only an important risk for ruminant-associated campylobacteriosis (17%). Different results were observed in Finland (de Haan et al. 2010), however, where it was shown that poultry and cattle were equally important reservoirs of human *C. jejuni* infection (45.4% and 44.3%, respectively). In the USA, a seroprevalence survey showed that farmers (mainly pig and cattle farmers) had significantly higher levels of anti-*C. jejuni* IgA and IgG antibodies than non-farmers (Vegosen et al. 2015). A study in Washington state (USA), estimated that living or working on a dairy farm produced the strongest significant associations with campylobacteriosis and concluded that in areas with high concentrations of dairy cattle, exposure to animals could be a more important risk of campylobacteriosis than exposure to poultry food products (Davis et al. 2013). This was in agreement with a study carried out in New Zealand (Gilpin et al. 2008), where the relative risk of a dairy farm worker being diagnosed with campylobacteriosis was estimated to be 1.88 (95% CI = 1.6–2.2). This study identified the primary means of transmission to farm workers and children to be cattle faeces. From 2002 to 2003, contact with farm animals was also a predominant risk factor for *C. jejuni* enteritis in young children (0–4 years old) in the Netherlands (Doorduyn et al. 2010). An outbreak of campylobacteriosis was documented in children attending a farm day camp in Minnesota state (USA) and was associated with caring for an ill calf and visibly getting manure on their hands (Smith et al. 2004). One review (Taylor et al. 2013) aiming to find sources of campylobacteriosis outbreaks in the USA from 1997 to 1998, showed that seven (3%) outbreaks were associated with animal contact. All the outbreaks except one were due to farm animals (calves, chickens, turkeys or pigs). The study underlined that children display more frequent hand-to-mouth activities, resulting in a greater potential to ingest pathogens present in the environment.

Campylobacter species other than *C. coli* and *C. jejuni* can also be transmitted from animals to humans. For example, a case of cellulitis caused by *C. fetus*, probably transmitted from cattle to a rural worker, was described in Uruguay (Iraola et al. 2015), and bacteriemia with acute febrile illness due to *C. fetus* was also reported in a cattle farmer (Zonios et al. 2005). One review (Wagenaar et al. 2014) reported five other cases of farmers infected by zoonotic *C. fetus*, with diverse clinical symptoms such as meningitis, coughing, cellulitis and diarrhoea.

In conclusion, we observe that the relative contribution of ruminant campylobacter to farmers' infections varies geographically, but a non-negligible proportion of human infection is due to reservoirs in cattle. Occupational exposure to pigs, however, does not seem to be a recurrent risk factor for campylobacteriosis.

18.2.2.3 Specific Prevention

In cases of subclinical human infection and the asymptomatic carriage of gastrointestinal infection by farmers, pathogens are excreted continuously. Therefore, in certain situations, it could be useful to identify and treat these asymptomatic individuals in order to avoid spreading pathogens throughout the community (Quilliam et al. 2013).

18.2.3 Salmonellosis

18.2.3.1 General Information

Salmonellosis is a common worldwide disease caused by *Salmonella enteritica*. Typhimurium and Enteritidis are the most common serological variants (serovars) of the *enterica* subspecies; they are enterobacteriaceae Gram-negative bacilli found in animal and human intestines and are shed through faeces. Humans usually become infected via contaminated water or food. However, non-foodborne infection may be transmitted through direct contact with infected animals or indirect contact with the contaminated environment (Schiellerup et al. 2001; Wall et al. 1995). Moreover, it has been shown that houseflies can be vectors for salmonella transmission between farms. Indeed, flies can acquire and harbour the bacteria and transport them over significant distances (Wang et al. 2011).

Human infection can be asymptomatic. Symptomatic individuals develop a gastroenteritis with diarrhoea, fever and abdominal cramps within 12–72 h of exposure. Healthy people usually recover within 4–7 days without treatment.

According to the CDC, in some cases, the diarrhoea associated with salmonella infection can be so severe that hospitalisation is required. Life-threatening complications may develop if the infection spreads from the intestines to the blood stream. Young children, older adults and people with weakened immune systems are the most likely to suffer severe infections.

The CDC estimates that non-typhoidal *Salmonella* causes approximately 1.2 million cases of illness and 450 deaths annually in the USA.

18.2.3.2 Reported Infections in Farmers

One review (Steinmuller et al. 2006) aiming to describe enteric disease outbreaks associated with animals in public settings – including farms accepting visitors – in USA from 1991 to 2005, showed that most of the 55 outbreaks were caused by *E. coli* O157 (32 cases) or *Salmonella* spp. (12 cases) after either direct or indirect contact with animals and their environment. All but three outbreaks were due to farm animals (calves, dairy cattle, chickens, turkeys, goats, sheep or pigs). Another review (Hoelzer et al. 2011) reported 16 other cases/outbreaks of Salmonella transmission from cattle to humans in different countries (USA, UK, Netherlands, Canada and New Zealand), three cases from sheep to humans and two cases from pigs to humans in animal farms, mostly associated with occupational exposure.

A case of farm animal (pigs or calves) to human transmission of *Salmonella enterica* serovar Typhimurium was reported in the Netherlands (Hendriksen et al. 2004).

A Dutch analysis of risk factors showed that pigs and cattle were the primary reservoirs of salmonellosis for children, people living in urban areas and during autumn and winter, whereas poultry was the primary reservoir for adults, for people living in rural areas, and during spring and summer (Mughini-Gras et al. 2014). A Danish study reported that women with occupational exposure to animals had significantly higher IgG antibody concentrations against Campylobacter, Salmonella and Yersinia, thus confirming that they were frequently infected with these bacteria but without clinical symptoms (Kantso et al. 2014).

18.2.3.3 Specific Recommendations

The UK's National Animal Diseases Information Service (www.nadis.org.uk) gives, among other, the following recommendations:

- Sick animals should be isolated in dedicated isolation boxes.
- Buildings should be cleaned and disinfected between occupancies; good drainage and waste removal should be provided.
- Good fences should be maintained to prevent straying stock.
- All feed stores should be protected from vermin, including birds.
- Whenever possible, slurry should only be spread on arable land; leave all grazing land empty for at least three weeks after spreading slurry.

18.2.4 Brucellosis

18.2.4.1 General Information

According to the CDC, brucellosis – also called Bang's disease, Malta fever or undulant fever is caused by the bacterial genus *Brucella*, which is a small, aerobic, facultative, intracellular Gram-negative coccobacillus. Four of the eight identified animal species of *Brucella* exhibit significant human pathogenicity: *B. melitensis* (from sheep and goats), *B. abortus* (from cattle), *B. canis* (from dogs) *and B. suis* (from pigs). In infected animals, the symptoms are chronic lesions of the reproductive organs, leading to abortion in pregnant females or sterility and orchitis in males. Bacteria are shed in large numbers in the animal's urine, milk, placental fluid and other fluids and are mainly transmitted from animal to animal by the ingestion of infected tissue or fluid, via semen during breeding and by suckling infected mothers. The transmission of *B. melitensis* and *B. abortus* from reservoir animals to humans mainly occurs by the ingestion of raw dairy products or undercooked meat, whereas occupational exposure to contaminated secretions (via abrasions to the skin) or inhalation of bioaerosols are also frequent routes of transmission. Indeed, occupational exposure is the main route of transmission for *B. suis*. Accidental exposure to a vaccine (*B. abortus*) during injection can also cause human infection (Kutlu et al. 2014). Brucellosis is a major worldwide public health issue, with more than 500,000 infections per year (Pappas et al. 2006). Infected humans can be asymptomatic (Zhen et al. 2013) or can develop flu-like symptoms with inconstant fevers, sweating and joint or muscular pain. Symptoms usually appear within 5–30 days of contact with the bacteria and may disappear for weeks or months and then return. Chronic symptoms include fatigue, fevers, arthritis and spondylitis. The intensity of symptoms can depend on the intensity of animal contact and on the species involved. *B. melitensis* is the most virulent species, followed by *B. suis*; *B. abortus* is a low-virulent species. According to the USA's Center for Food Security and Public Health (2007), *Brucella* species have varied geographical distributions. *B. abortus* is found worldwide in cattle-raising regions, except in Japan, Canada, some European countries, Australia, New Zealand

and Israel, where it has been eradicated. Eradication from domesticated herds is nearly complete in the USA. *B. melitensis* is particularly common around the Mediterranean. It also occurs in the Middle East and Central Asia, around the Persian Gulf and in some Central American countries. In the past, *B. suis* was found worldwide in swine-raising regions. This organism has been eradicated from domesticated pigs in the USA, Canada, many European countries and some other nations. However, it persists in wild or feral swine populations in some areas, including the USA, Europe and Queensland, Australia. Sporadic outbreaks are reported in domesticated herds or humans due to transmission from this source. *B. suis* continues to occur in domesticated herds in some countries in South and Central America (including Mexico) and Asia.

18.2.4.2 Reported Infections in Farmers

Pig Farmers

Cases of human brucellosis due to occupational exposure to *B. suis* are scarce in the literature. In 2006, an outbreak involving pig farm workers was reported in Brazil. Of the three workers showing a significant antibody titer (>60 IU/mL using an agglutination test) only one female worker, who had had direct contact with aborted foetuses, displayed clinical signs of infection (Meirelles-Bartoli et al. 2012).

Cattle and Goat Farmers

A study carried out in Portugal (Diez and Coelho 2013) reported that 4.5% of 154 cattle farmers had suffered from brucellosis. In Bangladesh, one review (Islam et al. 2013) reported that seroprevalence of brucellosis was found in 2.6% of livestock farmers and 18.6% of animal milkers. Moreover, goat farmers had a higher prevalence (8.5%) than other farmers (Rahman et al. 2012). In Pakistan, polymerase chain reaction assays detected Brucella species in the blood of 17.3% (9/48) of symptomatic livestock farmers (Asif et al. 2014).

In Spain, human brucellosis was shown to be significantly related to occupation and contact with cattle, sheep, goats or pigs (60.9% of seropositive farmers). However, it is not clear whether or not the cases reported were caused by the ingestion of contaminated milk or direct contact with animal secretions (Ascencio et al. 2015).

In Turkey, one study estimated that 11.8% (84/712) of veterinarians had occupational brucellosis. Of these, 51 had had direct contact with infected animal excreta or had inhaled contaminated aerosols. Cattle were the main source of exposure. The study did not test farmers, but we can assume that they too were exposed to these bacteria, even though direct contact with animals was less intense (Kutlu et al. 2014).

Due to its non-specific, flu-like clinical symptoms, occupational brucellosis is probably under-reported. Its true incidence remains unknown in most developing and developed countries.

18.2.4.3 Specific Prevention

Educating people exposed to farm animals is an important means of preventing brucellosis transmission. A study in Senegal reported that none of the 222 farmers questioned had heard of brucellosis and thus none used preventive measures, despite 70% of them regularly assisting their animals during parturition or abortion (Tebug et al. 2015). In Malawi, only 2.9% of the 140 dairy farm smallholders questioned had heard of brucellosis (Tebug et al. 2014). In contrast, a study to evaluate cattle farmers' knowledge of bovine brucellosis in Portugal showed that 74.7% of them knew that brucellosis was a zoonotic disease. In Jordan, although 100% of the livestock keepers interviewed were aware of the risk of zoonotic brucellosis, only 19% and 13% were aware of the risks of infection via direct contact with foetal membranes or physical contact with infected livestock, respectively.

18.2.5 Leptospira sp.

18.2.5.1 General Information

Leptospirosis is an important re-emerging zoonotic disease that affects humans and animals. It is caused by the bacterium *Leptospira interrogans* which possesses 24 different serogroups and over 200 serovars (Okello et al. 2015). Symptoms are non-specific and include high fever, headache, chills, muscle pain, vomiting, jaundice and diarrhoea. Infection in humans can also be asymptomatic. Leptospirosis can cause kidney damage, meningitis, hepatic failure, respiratory distress and even death (Campagnolo et al. 2000).

According to the CDC, *Leptospira sp.* are spread through the urine of infected animals (which are often asymptomatic and can excrete the bacteria continuously from a few months up to several years); it can get into water or soil and can survive there for weeks to months. Many different wild and domestic animal species are carriers of *Leptospira sp.* (rodents, dogs, horses, pigs, sheep, goats and cattle). Humans can be infected through contact with the urine or amniotic fluids of infected animals or water or soil contaminated by that urine. The bacteria can enter the body through broken skin or mucous membranes (eyes, nose or mouth), especially if the skin has a cut or scratch. Drinking contaminated water can also cause infection and leptospirosis outbreaks are usually caused by exposure to contaminated water, such as floodwaters.

Leptospirosis occurs worldwide but is most common in temperate or tropical climates. It is an occupational hazard for many people working outdoors or with animals, such as farmers.

Overall, leptospirosis has been estimated to cause 1.03 million cases and 58,900 deaths worldwide each year. These estimates make leptospirosis a leading zoonotic cause of morbidity and mortality. Furthermore, morbidity and mortality are greatest in the poorest regions of the world, in areas where surveillance is not routinely performed (Costa et al. 2015).

18.2.5.2 Reported Infections in Farmers

A recent systematic review of 18 studies aimed to assess and specify quantitative evidence of the environmental risks of leptospirosis transmission. It revealed that livestock production was associated with an increased risk of infection, with the exception of one study in Laos (Mwachui et al. 2015). Another recent study in Fiji revealed that leptospirosis was still an important zoonosis there. The authors conducted a cross-sectional seroprevalence study and used an eco-epidemiological approach to characterise the risk factors and drivers for human leptospirosis infection (Lau et al. 2016). Their results showed that working outdoors (87% of outdoor workers were full- or part-time farmers), having pigs in the community or at home, and a high density of cattle in the district were all associated with the presence of *Leptospira* antibodies in the sample population investigated (2152 participants from 81 communities on three main islands).

In northern India, a study observed that 56% (48/86) of the leptospirosis patients diagnosed by the region's Microbiological Institute from 2004–2008 had contact with cattle (Sethi et al. 2010).

In north-eastern Australia, the infection rate in feral pigs was close to 40% for certain serovar, from 1999 to 2000; 71 human cases were reported in that region during that period (Heise-Pavlov and Heise-Pavlov 2003). This paper did not mention the infected humans' occupations, however.

In New Zealand, Schollum and Blackmore (1982) reported that 31% of 70 people working or living on pig farms had positive serological tests for the serovar Pomona, the pig-associated serovar. A later report mentioned that working in an abattoir or livestock farming were the most frequently reported occupations in leptospirosis cases (The Institute of Environmental Sciences and Research Ltd 2013). Another study from New Zealand (Sanhueza et al. 2015) showed that 5.1% of veterinarians who spent at least a quarter of their time working with dairy cattle were seropositive. Even though farmers were not included in the latter survey, together these studies imply that they face far higher risks of contracting leptospirosis than the general population.

18.2.5.3 Specific Prevention

A study carried out in Missouri, USA, showed that hand washing during and after work with pigs had a significant protective effect, whereas smoking cigarettes and drinking beverages while working with infected pigs were found to be significant risk factors (Campagnolo et al. 2000).

The UK's National Animal Diseases Information Service (www.nadis.org.uk) gives the following recommendations:

- Leptospirosis can be controlled by an initial vaccination comprising two doses 3–4 weeks apart and followed by booster vaccinations at 12-month intervals.
- All newly purchased cattle should be vaccinated upon arrival and then isolated for 4 weeks.

- Surface water/streams and rivers should be fenced off and mains water supplies should be used wherever possible.
- Cows that have aborted should be isolated.
- All products of abortion should be correctly disposed of (buried, burnt or collected).

18.2.6 Zoonotic Tuberculosis

18.2.6.1 General Information

Mycobacterium bovis, an acid-fast bacilli classified as a biosafety level 3 pathogen for public health, is the causative agent of mammalian tuberculosis, also called zoonotic tuberculosis, that can infect cattle and humans, but also a broad host range including goats, pigs, sheep, horses, dogs and cats (O'Reilly and Daborn 1995). Several wild animals are reservoirs, particularly the European badger, which is known to introduce bovine tuberculosis into herds (Ward et al. 2010). In developed countries, eradication programmes have significantly reduced this disease's prevalence, but reservoirs in wildlife make complete eradication difficult (Thoen et al. 2006) and outbreaks in cattle herds are still reported (O'Hagan et al. 2016a, 2016b). The situation in some developing countries is different since the combination of a high prevalence of immuno-suppressed HIV patients and a high prevalence of tuberculosis in animals make the eradication of *M. bovis* very problematic. Infection in humans presents the same symptoms as classic pulmonary tuberculosis caused by *M. tuberculosis*, which is the much more prevalent form of human tuberculosis. Extra-pulmonary lesions may also occur when bacilli are ingested.

Human infection with *M. bovis* accounted for a small proportion (0.5–7.2%) of all patients with a bacteriologically confirmed diagnosis of tuberculosis in industrialised countries (de la Rua-Domenech, 2006). For example, in the UK, 129 cases of *M. bovis* infection in humans were reported from 2005 to 2008 (Mandal et al. 2011), while during that same period > 21,000 cases of *M. tuberculosis* cases were isolated. It was also reported that mixed infection (*M. tuberculosis* and *M. bovis*) is not rare (8.2% of human cases) in India (Prasad et al. 2005). The global occurrence of zoonotic tuberculosis over the last two decades was estimated in a systematic review (Müller et al. 2013): data from 61 countries suggested a very low global disease incidence (7 zoonotic cases/100,000 population/year). In cattle, *M. bovis* infects the respiratory tract and bacilli are excreted in aerosol form and can be transmitted to humans by inhalation, ingestion of unpasteurised contaminated milk or, more rarely, by direct contact with mucous membranes and a skin abrasion (Ashford et al. 2001). Person-to-person transmission has also been observed in the UK (Evans et al. 2007), but this seems to be very rare. Occupational airborne exposure to infected animals is currently the most common route of *M. bovis* transmission from cattle to humans (Collins 2000).

18.2.6.2 Reported Infections in Farmers

Cases in cattle farmers are still being reported in Ireland (O'Hagan et al. 2016a, 2016b), the USA (Rodwell et al. 2008) and in the UK (Mandal et al. 2011).

In Mexico, a high prevalence of latent and pulmonary tuberculosis among workers exposed to cattle infected with *M. bovis* was documented (Torres-Gonzales et al. 2013). Evidence of occupational infection with *M. bovis* in livestock farmers was also documented in Nigeria, Pakistan, Tanzania and south-east Ethiopia (Adesokan et al. 2012; Khattak et al. 2016; Katale et al. 2012; Gumi et al. 2012).

18.2.6.3 Specific Prevention

A study carried out in Northern Ireland showed that the risks of several significant factors associated with being infected with bovine tuberculosis on a farm could be reduced by adopting specific biosecurity measures, particularly in relation to newly purchased cattle (e.g. keeping a closed herd and pre-movement testing) and badger control (O'Hagan et al. 2016b).

18.2.7 Listeriosis

18.2.7.1 General Information

Listeria spp are Gram-positive bacilli living almost everywhere in soil, manure piles and grass. Listeriosis infection is common in cattle, sheep and goats and can occur in pigs, dogs and cats, some wild animals and humans. Animals infected with listeria can show signs restlessness, loss of appetite, fever, nervous system disorders and abortion. However, animal infections can also be asymptomatic. The reservoirs of infection are the soil and the body fluids of asymptomatic animals.

According to the CDC, infected animals can shed *L. monocytogenes* in faeces, milk and uterine discharges. It is also found in aborted foetuses and occasionally in the nasal discharges and urine of symptomatic animals. Soil or faecal contamination results in its presence on plants and in silage. Most infections are acquired by ingestion, but listeria can also spread by inhalation or direct contact. Symptoms vary in infected people, ranging from mild flu-like symptoms, such as nausea, vomiting and diarrhoea, to more serious infections such as meningitis and other potentially life-threatening complications. The elderly, pregnant women, new-born infants and people with weak immune systems are more susceptible to *Listeria spp* infections.

Cutaneous infections can occur and manifest as rashes of papules or pustules with fever.

18.2.7.2 Reported Cases in Farms

In the USA, analyses of enteric illnesses due to contact with animals and their environments revealed that < 1% of listeriosis cases were attributable to animal contact and no animal contact outbreaks were reported to the CDC from 2000–2011. Thus, the risk of contracting occupational enteritis is very limited. However, the cutaneous listeriosis is reported in the literature.

One review paper (Godshall et al. 2013) described at least 16 cases of cutaneous listeriosis after occupational exposure to bovine or sheep products of conception from 1957–2009 in the USA. In these cases, *L. monocytogene* infection caused the abortion of an animal foetus and was then transmitted via direct inoculation to the person who assisted the delivery, usually a farmer or veterinarian who had not worn protective gloves.

18.3 Zoonotic Agents Presenting Occupational Risks Specific to Pig Farms

18.3.1 Streptococcus suis

18.3.1.1 General Information

Streptococcus suis is an encapsulated, Gram-positive bacterial coccus that can occur singly, in pairs or in short chains (Wertheim et al. 2009) and until today, 35 serotypes have been identified to date based on the antigenicity of their respective capsular polysaccharides (Wertheim et al. 2009). The bacteria are best adapted to domestic pigs and multiple serotypes can form a part of their normal microflora (Lun et al. 2007) and some serotypes have occasionally been recovered from wild boar, horses, dogs and cats (Goyette-Desjardins et al. 2014; Staats et al. 1997). The most frequent manifestation of infection in pigs is meningitis, followed by arthritis, endocarditis, pneumonia, rhinitis, abortion and vaginitis (Gottschalk et al. 2007; Sanford and Tilker, 1982). The most likely route of transmission from pigs to humans is through broken skin when handling infected pig or pork products. Even though entry through the gastrointestinal tract after ingestion of uncooked infected pork has been suggested as another route of transmission, it has to be proven yet (Huong et al. 2014). In Europe and North America, therefore, *S. suis* infection is considered an occupational disease, as it is nearly exclusively reported in individuals with direct exposure to infected or carrier pigs, or infected raw or undercooked pork products, e.g. farmers, veterinarians, butchers and food processing workers (Gottschalk et al. 2007).

In the case of an infection in humans, symptoms can vary significantly. The most common symptom is meningitis, making up 70% of reported cases; other symptoms are sepsis, arthritis, endocarditis, endophthalmitis and spondylodiscitis. A subjective hearing loss appears in 50% of cases and up to 30% also show skin lesions. In a minority of cases (almost all in Asia), increased disease severity manifests in shorter incubation times, more rapid disease progression, the appearance of streptococcal toxic-shock syndrome and a higher mortality rate (Lun et al. 2007; Huong et al. 2014; Feng et al. 2010). Many human diagnostic laboratories are less aware of this pathogen in cases of bacterial infection and a misidentification of *S. suis* is not uncommon (Gottschalk et al. 2010). Molecular techniques have therefore been developed to identify *S. suis* strains easily, such as a PCR assay targeting the gene encoding for glutamate dehydrogenase (Goyette-Desjardins et al. 2014; Okwumabua et al. 2003).

18.3.1.2 Reported Cases in Farmers

S. suis infection in humans is considered a global problem, and the number of cases reported in the literature has significanty increased in the last few years. From 2009–2013, confirmed cases increased from about 700 to over 1,600 (Wertheim et al. 2009). About 90% (1,481 cases) were reported in Asia, followed by Europe with 8.5% (140 cases). Most of these European cases are reported in countries with a highly developed pig industry: Netherlands (51 cases), France (19 cases), UK (19 cases) and Spain (13 cases) (Goyette-Desjardins et al. 2014; Huong et al. 2014). The low number of cases in North America (three) could be explained by an under-diagnosis of *S. suis* strains (Gottschalk et al. 2010). Two notable outbreaks occurred in 1998 and 2005 in Sichuan, China, with 25 and 215 reported cases and a total of 14 and 39 deaths, respectively. These are the only known major outbreaks and they were characterised by severe disease progression and unusually high mortality rates (Gottschalk et al. 2010; Normile, 2005).

18.3.1.3 Specific Prevention

Because the main infectious sources are sick pigs, prevention of transmission to human beings depends on the management of sick animals. Thus, various types of vaccines have been developed for pigs, and the one commonly used in the pig industry is an inactivated autogenous vaccine generated from virulent strains isolated from sick pigs. As there is as yet no vaccine for humans, raising awareness of the disease within populations that are at high risk of encountering it is expected to help avoid human infection.

18.3.2 Yersiniosis

18.3.2.1 General Information

Yersinia are Gram-negative bacilli bacteria that can cause diseases in humans. *Y. enterolitica* are widespread in the natural environment and are the most common species causing human enteric yersiniosis. Pigs are the major animal reservoir for the few strains of *Y. enterocolitica* that cause human illness, but according to the CDC, rodents, rabbits, sheep, cattle, horses, dogs and cats can also carry strains that cause human illness. Contaminated pigs are usually asymptomatic, and the prevalence of *Y. enterocolitica* carriers varies widely between fattening units (between 0%–100% depending on the country) (Vilar et al. 2013; Nathues et al. 2013; Guertler et al. 2005; Fondrevez et al. 2014). Yersiniosis is the third most reported zoonotic disease in the European Union (EFSA, 2012).

According to the CDC, yersiniosis symptoms can vary depending on the infected person's age. In young children, common symptoms are fever, abdominal pain and diarrhoea, which is often bloody. Symptoms typically develop 4–7 days after exposure and may last 1–3 weeks or longer. Symptoms in older children and adults may include fever and pain on the right side of the abdomen, which may be confused with appendicitis. Complications are rare, but can include skin rashes, joint pains or the spread of bacteria to the bloodstream.

18.3.2.2 Reported Cases in Farmers

The organism is mainly acquired via insufficiently cooked pork or contaminated water, meat or milk (Fredriksson-Ahomaa et al. 2006), but direct contact with contaminated animals or their environment has been suggested (le Guern et al. 2016). *Y. enterocolitica* was also isolated from the flies in pig farms, indicating that bacteria can be transmitted between insects, animals and humans (Fukushima et al. 1979; Valentin-Weigand et al. 2014; Nathues et al. 2013). However, case reports confirming occupational infections due to direct contact with contaminated pigs are very sparse, and only a very small proportion of *Y. enterocolitica* illnesses has been attributed to animal contact (Schmitz and Tauxe 2009).

18.3.2.3 Specific Prevention

A Finnish study (Vilar et al. 2013) revealed that rearing pigs in pens (with or without sparse amounts of bedding) and buying piglets from more than one farm were the variables that contributed most to the occurrence of *Y. enterocolitica*. It also showed that using an all-in/all-out stock management system and using municipal water supplies were factors that could reduce the prevalence of *Y. enterocolitica* and thus the risks of transmission in pig farms.

18.3.3 Erysipeloid

18.3.3.1 General Information

Erysipelothrix rhusiopathiae is a Gram-positive bacillus occurring worldwide. It is primarily considered an animal pathogen, causing a skin disease known as erysipelas in a variety of animals and erysipeloid in humans. Pigs are most commonly affected, but cases have been reported in birds (turkeys), sheep, fish and reptiles. Bacteria can survive in soil for several weeks (Brooke and Riley 1999). In pig faeces, this bacterium's survival period ranges from 1 to 5 months. Humans acquire erysipeloid after direct contact with infected animals, their products or environmental sources. Occupational cases are most common among farmers, veterinarians, slaughterhouse workers and butchers. Infection is more likely to occur in the summer or early autumn and is usually localised on the back of the hand or fingers. Erysypeloid may be under-diagnosed because of its resemblance to other infections and the problems that can be encountered in isolating and identifying the bacteria (Brooke and Riley 1999). The palms, forearms, arms, face and legs are rarely affected. The lesion may be asymptomatic or accompanied by mild pruritus, pain and fever. Less commonly, erysipeloid can result in sepsis – a scenario often associated with endocarditis.

18.3.3.2 Reported Cases in Farmers

Very few cases of erysipeloid associated with farm animals have been reported; in a short review, Principe et al. (2016) found five cases, of which four occurred before 1992. The present research only found four cases of septic arthritis caused by *E. rusiopathiae* reported in the literature (Traer et al. 2008; Wong et al. 2003; Frolov and Baklanov, 1995; Allianatos et al. 2003). Of these, only one documented exposure to farm animals, including pigs. Another case report mentioned an instance of *E. rusiopathiae* pneumonia in an immunocompetent patient who had worked on a cattle farm (Meric and Ozcan 2012).

18.4 General Preventive Measures

18.4.1 Which Important Measures Should Be Taken?

The most effective preventive measure for avoiding animal to human transmission of microorganisms is keeping cattle and pigs free of zoonotic pathogens. However, a lot of biosecurity measures have to be taken in order to obtain "healthy herds." These biosecurity measures should not only prevent the transmission of zoonotic diseases between or within farms, but also the transmission of all non-zoonotic animal infections (Anonymous, project OZ0144, 2003). Lots of valuable information concerning biosecurity measures for farms can be found on the websites of different national governmental agencies or departments of agriculture. The most important preventive measures are summarised in Table 18.2.

Farmers' adherence to these preventive measures is not the rule, however. In 2003, a project in the UK investigated constraints to the implementation of improved biosecurity on cattle and sheep farms (Anonymous, project OZ0144, 2003). The project's main conclusions were that biosecurity and its implementation were complex and potentially value-laden issues for both farmers and veterinarians. Authors highlighted that overall farm biosecurity is a combination of complicated and simple measures that complement and sometimes contradict each other. They added that it is crucial to ensure that all the measures recommended to farmers are proven and validated in a transparent manner that takes into account practical farm-level issues. It must also be accepted that, on some farms, certain measures will be impossible to implement, and in such cases other specific solutions should be proposed. Moreover, it appears that the provision of certain types of information is clearly inadequate, including the hierarchical cost/benefit ratios of different farm-level biosecurity measures. This report underlined also the need to address the perceived barriers of "Why us?" among farmers (i.e the farmer perception that only farmers are expected to do something about biosecurity and that farmer efforts will be useless due to lack of action or wrong action by other). A 2007 survey highlighted that 34% of farmers questioned stated that biosecurity was almost non-existent on their farms (Anonymous, 2007). A 2012 questionnaire administered to 56 cattle farmers about their on-farm biosecurity practices showed that basic preventive measures were not systematically adhered to (Brennan and Christley 2012). For example, 30% of the farmers who purchased stock from other farms did not inquire about the selling farm's disease history, and most farmers did not isolate new animals arriving from other farms. Moreover, only 50% of farmers who shared equipment with other farmers disinfected it before or after use. In one third of farms, external contractors parked their vehicles in areas frequented by animals and most of them (90%) never cleaned their vehicles after visiting farms. It was also observed that 93% of farmers or their employees carried out no personal biosecurity measures, such as cleaning boots or changing overalls, between tasks with different animal groups.

18.4.2 Which Important Measurements Should Be Taken?

With regard to what farmers should measure in their barns in order to avoid occupational zoonotic diseases, the standard sanitary surveillance of their herds with serological or microbiological tests should be sufficient. If a farmer or an employee develops non-specific flu-like symptoms, he must consult a physician and report his potential exposure to zoonotic agents. This could help to identify the causal agent and apply effective treatment more quickly.

18.5 Conclusions

Occupational contact with farm animals is responsible for a number of human bacterial diseases. However, infected animals are often asymptomatic and farmers are consequently unable to identify the signs of specific hazards. Thus, all animals must be considered as potential sources of contamination and a routine surveillance as well as preventive measures should always be implemented rigorously. If signs of infection are visible either in animals or humans, veterinarian or physician should be called immediately. Moreover, educating farmers and their employees on the sources of contamination and pathogen transmission routes is essential to ensuring that everybody working closely with animals is aware of the potential dangers and adapts their behaviour so as to avoid any contamination. Simple, proven and efficient measures should be preferred over disproportionate measures whose efficacy has yet to be proved. The costs of implementing measures should be taken into account and governmental support should be considered in certain situations. The veterinarian should always inform the farmer about the current risks present on his farm. In general, veterinarians and all other people coming into contact with animal farm environments should strictly adhere to biosecurity measures.

Raising workers' awareness and good hygiene practices are efficient measures for mitigating risks. As instance, frequent hand-washing alone could probably prevent a substantial proportion of human infections every year (Hoelzer et al. 2011).

Moreover, in the context of zoonoses, the collaborative effort of specialists in environmental microbiology, human and veterinary medicine is needed for the surveillance of the diseases and for the development and implementation of adequate control measures. Indeed, it is now well recognised that a One-Health approach, linking human, animal and environmental health is essential to fight against zoonotic diseases.

References

Adesokan HK, Jenkins AO, van Soolingen D et al (2012) Mycobacterium bovis infection in livestock workers in Ibadan, Nigeria: evidence of occupational exposure. Int Tuberc Lung Dis 16(10):1388–1392

Allianatos PG, Tilentzoglou AC, Koutsoukou AD (2003) Septic arthritis caused by Erysipelothrix rhusiopathiae infection after arthroscopically assisted anterior cruciate ligament reconstruction. Arthroscopy 19:1–4

Angelakis E, Raoult D (2010) Q fever. Vet Microbiol 140(3–4):297–309

Anonymous (2003) University of Reading, Scottish Agricultural College. Constraints to uptake of adequate biosecurity on UK cattle and sheep farms, with special reference to zoonotic diseases. Department for Environment, Food and Rural Affairs, p 1–25

Anonymous (2007) Survey reveals lack of biosecurity on farms. Vet Rec 161:324–324

Asencio MA, Herraez O, Tenias JM et al (2015) Seroprevalence survey of Zoonoses in Extremadura, Southwestern Spain, 2002–2003. Jap Infect Dis 68(2):106–112

Ashford DA, Whitney E, Raghunathan et al (2001) Epidemiology of selected mycobacteria that infect humans and other animals. Revue Scientifique Et Technique De L'Office International Des Epizooties 20(1):325–337

Asif M, Waheed U, Farooq M et al (2014) Frequency of brucellosis in high risk human groups in Pakistan detected through polymerase chain reaction and its comparison with conventional slide agglutination test. Int Agr Biol 16(5):986–990

Bond KA, Vincent G, Wilks CR et al (2016) One health approach to controlling a Q fever outbreak on an Australian goat farm. Epidemiol Infect 144(6):1129–1141

Bosnjak E, Hvass A, Villumsen S et al (2010) Emerging evidence for Q fever in humans in Denmark: role of contact with dairy cattle. Clin Microbiol Infect 16(8):1285–1288

Brennan ML, Christley RM (2012) Biosecurity on cattle farms: a study in North-West England. PloS One 7(4):0028139

Brooke CJ, Riley TV (1999) Erysipelothrix rhusiopathiae: bacteriology, epidemiology and clinical manifestations of an occupational pathogen. Med Microbiol 48(9):789–799

Campagnolo ER, Warwick MC, Marx HL et al (2000) Analysis of the 1998 outbreak of leptospirosis in Missouri in humans exposed to infected swine. Am Vet Med Ass 216(5): 676–682

Center for Food Security and Public Health. Spickler, Anna Rovid. "Brucellosis" "2007 (2009 last update)." http://www.cfsph.iastate.edu/DiseaseInfo/factsheets.php

Collins CH (2000) The bovine tubercle bacillus. Brit J Biomed Sci 57(3):234–240

Costa F, Hagan JE, Calcagno J et al (2015) Global morbidity and mortality of leptospirosis: a systematic review. Plos Negl Trop Dis 9(9):0003898

Davis MA, Moore DL, Baker KNK et al (2013) Risk factors for Campylobacteriosis in two Washington State Counties with high numbers of dairy farms. Clin Microbiol 51(12):3921–3927

de Haan CPA, Kivisto RI, Hakkinen M et al (2010) Multilocus sequence types of Finnish bovine Campylobacter jejuni isolates and their attribution to human infections. BMC Microbiol 10:200

de la Rua-Domenech R (2006) Human Mycobacterium bovis infection in the United Kingdom: incidence, risks, control measures and review of the zoonotic aspects of bovine tuberculosis. Tuberc 86(2):77–109

De Lange MMA, Schimmer B, Vellema P et al (2014) Coxiella burnetii seroprevalence and risk factors in sheep farmers and farm residents in The Netherlands. Epidemiol Infect 142(6):1231–1244

Diez JG, Coelho AC (2013) An evaluation of cattle farmers' knowledge of bovine brucellosis in northeast Portugal. J Infect Public Health 6(5):363–369

Denis M, Chidaine B, Laisney MJ et al (2009) Comparison of genetic profiles of Campylobacter strains isolated from poultry, pig and Campylobacter human infections in Brittany, France. Pathologie Biologie 57(1):23–29

Dijkstra F, van der Hoek W, Wijers N et al (2012) The 2007-2010 Q fever epidemic in the Netherlands: characteristics of notified acute Q fever patients and the association with dairy goat farming. FEMS Immunol Med Microbiol 64(1):3–12

Doorduyn Y, Van den Brandhof WE, Van Duynhoven Y et al (2010) Risk factors for indigenous Campylobacter jejuni and Campylobacter coli infections in The Netherlands: a case-control study. Epidemiol Infect 138(10):1391–1404

Dorko E, Rimarova K, Pilipcinec E (2012) Influence of the environment and occupational exposure on the occurrence of q fever. Cent Eur J Public Health 20(3):208–214

Evans JT, Smith EG, Banerjee A et al (2007) Cluster of human tuberculosis caused by Mycobacterium bovis: evidence for person-to-person transmission in the UK. Lancet 369 (9569):1270–1276

Feng Y, Zhang H, Ma Y et al (2010) Uncovering newly emerging variants of Streptococcus suis, an important zoonotic agent. Trends Microbiol 18(3):124–131

Fenga C, Gangemi S, de Luca A et al (2015) Seroprevalence and occupational risk survey for Coxiella burnetii among exposed workers in Sicily, Southern Italy. Int Occup Med Environ Health 28(5):901–907

Fondrevez M, Minvielle B, Labbe A et al (2014) Prevalence of pathogenic Yersinia enterocolitica in slaughter-aged pigs during a one-year survey, 2010–2011, France. Int J Food Microbiol 174:56–62

Fredriksson-Ahomaa M, Stolle A, Siitonen A et al (2006) Sporadic human Yersinia enterocolitica infections caused by bioserotype 4/O : 3 originate mainly from pigs. J Med Microbiol 55(6):747–749
Frolov VM, Baklanov AN (1995) Occupational causes of erysipeloid morbidity among the workers engaged in meat-processing industry in the Lugansk region. Med Tr Prom Ekol 8:18–20
Fukushima H, Ito Y, Saito K et al (1979) Role of the fly in the transport of yersinia-enterocolitica. Appl Environ Microbiol 38(5):1009–1010
Gidding HF, Wallace C, Lawrence GL et al (2009) Australia's national Q fever vaccination program. Vaccine 27(14):2037–2041
Gilpin BJ, Scholes P, Robson B et al (2008) The transmission of thermotolerant Campylobacter spp. To people living or working on dairy farms in New Zealand. Zoon Public Health 55(7):352–360
Godshall CE, Suh G, Lorber B (2013) Cutaneous Listeriosis. Clin Microbiol 51(11):3591–3596
Gottschalk M, Segura M, Xu J (2007) Streptococcus suis infections in humans: the Chinese experience and the situation in North America. Anim Health Res Rev 8(1):29–45
Gottschalk M, Xu J, Calzas C et al (2010) Streptococcus suis: a new emerging or an old neglected zoonotic pathogen? Future Microbiol 5(3):371–391
Goyette-Desjardins G, Auger J-P, Xu J et al (2014) Streptococcus suis, an important pig pathogen and emerging zoonotic agent – an update on the worldwide distribution based on serotyping and sequence typing. Emerg Microbes Infect 3:e45
Gras LM, Smid JH, Wagenaar JA et al (2012) Risk Factors for Campylobacteriosis of chicken, ruminant, and environmental origin: a combined case-control and source attribution analysis. PloS One 7(8):0042599
Gullan LN, Cowie BC, Franklin LJ (2012) Australia's national Q fever vaccination program. Vaccine 27:2037–2041
Gumi B, Schelling E, Berg S et al (2012) Zoonotic transmission of tuberculosis between pastoralists and their livestock in South-East Ethiopia. Ecohealth 9(2):139–149
Gurtler M, Alter T, Kasimir S et al (2005) Prevalence of Yersinia enterocolitica in fattening pigs. J Food Prot 68(4):850–854
Havelaar AH, van Pelt W, Ang CW et al (2009) Immunity to Campylobacter: its role in risk assessment and epidemiology. Crit Rev Microbiol 35(1):1–22
Heederik D, Sigsgaard T, Thorne PS et al (2007) Health effects of airborne exposures from concentrated animal feeding operations. Environ Health Persp 115(2):298–302
Heise-Pavlov PM, Heise-Pavlov SR (2003) Feral pigs in tropical lowland rainforest of northeastern Australia: ecology, zoonoses and management. Wildlife Biol 9:21–27
Hendriksen SWM, Orsel K, Wagenaar JA et al (2004) Animal-to-human transmission of Salmonella Typhimurium DT104A variant. Emerg Infect Dis 10(12):2225–2227
Hoelzer K, Switt AIM, Wiedmann M (2011) Animal contact as a source of human non-typhoidal salmonellosis. Vet Res 42:34
Huong VT, Ha N, Huy NT et al (2014) Epidemiology, clinical manifestations, and outcomes of Streptococcus suis infection in humans. Emerg Infect Dis 20(7):1105–1114
Inglis GD, McAllister TA, Larney FJ et al (2010) Prolonged survival of Campylobacter species in bovine manure compost. Appl Environ Microbiol 76(4):1110–1119
Iraola G, Betancor L, Calleros L et al (2015) A rural worker infected with a bovine-prevalent genotype of Catnpylobacter fetus subsp fetus supports zoonotic transmission and inconsistency of MLST and whole-genome typing. Eur J Clin Microbiol Infect Dis 34(8): 1593–1596
Islam MA, Khatun MM, Werre SR et al (2013) A review of Brucella seroprevalence among humans and animals in Bangladesh with special emphasis on epidemiology, risk factors and control opportunities. Vet Microbiol 166(3–4):317–326

Jonas R, Kittl S, Overesch G et al (2015) Genotypes and antibiotic resistance of bovine Campylobacter and their contribution to human campylobacteriosis. Epidemiol Infect 143(11): 2373–2380

Kantso B, Andersen AMN, Molbak K et al (2014) Campylobacter, Salmonella, and Yersinia antibodies and pregnancy outcome in Danish women with occupational exposure to animals. Int J Infect Dis 28:74–79

Katale BZ, Mbugi EV, Kendal S et al (2012) Bovine tuberculosis at the human-livestock-wildlife interface: is it a public health problem in Tanzania? A review. Onderstepoort Jof Vet Res 79(2):463

Khattak I, Mushtaq MH, Ahmad MUD et al (2016) Risk factors associated with Mycobacterium bovis skin positivity in cattle and buffalo in Peshawar, Pakistan. Trop Anim Health Product 48(3):479–485

Kutlu M, Ergonul O, Sayin-Kutlu S et al (2014) Risk factors for occupational brucellosis among veterinary personnel in Turkey. Prev Vet Med 117(1):52–58

Lau CL, Watson CH, Lowry JH et al (2016) Human leptospirosis infection in Fiji: an eco-epidemiological approach to identifying risk factors and environmental drivers for transmission. PloS Neglect Trop Dis 10:0004405

Le Guern AS, Martin L, Savin C et al (2016) Yersiniosis in France: overview and potential sources of infection. Int J Infect Dis 46:1–7

Lun ZR, Wang QP, Chen XG et al (2007) Streptococcus suis: an emerging zoonotic pathogen. Lancet Infect Dis 7(3):201–209

Macellaro A, Akesson A, Norlander L (1993) A survey of q-fever in Sweden. Eur J Epidemiol 9(2):213–216

Mandal S, Bradshaw L, Anderson LF et al (2011) Investigating transmissio of Mycobacterium bovis in the United Kingdom in 2005 to 2008. Clin Microbiol 49(5):1943–1950

Meirelles-Bartoli RB, Mathias LA, Samartino LE (2012) Brucellosis due to Brucella suis in a swine herd associated with a human clinical case in the State of Sao Paulo, Brazil. Trop Anim Health Prod 44(7):1575–1579

Meric M, Ozcan SK (2012) Erysipelothrix rhusiopathiae pneumonia in an immunocompetent patient. J Med Microbiol 61(3):450–451

Miller M et al (2005) Australia's notifiable diseases status, 2003 annual report of the National Notifiable Diseases Surveillance System. Communicable Dis Intel Q Rep 29:1–61

Monno R, Fumarola L, Trerotoli P et al (2009) Seroprevalence of Q fever, brucellosis and leptospirosis in farmers and agricultural workers in Bari, Southern Italy. Ann Agr Environ Med 16(2):205–209

Mossong J, Mughini-Gras L, Penny C et al (2016) Human Campylobacteriosis in Luxembourg, 2010-2013: a case-control study combined with multilocus sequence typing for source attribution and risk factor analysis. Sci Rep 7(7):20939

Mughini-Gras L, Enserink R, Friesema I et al (2014) Risk factors for Human Salmonellosis originating from pigs, cattle, broiler chickens and egg laying hens: a combined case-control and source attribution analysis. PloS One 9(2):0087933

Mughini-Gras L, Smid JH, Wagenaar JA et al (2012) Risk factors for campylobacteriosis of chicken, ruminant, and environmental origin: a combined case-control and source attribution analysis. PloS One 7(8):0042599

Müller B, Dürr S, Alonso S et al (2013) Zoonotic Mycobacterium bovis-induced tuberculosis in humans. Emerg Infect Dis 19(6):899–908

Mwachui MA, Crump L, Hartskeerl R et al (2015) Environmental and behavioural determinants of leptospirosis transmission: a systematic review. Plos Neglect Trop Dis 9(9):0003843

Nathues C, Gruning P, Fruth A et al (2013) Campylobacter spp., Yersinia enterocolitica, and Salmonella enterica and their simultaneous occurrence in German fattening pig herds and their environment. J Food Prot 76(10):1704–1711

Njeru J, Henning K, Pletz MW et al (2016) Q fever is an old and neglected zoonotic disease in Kenya: a systematic review. BMC Public Health 16:297

Normile D (2005) Infectious diseases. WHO probes deadliness of China's pig-borne disease. Science 309(5739):1308–1309

O'Hagan MJH, Matthews DI, Laird C et al (2016a) Herd-level risk factors for bovine tuberculosis and adoption of related biosecurity measures in Northern Ireland: a case-control study. Vet J 213:26–32

O'Hagan MJH, Matthews DI, Laird C et al (2016b) Farmers' beliefs about bovine tuberculosis control in Northern Ireland. Vet J 212:22–26

Okello AL, Burniston S, Conlan JV et al (2015) Prevalence of endemic pig-associated zoonoses in Southeast Asia: a review of findings from the Lao People's Democratic Republic. Am J Trop Med Hyg 92(5):1059–1066

Okwumabua O, O'Connor M, Shull E (2003) A polymerase chain reaction (PCR) assay specific for Streptococcus suis based on the gene encoding the glutamate dehydrogenase. FEMS Microbiol Lett 218(1):79–84

O'Reilly LM, Daborn CJ (1995) The epidemiology of mycobacterium-bovis infections in animals and man – a review. Tubercle Lung Dis 76:1–46

Pappas G, Papadimitriou P, Akritidis N et al (2006) The new global map of human brucellosis. Lancet Infect Dis 6(2):91–99

Prasad HK, Singhal A, Mishra A et al (2005) Bovine tuberculosis in India: potential basis for zoonosis. Tuberculosis 85(5–6):421–428

Principe L, Bracco S, Mauri C et al (2016) Erysipelothrix rhusiopathiae bacteremia without endocarditis: rapid identification from positive blood culture by MALDI-TOF mass spectrometry. A case report and literature review. Infect Dis Rep 8(1):8–11

Quilliam RS, Cross P, Williams AP et al (2013) Subclinical infection and asymptomatic carriage of gastrointestinal zoonoses: occupational exposure, environmental pathways, and the anonymous spread of disease. Epidemiol Infect 141(10):2011–2021

Rahman A, Dirk B, Fretin D et al (2012) Seroprevalence and risk factors for Brucellosis in a high-risk group of individuals in Bangladesh. Foodborne Path Dis 9(3):190–197

Reid A, Malone J (2004) Q fever in Ireland. A seroprevalence study of exposure to Coxiella burnettii among Department of Agriculture workers. Occup Med 54(8):544–547

Rodwell TC, Moore M, Moser KS et al (2008) Tuberculosis from Mycobacterium bovis in binational communities, United States. Emerg Infect Dis 14(6):909–916

Sanford SE, Tilker ME (1982) Streptococcus suis type II-associated diseases in swine: observations of a one-year study. Am Vet Med Ass 181(7):673–676

Sanhueza JM, Heuer C, Wilson PR et al (2015) Prevalence and risk factors for Leptospira exposure in New Zealand veterinarians. Epidemiol Infect 143(10):2116–2125

Schiellerup P, Abdul-Redha RJ, Baggesen DL et al (2001) Five cases of gastroenteritis with multiresistant Salmonella enteritica serovar Typhimurium DT104 related to farm animals in Denmark. Ugeskr Laeger 163:5677–5678

Schimmer B, Lenferink A, Schneeberger P et al (2012) Seroprevalence and risk factors for Coxiella burnetii (Q fever) seropositivity in dairy goat farmers' households in the Netherlands, 2009-2010. PloS One 7(7):0042364

Schimmer B, Schotten N, van Engelen E et al (2014) Coxiella burnetii seroprevalence and risk for humans on dairy cattle farms, the Netherlands, 2010–2011. Emerg Infect Dis 20(3): 417–425

Schmitz AM, Tauxe RV (2009) Yersinia enterocolitica Infections. In: Brachman PS, Abrutyn E (eds). Bacterial infections of humans. Springer, New York, p 939–957

Schneeberger PM, Wintenberger C, van der Hoek W et al (2014) Q fever in the Netherlands-2007-2010: what we learned from the largest outbreak ever. Medecine et Maladies Infectieuses 44(8):339–353

Schollum LM, Blackmore DK (1982) Leptospirosis of pig farmers – the results of a serological survey. New Zealand Med J 95(707):299–301
Sethi S, Sharma N, Kakkar N et al (2010) Increasing trends of leptospirosis in Northern India: a clinico-epidemiological study. Plos Negl Trop Dis 4(1):0000579
Smith KE, Stenzel SA, Bender JB et al (2004) Outbreaks of enteric infections caused by multiple pathogens associated with calves at a farm day camp. Pediatr Infect Dis J 23(12):1098–1104
Staats JJ, Feder I, Okwumabua O et al (1997) Streptococcus suis: past and present. Vet Res Communic 21(6):381–407
Steinmuller N, Demma L, Bender JB et al (2006) Outbreaks of enteric disease associated with animal contact: not just a foodborne problem anymore. Clin Infect Dis 43(12):1596–1602
Szymanska-Czerwinska M, Niemczuk K et al (2014) Prevalence of Coxiella burnetii in dairy herds – diagnostic methods and risk to humans – a review. Bull Vet Inst Pulawy 58(3): 337–340
Tabibi R, Baccalini R, Barassi A et al (2013) Occupational exposure to zoonotic agents among agricultural workers in Lombardy Region, Northern Italy. Ann Agr Environ Med 20(4): 676–681
Taylor EV, Herman KM, Ailes EC et al (2013) Common source outbreaks of Campylobacter infection in the USA, 1997–2008. Epidemiol Infect 141(5):987–996
Tebug SF, Kamga-Waladjo AR, Ema PJN et al (2015) Cattle farmer awareness and behavior regarding prevention of zoonotic disease transmission in Senegal. J Agromed 20(2):217–224
Tebug SF, Njunga GR, Chagunda MGG et al (2014) Risk, knowledge and preventive measures of smallholder dairy farmers in northern Malawi with regard to zoonotic brucellosis and bovine tuberculosis. Onderstepoort J Vet Res 81(1):594
The Institute of Environmental Science and Research Ltd. Notifiable and Other Diseases in New Zealand: Annual Report 2013 Porirua, New Zealand
Thoen C, LoBue P, de Kantor I (2006) The importance of Mycobacterium bovis as a zoonosis. Vet Microbiol 112(2–4):339–345
Torres-Gonzalez P, Soberanis-Ramos O, Martinez-Gamboa A et al (2013) Prevalence of latent and active tuberculosis among dairy farm workers exposed to cattle infected by Mycobacterium bovis. Plos Negl Trop Dis 7(4):0002177
Traer EA, Williams MR, Keenan JN (2008) Erysipelothrix rhusiopathiae infection of a total knee arthroplasty – an occupational hazard. J Arthroplasty 23(4):609–611
Valentin-Weigand P, Heesemann J, Dersch P (2014) Unique virulence properties of Yersinia enterocolitica O:3-An emerging zoonotic pathogen using pigs as preferred reservoir host. Int J Med Microbiol 304(7):824–834
Vegosen L, Breysse PN, Agnew J et al (2015) Occupational exposure to swine, poultry, and cattle and antibody biomarkers of Campylobacter jejuni exposure and autoimmune peripheral neuropathy. PloS One 10(12):0143587
Verso MG, Vesco G, Villari S et al (2016) Analysis of seroprevalence against Coxiella burnetii in a sample of farm workers in Western Sicily. Ann Agr Environ Med 23(1):71–74
Vilar MJ, Virtanen S, Heinonen M et al (2013) Management practices associated with the carriage of Yersinia enterocolitica in pigs at farm level. Foodborne Pathogens Dis 10(7):595–602
Wagenaar JA, van Bergen MAP, Blaser MJ et al (2014) Campylobacter fetus infections in humans: exposure and disease. Clin Infect Dis 58(11):1579–1586
Wall PG, Morgan D, Lamden K et al (1995) Transmission of multiresistant strains of Salmonella-typhimurium from cattle to man. Vet Rec 136(23):591–592
Walsh MG (2012) Assessing Q fever in a representative sample from the United States population: identification of a potential occupational hazard. Epidemiol Infect 140(1):42–46
Wang YC, Chang YC, Chuang HL et al (2011) Transmission of Salmonella between swine farms by the housefly (Musca domestica). J Food Protect 74(6):1012–1016
Ward AI, Judge J, Delahay RJ (2010) Farm husbandry and badger behaviour: opportunities to manage badger to cattle transmission of Mycobacterium bovis? Prev Vet Med 93(1):2–10

Wertheim HF, Nghia HD, Taylor W et al (2009) Streptococcus suis: an emerging human pathogen. Clin Infect Dis 48(5):617–625

Wong RCC, Kong KO, Lin R et al (2003) Chronic monoarthritis of the knee in systemic lupus erythematosus. Lupus 12(4):324–326

Zhen Q, Lu Y, Yuan X et al (2013) Asymptomatic brucellosis infection in humans: implications for diagnosis and prevention. Clin Microbiol Infect 19(9):831–831

Zonios DI, Panayiotakopoulos GD, Kabletsas EO et al (2005) Campylobacter fetus bacteraemia in a healthy individual: clinical and therapeutical implications. J Infect 51(4):329–332

Part VI
Antimicrobial Resistances in Specific Environments

Chapter 19
Antifungal Resistances

Raquel Sabino

Abstract The increasing number of invasive procedures and clinical therapies has led to an increase number of patients at-risk of suffering invasive fungal infections. The prophylaxis with antifungals is now broader used in specific groups of patients. Therefore, it is not surprising the emergence of antifungal resistance. In the following chapter this issue will be discussed, raising awareness to possible sources as well as modes of transmission of resistant isolates in specific environmental settings.

Keywords Resistance · antifungals · fungi · *Candida* · *Aspergillus*

19.1 Introduction

In the last decades, the problematic of the emergence of microorganisms resistant to antimicrobial substances has been rising. Although antibiotic-resistant bacterial infections are a widely-recognized public health threat, less is known about the effects of antifungal resistance and the burden of drug-resistant fungal infections.

The incidence of fungal infections has increased together with the increasing number of available therapies and clinical procedures, and the population of patients at risk has expanded to include those with a broad list of medical conditions, such as solid-organ and hematopoietic stem cell transplantation (HSCT), cancer, receipt of immunosuppressive therapy, AIDS, premature birth, advanced age, and major surgery (Nucci and Marr 2005). Further, the spectrum of fungal pathogens causing infections in immunocompromised hosts is also growing (Low and Rotstein 2011). In the last decades, the development of antifungal substances led to major improvements in the mortality rates of patients with fungal infections. The growth of populations vulnerable to fungal infections has been accompanied by increased use of azoles as prophylaxis, as well as for empirical therapy, and for

R. Sabino (✉)
Reference Unit for Parasitic and Fungal Infections, Infectious Diseases Department, National Institute of Health Dr. Ricardo Jorge, Lisbon, Portugal
e-mail: raquel.sabino@insa.min-saude.pt

C. Viegas et al. (eds.), *Exposure to Microbiological Agents in Indoor and Occupational Environments*, DOI 10.1007/978-3-319-61688-9_19

treatment of proven infections. The use of antifungal substances also in other fields, such as agriculture and construction, conducted to big improvements in the quality and amount of the obtained harvests as well as to higher quality raw-materials. Thus, if in one hand we have limited available therapeutic options for pathogens that are resistant to more than class of antifungals, in other hand, all these antifungal applications also cause an emergence of less susceptible strains or species. The exposure of immunocompromised patients or persons with a hyper reactive immune system to these resistant strains may lead to serious invasive fungal infections, difficult to manage due to the lack of response to clinical antifungals.

In the following pages, antifungal resistance will be discussed, raising awareness to this emerging problem.

19.2 Antifungal Classes

Antifungals can be grouped into four main classes based on their action: azoles, which inhibit the synthesis of ergosterol (the main fungal sterol); polyenes, which interact with fungal membrane sterols, binding to ergosterol and leading to the formation of pores that cause osmotic desregulation; echinocandins, which inhibit the synthesis of β-D-glucan in fungal cell wall; and fluoropyrimidines, which inihibits fungal DNA synthesis.

19.3 Detection of Antifungal Resistances

Several methods were developed and are available in order to allow the determination of the antifungal susceptibility pattern, and consequently, the detection of antifungal resistances. The main goal of all those methodologies is to select the most active antifungal against a specific fungal infection. For this purpose, several methods are available (Johnson 2008, Cuenca-Estrella et al. 2010; Alastruey-Izquierdo et al. 2015, Perkhofer et al. 2010): agar-based test methods (disk diffusion, gradient strips – E-test, and agar supplemented with antifungals-resistant screening media), broth-based test methods (macro and microdilutions), semi-automated methods (Sensititre YeastOne, Vitek2, and others), and also alternative methodologies such as MTT, XTT or flow cytometry.

With more specific methodologies, it is also possible to detect which underlying molecular mechanism is responsible for the resistance of a certain strain.

It is not advisable to perform antifungal susceptibility testing to all clinical fungal isolates. Nevertheless, there are specific cases when they should be performed (http://www.eucast.org/ Kahlmeter G, EUCAST, ESCMID):

- Recurrent superficial infections;
- In cases of deep seated *Candida* infections (e.g., *C. glabrata* with variable susceptibility profile to azoles and echinocandins)

- In clinically relevant *Aspergillus* (deep infections, therapeutic failure, previous prophylaxis with azoles and/or recurrent isolations) or if the *Aspergillus* species is characterized by low susceptibility to certain antifungals (e.g., *Aspergillus lentulus)*
- If there are no cutoffs for a specific fungal group, the susceptibility testing can help in choosing the correct antifungal (if the strain is completely resistant in all dilutions, that antifungal probably is not the first to be chosen as therapy)
- In order to understand the local/national/international epidemiology, namely to know the frequency of resistance of a certain species to a certain antifungal

19.4 Modes of antifungal resistance

In fungi, the known mechanisms of resistance include changes in the cell wall or plasma membrane leading to an impaired uptake of antifungals, efflux pumps that take antifungals outside the cell, overexpression of the antifungal targets, mutations of the in gene encoding target proteins, activation of alternative pathways that increase the metabolism of the antifungal, sequestration of the antifungal in organelle-like vacuoles, or chromosomal changes to increase the number of copies of the required gene (Eggiman et al. 2003). Horizontal gene transfer, common in bacteria, is not known in fungi (Hof 2008).

When a resistant fungal strain is detected, different antifungal resistance patterns should be considered (Loeffler and Stevens 2003):

- A primary (intrinsic) resistance, found naturally among certain fungi without prior exposure to the drug, and emphasizes the importance of identification of fungal species from clinical specimens.

In this case, we have examples such as *Candida krusei*, resistant to fluconazole (Orozco et al. 1998), some or specific *Aspergillus* (Balagee et al. 2005) *Fusarium* (Carneiro et al. 2011) or *Lomentospora prolificans* (syn. *Scedosporium prolificans*), resistant to amphotericin B (Alastruey et al. 2008, Cuenca et al. 1999). Also, the order mucorales comprises several pathogenic species that are resistant to voriconazole (Alastruey et al. 2009). Species of the genera *Fusarium* and *Scedosporium* also show low susceptibility to several antifungals (Alastruey-Izquierdo et al. 2007, 2008, Cuenca-Estrella et al. 1999). In addition, multi-resistant species are also present as human pathogens. *Lomentospora prolificans* is resistant to all azoles, echinocandins and amphotericin B, and has been associated with poorer outcomes (Rodriguez-Tudela et al. 2009).

These species have coded in their genome molecular mechanisms that enable them to survive in presence of those antifungals.

- A secondary (acquired) resistance, also defined as acquired resistance, which arises when initially susceptible microorganisms develop resistance after exposure of the organism to the drug.

A distinction should be made between fungal resistance to an inhibitor and clinical treatment failure with an antifungal agent. A person whose infection persists despite treatment with an antifungal drug might be reasonably described as "clinically resistant" to the therapeutic agent (Sanglard and Odds 2002). Clinical resistance is thus defined as the failure to eradicate a fungal infection despite the administration of an antifungal agent with *in vitro* activity against the organism (Kanafani and Perfect 2008). Such failures can be attributed to a combination of factors, namely: characteristics of the antifungal and its ability to diffuse in the tissue/cells of the patient; presence of serum proteins to which the antifungal may bind; interactions with other drugs; underlying disease and patient immune response; factors associated with the infecting fungal strain (virulence factors); local and severity of the infection; interaction patient/fungi/drug. The discordance between in vivo and in vitro data is illustrated by the "90–60 rule," which maintains that infections due to susceptible strains respond to appropriate therapy in 90% of cases, whereas infections due to resistant strains respond in about 60% of cases (Rex and Pfaller 2002).

19.5 Sources of Antifungal Resistance

A fungal infection may rise when there is an imbalance between the immunological system of the patient and their surrounding environment. The exposure to a specific environmental setting with high fungal loads, more virulent or less antifungal susceptible species or strains will enhance the chances of those patients acquire an invasive fungal infection difficult to treat. Environmental reservoirs of both antibiotic resistant bacteria (Francino 2012; Wellington et al. 2013) and antifungal resistant fungi (Chowdhary et al. 2013; Snelders et al. 2008) have been emerging. The causes may be associated with the release of antibiotic and antifungal residues, from agriculture, animal feeding, aquaculture, and also from hospital sources (Solo Gabriele et al. 2016). Hospital environment displays a very important role, and one should consider, as source of transmission, the person to person transmission, ventilation and water systems, cleaning activities, potted plants, contaminated surfaces, construction work, among others (Sautour et al. 2009).

Among this major concern of resistance are the fungal genera *Candida* and *Aspergillus*. Some species of *Candida* are becoming increasingly resistant to first-line and second-line antifungals – namely, echinocandins and fluconazole. Azole-resistant *Aspergillus* isolates are also arising as public health concern. *Candida* and *Aspergillus* resistance will thus be discussed in more detail in the following pages.

19.5.1 Candida Resistance

Candida spp. are part of the normal endogenous flora and can be found temporary or permanently in the gastrointestinal tract of 40–50% of the healthy adult population (Eggimann et al. 2003) and in the genital tract of about 20% of healthy women (Sobel 2005). Although *Candida* species are not part of the normal flora of the skin, they may colonize fingers or body folds transiently (Trofa et al. 2008). Only 5–15% of hospitalized patients are already colonized at entry, but this proportion increases with time and exposure to risk factors and it has been reported that, as many as 50–86% of critically ill patients may become colonized with *Candida* spp. during prolonged ICU stay (Egimann et al. 2003). possible that inanimate surfaces in hospital environment and the colonized patients themselves constitute a reservoir from which other patients or even the same patient acquire the fungus and in some cases develop candidemia (San-Miguel et al. 2004) and the hands of healthcare workers may be the predominant environmental source (Khun et al. 2004).

The increased use of azole antifungal agents for prophylaxis has decreased the incidence of invasive infections with *Candida* species in some populations (e.g., bone marrow transplant recipients) (Slavin et al. 1995; Marr et al. 2000, Sanglard and Odds 2002). Conversely, increased azole usage may be associated with the emergence of azole-resistant *C. albicans* or a shift in species distribution presenting resistance mechanisms that have compromised the spectrums of both fluconazole and itraconazole among certain groups of patients at high risk for *Candida* bloodstream infections (Girmenia and Martino 1998; White et al. 1998; Trick et al. 2002). Resistance to the azoles arises within the commensal population of the treated individual, primarily because azoles are fungistatic (inhibit growth but do not kill) (Ford et al. 2015). Epidemiological data suggest that the intensity of fluconazole use is driving the appearance of resistant isolates (Pfaller et al. 1999) such as *Candida krusei, C. norvegensis* and *C. inconspicua*, that are intrinsically resistant to some triazoles. *Candida glabrata* can be resistant to usual doses of triazoles, but sensitive to higher doses (Eggiman et al. 2003; Tortorano et al. 2006). Cross-resistance among azoles is expected as the target of action on fungi is similar. In HIV positive patients, a high level of cross resistance to itraconazole was observed in fluconazole resistant *C. glabrata* and *C. tropicalis*. Cross-resistance in *C. glabrata* strains was due to increased expression of CgCDR1, CgCDR2 genes and CDR efflux pumps (Chowdhary et al. 2013).

Structural changes in the sterol content of the cell wall are associated with the ability of some *Candida* strains to resist polyenes. The lack of ergosterol, replaced by more saturated forms, results in a reduced binding of liposomal amphotericin B and nystatin to *C. lusitaniae*, but those strains remain, however, sensitive to azoles.

More recently, several outbreaks caused by the species *Candida auris* have been described. This is anemerging multidrug-resistant (MDR) yeast causing invasive healthcare-associated infections with high mortality. Some strains of *C. auris* have elevated minimum inhibitory concentrations to the three major classes of antifungals, severely limiting treatment options (Chatterjee et al. 2015).

Infection by *Candida* frequently involves biofilm growth. This intricate and complex structure allows *Candida* to adhere to surfaces (catheters and other commonly used medical devices) but leads to an extraordinary resistance to the administrated antifungal drugs. This resistance appears to be multifactorial, involving mechanisms that are active in antifungal resistance of planktonic cells (such as increased efflux pump activity), as well as mechanisms specific to the biofilm lifestyle (Taff et al. 2013; Nett 2014).

19.5.2 *Aspergillus* Resistance to Azoles

More than 344 species of *Aspergillus* have been described (Frisvad and Larsen 2016). Molecular studies classify *Aspergillus* into five "species complexes": *Fumigati*, *Flavi*, *Nidulantes*, *Usti* and *Terrei*. Each of these complexes comprises many related species, termed as "cryptic" and were previously classified as single morpho-species with rather different antifungal susceptibility profiles (Balajee et al. 2007). For species delimitation, identification based upon comparative sequence of a protein encoding locus, such as the β-tubulin, calmodulin, and others, are now used for *Aspergillus* cryptic species' detection (Balajee et al. 2007). Importantly, a number of these rare *Aspergillus* species are resistant both to azoles and echinocandins, presenting clinicians with obvious challenges for patient management. It is therefore important to understand the biodiversity of correctly identified *Aspergillus* species in hospital settings to guide therapeutically decisions.

Besides this intrinsic resistance, secondary resistance mechanisms are also described in *Aspergillus*, becoming one of the actual major fungal, with high rates of resistance to azoles, especially in Europe (Snelders et al. 2008).This situation may occur though the selection of resistant strains during antifungal therapy/profilaxy. Nevertheless, several studies showed that resistant strains may originate from the environment, where the use of antifungal substances such as pesticides may select positively those resistant isolates, causing an increase of that resistant population.

This problem has been studied since a new molecular mechanism of resistance to azoles was found in *A. fumigatus* environmental strains (Chowdhary et al. 2013; Snelders et al. 2008; Mortensen et al. 2010; Verweij et al. 2009a). Snelders et al. (2008) was able to identify a mutation in the azole target *Cyp51A* (a L98H substitution), which was associated with a 34-bp tandem repeat (TR34) in the gene promoter. This mutation results in resistance to all medical azoles (pan-azole resistance). Other *Cyp51A* mutations than the L98H/TR34 are now also reported from environmental isolates, including TR46/Y121F/T289A (Snelders et al. 2012).

These mechanisms (TR34/L98H and TR46/Y121F/T289A mutations of the CYP51A gene) were found in clinical samples from naïve patients, not subjected to previous antifungal profilaxy/therapy, which lead to the hypothesis of acquisition of resistant strains directly from the environment. The consequence of this is that patients would inhale both susceptible and resistant conidia, but the resistant conidia may have a selective advantage (Snelders et al. 2008), thus allowing their germination in the lungs and subsequent invasive disease, especially in patients at risk. These new mechanisms of resistance have been increasing in frequency over the last years and are one of the biggest recent concerns of the mycological community.

In the European Union, half of the total land of cereals and grapevine is treated with azole fungicides. Notably, much smaller quantities of azoles are used in the United States for plant protection (Chambers et al. 2014) and, coincidently, few azole-resistant *Aspergillus* isolates have been found so far.

Triazole pesticides (also called DMI- sterol dimetilation inhibitors) used in agriculture show a very similar structure to the ones used in clinical practice. Azoles exert their action by inhibiting the C14α demethylation of lanosterol in fungi, which interferes with the synthesis of ergosterol in the fungal cell membrane. Apart from being selected due to local use of azole fungicides, azole-resistant *Aspergillus* spp. might be introduced in the environment via the use of imported azole-treated commercial compost, an ecological niche that has been suggested as a key component in resistance development. Moreover, *Aspergillus* isolates are frequently recovered from poultry farms; in these settings, fungi can be exposed to antifungal drugs regularly used for the prevention of avian mycosis (Sabino et al. 2016).

The true prevalence of the environmental type resistant isolates is largely unknown. In the Netherlands they accounted for approx. 10% in 2007, in Danish soil samples for 8% and in Danish patients approximately 4% in 2009–2011 (Mortensen et al. 2010, 2011; Verweij et al. 2009b). Azole resistance among *A. fumigatus* isolates was not only found in Dutch hospitals (Snelders et al. 2008) resistant isolates were also detected in several European countries, such as Spain (Mellado et al. 2007), Belgium (Lagrou et al. 2008), Denmark (Arendrup et al. 2008), Sweden (Chryssanthou 1997) and France (Dannaoui et al. 2001).

The problematic of the emergence of resistant *Aspergillus fumigatus* is now considered as a growing public Health concern with global dimensions. The European Center for Disease Prevention and Control (ECDC) published a report integrating all evidence for a causal role of fungicides in resistance in *A. fumigatus* in February 2013 (ECDC 2013).

One of the recommendations given in this report is the increase of the environmental surveillance, with extensive and continued environmental field studies, investigating the environmental origin of resistant strains.

References

Alastruey-Izquierdo A, Castelli MV, Cuesta I et al (2009) Activity of posaconazole and other antifungal agents against Mucorales strains identified by sequencing of internal transcribed spacers. Antimicrob Agents Chemother 53:1686–1689

Alastruey-Izquierdo A, Cuenca-Estrella M, Monzón A et al (2007) Prevalence and susceptibility testing of new species of *Pseudallescheria* and *Scedosporium* in a collection of clinical mold isolates. Antimicrob Agents Chemother 51:748–751

Alastruey-Izquierdo A, Cuenca-Estrella M, Monzon A et al (2008) Antifungal susceptibility profile of clinical *Fusarium* spp. isolates identified by molecular methods. J Antimicrob Chemother 61:805–809

Alastruey-Izquierdo A, Melhem MSC, Bonfietti LX et al (2015) Susceptibility test for fungi: clinical and laboratorial correlations in medical mycology. Rev Inst Med Trop S Paulo 57:57–64. doi:10.1590/S0036-46652015000700011

Arendrup MC, Perkhofer S, Howard SJ et al (2008) Establishing in vitro–in vivo correlations for *Aspergillus fumigatus*: the challenge of azoles vs echinocandins. Antimicrob Agents Chemother 52:3504–3511. doi:10.1128/AAC.00190-08

Balajee S, Houbraken J, Verweij P et al (2007) *Aspergillus* species identification in the clinical setting. Stud Mycol 59:39–46

Balajee SA, Gribskov JL, Hanley E et al (2005) *Aspergillus lentulus* sp. nov., a new sibling species of A. fumigatus. Eukaryot Cell 4:625–632

Carneiro HA, Coleman JJ, Restrepo A et al (2011) *Fusarium* infection in lung transplant patients report of 6 cases and review of the literature. Medicine (Baltimore) 90:69–80

Chambers JE, Greim H, Kendall RJ et al (2014) Human and ecological risk assessment of a crop protection chemical: a case study with the azole fungicide epoxiconazole. Crit Rev Toxicol 44:176–210. doi:10.3109/10408444.2013.855163

Chatterjee S, Alampalli SV, Nageshan RK et al (2015) Draft genome of a commonly misdiagnosed multidrug resistant pathogen *Candida auris*. BMC Genomics 16:686. doi:10.1186/s12864-015-1863-z

Chowdhary A, Kathuria S, Xu J et al (2013) Emergence of azole-resistant *Aspergillus fumigatus* strains due to agricultural azole use creates an increasing threat to human health. PLoS Pathog 9:e1003633. doi:10.1371/journal.ppat.1003633

Chryssanthou E (1997) In vitro susceptibility of respiratory isolates of *Aspergillus* species to itraconazole and amphotericin B acquired resistance to itraconazole. Scand J Infect Dis 29:509–512. doi:10.3109/00365549709011864

Cuenca-Estrella M, Gomez-Lopez A, Alastruey-Izquierdo A et al (2010) Comparison of the Vitek2 antifungal susceptibility system with the Clinical and Laboratory Standards Institute (CLSI) and European Committee on Antimicrobial Susceptibility Testing (EUCAST) broth microdilution reference methods and with the Sensititre YeastOne and Etest Techniques for in vitro detection of antifungal resistance in yeast isolates. Clin Microbiol 48:1782–1786. doi:10.1128/JCM.02316-09

Cuenca-Estrella M, Ruiz-Diez B, Martinéz-Suaréz JV et al (1999) Comparative in-vitro activity of voriconazole (UK-109,496) and six other antifungal agents against clinical isolates of *Scedosporium prolificans* and *Scedosporium apiospermum*. J Antimicrob Chemother 43:149–151

Dannaoui E, Borel E, Monier MF et al (2001) Acquired itraconazole resistance in *Aspergillus fumigatus*. J Antimicrob Chemother 47:333–340. doi:10.1093/jac/47.3.333

Eggimann P, Garbino J, Pittet D (2003) Epidemiology of *Candida* species infections in critically ill non-immunosuppressed patients. Lancet Infec Dis 3:685–700

European Centre for Disease Prevention and Control (2013) Risk assessment on the impact of environmental usage of triazoles on the development and spread of resistance to medical triazoles in Aspergillus species. ECDC, Stockholm

Ford CB, Funt JM, Abbey D et al (2015) The evolution of drug resistance in clinical isolates of *Candida albicans*. Elife 4:e00662. doi:10.7554/eLife.00662

Francino PM (2012) The ecology of bacterial genes and the survival of the new. Int J Evol Biol. doi:10.1155/2012/394026

Frisvad JC, Larsen TO (2016) Extrolites of *Aspergillus fumigatus* and other pathogenic species in *Aspergillus* section *Fumigati*. Front Microbiol 6:1485. doi:10.3389/fmicb.2015.01485

Girmenia C, Martino P (1998) Fluconazole and the changing epidemiology of candidemia. Clin Infect Dis 27:232–234

Hof H (2008) Will resistance in fungi emerge on a scale similar to that seen in bacteria? Eur J Clin Microbiol Infect Dis 27:327–334

Johnson EM (2008) Issues in antifungal susceptibility testing. J Antimicrob Chemother 61:i13–i18. doi:10.1093/jac/dkm427

Kanafani ZA, Perfect JR (2008) Resistance to antifungal agents: mechanisms and clinical impact. Clin Infect Dis 46:120–128. doi:10.1086/524071

Kuhn DM, Mukherjee PK, Clark TA et al (2004) *Candida parapsilosis* characterization in an outbreak setting. Emerg Infect Dis 10:1074–1081

Lagrou K, De Vleeschouwer J, Meerseman W, et al. (2008) Triazole resistance among clinical *Aspergillus fumigatus* isolates. Proceedings of the 3th Advances Against Aspergillosis, Miami Beach, FL, Jan 16–19.

Loeffler J, Stevens DA (2003) Antifungal drug resistance. Clin Infect Dis 36:S31–S41

Low C-Y, Rotstein C (2011) Emerging fungal infections in immunocompromised patients. F1000 Med Rep 3:14. doi:10.3410/M3-14

Marr KA, Seidel K, Slavin MA et al (2000) Prolonged fluconazole prophylaxis is associated with persistent protection against candidiasis-related death in allogeneic marrow transplant recipients: long-term follow-up of a randomized, placebo-controlled trial. Blood 96:2055–2061

Mellado E, Garcia-Effron G, Alcázar-Fuoli l et al (2007) A new *Aspergillus fumigatus* resistance mechanism conferring in vitro cross-resistance to azole antifungals involves a combination of cyp5A alterations. Antimicrob Agents Chemother 51:1897–1904. doi:10.1128/AAC.01092-06

Mortensen KL, Jensen RH, Johansen HK et al (2011) *Aspergillus* species and other molds in respiratory samples from patients with cystic fibrosis: a laboratory-based study with focus on Aspergillus fumigatus azole resistance. J Clin Microbiol 49:2243–2251

Mortensen KL, Mellado E, Lass-Florl C et al (2010) Environmental study of azole-resistant *Aspergillus fumigatus* and other aspergilli in Austria, Denmark, and Spain. Antimicrob Agents Chemother 54:4545–4549

Nett JE (2014) Future directions for anti-biofilm therapeutics targeting *Candida*. Expert Rev Anti Infect Ther 12:375–382

Nucci M, Marr KA (2005) Emerging fungal diseases. Clin Infect Dis 41:521–526. doi:10.1086/432060

Orozco AS, Higginbotham LM, Hitchcock CA (1998) Mechanism of fluconazole resistance in *Candida krusei*. Antimicrob Agents Chemother 42:2645–2649

Perkhofer S, Mrazek C, Hartl L et al (2010) In vitro susceptibility testing in fungi: what is its role in clinical practice? Curr Infect Dis Rep 12:401. doi:10.1007/s11908-010-0134-z

Pfaller MA, Jones RN, Doern GV et al (1999) International surveillance of blood stream infections due to *Candida* species in the European SENTRY Program: species distribution and antifungal susceptibility including the investigational triazole and echinocandin agents. SENTRY Participant Group (Europe). Diagn Microbiol Infect Dis 35:19–25

Rex J, Pfaller M (2002) Has antifungal susceptibility testing come of age? Clin Infect Dis 35:982–989

Rodriguez-Tudela JL, Berenguer J, Guarro J et al (2009) Epidemiology and outcome of *Scedosporium prolificans* infection, a review of 162 cases. Med Mycol 47:359–370

Sabino R, Carolino E, Verissimo C et al (2016) Antifungal susceptibility of 175 *Aspergillus* isolates from various clinical and environmental sources. Med Mycol 54:740–756. doi:10.1093/mmy/myw024. Epub 2016 May 3

Sanglard D, Odds FC (2002) Resistance of *Candida* species to antifungal agents: molecular mechanisms and clinical consequences. Lancet Infect Dis 2:73–85

San-Miguel LG, Pla J, Cobo J et al (2004) Morphotypic and genotypic characterization of sequential *Candida parapsilosis* isolates from an outbreak in a pediatric intensive care unit. Diagn Microbiol Infect Dis 49:189–196

Sautour M, Dalle F, Olivieri C et al (2009) A prospective survey of air and surface fungal contamination in a medical mycology laboratory at a tertiary care university hospital. Am J Infect Control 37:189–194
Slavin MA, Osborne B, Adams R et al (1995) Efficacy and safety of fluconazole prophylaxis for fungal infections after bone marrow transplantation: a prospective, randomized, double-blind study. J Infect Dis 171:1545–1552
Snelders E, Camps SM, Karawajczyk A et al (2012) Triazole fungicides can induce cross-resistance to medical triazoles in *Aspergillus fumigatus*. PLoS One 7:e31801
Snelders E, van der Lee H, Kuijpers J et al (2008) Emergence of azole resistance in *A. fumigatus* and spread of a single resistance mechanism. PLoS Med 5:e219
Sobel JD (2005) Genital candidiasis. Medicine 33:62–65
Solo-Gabriele HM, Harwood VJ, Kay D et al (2016) Beach sand and the potential for infectious disease transmission: observations and recommendations. J Mar Biol Assoc UK 96:101–120
Taff HT, Mitchell KF, Edward JA, et al (2013) Mechanisms of *Candida* biofilm drug resistance. Future Microbiol 8:1325–1337. doi:10.2217/fmb.13.101
Tortorano AM, Kibbler C, Peman J et al (2006) Candidemia in Europe: epidemiology and resistance. Int J Antimicrob Agents 27:359–366
Trick WE, Fridkin SK, Edwards JR et al (2002) Secular trend of hospital-acquired candidemia among unit patients in the United States during 1989–1999. Clin Infect Dis 35:627–630
Trofa D, Gácser A, Nosanchuk J (2008) *Candida parapsilosis*, an emerging fungal pathogen. Clin Microbiol Rev 21:606–625
Verweij PE, Howard SJ, Melchers WJ et al (2009b) Azole-resistance in *Aspergillus*: proposed nomenclature and breakpoints. Drug Resist Updat 12:141–147
Verweij PE, Snelders E, Kema GJH et al (2009a) Azole resistance in *Aspergillus fumigatus*: a side-effect of environmental fungicide use? Lancet Infect Dis 9:789–795
Wellington EMH, Boxall ABA, Cross P et al (2013) The role of the natural environment in the emergence of antibiotic resistance in Gram-negative bacteria. Lancet Infect Dis 13:155–165
White TC, Marr KA, Bowden RA (1998) Clinical, cellular, and molecular factors that contribute to antifungal drug resistance. Clin Microbiol Rev 11:382–402

Chapter 20
Bacterial Resistances

Vera Manageiro, Vanessa Salgueiro, Eugénia Ferreira and Manuela Caniça

Abstract Here we reviewed several factors involved in the emergence of antibiotic resistance. They are numerous, and the constant adaptation of microorganisms to the selective pressure exerted by antibiotics is extraordinary. The monitoring systems to assess antibiotic resistance levels and the extent of dissemination were highlighted. In addition, the success of spread of certain bacterial lineages and resistant mechanisms remains sometimes difficult to determine. The need to enlarge research in the area of antibiotic resistance was also stated, not only to better understand the dynamics of dissemination of resistance between different bacteria and different ecosystems, but also to enlarge the pharmaceutical pipeline of antibacterials against multidrug-resistant pathogens. It is manifest the severe consequences of antibiotic resistances to humans, animals and environment, constituting a global public health priority. In consequence, it should be tackled on all fronts in view to the essential concept of "One World-One Medicine-One Health."

Keywords Antibiotic resistance · one-health · healthcare · animal · food-chain · environment

---

V. Manageiro · E. Ferreira · M. Caniça (✉)
National Reference Laboratory of Antibiotic Resistances and Healthcare Associated Infections (NRL-AMR/HAI), Department of Infectious Diseases, National Institute of Health Doutor Ricardo Jorge, Lisbon, Portugal

Centre for the Studies of Animal Science, Institute of Agrarian and Agri-Food Sciences and Technologies, Oporto University, Oporto, Portugal
e-mail: manuela.canica@insa.min-saude.pt

V. Salgueiro
National Reference Laboratory of Antibiotic Resistances and Healthcare Associated Infections (NRL-AMR/HAI), Department of Infectious Diseases, National Institute of Health Doutor Ricardo Jorge, Lisbon, Portugal

C. Viegas et al. (eds.), *Exposure to Microbiological Agents in Indoor and Occupational Environments*, DOI 10.1007/978-3-319-61688-9_20

20.1 Tackling the Emergence of Antibiotic Resistance

20.1.1 The Problem of Antibiotic Resistances

Antibiotic resistance has had a real explosion at national and international level, representing a global concern (Fair and Tor 2014). This problem has serious consequences to infection control in hospital and in the community, since some bacteria became totally resistant to all known antibiotics. So, in these cases, antibiotics and respective targets no longer recognize each other. Indeed, there are around 50,000 deaths in Europe and the US alone each year due to the difficulty of treating infections with the existing antibiotic pipeline, to which we can add many hundreds of thousands more dying in other areas of the globe (O'Neill 2014).

Of note is that antibiotic resistance is increasing and should not be overlooked; it does not have a temporary importance, since it is expected that, by 2050, deaths due to infections caused by antibiotic resistant microorganisms will reach about 10 million people (O'Neill 2014). If we do not act today, the problem will be more serious in the future. However, it cannot be handled in each place separately, but within an integrated plan (Brogan and Mossialos 2016; WHO 2015).

20.1.2 Facing the Problem

Rethinking the decrease in resistance to antibiotics still means preventing the disease, or rather, its spread, especially when caused by resistant bacteria. At this point, we consider mainly sites with a large number of people, especially patients, immunocompromised (e.g. after surgeries) and/or elderly, where the easiness of acquisition and transmission of bacteria with antibiotic resistance is facilitated. These sites vary widely in physical and functional characteristics, ranging from hospitals, long-term care facilities to specialty units (e.g. intensive care units, burn units, and neonatal intensive care units), integrated care units, community health units without hospitalization, tertiary care facilities, and nursing homes, among others. In fact, the severity and extent of disease caused by antibiotic resistant pathogens varies with the environment(s) in which they are found and by the population(s) affected (Charani et al. 2017; Roca et al. 2015; Rogers et al. 2011; Tacconelli et al. 2014).

WHO and medical doctors recognize that the fight against antibiotic resistance depends on three main areas: stewardship education (including the appropriate use of diagnostic), improvement of clinical microbiology and reporting, and regulatory interventions, namely at a pharmaceutical level (FAO/OIE/WHO 2016a; Robilotti et al. 2017).

Clinical decision support systems for antibiotic management have been reproduced in last years with the rise of antibiotic stewardship, but it is needed to evaluate if they have been designed for maximum benefit and assessed correctly. Recently, Rawson et al. (2017) did a systematic review sustaining the responsibility for clinical decision support systems to optimize antibiotic therapy prescribing, but suggesting substantial improvement in how these systems are developed, evaluated, and used. The authors considered that interventions must be multi-modal so that potential synergistic effects can be explored to make certain that interventions are utilized (Rawson et al. 2017). The education and awareness of the population in general, with the appropriate information in each case, are still fundamental.

Furthermore, the movement of both people, such as international travels and the flow of patients between hospitals, and food, among different regions or countries, have to be taken into account, since patients might be contaminated or carriers of resistant bacteria, and food might be contaminated with bacteria, especially those resistant to antibiotics. These traits make the dissemination of antibiotic resistance more difficult to contain and control, becoming this not a national but also an international problem (Hawkey 2015). In fact, several studies have shown that intercontinental travels create new opportunities for antimicrobial-resistant pathogens to be spread globally (van der Bij and Pitout 2012; Johnson and Woodford 2013; Rogers et al. 2011; Zanger 2014). No country can therefore successfully tackle antibiotic resistance by acting alone: it will not be possible to break the chain of dissemination of antibiotic resistance if there is no concerted action among the different stakeholders (FAO/OIE/WHO 2016b). Certainly, for instance, if the combat of this plague is performed separately by human doctors, without the concomitant action of the veterinarian and environmental parts, the result will not be effective.

The use of antibiotics in human and veterinary medicine is just the tip of the iceberg. In fact, reducing the use of antibiotics in veterinary that are of vital importance for the treatment of serious human diseases, is only one way of combating the problem. Taking responsibility for the prescription of antibiotics, requesting a laboratory examination prior to prescribe it is also crucial. The use of antibiotics for prevention of disease should be avoided, except in cases absolutely necessary. New studies are still essential in order to find products that are alternative to antibiotics and that can mingle bacteria, preventing or bypassing the development of bacterial infections (Strachan and Davies 2017; Wright and Sutherland 2007).

In other way, the environmental component plays an equally important role, since current news gives us detailed information on frequent discharges from factory activity or wastewater treatment plants to river beds, agricultural land, among others (Berglund 2015; Marti et al. 2013). Intensive farms in the veterinary area also make an important contribution to the contamination of soils and the environment in general (Jones-Dias et al. 2016a). In consequence, FAO published an action plan on antimicrobial resistance for 2016–2020, which aims to support also the food and agriculture sectors in implementing the Global Action Plan on Antimicrobial Resistance to minimize the impact of antibiotic resistance (FAO/OIE/WHO 2016b).

20.1.3 Dynamics of Antibiotic Resistance in Bacteria

The approaches to the prevention and control of antibiotic resistance pathogens need to be adapted to the specific needs of each population and individual society, region, and local. As mentioned above, the prevention and control of antibiotic resistance, namely multidrug resistance (MDR), is a national and international priority.

Accordingly, many countries, including Portugal, increased responsiveness in combating antibiotic resistance, at different levels. Nevertheless, the dynamics of the bacteria that produce antibiotic resistance is voracious with increasingly and sophisticated new resistance mechanisms. For instance, a single gene, such as *aac(6')-Ib-cr*, confers resistance to aminoglycosides and fluoroquinolones. In the same way, efflux pumps, which confer a low-level of MDR in Gram-negative bacteria, in coproduction with other resistance mechanisms (β-lactamase production, loss of porins and/or impermeability), might increase the level of antibiotic resistance (Poole 2007).

The acronym ESKAPE(E), which encompasses *Enterococcus faecium*, *Staphylococcus aureus*, *Klebsiella pneumoniae*, *Acinetobacter baumannii*, *Pseudomonas aeruginosa*, and *Enterobacter cloacae* (*Escherichia coli*), has allowed us to congregate the most frequent strains with resistance to the most worrisome antibiotics (Pendleton et al. 2013). Recently, WHO drew attention to the current extent of the antibiotic resistance problem and to the need of effectively monitor it internationally, by presenting a list of more important antibiotic resistant bacteria posing a menace to treat the infections they cause (WHO 2017). WHO also highlighted that there is an urgent need for research and development of new antibiotics, opening the border for new treatments, essentially against multidrug- and extensively drug-resistant Gram-negative bacteria (WHO 2017). The list of priorities for selection of those bacteria was based on the fact that all are associated to mortality, covering health and community levels, with an important prevalence of resistance and drawing a resistance trend during 10 years; those bacteria has also broad transmission capacity and the possibility to be prevented in hospital and community context, as well as a possibility of treatment; the list had also taken into account the existing pipeline for treatment (Table 20.1).

Table 20.1 Global priority list of antibiotic resistant bacteria to guide research, discovery, and development of new antibiotics (WHO 2017)

Priority 1: Critical[a]
Acinetobacter baumannii and *Pseudomonas aeruginosa* with resistance to carbapenems; *Enterobacteriaceae* (including *Klebsiella pneumoniae*, *Escherichia coli*, *Enterobacter* spp., *Serratia* spp., *Proteus* spp., *Providencia* spp., and *Morganella* spp.) with resistance to carbapenems and 3rd generation cephalosporins.
Priority 2: High
Enterococcus faecium with resistance to vancomycin; methicillin-resistant *Staphylococcus aureus* and *S. aureus* with intermediate susceptibility or resistance to vancomycin; *Helicobacter pylori* with resistance to clarithromycin; *Campylobacter*, *Salmonella* spp. and *Neisseria gonorrhoeae* with resistance to fluoroquinolones; *Neisseria gonorrhoeae* with resistance to 3rd generation cephalosporins.
Priority 3: Medium
Streptococcus pneumoniae non-susceptible to penicillin; *Haemophilus influenzae* with ampicillin resistance; *Shigella* spp. with resistance to fluoroquinolones.

[a]Mycobacteria causing human tuberculosis was not included in this prioritization because WHO considered that it is already a globally established priority for which innovative new treatments are urgently needed.

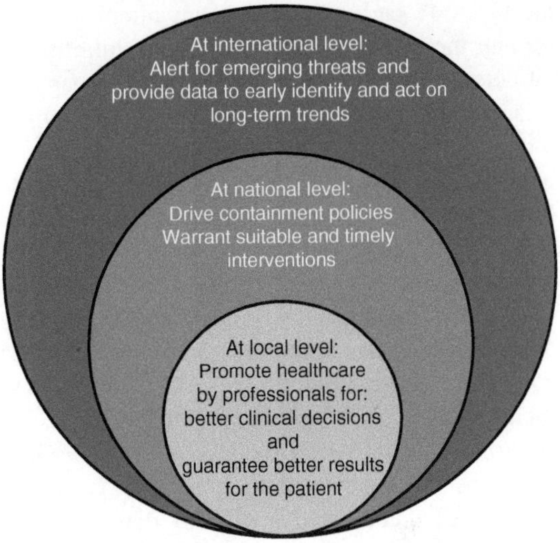

Fig. 20.1 How surveillance can improve the results of the emergence of resistance (Adapted from O'Neill 2017).

The constant assessment of the levels of susceptibility to antibiotics in bacteria is of the utmost importance in order to be able to act. Indeed, actions are needed to prevent infection and to design antibiotic prescribing schemes, to detect outbreaks, and to plan how to circumvent paths of dissemination that are signalled (Fig. 20.1).

20.2 Detection of Antibiotic Resistance

Healthcare-associated infections (HAI) are related, in a high percentage, to the circulation of antibiotic resistant bacteria, increasingly resistant to all antibiotics from the therapeutic set (pandrug-resistant bacteria). Therefore, the National Reference Laboratories of Antibiotic Resistances (NRL-ARs) plays a key role.

The implementation of antibiotic resistance monitoring schemes at national and international level allows the effective prevention of the emergence and spread of microorganisms with intermediate susceptibility or resistance to antibiotics. The mandatory notification of isolates with different susceptibility patterns to the NRL-ARs aims to support the modification or modulation of antibiotic prescribing and consumption policy locally, preventing the emergence or transmission of resistant microorganisms and infections by applying good practices of infection control. The notification of priority antibiotic resistant bacteria should be independently of the type of samples in which they are isolated (Fig. 20.1): samples used to the diagnosis of the infection, to evaluate asymptomatic colonization or to the environmental assessment in healthcare units. It is important to use standard methodologies intended for the determination of antibiotic susceptibility of bacteria. For example, some β-lactamases that inactivate the β-lactam ring of carbapenems, the OXA-48 carbapenemases, are often not detected by antibiotic susceptibility testing; the use of temocillin and molecular methods are an asset for its laboratorial detection (Robert et al. 2017). The carbapenemase NDM (New Delhi Metalo-β-lactamase) can also go unnoticed using only phenotypic methods since can confer only reduced susceptibility to carbapenems; in this case, it is very useful to include faropenem in the antibiotic susceptibility testing for its detection. Faropenem, when in the presence of this or another carbapenemase, is forthrightly sensitive to alert the microbiologist for the presence of a carbapenemase-producing *Enterobacteriaceae* (Hu et al. 2014). These approaches, along with the use of new generation methods, such as whole-genome sequencing (WGS), has been shown to provide the required power of discrimination to track transmission networks for many bacterial pathogens, as well as antimicrobial resistance (Schürch and van Schaik 2017; Shimose et al. 2015). In HAI outbreaks and other environmental settings associated with antibiotic resistance dissemination, WGS is of utmost importance to look for new ways to deal with antibiotic resistance superbugs. Accordingly, NRL-ARs play a key role in providing early warning of emerging threats and data to identify and act on trends, locally and globally.

20.3 Dissemination of Antibiotic Resistance

The ability of pathogens to infect a wide range of hosts provides them with enormous opportunities to share their genetic material with each other, and then creating new resistant strains at an unparalleled swiftness. Plasmids and other mobile genetic elements (MGE), present in Gram-negative bacteria, are able to be transmitted among bacteria, with the aggravating circumstance that they may also encode for resistance for many antibiotics. This is a phenomenon which occurs universally in different reservoirs and situations, such as globalization and other migratory movements, in the environment, human and animal guts and food chain, reaching commensal and pathogenic bacteria (Caniça et al. 2015). Indeed, there might be innumerous of such events at any moment, around the world.

20.3.1 In Healthcare Settings

Infections caused by multidrug-resistant bacteria are increasing among the elderly people living in nursing homes or other community care facilities. Unavoidably, antibiotic resistant bacteria are frequently being recovered in these settings (Bonomo 2000; Nicolle et al. 1996; van Buul et al. 2012). Thus, the understanding and monitoring of such phenomena is essential to avoid further dissemination. In a recent study, Salgueiro et al. (2016) analyzed indoor air samples collected from bedrooms, living rooms and outdoor of four nursing homes located in Lisbon, Portugal. Non-susceptibility to β-lactam antibiotics was found in

Staphylococcus capitis (cefoxitin) due to the presence of the *mecA* gene, in *Klebsiella oxytoca* (ampicillin, piperacillin and piperacillin-tazobactam) linked to the presence of $bla_{\text{OXY-5-type}}$ gene and in *Pseudomonas putida* (meropenem and piperacillin-tazobactam). Regarding quinolones, non-susceptibility was detected in *S. capitis*, *Staphylococcus haemolyticus* and *K. oxytoca*. Other bacteria isolated (*Acinetobacter* spp., *Pantoea* spp., *Sphingomonas paucimobilis*, *Micrococcus luteus*) were susceptible to antibiotics. The genetic relatedness of *K. oxytoca* from nursing homes and clinical isolates recovered within the same region, allowed to conclude that they were not genetically related. The authors concluded that the molecular characterization of the *mecA*-positive *S. capitis* from the nursing homes and from a hospital within the same region might suggest a potential dissemination of strains between these two environments. Globally, nursing homes environments may act as complementary reservoirs of antibiotic resistant bacteria and antibiotic resistance genes. Thus, a better understanding of the antibiotic resistance mechanisms and dissemination pathways in other reservoirs than human is essential to control its emergence and spread.

The horizontal transfer of $bla_{\text{CTX-M}}$ genes mediated by plasmids and/or other MGE contributes to the dissemination of CTX-M enzymes to community and hospital environments, which belongs mainly to the lineage ST131 (Mendonça et al. 2007; Nicolas-Chanoine et al. 2014). However, recent findings suggest that prevalence of ST131 *fim*H30-Rx *E. coli* in non-clinical settings is maintained by the constant exchange of isolates throughout the time (Mathers et al. 2015). Indeed, ST131 lineage is continuously in contact with a live audience, constituting a public health concern. These clinically relevant multidrug-resistant *E. coli* isolates have been on the rise for years (Nicolas-Chanoine et al. 2014), contributing to the change in strategies to treat infections caused by third-generation-resistant Gram-negative bacteria, replacing these antibiotics by carbapenems. The consequence of this alteration in treatments culminated in the very serious increase of isolates resistant to carbapenems (Morrill et al. 2015). Clearly, the spread of antibiotic resistance has made bacterial infections gradually more difficult to treat (Blair et al. 2015).

The emerging problem of carbapenem resistance in Gram-negative bacteria, such as *A. baumannii*, *P. aeruginosa* and *Enterobacteriaceae*, with particular emphasis on *K. pneumoniae* in HAI and other healthcare settings, should be highlighted (Morrill et al. 2015). These bacteria produce carbapenemases, which may have different names according to the bacteria and/or substrate in which they act. The carbapenemase KPC (*Klebsiella pneumoniae* carbapenemase) is the most frequent in Portugal, especially associated to HAI (Manageiro et al. 2015a). The silent, but real, spread of NDM in Portugal is beginning to be equally worrying (Manageiro et al. 2015b). Of note is the gradual but consistent emergence of OXA-48. By instance, Manageiro et al. (2015c) reported a 63,584-bp conjugative IncL plasmid from an *Enterobacter cloacae* clinical isolate, which carried the Tn*1999.2* composite transposon that enclosed a $bla_{\text{OXA-48}}$ gene, as well as integrase- and transposases-encoding genes. This is an example of how is important the complete sequence of a plasmid to understand its evolution and differentiation given the clinical and epidemiological relevance of these MGEs. In addition, the availability of complete plasmid sequences from different countries can support the global epidemiological surveillance of antibiotic resistance spread. In fact, plasmids and other MGE are also potentially involved in the maintenance of ecosystem functions, and are extremely stable even without evident survival advantage to the host. These apparently stable MGEs can encode traits such as antibiotic or heavy metal resistance, in the absence of direct selective pressure (Kazimierczal et al. 2009; Jones et al. 2010).

Of notice, is the recent and alarming emergence of the plasmid-mediated resistance to colistin in several countries, due to the spread of *mcr-1* gene (Nordmann and Poirel 2016). Colistin is an antibiotic used as a last resort in the absence of response to the treatment of multidrug-resistant bacterial infections, including resistance to carbapenems and to most of the other available classes of antibiotics (Watkins et al. 2016). The resistance to colistin was previously known to be mediated only through the chromosome, but in 2016 a plasmid-mediated *mcr-1* gene was reported (Liu et al. 2016). This gene is easily disseminated by at least one type of MGE, the plasmids. The progression of dissemination at a high scale is then facilitated, and it is up to us to assess the degree and conditions of dissemination and how to control it, particularly through the hospital environment, but also in other reservoirs, such as in animals and food chain (Catry et al. 2015; Nordmann and Poirel 2016).

20.3.2 In Animal and Food-Chain Settings

Although *E. coli* is a common inhabitant of the gastrointestinal tract of humans and animals, the detection of transposons, plasmids and bacteriophages are essential to the acquisition of pathogenicity factors that enlarge their ability to adapt to new niches, allowing bacteria to increase the capacity to cause a broad spectrum of diseases (Bien et al. 2012). In a recent study, all isolates from a collection of *Salmonella enterica* and *E. coli* recovered from animals of different origins displayed genomic factors that may be decisive to cause a zoonotic infection reflecting high probabilities for the isolates to be human pathogens (>93%) (Jones-Dias et al. 2016b). In fact this showed the intrinsic potential for transboundary dissemination among bacteria from different environments. The isolates presented a set of genetic features crucial to support their own successful dissemination, such as: multiple antibiotic resistance genes carried by well-known MGE, virulence factors adequate to zoonotic transmission and numerous other pathogenicity factors.

Furthermore, new resistance mechanisms are emerging in isolates from human but also in other reservoirs as in animals. A recent example is the new CTX-M-166-harbouring O6:H16 ST48-*fimH*34 *E. coli*, recovered from a *Gallus gallus* broiler flock from an industrial poultry unit, carrying a new amino acid substitution, when compared with CTX-M-1. This isolate revealed the presence of an IncI1/ST103-IS*Ecp1*-*bla*$_{\text{CTX-M-166}}$-*orf477* plasmid region and of diverse antibiotic and virulence acquired-genes (Manageiro et al. 2016).

The growing concern about the dissemination of acquired resistance genes by MGE in commensal and/or potential pathogens found in the food chain, e.g. in vegetable microbiomes, has also been demonstrated (Berg et al. 2014). In a recent study, the authors detected in vegetable samples integrons of three different classes (class 1, 2 and 3), associated with other clinically relevant MGE, reinforcing the mobilization potential of antibiotic resistance in Gram-negative bacteria (Jones-Dias et al. 2016c). The diverse integron backbones were associated with transposable elements (e.g. Tn*402*, Tn*7*, *ISCR*1, Tn*2**, IS*26*, IS*1326* and IS*3*) that conferred greater mobility. Moreover, clinically relevant acquired antibiotic resistance genes, such as the new *mcr-1* gene, *qnrA1*, $bla_{\text{GES-11}}$, *mphA* and *oqxAB* were also identified. Although continuous monitoring of products of vegetable and animal origin is essential, the decrease or cessation of certain antibiotics in manure and agriculture is strongly advised to preserve drug molecules for clinical treatment of infections (Thanner et al. 2016).

We also emphasize a worrying antibiotic multidrug resistance rate of human, food and animal *Campylobacter* spp. and the emergence of strains resistant to antibiotics of human use (Duarte et al. 2014).

20.3.3 In Environmental Settings

Specific regulations on how to evaluate, to manage, or minimize the potential human health risks associated with the presence of antibiotics and antibiotic resistant bacteria and the diverse possibilities of human exposures via a variety of water, sanitation and hygiene paths and transmission routes, are needed particularly when subjected to anthropogenic pressures (Szczepanowski et al. 2009; WHO 2014). This necessity was highlighted in a report of the $bla_{\text{GES-5}}$ gene, coding for a carbapenemase, among a diversity of other β-lactamase-encoding resistance mechanisms detected in ubiquitous bacteria isolated from aquatic environment samples (Manageiro et al. 2014). The $bla_{\text{GES-5}}$ gene was also previously recovered from an activated sludge bacterial community of a municipal wastewater treatment plant in Germany, which was reliable with a potential emergence of this carbapenemase in the environment (Girlich et al. 2012).

In other reservoirs, such as in natural world, there are also resistance mechanisms that appear in clinical settings, which show the promiscuity of antibiotic resistance spread. An example is the production of CTX-M, SHV, CMY and TEM β-lactamases by *E. coli* isolated in intensive farming soils, reported by Jones-Dias et al. (2016a). In that study the authors also demonstrated the first appearance of IncI1/ST28-harbouring $bla_{\text{CTX-M-1}}$ gene. Indeed, it was highlighted the need to monitor the spread of that plasmid to prevent its establishment as other successfully dispersed plasmids. This research may help to reveal paths of contamination by mobile antibiotic resistance determinants, and the risks for their dissemination.

20.4 Conclusion

By linking human, animal, and environment health we may recognize the importance of the One-Health approach to address emerging infectious diseases. In fact, this emphasizes that the health of humans is connected to that of animals and their shared environment. In these settings MGEs are crucial to the dissemination of antibiotic resistance and to increase the number of human bacterial infections without treatment.

Research in this area must be enlarged, not only to better understand the dynamics of dissemination of resistance between different bacteria and different settings, but also to enlarge the pharmaceutical pipeline of antibacterials against multidrug-resistant pathogens and/or obtain molecules that can inactivate bacteria or dribbling its action.

Acknowledgements The authors thank Fundação para a Ciência e a Tecnologia (FCT) for project grant PEst-OE/AGR/UI0211/2011-2014, Strategic Project UI211-2011-2014. V. Manageiro was supported by FCT fellowship (grant SFRH/BPD/77486/2011), financed by the European Social Funds (COMPETE-FEDER) and national funds of the Portuguese Ministry of Education and Science (POPH-QREN).

References

Berg G, Erlacher A, Smalla K et al (2014) Vegetable microbiomes: is there a connection among opportunistic infections, human health and our 'gut feeling'? Microb Biotechnol 7:487–495

Berglund B (2015) Environmental dissemination of antibiotic resistance genes and correlation to anthropogenic contamination with antibiotics. Infect Ecol Epidemiol 5:28564

Bien J, Sokolova O, Bozko P (2012) Role of uropathogenic *Escherichia coli* virulence factors in development of urinary tract infection and kidney damage. Int J Nephrol 2012:681473

Blair JM, Webber MA, Baylay AJ et al (2015) Molecular mechanisms of antibiotic resistance. Nat Rev Microbiol 13:42–51

Bonomo RA (2000) Multiple antibiotic resistant bacteria in long-term-care facilities: an emerging problem in the practice of infectious diseases. Clin Infect Dis 31:1414–1422

Brogan DM, Mossialos E (2016) A critical analysis of the review on antimicrobial resistance report and the infectious disease financing facility. Global Health 12:8

Caniça M, Manageiro V, Jones-Dias D et al (2015) Current perspectives on the dynamics of antibiotic resistance in different reservoirs. Res Microbiol 166:594–600

Catry B, Cavaleri M, Baptiste K et al (2015) Use of colistin-containing products within the European Union and European Economic Area (EU/EEA): development of resistance in animals and possible impact on human and animal health. Int J Antimicrob Agents 46:297–306

Charani E, Tarrant C, Moorthy K et al (2017) Understanding antibiotic decision making in surgery – a qualitative analysis. Clin Microbiol Infect pii:51198-743X(17)30182-9

Duarte A, Santos A, Manageiro V et al (2014) Human, food and animal *Campylobacter* spp. isolated in Portugal: high genetic diversity and antibiotic resistance rates. Int J Antimicrob Agents 44:306–313

Fair RJ, Tor Y (2014) Antibiotics and bacterial resistance in the 21st century. Perspect Medicin Chem 6:25–64

FAO/OIE/WHO (Food and Agriculture Organization of the United Nations, World Organisation for Animal Health, World Health Organization) (2016a) Antimicrobial resistance: a manual for developing national action plans. WHO, Geneva

FAO/OIE/WHO (Food and Agriculture Organization of the United Nations, World Organisation for Animal Health, World Health Organization) (2016b) The FAO action plan on antimicrobial resistance 2016-2020: Supporting the food and agriculture sectors in implementing the Global Action Plan on Antimicrobial Resistance to minimize the impact of antimicrobial resistance. FAO, Rome

Girlich D, Poirel L, Szczepanowski R et al (2012) Carbapenem-hydrolyzing GES-5-encoding gene on different plasmid types recovered from a bacterial community in a sewage treatment plant. Appl Environ Microbiol 78:1292–1295

Hawkey PM (2015) Multidrug-resistant Gram-negative bacteria: a product of globalization. J Hosp Infect 89:241–247

Hu F, Ahn C, O'Hara JA et al (2014) Faropenem disks for screening of *Klebsiella pneumoniae* carbapenemase-producing *Enterobacteriaceae*. J Clin Microbiol 52:3501–3502

Johnson AP, Woodford N (2013) Global spread of antibiotic resistance: the example of New Delhi metallo-β-lactamase (NDM)-mediated carbapenem resistance. J Med Microbiol 62:499–513

Jones BV, Sun F, Marchesi JR (2010) Comparative metagenomic analysis of plasmid encoded functions in the human gut microbiome. BMC Genomics 11:46

Jones-Dias D, Manageiro V, Caniça M (2016a) Influence of agricultural practice on mobile *bla* genes: IncI1-bearing CTX-M, SHV, CMY and TEM in *Escherichia coli* from intensive farming soils. Environ Microbiol 18:260–272

Jones-Dias D, Manageiro V, Graça R et al (2016b) QnrS1- and Aac(6')-Ib-cr-producing *Escherichia coli* among isolates from animals of different sources: susceptibility and genomic characterization. Front Microbiol 7:671

Jones-Dias D, Manageiro V, Ferreira E et al (2016c) Architecture of class 1, 2, and 3 integrons from Gram-negative bacteria recovered among fruits and vegetables. Front Microbiol 7:1400

Kazimierczal KA, Scott KP, Kelly D et al (2009) Tetracycline resistome of the organic pig gut. Appl Environ Microbiol 75:1717–1722

Liu YY, Wang Y, Walsh TR et al (2016) Emergence of plasmid-mediated colistin resistance mechanism MCR-1 in animals and human beings in China: a microbiological and molecular biological study. Lancet Infect Dis 16:161–168

Manageiro V, Clemente L, Duarte S et al (2016). Draft genome sequence of an *Escherichia coli* strain isolated from a *Gallus gallus* broiler producing the novel CTX-M-166 variant. Genome Announc 4(5):e01029-16

Manageiro V, Ferreira E, Caniça M et al (2014) GES-5 among the β-lactamases detected in ubiquitous bacteria isolated from aquatic environment samples. FEMS Microbiol Lett 351:64–69

Manageiro V, Pinto M, Caniça M (2015c) Complete sequence of a bla_{OXA-48}-harboring IncL plasmid from an *Enterobacter cloacae* clinical isolate. Genome Announc 3(5):e01076–15

Manageiro V, Ferreira E, Almeida J et al (2015a) Predominance of KPC-3 in a survey for carbapenemase-producing *Enterobacteriaceae* in Portugal. Antimicrob Agents Chemother 59:3588–3592

Manageiro V, Sampaio DA, Pereira P et al (2015b) Draft genome sequence of the first NDM-1-producing *Providencia stuartii* isolated in Portugal. Genome Announc 3(5):e01077–15

Marti E, Jofre J, Balcazar JL (2013) Prevalence of antibiotic resistance genes and bacterial community composition in a river influenced by a wastewater treatment plant. PLoS One 8: e78906

Mathers AJ, Peirano G, Pitout JD (2015) The role of epidemic resistance plasmids and international high-risk clones in the spread of multidrug-resistant Enterobacteriaceae. Clin Microbiol Rev 28:565–591

Mendonça N, Leitão J, Manageiro V et al (2007) Spread of extended-spectrum β-lactamase CTX-M-producing *Escherichia coli* clinical isolates in community and nosocomial environments in Portugal. Antimicrob Agents Chemother 51:1946–1955

Morrill HJ, Pogue JM, Kaye KS, LaPlante KL (2015) Treatment options for carbapenem-resistant *Enterobacteriaceae* infections. Open Forum Infect Dis 2:ofv050
Nicolas-Chanoine MH, Bertrand X, Madec JY (2014) Escherichia coli ST131, an intriguing clonal group. Clin Microbiol Rev 27:543–574
Nicolle LE, Strausbaugh LJ, Garibaldi RA (1996) Infections and antibiotic resistance in nursing homes. Clin Microbiol Rev 9:1–17
Nordmann P, Poirel L (2016) Plasmid-mediated colistin resistance: an additional antibiotic resistance menace. Clin Microbiol Infect 22:398–400
O'Neill J (2014) The review on antimicrobial resistance. Antimicrobial resistance: tackling a crisis for the health and wealth of nations. http://www.jpiamr.eu/wp-content/uploads/2014/12/AMR-Review-Paper-Tackling-a-crisis-for-the-health-and-wealth-of-nations_1-2.pdf
O'Neill J (2017) The review on antimicrobial resistance. tackling drug-resistant infections globally: final report and recommendations. https://amr-review.org/sites/default/files/160518_Final%20paper_with%20cover.pdf
Pendleton JN, Gorman SP, Gilmore BF (2013) Clinical relevance of the ESKAPE pathogens. Expert Rev Anti Infect Ther 11:297–308
Poole K (2007) Efflux pumps as antimicrobial resistance mechanisms. Ann Med 39:162–176
Rawson TM, Moore LS, Hernandez B et al (2017) A systematic review of clinical decision support systems for antimicrobial management: are we failing to investigate these interventions appropriately? Clin Microbiol Infect pii: S1198-743X(17)30125-8
Robert J, Pantel A, Merens A et al (2017) Development of an algorithm for phenotypic screening of carbapenemase producing *Enterobacteriaceae* in the routine laboratory. BMC Infect Dis 17:78
Robilotti E, Holubar M, Nahrgang S et al (2017) Educating front-line clinicians about antimicrobial resistance. Lancet Infect Dis 17:257–258
Roca I, Akova M, Baquero F et al (2015) The global threat of antimicrobial resistance: science for intervention. New Microbes New Infect 6:22–29
Rogers BA, Aminzadeh Z, Hayashi Y et al (2011) Country-to-country transfer of patients and the risk of multi-resistant bacterial infection. Clin Infect Dis 53:49–56
Salgueiro V, Manageiro V, Jones-Dias D et al (2016) Antibiotic resistant bacteria from air samples collected in nursing homes. 26th SETAC (Society of Environmental Toxicology and Chemistry) Europe Annual Meeting, Nantes, France
Schürch AC, van Schaik W (2017) Challenges and opportunities for whole-genome sequencing-based surveillance of antibiotic resistance. Ann N Y Acad Sci 1388:108–120
Shimose LA, Doi Y, Bonomo RA et al (2015) Contamination of ambient air with *Acinetobacter baumannii* on consecutive inpatient days. J Clin Microbiol 53:2346–2348
Strachan CR, Davies J (2017) The whys and wherefores of antibiotic resistance. Cold Spring Harb Perspect Med 7:a025171
Szczepanowski R, Linke B, Krahn I et al (2009) Detection of 140 clinically relevant antibiotic-resistance genes in the plasmid metagenome of wastewater treatment plant bacteria showing reduced susceptibility to selected antibiotics. Microbiology 155:2306–2319
Tacconelli E, Cataldo MA, Dancer SJ et al (2014) ESCMID guidelines for the management of the infection control measures to reduce transmission of multidrug-resistant Gram-negative bacteria in hospitalized patients. Clin Microbiol Infect 20(S1):1–55
Thanner S, Drissner D, Walsh F (2016) Antimicrobial resistance in agriculture. MBio 7: e02227-15
van Buul LW, Steen JT, Veenhuizen RB et al (2012) Antibiotic use and resistance in long term care facilities. J Am Med Dir Assoc 13:568. e1–e113
van der Bij AK, Pitout JDD (2012) The role of international travel in the worldwide spread of multiresistant *Enterobacteriaceae*. J Antimicrob Chemother 67:2090–2100
Watkins RR, Smith TC, Bonomo RA (2016) On the path to untreatable infections: colistin use in agriculture and the end of 'last resort' antibiotics. Expert Rev Anti Infect Ther 14:785–788

WHO (2014) Antimicrobial resistance: an emerging water, sanitation and hygiene issue. http://apps.who.int/iris/bitstream/10665/204948/1/WHO_FWC_WSH_14.7_eng.pdf?ua=1

WHO (2015) Global action plan on antimicrobial resistance. http://www.wpro.who.int/entity/drug_resistance/resources/global_action_plan_eng.pdf

WHO (2017) Global priority list of antibiotic resistant bacteria to guide research, discovery, and development of new antibiotics. http://www.who.int/medicines/publications/WHO-PPL-Short_Summary_25Feb-ET_NM_WHO.pdf

Wright GW, Sutherland AD (2007) New strategies for combating multidrug-resistant bacteria. Trends Mol Med 13:260–267

Zanger P (2014) Methicillin-resistant *Staphylococcus aureus* and intercontinental travel – "bad bugs on the move!". J Travel Med 21:225–227

Kolibri 360 GmbH
Ferdinand-Jühlke-Straße 7
99095 Erfurt, Deutschland
produktsicherheit@kolibri360.de